WITHDRAWN

Renewable Fuel Standard

POTENTIAL ECONOMIC AND ENVIRONMENTAL EFFECTS OF
U.S. Biofuel Policy

Committee on Economic and Environmental Impacts of Increasing Biofuels Production

Board on Agriculture and Natural Resources
Division on Earth and Life Studies

Board on Energy and Environmental Systems
Division on Engineering and Physical Sciences

NATIONAL RESEARCH COUNCIL
OF THE NATIONAL ACADEMIES

THE NATIONAL ACADEMIES PRESS
Washington, D.C.
www.nap.edu

THE NATIONAL ACADEMIES PRESS   **500 Fifth Street, N.W.**   **Washington, DC 20001**

NOTICE: The project that is the subject of this report was approved by the Governing Board of the National Research Council, whose members are drawn from the councils of the National Academy of Sciences, the National Academy of Engineering, and the Institute of Medicine. The members of the committee responsible for the report were chosen for their special competences and with regard for appropriate balance.

This study was funded by the Department of Treasury under Award TOS-09-051. Any opinions, findings, conclusions, or recommendations expressed in this publication are those of the author(s) and do not necessarily reflect the views of the organizations or agencies that provided support for the project.

International Standard Book Number-13: 978-0-309-18751-0
International Standard Book Number-10: 0-309-18751-6

Additional copies of this report are available from the National Academies Press, 500 Fifth Street, N.W., Lockbox 285, Washington, DC 20055; (800) 624-6242 or (202) 334-3313 (in the Washington metropolitan area); Internet, http://www.nap.edu.

Copyright 2011 by the National Academy of Sciences. All rights reserved.

Printed in the United States of America

THE NATIONAL ACADEMIES
Advisers to the Nation on Science, Engineering, and Medicine

The **National Academy of Sciences** is a private, nonprofit, self-perpetuating society of distinguished scholars engaged in scientific and engineering research, dedicated to the furtherance of science and technology and to their use for the general welfare. Upon the authority of the charter granted to it by the Congress in 1863, the Academy has a mandate that requires it to advise the federal government on scientific and technical matters. Dr. Ralph J. Cicerone is president of the National Academy of Sciences.

The **National Academy of Engineering** was established in 1964, under the charter of the National Academy of Sciences, as a parallel organization of outstanding engineers. It is autonomous in its administration and in the selection of its members, sharing with the National Academy of Sciences the responsibility for advising the federal government. The National Academy of Engineering also sponsors engineering programs aimed at meeting national needs, encourages education and research, and recognizes the superior achievements of engineers. Dr. Charles M. Vest is president of the National Academy of Engineering.

The **Institute of Medicine** was established in 1970 by the National Academy of Sciences to secure the services of eminent members of appropriate professions in the examination of policy matters pertaining to the health of the public. The Institute acts under the responsibility given to the National Academy of Sciences by its congressional charter to be an adviser to the federal government and, upon its own initiative, to identify issues of medical care, research, and education. Dr. Harvey V. Fineberg is president of the Institute of Medicine.

The **National Research Council** was organized by the National Academy of Sciences in 1916 to associate the broad community of science and technology with the Academy's purposes of furthering knowledge and advising the federal government. Functioning in accordance with general policies determined by the Academy, the Council has become the principal operating agency of both the National Academy of Sciences and the National Academy of Engineering in providing services to the government, the public, and the scientific and engineering communities. The Council is administered jointly by both Academies and the Institute of Medicine. Dr. Ralph J. Cicerone and Dr. Charles M. Vest are chair and vice chair, respectively, of the National Research Council.

www.national-academies.org

COMMITTEE ON ECONOMIC AND ENVIRONMENTAL IMPACTS OF INCREASING BIOFUELS PRODUCTION

LESTER B. LAVE, *Chair (until May 9, 2011)*, IOM,[1] Carnegie Mellon University, Pittsburgh, Pennsylvania
INGRID (INDY) C. BURKE, *Cochair (from May 9, 2011)*, University of Wyoming, Laramie
WALLACE E. TYNER, *Cochair (from May 9, 2011)*, Purdue University, West Lafayette, Indiana
VIRGINIA H. DALE, Oak Ridge National Laboratory, Tennessee
KATHLEEN E. HALVORSEN, Michigan Technological University, Houghton
JASON D. HILL, University of Minnesota, St. Paul
STEPHEN R. KAFFKA, University of California, Davis
KIRK C. KLASING, University of California, Davis
STEPHEN J. MCGOVERN, PetroTech Consultants, Mantua, New Jersey
JOHN A. MIRANOWSKI, Iowa State University, Ames
ARISTIDES (ARI) PATRINOS, Synthetic Genomics, Inc., La Jolla, California
JERALD L. SCHNOOR, NAE,[2] University of Iowa, Iowa City
DAVID B. SCHWEIKHARDT, Michigan State University, East Lansing
THERESA L. SELFA, State University of New York – College of Environmental Science and Forestry, Syracuse
BRENT L. SOHNGEN, Ohio State University, Columbus
J. ANDRES SORIA, University of Alaska, Fairbanks

Project Staff

KARA N. LANEY, Study Codirector
EVONNE P.Y. TANG, Study Codirector
KAMWETI MUTU, Research Associate
KAREN L. IMHOF, Administrative Coordinator
ROBIN A. SCHOEN, Director, Board on Agriculture and Natural Resources
JAMES ZUCCHETTO, Director, Board on Energy and Environmental Systems

Editor

PAULA TARNAPOL WHITACRE, Full Circle Communications, LLC

---

[1]Institute of Medicine.
[2]National Academy of Engineering.

BOARD ON AGRICULTURE AND NATURAL RESOURCES

NORMAN R. SCOTT, *Chair,* NAE,[1] Cornell University, Ithaca, New York
PEGGY F. BARLETT, Emory University, Atlanta, Georgia
HAROLD L. BERGMAN, University of Wyoming, Laramie
RICHARD A. DIXON, NAS,[2] Samuel Roberts Noble Foundation, Ardmore, Oklahoma
DANIEL M. DOOLEY, University of California, Oakland
JOAN H. EISEMANN, North Carolina State University, Raleigh
GARY F. HARTNELL, Monsanto Company, St. Louis, Missouri
GENE HUGOSON, Global Initiatives for Food Systems Leadership, St. Paul, Minnesota
MOLLY M. JAHN, University of Wisconsin, Madison
ROBBIN S. JOHNSON, Cargill Foundation, Wayzata, Minnesota
A.G. KAWAMURA, Solutions from the Land, Washington, DC
JULIA L. KORNEGAY, North Carolina State University, Raleigh
KIRK C. KLASING, University of California, Davis
VICTOR L. LECHTENBERG, Purdue University, West Lafayette, Indiana
JUNE BOWMAN NASRALLAH, NAS,[2] Cornell University, Ithaca, New York
PHILIP E. NELSON, Purdue University, West Lafayette, Indiana
KEITH PITTS, Marrone Bio Innovations, Davis, California
CHARLES W. RICE, Kansas State University, Manhattan
HAL SALWASSER, Oregon State University, Corvallis
ROGER A. SEDJO, Resources for the Future, Washington, DC
KATHLEEN SEGERSON, University of Connecticut, Storrs
MERCEDES VAZQUEZ-AÑON, Novus International, Inc., St. Charles, Missouri

Staff

ROBIN A. SCHOEN, Director
CAMILLA YANDOC ABLES, Program Officer
RUTH S. ARIETI, Research Associate
KAREN L. IMHOF, Administrative Coordinator
KARA N. LANEY, Program Officer
AUSTIN J. LEWIS, Senior Program Officer
JANET M. MULLIGAN, Senior Program Associate for Research
KATHLEEN REIMER, Senior Program Assistant
EVONNE P.Y. TANG, Senior Program Officer
PEGGY TSAI, Program Officer

---

[1]National Academy of Engineering.
[2]National Academy of Sciences.

BOARD ON ENERGY AND ENVIRONMENTAL SYSTEMS

ANDREW BROWN, JR., *Chair,* NAE,[1] Delphi Corporation, Troy, Michigan
RAKESH AGRAWAL, NAE,[1] Purdue University, West Lafayette, Indiana
WILLIAM BANHOLZER, NAE,[1] The Dow Chemical Company, Midland, Michigan
MARILYN BROWN, Georgia Institute of Technology, Atlanta
MICHAEL CORRADINI, NAE,[1] University of Wisconsin, Madison
PAUL A. DeCOTIS, New York State Energy R&D Authority, Albany, New York
CHRISTINE EHLIG-ECONOMIDES, NAE,[1] Texas A&M University, College Station
SHERRI GOODMAN, CNA, Alexandria, Virginia
NARAIN HINGORANI, NAE,[1] Independent Consultant, Los Altos Hills, California
ROBERT J. HUGGETT, Independent Consultant, Seaford, Virginia
DEBBIE A. NIEMEIER, University of California, Davis
DANIEL NOCERA, NAS,[2] Massachusetts Institute of Technology, Cambridge
MICHAEL OPPENHEIMER, Princeton University, New Jersey
DAN REICHER, Stanford University, California
BERNARD ROBERTSON, NAE,[1] Daimler-Chrysler (retired), Bloomfield Hills, Michigan
ALISON SILVERSTEIN, Consultant, Pflugerville, Texas
MARK H. THIEMENS, NAS,[2] University of California, San Diego
RICHARD WHITE, Oppenheimer & Company, New York City

Staff

JAMES ZUCCHETTO, Director
K. JOHN HOLMES, Associate Director
DANA CAINES, Financial Associate
ALAN CRANE, Senior Program Officer
JONNA HAMILTON, Program Officer
LANITA JONES, Administrative Coordinator
ALICE WILLIAMS, Senior Program Assistant
MADELINE WOODRUFF, Senior Program Officer
JONATHAN YANGER, Senior Project Assistant

---

[1]National Academy of Engineering.
[2]National Academy of Sciences.

In Memoriam

Lester B. Lave
(1939-2011)

The committee dedicates this report to Dr. Lester Lave, chair for the majority of the duration of the study until his passing. Dr. Lave was a supremely accomplished scholar and educator, who conducted work of international significance and dedicated much of his time to National Research Council and Institute of Medicine studies. Dr. Lave was an inspirational leader. He framed complex questions in tractable ways, stimulated productive discussion on critical topics, listened carefully, and provided a strong hand to focus the committee's work. This report and each member of the committee benefited from his commitment to excellence.

Preface

Prediction is very difficult, especially if it's about the future.
—Niels Bohr

In the United States, we have come to depend upon plentiful and inexpensive energy to support our economy and lifestyles. In recent years, many questions have been raised regarding the sustainability of our current pattern of high consumption of nonrenewable energy and its environmental consequences. Further, because the United States imports about 55 percent of the nation's consumption of crude oil, there are additional concerns about the security of supply. Hence, efforts are being made to find alternatives to our current pathway, including greater energy efficiency and use of energy sources that could lower greenhouse-gas (GHG) emissions such as nuclear and renewable sources, including solar, wind, geothermal, and biofuels. This study focuses on biofuels and evaluates the economic and environmental consequences of increasing biofuel production. The statement of task asked this committee to provide "a qualitative and quantitative description of biofuels currently produced and projected to be produced by 2022 in the United States under different policy scenarios. . . ."

The United States has a long history with biofuels. Recent interest began in the late 1970s with the passage of the National Energy Conservation Policy Act of 1978, which established the first biofuel subsidy, applied in one form or another to corn-grain ethanol since then. The corn-grain ethanol industry grew slowly from the early 1980s to around 2003. From 2003 to 2007, ethanol production grew rapidly as methyl tertiary butyl ether was phased out as a gasoline oxygenate and replaced by ethanol. Interest in providing other incentives for biofuels increased also because of rising oil prices from 2004 and beyond. The Energy Independence and Security Act of 2007 established a new and much larger Renewable Fuel Standard and set in motion the drive toward 35 billion gallons of ethanol-equivalent biofuels plus 1 billion gallons of biodiesel by 2022. This National Research Council committee was asked to evaluate the consequences of such a policy; the nation is on a course charted to achieve a substantial increase in biofuels, and there are challenging and important questions about the economic and environmental consequences of continuing on this path.

The committee brings together expertise on the many dimensions of the topic. In addition, we called upon numerous experts to provide their perspectives, research conclusions, and insight. Yet, with all the expertise available to us, our clearest conclusion is that there is very high uncertainty in the impacts we were trying to estimate. The uncertainties include essentially all of the drivers of biofuel production and consumption and the complex interactions among those drivers: future crude oil prices, feedstock costs and availability, technological advances in conversion efficiencies, land-use change, government policy, and more. The U.S. Department of Energy projects crude oil price in 2022 to range between $52 and $191 per barrel (in 2008 dollars), a huge range. There are no commercial cellulosic biofuel refineries in the United States today. Consequently, we do not know much about growing, harvesting, and storing such feedstocks at scale. We do not know how well the conversion technologies will work nor what they will cost. We do not have generally agreed upon estimates of the environmental or GHG impacts of most biofuels. We do not know how landowners will alter their production strategies. The bottom line is that it simply was not possible to come up with clear quantitative answers to many of the questions. What we tried to do instead is to delineate the sources of the uncertainty, describe what factors are important in understanding the nature of the uncertainty, and provide ranges or conditions under which impacts might play out.

Under these conditions, scientists often use models to help understand what future conditions might be like. In this study, we examined many of the issues using the best models available. Our results by definition carry the assumptions and inherent uncertainties in these models, but we believe they represent the best science and scientific judgment available.

We also examined the potential impacts of various policy alternatives as requested in the statement of work. Biofuels are at the intersection of energy, agricultural, and environmental policies, and policies in each of these areas can be complex. The magnitude of biofuel policy impacts depends on the economic conditions in which the policy plays out, and that economic environment (such as growth of gross domestic product and oil price) is highly uncertain. Of necessity, we made the best assumptions we could and evaluated impacts contingent upon those assumptions.

Biofuels are complicated. Biofuels are controversial. There are very strong advocates for and political supporters of biofuels. There are equally strong sentiments against biofuels. Our deliberations as a committee focused on the scientific aspects of biofuel production—social, natural, and technological. Our hope is that the scientific evaluation sheds some light on the heat of the debate, as we have delineated the issues and consequences as we see them, together with all the inherent uncertainty.

Ingrid C. Burke
Wallace E. Tyner
Cochairs, Committee on Economic and Environmental
Effects of Increasing Biofuels Production

Acknowledgments

This report is a product of the cooperation and contribution of many people. The members of the committee thank all the speakers who provided briefings to the committee. (Appendix C contains a list of presentations to the committee.) Members also wish to express gratitude to Nathan Parker, University of California, Davis, and Alicia Rosburg, Iowa State University, who provided input to the committee.

This report has been reviewed in draft form by persons chosen for their diverse perspectives and technical expertise in accordance with procedures approved by the National Research Council's Report Review Committee. The purpose of this independent review is to provide candid and critical comments that will assist the institution in making its published report as sound as possible and to ensure that the report meets institutional standards of objectivity, evidence, and responsiveness to the study charge. The review comments and draft manuscript remain confidential to protect the integrity of the deliberative process. We wish to thank the following individuals for their review of this report:

Robert P. Anex, University of Wisconsin, Madison
Dan L. Cunningham, University of Georgia
William E. Easterling, Pennsylvania State University
R. Cesar Izaurralde, Joint Global Change Research Institute and University of Maryland
James R. Katzer, ExxonMobil (retired)
Eric F. Lambin, Stanford University and University of Louvain
Bruce Lippke, University of Washington
Heather MacLean, University of Toronto
K. Ramesh Reddy, University of Florida
John Reilly, Massachusetts Institute of Technology
James C. Stevens, Dow Chemical Company
Scott Swinton, Michigan State University
William Ward, Clemson University

Although the reviewers listed above have provided many constructive comments and suggestions, they were not asked to endorse the conclusions or recommendations, nor did they see the final draft of the report before its release. The review of this report was overseen by Dr. Thomas E. Graedel, Yale University, appointed by the Division on Earth and Life Studies, and Dr. M. Granger Morgan, Carnegie Mellon University, appointed by the NRC's Report Review Committee. They were responsible for making certain that an independent examination of this report was carried out in accordance with institutional procedures and that all review comments were carefully considered. Responsibility for the final content of this report rests entirely with the authoring committee and the institution.

Contents

List of Tables, Figures, and Boxes

TABLES

FIGURES

BOXES

Summary

Biofuels that can be produced from renewable domestic resources offer an alternative to petroleum-based fuels. To encourage the production and consumption of biofuels in the United States, the U.S. Congress enacted the Renewable Fuel Standard (RFS) as part of the 2005 Energy Policy Act and amended it in the 2007 Energy Independence and Security Act (EISA). The RFS, as amended by EISA (referred to as RFS2 hereafter), mandated volumes of renewable fuels to be used in U.S. transportation fuel from 2008 to 2022 (Figure S-1; see Box S-1 for definitions of renewable fuels pertaining to RFS2). At the request of the U.S. Congress, the National Research Council convened a committee of 16 experts to provide an independent assessment of the economic and environmental benefits and concerns associated with achieving RFS2. The committee drew on its own expertise and solicited input from many experts in federal agencies, academia, trade associations, stakeholders' groups, and nongovernmental organizations in a series of open meetings and in writing to fulfill the statement of task. (See complete statement of task in Appendix A.)

The committee was asked to

- Describe biofuels produced in 2010 and projected to be produced and consumed by 2022 using RFS-compliant feedstocks primarily from U.S. forests and farmland. The 2022 projections were to include per-unit cost of production.
- Review model projections and other estimates of the relative effects of increasing biofuel production as a result of RFS2 on the prices of land, food and feed, and forest products; on the imports and exports of relevant commodities; and on federal revenue and spending.
- Discuss the potential environmental harm and benefits of biofuel production and the barriers to achieving the RFS2 consumption mandate.

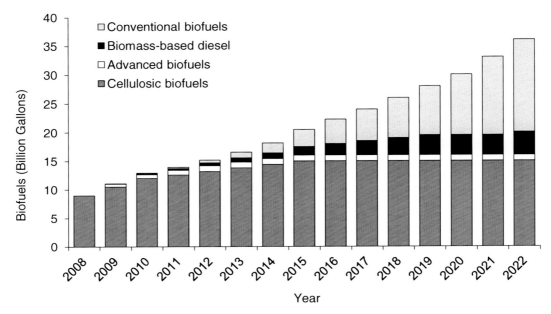

FIGURE S-1 Renewable fuel volume consumption mandated by RFS2.
NOTE: All volumes, except for volumes of biomass-based diesel, are shown in billions of gallons of ethanol-equivalent. The consumption mandate for biomass-based diesel is to be met on a biodiesel-equivalent basis.

BOX S-1
Definitions of Renewable Fuels in RFS2

RFS2 divides the total renewable fuel requirement into four categories:

- Conventional biofuel that is ethanol derived from corn starch and has a life-cycle greenhouse-gas (GHG) threshold of at least 20-percent reduction in emissions compared to petroleum-based gasoline and diesel.
- Biomass-based diesel that achieves life-cycle GHG reduction threshold of at least 50 percent.
- Advanced biofuels that are renewable fuels other than ethanol derived from corn starch and that achieve a life-cycle GHG reduction threshold of at least 50 percent. Advanced biofuels can include cellulosic biofuels and biomass-based diesel.
- Cellulosic biofuels derived from any cellulose, hemicellulose, or lignin from renewable biomass that achieve a life-cycle GHG reduction threshold of at least 60 percent.

KEY FINDINGS

FINDING: Absent major technological innovation or policy changes, the RFS2-mandated consumption of 16 billion gallons of ethanol-equivalent cellulosic biofuels is unlikely to be met in 2022.

The United States had the capacity to produce 14.1 billion gallons of ethanol per year from corn grain and 2.7 billion gallons of biodiesel per year from soybean oil, other vegetable oils, and animal fats at the end of 2010. That year, about 13.2 billion gallons of ethanol and

311 million gallons of biodiesel were produced in the United States. Therefore, adequate volumes are likely to be produced to meet the consumption mandates of 15 billion gallons of conventional biofuel and at least 1 billion gallons of biodiesel[1] by 2022. In contrast, whether and how the mandate for cellulosic biofuels will be met is uncertain. Although several studies suggested that the United States can produce adequate biomass feedstock for conversion to 16 billion gallons of ethanol-equivalent cellulosic biofuels to meet the consumption mandate, no commercially viable biorefineries exist for converting lignocellulosic biomass to fuels as of the writing of this report. Another report, *Liquid Transportation Fuels from Coal and Biomass: Technological Status, Costs, and Environmental Impacts*, estimated that aggressive deployment, in which the capacity build rate doubles the historic capacity build rate of corn-grain ethanol, is needed if 16 billion gallons of ethanol-equivalent cellulosic biofuels are to be produced by 2022. That estimate was based on the assumption that robust commercial-scale technology would be ready for deployment by 2015. Although the government guarantees a market for cellulosic biofuels regardless of price up to the level of the consumption mandate,[2] policy uncertainty and high cost of production might deter investors from aggressive deployment. Therefore, the capacity for producing cellulosic biofuels to meet the RFS2 consumption mandate will not be available unless innovative technologies are developed that unexpectedly improve the cellulosic biofuel production process, and technologies are scaled up and undergo several commercial-scale demonstrations in the next few years to optimize capital and operating costs.

FINDING: Only in an economic environment characterized by high oil prices, technological breakthroughs, and a high implicit or actual carbon price would biofuels be cost-competitive with petroleum-based fuels.

The committee used the Biofuel Breakeven Model to evaluate the costs and feasibility of a local or regional cellulosic biomass market for a variety of potential feedstocks. The model was used to estimate the minimum price that biomass producers would be willing to accept (WTA) for a dry ton of biomass delivered to the biorefinery gate and the maximum price that biorefineries would be willing to pay (WTP) to at least break even.

The price of crude oil, which is the chief competitor with biofuels, is a key determinant in the competitiveness of cellulosic biofuel and other advanced biofuels in the marketplace. Because crude oil prices are highly volatile, the difference between the WTP and WTA was calculated for three oil prices: $52, $111, and $191 per barrel, which are the low, reference, and high price projections for 2022 from the Department of Energy's Annual Energy Outlook in 2008$. Table S-1 shows that the price gap is positive for all potential cellulosic feedstocks if the oil price is $111 per barrel and policy incentives for biofuels do not exist. In this scenario, no cellulosic feedstock market is feasible without policy incentives.

A cellulosic feedstock market would be feasible under other circumstances, such as if the price of oil reaches $191 per barrel, if a carbon price makes the price of cellulosic biofuels more competitive, if government subsidy payments are high enough, or if government mandates are enforced at given levels of biofuel blending. Oil price affects both the processor's WTP through fuel revenues and the supplier's WTA through production, handling, and transport costs. The price gap is eliminated for several feedstocks when oil prices are

---

[1]The actual consumption mandate for biomass-based diesel is 1.0 billion gallons per year in 2012. Thereafter, the volume, no less than 1.0 billion gallons of biodiesel equivalent per year, is to be determined by EPA in a future rule making.

[2]RFS2 mandates that the production capacity of cellulosic biofuels be used to the extent that companies build it.

TABLE S-1 Estimated Unit Price That Biorefineries Are Willing to Pay (WTP) for Biofuel Feedstock and Estimated Unit Price That Suppliers Are Willing to Accept (WTA) for Cellulosic Biomass When Oil Is $111 per Barrel and No Policy Incentives Exist

	WTA	WTP	Price Gap (Per Dry Ton)	Price Gap (Per Gallon of Ethanol)
Corn stover in a corn-soybean rotation	$92	$25	$67	$0.96
Corn stover in a 4-year corn-alfalfa rotation	$92	$26	$66	$0.94
Alfalfa	$118	$26	$92	$1.31
Switchgrass in the Midwest	$133	$26	$106	$1.51
Switchgrass in Appalachia	$100	$26	$74	$1.06
Miscanthus in the Midwest	$115	$26	$89	$1.27
Miscanthus in Appalachia	$105	$27	$79	$1.13
Wheat straw	$75	$27	$49	$0.70
Short-rotation woody crops	$89	$24	$65	$0.93
Forest residues	$78	$24	$54	$0.77

NOTE: Conversion yield of biomass to ethanol is assumed to be 70 gallons per dry ton. These results are based on original modeling work by the committee that builds upon the work performed in *Liquid Transportation Fuels from Coal and Biomass: Technological Status, Costs, and Environmental Impacts* (NAS-NAE-NRC, 2009).

at or above $191 per barrel. Alternatively, a carbon price[3] of $118-$138 per tonne of CO_2 equivalent can close the gap between WTP and WTA at an oil price of $111 per barrel for some feedstocks given current technology. A subsidy of $1.01 per gallon of cellulosic biofuel blended with fossil fuel was established in 2008, but this payment is not sufficient to close the price gap at $111 per barrel of oil.[4]

RFS2 is decoupled from biofuel cost of production and economics. Although the economics may be a strong deterrent to developing capacity, cellulosic biofuels will have a government-mandated market to the extent that capacity is built. The future of RFS2 after it expires in 2022 is a source of uncertainty for investors.

FINDING: RFS2 may be an ineffective policy for reducing global GHG emissions because the effect of biofuels on GHG emissions depends on how the biofuels are produced and what land-use or land-cover changes occur in the process.

GHGs are emitted into the atmosphere or stored in soil during different stages of biofuel production—for example, CO_2 storage in biomass during growth and emissions from fossil fuel combustion in the manufacturing, transport, and application of agricultural inputs, from fermentation to ethanol, and from tailpipe emissions. Processes that affect GHG emissions of biofuels also include land-use and land-cover changes. If the expanded production involves removing perennial vegetation on a piece of land and replacing it with an annual commodity crop, then the land-use change would incur a one-time GHG emission from biomass and soil that could be large enough to offset GHG benefits gained by displacing petroleum-based fuels with biofuels over subsequent years. Furthermore, such land conversion may disrupt any future potential for storing carbon in biomass and soil.

---

[3]A carbon price can be enacted through a carbon tax credit provided to the biofuel producer (or feedstock supplier) per dry ton of cellulosic feedstock refined or as the market price for carbon credits if processors are allocated marketable carbon credits for biofuel GHG reductions relative to conventional gasoline.

[4]These conclusions are based on average prices for a cellulosic biofuel industry that is assumed to be commercially competitive and viable. Other studies have shown small quantities of biomass feedstocks could be available at significantly lower prices.

In contrast, planting perennial bioenergy crops in place of annual crops could potentially enhance carbon storage in that site.

Indirect land-use change occurs if land used for production of biofuel feedstocks causes new land-use changes elsewhere through market-mediated effects. The production of biofuel feedstocks can constrain the supply of commodity crops and raise prices. If agricultural growers anywhere in the world respond to the market signals (higher commodity prices) by expanding production of the displaced commodity crop, indirect land-use change occurs. This process might ultimately lead to conversion of noncropland (such as forests or grassland) to cropland. Because agricultural markets are intertwined globally, production of bioenergy feedstock in the United States will result in land-use and land-cover changes somewhere in the world, but the extent of those changes and their net effects on GHG emissions are uncertain.

Biofuels produced from residues or waste products, such as corn stover and municipal solid waste, will not contribute much GHG emissions from land-use or land-cover changes as long as adequate residue is left in the field to maintain soil carbon. However, it is not economically and environmentally feasible to produce enough biomass to meet RFS2 through crop residue or municipal solid waste. Therefore, dedicated energy crops will have to be grown to meet the mandate, which will likely require conversion of uncultivated cropland or the displacement of commodity crops and pastures. The extent of market-mediated land-use change and the associated GHG emissions as a result of increasing biofuels and dedicated bioenergy crop production in the United States are difficult to estimate and highly uncertain. Although RFS2 imposes restrictions to discourage bioenergy feedstock producers from land-clearing or land-cover change in the United States that would result in net GHG emissions, the policy cannot prevent market-mediated effects nor control land-use or land-cover changes in other countries. Therefore, the extent to which biofuel produced from dedicated energy crops will result in savings in GHG emissions compared to using petroleum is uncertain.

ECONOMIC EFFECTS OF INCREASING BIOFUEL PRODUCTION

Land Prices

FINDING: Absent major increases in agricultural yields and improvement in the efficiency of converting biomass to fuels, additional cropland will be required for cellulosic feedstock production; thus, implementation of RFS2 is expected to create competition among different land uses, raise cropland prices, and increase the cost of food and feed production.

Cropland acreage in the United States has been declining as it has in all developed countries. If the United States produces 16 billion gallons of ethanol-equivalent cellulosic biofuels by 2022, 30-60 million acres of land might be required for cellulosic biomass feedstock production, thereby creating competition among land uses. Although biofuels produced from crop and forest residues and from municipal solid wastes could reduce the amount of land needed for cellulosic feedstock production, those sources are inadequate to supply 16 billion gallons of ethanol-equivalent cellulosic biofuels, particularly if a proportion of crop and forest residues are left in the field to maintain soil quality.

Food and Feed Prices

FINDING: Food-based biofuel is one of many factors that contributed to upward price pressure on agricultural commodities, food, and livestock feed since 2007; other factors

affecting those prices included growing population overseas, crop failures in other countries, high oil prices, decline in the value of the U.S. dollar, and speculative activity in the marketplace.

To date, the agricultural commodities most affected by U.S. biofuel production are corn and soybean. The increased competition for these commodities created by an expanding biofuel market has contributed to upward pressure on their prices, but the increase has had a small effect on consumers' food retail prices, except livestock products, because corn and soybean typically undergo some processing before reaching consumers' food basket. The difference between the price of an unprocessed commodity and the retail price of processed food is typically large. The committee estimated that an increase of 20-40 percent in agricultural commodity prices would result in an increase in the retail price of most processed grocery food products (for example, breakfast cereal and bread) containing those commodities of only 1 to 2 percent.

Corn and soybean are used as animal feed, so the livestock market has experienced increased competition from the biofuels market. Some of this competition is alleviated by the ability of livestock producers to feed their animals dried distillers grain with solubles (DDGS), a coproduct of dry-milling corn grain into ethanol. However, there are limits to the amount of DDGS that can be used without impairing efficient production and the quality of the product. Moreover, increased commodity prices raise the production costs of livestock, and the animal producer's ability to pass increased production costs quickly on to consumers is limited because high prices decrease demand. The reproductive pipeline involved in livestock production makes it difficult for producers to adjust herd numbers quickly in response to increased feed costs.

Price of Woody Biomass

Wood is the most widely available cellulosic bioenergy feedstock in the United States at present, and it will be an important source of supply for cellulosic biofuel refineries if they become economically viable. If a commercial woody biomass refinery is built, it would require a large supply of dry biomass to operate efficiently (1,000-2,000 dry tons per day). Residues from forest harvesting operations could provide only a modest supply of cellulosic feedstock for such an operation due to the high marginal cost of harvesting these additional materials, the limited legal definition for accessing residues, and the uncertain nature of future federal subsidies. Although there are currently large supplies of milling residues in the wood processing industry, most of these residues are already committed to electricity production (in recent years, up to 132 million dry tons of roundwood equivalent[5]), and thus would be costly for cellulosic biofuel producers to purchase. Pulpwood is the closest marketable commodity that could enter woody biomass markets, but it is a higher value product (and thus more costly as a feedstock) than either forest harvest residues or milling residues. As a result, RFS2 is likely to have large effects on wood product prices. Some factors could mitigate these effects, including technological breakthroughs that reduce the cost of extracting forest residues, changes in the legal definition for accessing residues, and the size of subsidies for forest residues.

---

[5]This includes industrial roundwood used directly to produce energy as well as residues, black liquor from the pulping process, and fuelwood harvested from the forest.

Imports and Exports of Relevant Commodities

A growing biofuel industry was one factor that contributed to an increase in international commodity prices. However, exports of corn, soybean, and wheat held steady or even increased largely due to a huge decline in the value of the U.S. dollar between 2002 and 2008. With a lower value for the U.S. dollar, commodity prices did not increase nearly as much in other currencies such as the euro or yen. If commodity prices had not increased as a result of biofuel production and other factors and the U.S. dollar had still depreciated, exports likely would have increased more.

Increased animal product costs (for example, prices of meat and dairy) as a result of the simultaneous implementation of RFS2 and the European Union's biofuel mandates are expected to decrease the global value of livestock industries substantially, with one estimate being $3.7 billion between 2006 and 2015 (2006$). Most of this decrease will occur outside the United States, which will observe only a minor reduction ($0.9 billion) in its livestock and processed livestock products. The effect in the United States is buffered by the increasing availability of coproducts from corn-grain ethanol production, especially DDGS.

Current estimates suggest that the RFS2 mandate will likely increase wood imports into the United States. If wood currently used by the wood processing sector is diverted to meeting the RFS2 mandate, the shift in industrial wood from traditional uses to biofuels could cause the United States to import more industrial wood from elsewhere. The scale of this effect, however, cannot be precisely estimated at this time.

Achieving RFS2, along with increasing fuel efficiency standards in vehicles, can contribute to reducing the nation's dependence on oil imports. If RFS2 is to be achieved, domestically produced biofuels can displace 1.6 million barrels of petroleum-based fuels each day. (Consumption of petroleum-based transportation fuels in 2010 was 13.5 million barrels per day.) Even if part of the RFS2 consumption mandate is to be met by imported ethanol, a net reduction in the volume of imported oil is expected.

Federal Budget

FINDING: Achieving RFS2 would increase the federal budget outlays mostly as a result of increased spending on payments, grants, loans, and loan guarantees to support the development of cellulosic biofuels and forgone revenue as a result of biofuel tax credits.

Federal Spending

Agricultural Commodity Payments

Federal spending on agricultural commodity payments is not expected to change as a result of increasing biofuel production in the United States. Government payments to the producers of the major agricultural commodities primarily take one of two forms: direct payments and countercyclical payments. Direct payments are fixed payments provided to crop producers regardless of the market price received by crop producers. Thus, under no circumstances would RFS2 generate savings in the budget cost of the direct payment program. Countercyclical payments are paid when the market price for a crop is less than the effective target price of that crop. The effective target price of a crop is calculated as the legislated target price of that crop minus the direct payment for that crop. U.S. agricultural commodity prices are projected to exceed effective target prices from 2011 to 2021. If these projections hold true, then no countercyclical payments will be paid.

Conservation Reserve Program

The effect of biofuel production on the federal spending for conservation programs is uncertain. The Conservation Reserve Program (CRP) is the largest federal conservation program directed at agricultural land. Its objective is to provide "technical and financial assistance to eligible farmers and ranchers to address soil, water, and related natural resource concerns on their lands in an environmentally beneficial and cost-effective manner" (http://www.nrcs.usda.gov/programs/crp/). At the time this report was written, participants in the program received an average payment of $44 an acre. Federal outlays for fiscal year 2010 were estimated to cost $1.7 billion. If land is withdrawn from CRP for biofuel feedstock production and not replaced by new enrollment, the cost of CRP will decrease. However, CRP application acreage in a given year typically exceeds the maximum program acreage. The cost of the program will increase if enrollment applications are insufficient and if per-acre payment levels are increased to keep CRP competitive with crop or biofuel feedstock production and to incentivize producers to keep the most sensitive land in the program.

Nutritional and Other Income Assistance Programs

Nutritional and other income assistance programs are often adjusted for changes in the general price level as a means of protecting the real purchasing power of program recipients; therefore, if food retail prices increase, the program payments will typically be adjusted to reflect this change. Under such circumstances, expenses will increase not only for the Supplemental Nutrition Assistance Program and the Special Supplemental Assistance Program for Women, Infants, and Children,[6] but also for much larger income assistance programs, such as Social Security, military or civilian retirement programs, or Supplemental Security Income. Given that biofuels are only one of many factors affecting food retail prices, possible increases in the costs of these programs cannot be solely attributed to RFS2.

Grants, Loans, and Loan Guarantees

Grants, loans, and loan guarantees to support the production of feedstock, the cost of biofuel processing, and the development of cellulosic biofuel infrastructure have also been made. Biofuel production subsidies that reduce the cost of feedstock purchased by cellulosic biofuel refineries are typically provided in the form of payments per unit of feedstock purchased. Research into lowering the cost of biofuel processing can be aimed at many different areas in the production chain, including investment in increasing crop yields and in increasing the amount of biofuel produced per unit of biomass. Subsidies to reduce the capital investment cost of constructing cellulosic biofuel refineries are typically provided in the form of tax credits, grants, loans, or loan guarantees that provide a rate of interest below what investors could obtain from alternative financing sources.

Forgone Federal Revenue

Transportation fuels are taxed in the United States, but the structure of excise tax rates and exemptions varies by transportation mode and fuel type. Biofuel use is encouraged through a federal tax credit to fuel blenders. The 2008 farm bill set the Volumetric Ethanol Excise Tax Credit (VEETC) at $0.45 per gallon of ethanol blended with gasoline. Blenders receive a $1 per gallon tax credit for the use of biodiesel, and a $1.01 per gallon credit for the use of cellulosic biofuel. The value of payments made to blenders for the use of biodiesel

---

[6]Two nutritional assistance programs operated by the U.S. Department of Agriculture.

and cellulosic biofuel is less than $1 billion a year because these fuels are not produced in large volumes. However, forgone federal tax revenue as a result of VEETC was $5.4 billion in 2010 and is anticipated to increase to $6.75 billion in 2015 as corn-grain ethanol production approaches the mandate limit. The forgone revenue is much larger than any savings that could be gained from reduced CRP enrollment. As of the writing of this report, the biofuel subsidies were under review by Congress.

Impact with No Federal Subsidies

All biofuel tax credits will end in 2012 unless Congress takes action to extend them, but RFS2 will remain in effect. Without biofuel tax credits and with RFS2 in effect, the cost of biofuel programs is borne directly by consumers, as they are forced to pay a higher cost for the blended renewable fuel than for petroleum-based products. Otherwise, consumers bear the cost of biofuel programs indirectly through taxes paid.

ENVIRONMENTAL EFFECTS OF INCREASING BIOFUEL PRODUCTION

FINDING: The environmental effects of increasing biofuel production largely depend on feedstock type, site-specific factors (such as soil and climate), management practices used in feedstock production, land condition prior to feedstock production, and conversion yield. Some effects are local and others are regional or global. A systems approach that considers various environmental effects simultaneously and across spatial and temporal scales is necessary to provide an assessment of the overall environmental outcome of increasing biofuel production.

Although using biofuels holds potential to provide net environmental benefits compared to using petroleum-based fuels, the environmental outcome of biofuel production cannot be guaranteed because the key factors that influence environmental effects from bioenergy feedstock production are site specific and depend on the type of feedstocks produced, the management practices used to produce them, prior land use, and any land-use changes that their production might incur. In addition to GHG emissions, biofuel production affects air quality, water quality, water quantity and consumptive use, soil, and biodiversity. Thus, the environmental effects of biofuels cannot be focused on one environmental variable. Environmental effects of increasing biofuel production have to be considered across spatial scales because some effects are local and regional (for example, water quality and quantity) and others are global (for example, GHG emissions have the same global effect irrespective of where they are emitted). Planning based on landscape analysis could help integrate biofuel feedstock production into agricultural landscapes in ways that improve environmental outcomes and benefit wildlife by encouraging placement of cellulosic feedstock production in areas that can enhance soil quality or help reduce agricultural nutrient runoffs, anticipating and reducing the potential of groundwater overuse and enhancing wildlife habitats.

Air Quality

Air quality modeling suggests that production and use of ethanol as fuel to displace gasoline is likely to increase such air pollutants as particulate matter, ozone, and sulfur oxides. Published studies projected that overall production and use of ethanol will result in higher pollutant concentration for ozone and particulate matter than their gasoline counterparts on a national average. Unlike GHG effects, air-quality effects from corn-grain ethanol are largely localized. The potential extent to which the air pollutants harm human health depends on whether the pollutants are emitted close to highly populated areas and exposure.

Water Quality

An assessment of the effects of producing biofuels to achieve the RFS2 consumption mandate on water quality requires detailed information on where the bioenergy feedstocks would be grown and how they would be integrated into the existing landscape. The increase in corn production has contributed to environmental effects on surface and ground water, including hypoxia, harmful algal blooms, and eutrophication. Additional increases in corn production under RFS2 likely will have additional negative environmental effects (though production of corn-grain ethanol in 2010 was only 1 billion gallons less than the consumption mandate for years 2015 to 2022). Perennial and short-rotation woody crops for cellulosic feedstocks with low agrichemical inputs and high nutrient uptake efficiency hold promise for improving water quality under RFS2, particularly if integrated with food-based crops. Use of residues would not require much additional inputs so that they are not likely to incur much negative effects on water quality as long as enough residues are left in field to prevent soil erosion. The site-specific details of the implementation of RFS2, and particularly the balance of feedstocks and levels of inputs, will determine whether or not RFS2 will lead to improved or diminished water quality.

Water Quantity and Consumptive Water Use

Published estimates of consumptive water use over the life cycle of corn-grain ethanol (15-1,500 gallons per gallon of gasoline equivalent) and cellulosic biofuels (2.9-1,300 gallons per gallon of gasoline equivalent) are higher than petroleum-based fuels (1.9-6.6 gallons per gallon of gasoline equivalent), but the effects of water use depend on regional availability. An individual refinery might not pose much stress on a water resource, but multiple refineries could alter the hydrology in a region. In particular, biorefineries are most likely situated close to sources of bioenergy feedstock production, both of which draw upon local water resources. Yet, regional water availability was not always taken into account in the models that project cellulosic biorefinery locations.

Soil Quality and Biodiversity

Effects of biofuel production on soil quality and biodiversity primarily result from the feedstock production and removal stages, particularly on the rates of biological inputs and outputs and the levels of removal. The effects of achieving RFS2 on biodiversity currently cannot be readily quantified or qualified largely because of the uncertainty in the future. Bioenergy feedstock production can reduce or enhance biodiversity depending on the compatibility of feedstock type, management practices, timing of harvest, and input use with plants and animals in the area of production and its surroundings. Precise regional assessments at each site of feedstock production for biofuels are needed to assess the collective effects of achieving RFS2 on biodiversity.

BARRIERS TO ACHIEVING RFS2

FINDING: Key barriers to achieving RFS2 are the high cost of producing cellulosic biofuels compared to petroleum-based fuels and uncertainties in future biofuel markets.

RFS2 guarantees a market for cellulosic biofuels produced, but market uncertainties could deter private investment. Of the three crude oil prices tested in this study, the only one for which biofuels were economic without subsidies was $191. The breakeven crude oil price would be between $111 and $191. If the biofuel is ethanol, there also are infrastructure and blend wall[7] issues to surmount. Production of "drop-in"[8] fuels instead of ethanol would eliminate these additional downstream costs. Although RFS2 provides a market for the biofuels produced even at costs considerably higher than fossil fuels, uncertainties in enforcement and implementation of RFS2 mandate levels affect investors' confidence and discourage investment. EPA has the right to waive or defer enforcement of RFS2 under a variety of circumstances, and the agency is "required to set the cellulosic biofuel standard each year based on the volume projected to be available during the following year." In 2011, the RFS level of 250 million gallons of ethanol-equivalent cellulosic biofuel was reduced to 6.6 million gallons. As of 2011, biofuel production is contingent on subsidies, RFS2, and similar policies.

Opportunities to reduce costs of biofuels include decreasing the cost of bioenergy feedstock, which constitutes a large portion of operating costs, and increasing the conversion efficiency from biomass to fuels. Research and development to improve feedstock yield through breeding and biotechnology and conversion yield could reduce costs of biofuel production and potentially reduce the environmental effects per unit of biofuel produced.

---

[7]Most ethanol in the United States is consumed as a blend of 10-percent ethanol and 90-percent gasoline. If every drop of gasoline-type fuel consumed in U.S. transportation could be blended, then a maximum of about 14 billion gallons of ethanol could be blended.

[8]A nonpetroleum fuel that is compatible with existing pipelines and delivery mechanisms for petroleum-based fuels.

1

Introduction

Transportation is a critical component of sustained economic growth in industrialized societies. Globally, about 94 percent of transportation fuels used are derived from crude oil. As the largest consumer of crude oil in 2009, the United States was responsible for about 27 percent of oil used worldwide (EIA, 2010f). The large quantities of oil consumed in the United States (about 19 million to 21 million barrels per day in 2005-2009) contribute to two major problems: energy security and greenhouse-gas[1] (GHG) emissions. With respect to energy security, 52 to 60 percent of oil consumed in the United States was imported in 2005-2009 (EIA, 2009a). The use of petroleum-based fuel in transportation contributes about 30 percent of all carbon dioxide (CO_2) emissions in the United States (EPA, 2010a).

The U.S. Congress enacted the Energy Independence and Security Act (EISA) in 2007 (110 P.L. 140) "to move the United States toward greater energy independence and security, to increase the production of clean renewable fuels, to protect consumers, to increase the efficiency of products, buildings, and vehicles, to promote research on and deploy GHG capture and storage options, and to improve the energy performance of the Federal Government, and for other purposes." A subtitle within EISA entitled the Renewable Fuel Standard (RFS) mandates the amounts of biofuels to be consumed each year. At the request of the U.S. Congress, the National Research Council convened a committee to assess the economic and environmental impacts of increasing biofuel production. (See Appendix A for statement of task and Appendix B for committee membership.) In addition to drawing on its own expertise, the committee solicited input from many experts in federal agencies, academia, trade associations, and nongovernmental organizations in a series of open meetings and in writing to fulfill the statement of task. (See Appendix C for a list of presentations to the committee.) Box 1-1 shows how different

---

[1]Greenhouse gases are gases in the atmosphere that absorb and emit radiation within the thermal infrared range, and hence produce a warming effect. The Earth's natural "greenhouse effect" makes the surface temperature of the planet suitable for living organisms. However, a multitude of evidence shows that emissions of large quantities of greenhouse gases from human activities (for example, burning fossil fuels) have significantly intensified the greenhouse effect (NRC, 2010a). The main greenhouse gases in the Earth's atmosphere are water vapor, CO_2, methane, nitrous oxide, and ozone.

BOX 1-1
Structure of the Report

1. Introduction

2. Biofuel supply chain

3. Projected supply of cellulosic biomass

4. Economics and economic effects of biofuel production   5. Environmental effects and tradeoffs of biofuels

6. Barriers to achieving RFS2

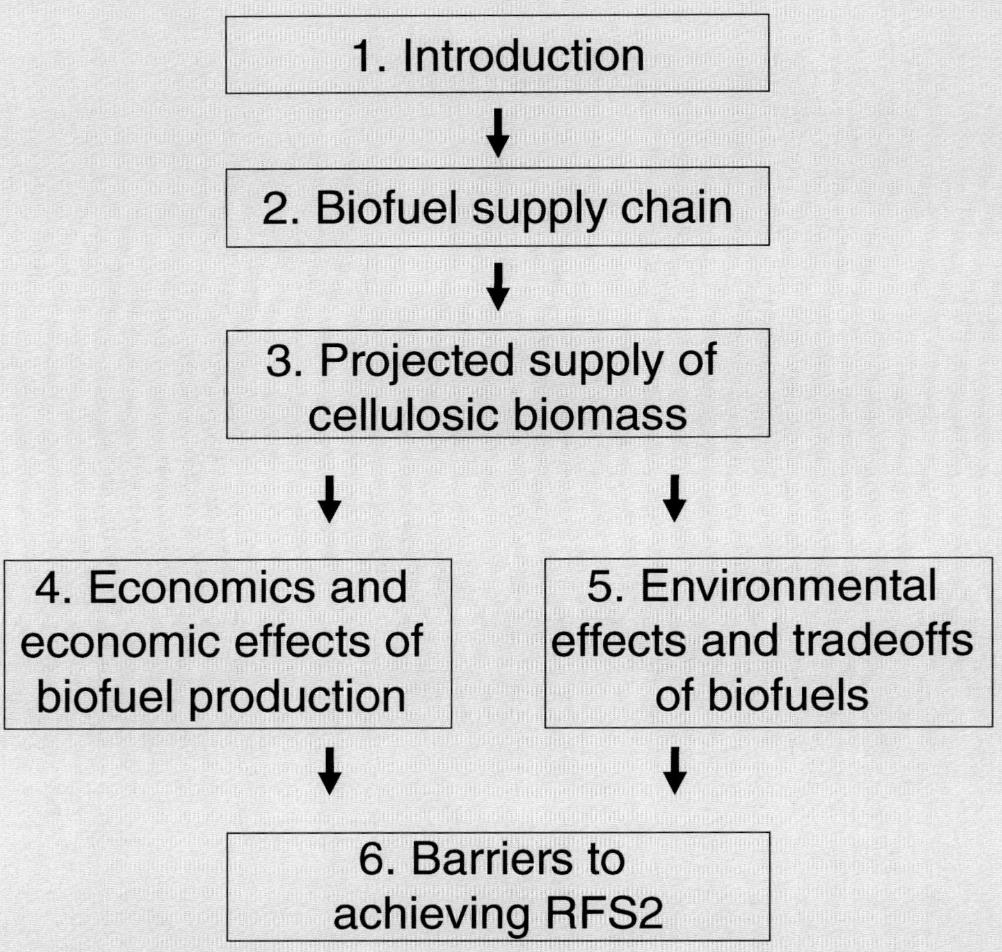

FIGURE 1-1 Structure of the report.

CHAPTER 1
• Introduction.

CHAPTER 2
• A quantitative and qualitative description of biofuels currently produced.

CHAPTER 3

- A qualitative and quantitative description of biofuels that could be produced in different regions of the United States, including a review of estimates of potential biofuel production levels using RFS-compliant feedstocks from U.S. forests and farmland.

CHAPTER 4

- Estimates of the per-unit costs of biofuel feedstock production.
- A quantitative description of biofuels projected to be produced and consumed by 2022 in the United States under different policy scenarios, including scenarios with and without current Renewable Fuel Standard (RFS) and biofuel tax and tariff policies, and considering a range of future fossil energy and biofuel prices, the impact of a carbon price, and advances in technology.
- An assessment of the effects of current and projected levels of biofuel production, and the incremental impact of additional production, on the number of U.S. acres used for crops, forestry, and other uses, and the associated changes in the price of rural and suburban land.
- A review of economic model results and other estimates of the relative effects of the RFS, biofuel tax and tariff policy, production costs, and other factors, alone and in combination, on biofuel and petroleum refining capacity, and on the types, amounts, and prices of biofuel feedstocks, biofuels, and petroleum-based fuels (including finished motor fuels) produced and consumed in the United States.
- An analysis of the effects of current and projected levels of biofuel production, and the incremental impact of additional production, on U.S. exports and imports of grain crops, forest products, and fossil fuels and on the price of domestic animal feedstocks, forest products, and food grains.
- An analysis of the effect of projected biofuel production on federal revenue and spending, through costs or savings to commodity crop payments, biofuel subsidies, and tariff revenue.
- An analysis of the pros and cons of achieving legislated RFS levels, including the impacts of potential short-falls in feedstock production on the prices of animal feed, food grains, and forest products, and including an examination of the impact of the cellulosic biofuel tax credit established by Sec. 15321 of the Food, Conservation, and Energy Act of 2008 on the regional agricultural and silvicultural capabilities of commercially available forest inventories. This analysis explores policy options to maintain regional agricultural and silvicultural capacity in the long term, given RFS requirements for annual increases in the volume of renewable fuels, and includes recommendations for the means by which the federal government could prevent or minimize adverse impacts of the RFS on the price and availability of animal feedstocks, food, and forest products, including options available under current law.

CHAPTER 5

- An analysis of the effect of current and projected future levels of biofuel production and use, and the incremental impact of additional production, on the environment. The analysis considers impacts due to changes in land use, fertilizer use, runoff, water use and quality, GHG and local pollutant emissions from vehicles utilizing biofuels, use of forestland biomass, and other factors relevant to the full life-cycle of biofuel production and use. The analysis summarizes and evaluates various estimates of the indirect effects of biofuel production on changes in land use and the environmental implications of those effects. A comparison of corn ethanol versus other biofuels and renewable energy sources for the transportation sector based on life-cycle analyses, considering cost, energy output, and environmental impacts, including GHG emissions.

CHAPTER 6

- An analysis of barriers to achieving the RFS requirements.

components of the statement of task are addressed in the six chapters of the report and the interconnectedness of the different chapters. This chapter provides a brief history of U.S. biofuel policies, goals, and production. It highlights some of the economic and environmental opportunities and concerns raised regarding U.S. biofuel policy.

Throughout this report, the committee uses the term "biofuels" to specify liquid fuels for transportation derived from biological sources. "Bioenergy," which encompasses all forms of energy for electricity or heat generation and for transportation produced from biological sources, is a broader term than biofuels. Recognizing that biomass can be used to produce various forms of energy, the term "bioenergy feedstock" is used throughout the report. Quantities of bioenergy feedstock are reported in dry weight. A glossary of terms and list of select acronyms and abbreviations are provided in Appendixes D and E. With the exception of GHG emissions, standard units that are commonly used in the United States are used throughout the report. Conversion factors to International Systems of Units are in Appendix F. GHG emissions are expressed in tonnes of CO_2 equivalent (CO_2 eq), the unit commonly used by the Intergovernmental Panel on Climate Change.

INTEREST IN BIOFUELS

Changing Demand for Transportation Energy

Since World War II, petroleum consumption from transportation in the United States has increased by at least four-fold. The transportation sector required just over 3 million barrels per day of petroleum in 1949; by 2009, it consumed about 13 million barrels per day (EIA, 2010b). The Energy Information Administration (EIA) projected that U.S. demand for transportation fuel will reach about 15 million barrels per day by 2022 and 16 million barrels per day by 2035 (EIA, 2010a). The projections suggest that increase in demand is expected to slow down. Petroleum consumption is not likely to increase in lockstep with population growth because of projected improvements in energy efficiency and projected increases in the use of biofuels as transportation fuel (EIA, 2010a). Therefore, although motor gasoline consumption increased from 7.4 million barrels per day in 1978 to 9.0 million barrels per day in 2008, EIA projected that consumption will peak at 9.5 million barrels per day in 2012 and eventually decline to 9.1 million barrels per day by 2035 (EIA, 2009b, 2010a).

Increase in U.S. demand for transportation fuel is projected to slow down in the next 25-30 years, but global demand is likely to continue to grow. World demand for petroleum increased from 63 million barrels per day in 1980 to 85.8 million barrels per day in 2008 (EIA, 2010e). Much of that growth took place in developing countries. Demand in OECD (Organisation for Economic Co-operation and Development) countries[2] increased by less than 6 million barrels per day, while demand in non-OECD countries nearly doubled, from 21 million barrels per day to 38 million barrels per day. In 1980, non-OECD countries' petroleum consumption was equivalent to just over 50 percent of that in OECD countries; by 2008, it had reached 80 percent. EIA projected transportation energy use to increase by 2.6 percent per year from 2007 to 2035 in non-OECD countries, surpassing OECD transportation energy consumption in 2025. World demand for petroleum in 2035 is projected to be 111 million barrels per day (EIA, 2010e).

---

[2]OECD member countries as of March 2011 were Australia, Austria, Belgium, Canada, Chile, Czech Republic, Denmark, Estonia, Finland, France, Germany, Greece, Hungary, Iceland, Ireland, Israel, Italy, Japan, Korea, Luxembourg, Mexico, the Netherlands, New Zealand, Norway, Poland, Portugal, Slovak Republic, Slovenia, Spain, Sweden, Switzerland, Turkey, the United Kingdom, and the United States.

Historic Interest in Biofuels

Petroleum is the dominant source of motor fuel today, but this was not the case when internal combustion engines and automobiles were first invented. Indeed, biofuels have been an energy source for vehicle engines since the development of the automobile. Henry Ford's first vehicle in 1896 ran on pure ethanol, and his Model T, produced in 1908, could use ethanol, gasoline, or a blend of the two (EIA, 2008). Early models of the diesel engine also could operate on vegetable oil (Knothe, 2001). The United States used 50-60 million gallons of ethanol per year as motor fuel while engaged in World War I (EIA, 2008). During World War II, the U.S. Army built an ethanol plant in Omaha, Nebraska, to supplement its fuel needs (EIA, 1995), and experiments with blends of petroleum diesel and diesel from vegetable and cottonseed oils were conducted (Knothe, 2001).

In the nascent years of the automobile industry, before the mass production of vehicles for personal use, ethanol contended to be the motor fuel of choice. Unlike gasoline, ethanol did not cause engine knock, and ethanol engines had higher compression ratios (Carolan, 2009). However, gasoline had several advantages over ethanol that caused it to be more successful before World War I and to be wholly dominant in the marketplace by the 1930s. Gasoline was a byproduct of kerosene production, an industry that was already well developed but that was beginning to lose its market share to electrical lighting powered by coal. There were over 100 kerosene refineries in the United States at the beginning of the 20th century with an established, decentralized distribution network (Melaina, 2007). As a less valuable byproduct of kerosene, gasoline could use this network. There was an abundance of gasoline—7 million barrels of gasoline were produced in 1905—but only 600,000 barrels were used to fuel automobiles. Gasoline was often used as a solvent but was frequently disposed of in rivers if it was not economical to distribute (Melaina, 2007). Therefore, there was a plentiful supply of gasoline fuel to meet growing demand when Ford began to produce Model T cars quickly and cheaply; though rudimentary, the infrastructure to deliver the fuel was already in place. Later, the alleviation of engine knock through the inclusion of tetraethyl lead in gasoline and an increase in compression ratios removed the few remaining technological barriers to using gasoline in vehicle engines (Dimitri and Effland, 2007; Carolan, 2009).

In contrast, fuel from corn was not as abundant as gasoline, and the distribution system for grain commodities was not congruent with fuel distribution (Dimitri and Effland, 2007). High corn prices in much of the early 20th century also made ethanol less cost competitive with gasoline (Giebelhaus, 1980; Dimitri and Effland, 2007). Later agricultural mechanization and favorable subsidies expanded corn acres at the expense of other possible ethanol feedstocks (Carolan, 2009). Ethanol's competitiveness also was reduced by alcohol tax during and following the Civil War until the tax was repealed in 1906 (Giebelhaus, 1980; Dimitri and Effland, 2007; Carolan, 2009). A movement emerged in the 1930s to produce fuel and other industrial products not only from corn but also from other crops high in sugar and starch such as sugarcane, Jerusalem artichokes, and sweet potatoes (Giebelhaus, 1980; Finlay, 2003). However, gasoline was firmly established as the primary abundant and cost-effective vehicle fuel by that time.

From the 1940s through the 1960s, the United States continued to increase petroleum production, which met most of domestic demand. Biofuel production was largely abandoned. However, domestic petroleum production peaked in 1970 at 9.6 million barrels per day while demand continued to increase (EIA, 2009b). Though just over 20 percent of U.S. consumption was met with imports in 1970, the United States was importing over 40 percent of its petroleum needs by the second half of the decade (EIA, 2009b). That reliance on foreign oil was acutely felt in the form of gasoline shortages and increased retail prices when the Organization of the Petroleum Exporting Countries (OPEC) embargoed oil in 1973 and when political unrest curtailed oil production in Iran in 1978. Those disruptions led to increased exploration for domestic fossil fuel reserves and also renewed interest and investment in biofuels (Duffield et al., 2008).

Policies to Encourage Biofuel Production

Spurred by concerns about energy security, the federal government included a tax incentive in the Energy Tax Act of 1978 (95 P.L. 619) in the form of an exemption for ethanol blends of at least 10-percent ethanol by volume from the $0.04 per gallon federal motor fuels tax (Table 1-1). Because ethanol was only 10 percent of the total volume, the $0.04 tax exemption on the blend amounted to a subsidy of $0.40 per gallon of ethanol. The Energy Security Act of 1980 offered insured loans to small ethanol plants (96 P.L. 294). This act also instructed the Secretary of Agriculture and the Secretary of Energy to develop a plan to increase ethanol production to the equivalent of 10 percent of the total U.S. gasoline consumption by 1990 (Duffield et al., 2008). Although interest in biofuels waned when the

TABLE 1-1 History of U.S. Ethanol and Biofuel Legislation

Year	Legislation	Provision
1978	Energy Tax Act of 1978	$0.40 per gallon of ethanol tax exemption on the $0.04 gasoline excise tax
1980	Crude Oil Windfall Profit Tax Act and the Energy Security Act	Promoted energy conservation and domestic fuel development
1982	Surface Transportation Assistance Act	Increased tax exemption to $0.50 per gallon of ethanol and increased the gasoline excise tax to $0.09 per gallon
1984	Tax Reform Act	Increased tax exemption to $0.60 per gallon of ethanol
1988	Alternative Motor Fuels Act	Created research and development programs and provided fuel economy credits to automakers
1990	Omnibus Budget Reconciliation Act	Ethanol tax incentive extended to 2000 but decreased to $0.54 per gallon of ethanol
1990	Clean Air Act amendments	Acknowledged contribution of motor fuels to air pollution – oxygen requirements for motor fuels
1992	Energy Policy Act	Tax deductions allowed on vehicles that could run on E85
1998	Transportation Efficiency Act of the 21st Century	Ethanol subsidies extended through 2007 but reduced to $0.51 per gallon of ethanol by 2005
2004	American Jobs Creation Act	Changed the mechanism of the ethanol subsidy to a blender tax credit instead of the previous excise tax exemption Extended the ethanol tax exemption to 2010
2005	Energy Policy Act	Established the Renewable Fuel Standard starting at 4 billion gallons in 2006 and rising to 7.5 billion in 2012 Eliminated the oxygen requirement for gasoline, but failed to provide MTBE legal immunity
2007	Energy Independence and Security Act	Established a Renewable Fuel Standard totaling 35 billion gallons of ethanol-equivalent biofuels and 1 billion gallons of biodiesel by 2022

SOURCE: Tyner (2008). Reprinted with permission from the American Institute of Biological Sciences.

price volatility of oil lessened in the 1980s, blenders' tax exemptions were modified and continued throughout the decade, and a tariff was established to prevent foreign ethanol producers from taking advantage of the credits. Despite these policies, the use of ethanol in motor fuels did not achieve the 10-percent target.

In the 1990s, the federal government began to look to ethanol as a means of combating air pollution from vehicle emissions. The Clean Air Act amendments of 1990 created mandates for oxygenates in gasoline to address carbon monoxide and ozone problems in urban areas (101 P.L. 549). This mandate increased demand for ethanol, but a petroleum derivative, methyl tertiary butyl ether (MTBE), was a more cost-effective oxygenate (Duffield et al., 2008). The Energy Policy Act of 1992 (102 P.L. 486) amended the motor fuels tax exemption and the blenders' credit to improve ethanol's ability to compete with MTBE. From 1992 to 1999, annual U.S. consumption of ethanol in gasoline-equivalent gallons increased from 719 million to 979 million. Almost all of the consumption and the growth in consumption was from using ethanol as an oxygenate (EIA, 2009b), which continued after the phase-out of MTBE use that began in the late 1990s because of concerns about water-quality contamination and state regulations prohibiting its use in motor fuels.

In the 2000s, federal legislation began to explicitly support biofuels to provide opportunities for agricultural and rural development. The Farm Security and Rural Investment Act of 2002 (107 P.L. 171) was the first farm bill to contain a title devoted to energy. It authorized a number of programs in support of biofuels, including grants for converting biomass into energy, programs to encourage farmers to increase use of renewable energy, and grants to promote public outreach about biodiesel. The 2002 farm bill also authorized continued funding for the Biomass Research and Development Initiative, which had been approved in an earlier bill, and codified the Bioenergy Program within the U.S. Department of Agriculture's Commodity Credit Corporation. The Bioenergy Program made payments available to biofuel producers who increased their level of production over the previous year (Duffield et al., 2008).

The 2008 farm bill, called The Food, Conservation and Energy Act (110 P.L. 246), included several provisions that encourage biomass production for fuels. Under the farm bill, the Biomass Crop Assistance Program (BCAP) and the Bioenergy Program for Advanced Biofuels were established. BCAP provides financial assistance for crop establishment, annual payments for crop production, and subsidies for collecting and delivering biomass material to production facilities. Biomass production has to be within an economically practical distance from conversion facilities to receive payments. The Bioenergy Program for Advanced Biofuels aims "to support and ensure an expanding production of advanced biofuels by providing payments to eligible advanced biofuel producers in rural areas" (USDA-RD, 2010). The program provides payments to eligible producers of fuel derived from renewable biomass, other than corn grain. The farm bill also includes a provision for the U.S. Department of Agriculture and the U.S. Department of Energy to award grants competitively to eligible entities to research, develop, and demonstrate biomass projects through the Biomass Research and Development Initiative.

In addition to support for biomass feedstock and biofuel production, the farm bill establishes a producer credit of $1.01 for each gallon of cellulosic biofuel until December 31, 2012. It modifies the Volumetric Ethanol Excise Tax Credit (VEETC) of $0.51 per gallon of ethanol blended into gasoline to companies that blend gasoline with ethanol; the VEETC, which was established by the American Jobs Creation Act of 2004, replaced the original excise tax exemption program from 1978 that was limited to specific blends with a tax credit based on the volume of ethanol consumed (a volume-based credit for biodiesel was also introduced) (Koplow, 2007; Solomon et al., 2007). This farm bill reduced the VEETC to $0.45 per gallon of ethanol blended into gasoline in the first calendar year that EPA certifies that 7.5 billion gallons of renewable fuel have been blended into gasoline. The farm bill also extended the $0.54 per gallon duty on imported ethanol through December 31, 2010. The expiration date of the import duty was pushed back to December 31, 2011, by the Tax Relief, Unemployment Insurance Reauthorization, and Job Creation Act of 2010 (111 P.L. 312).

RENEWABLE FUEL STANDARD

The steep rise in oil prices, growing concerns over energy security and GHG emissions, and the desire to support domestic farm and rural economies combined to reinvigorate support for biofuels in the mid-2000s. The Energy Policy Act (EPAct) of 2005 (109 P.L. 58) established a national Renewable Fuel Standard (RFS) that mandated an increased use of renewable fuels from 4.0 billion gallons per year in 2006 to 7.5 billion gallons per year in 2012. EISA of 2007 included a subtitle that amended RFS in EPAct and increased the volumes of renewable fuels to be phased in substantially. (RFS under EISA is referred to as RFS2 hereafter.) Mandated volumes of renewable fuel consumption began at 9 billion gallons in 2008 and will reach 36 billion gallons of biofuels (35 billion gallons of ethanol equivalent and 1 billion gallons of biodiesel) in 2022 (Figure 1-2). The term "renewable fuel" was defined as "fuel that is produced from renewable biomass and that is used to replace or reduce the quantity of fossil fuel present in a transportation fuel" (110 P.L. 140). RFS2 is an energy-equivalent standard because EPA interpreted the mandated volumes as ethanol-equivalent in its final rule (EPA, 2010a) to account for different energy contents of various possible biofuels (Table 1-2).

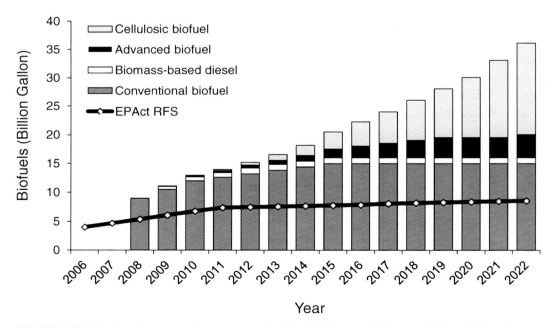

FIGURE 1-2 Mandated consumption target under the 2005 Energy Policy Act (EPAct RFS) and mandated consumption targets for different categories of biofuels under the 2007 Energy Independence and Security Act (EISA).

TABLE 1-2 Equivalence Values Assigned to Renewable Fuels

Fuel Type	Equivalence Value
Ethanol	1.0
Biodiesel (mono-alkyl ester)	1.5
Butanol	1.3
Nonester renewable diesel	1.7
Biogas	1.0
Electricity	1.0

NOTE: The energy content of pure ethanol is about 76,000 Btu per gallon (lower heating value).
SOURCE: EPA (2010b).

In addition to increasing the required volumes, RFS2 divides the total renewable fuel requirement into four categories:

- Conventional biofuel that is ethanol derived from corn starch. Conventional biofuel produced from facilities that commenced construction after December 19, 2007, would have to achieve a threshold of at least 20-percent reduction in GHG emissions compared to petroleum-based gasoline and diesel to qualify as a renewable fuel under RFS2.
- Advanced biofuels that are renewable fuels other than ethanol derived from corn starch and that achieve life-cycle GHG reduction threshold of at least 50 percent. Advanced biofuels can include ethanol and other types of biofuels derived from such renewable biomass as cellulose, hemicellulose, lignin, sugar, or any other starch other than corn starch, biomass-based diesel, and coprocessed renewable diesel.[3]
- Cellulosic biofuels that are renewable fuels derived from any cellulose, hemicellulose, or lignin from renewable biomass and that achieve life-cycle GHG reduction threshold of at least 60 percent. In general, cellulosic biofuels also qualify as renewable fuels and advanced biofuels.
- Biomass-based diesel, including biodiesel made from vegetable oils or animal fats and cellulosic diesel, that achieves life-cycle GHG reduction threshold of at least 50 percent—for example, soybean biodiesel and algal biodiesel. Coprocessed renewable diesel is excluded from this category.

The four renewable-fuel categories are nested within an overall mandate. There actually is no mandate for corn-grain ethanol, but a maximum quantity of conventional biofuels that can be filled with corn-grain ethanol. If any advanced or cellulosic biofuel become less expensive than corn-grain ethanol, the mandate for conventional biofuel could be filled entirely with advanced or cellulosic biofuel. RFS2 also requires that all renewable fuels be made from feedstocks that meet a new definition of renewable biomass. EISA's definition of renewable biomass incorporates land restrictions for planted crops, crop residue, planted trees and tree residue, slash and precommercial thinnings, and biomass from wildfire areas. Detailed definitions and EPA's interpretations of the terms are found in *Regulation of Fuels and Fuel Additives: Changes to Renewable Fuel Standard Program; Final Rule* (EPA, 2010b, pp. 14691-14697). A brief version of EISA's renewable biomass definition and land restrictions is included below.

---

[3]Coprocessed renewable diesel refers to diesel made from renewable material mixed with petroleum during the hydrotreating process.

- Planted crops or crop residues that were cultivated at any time prior to December 19, 2007, on land that is either actively managed or fallow, and nonforested.
- Planted trees and tree residue from actively managed tree plantations on nonfederal land cleared at any time prior to December 19, 2007, including land belonging to an Indian tribe or an Indian individual, that is held in trust by the United States or subject to a restriction against alienation imposed by the United States.
- Slash and precommercial thinnings from nonfederal forestlands, including forestlands belonging to an Indian tribe or an Indian individual, that are held in trust by the United States or subject to a restriction against alienation imposed by the United States.
- Biomass obtained from the immediate vicinity of buildings and other areas regularly occupied by people, or of public infrastructure, at risk from wildfire.
- Algae.

Figures 1-3 and 1-4 show the influence of energy policies on the projected production of biofuels over time. Before the enactment of EPAct in 2005, production of fuel ethanol was projected to reach a plateau by 2005. In 2006, EIA projected production of fuel ethanol to increase from 4 billion gallons in 2005 to about 11 billion gallons in 2012. As of 2009, the amounts of fuel ethanol produced each year from biomass closely approximated EIA's 2009 projections (EIA, 2009a,b). In 2010, 13.2 billion gallons of ethanol, mostly from corn grain, were produced and consumed in the United States (EIA, 2010d). Biodiesel production from vegetable oil or animal fats peaked in 2008 at 678 million gallons (EIA, 2010b) but fell to 330 million gallons in 2010 (EIA, 2010c).[4] As in the case of EPAct, EISA included provisions for biofuels research and development and for research, development, and demonstration related to biofuel distribution and advanced biofuel infrastructure. In addition to federal energy policies, some states mandate a percentage of ethanol be blended in all gasoline sold (Duffield et al., 2008).

IMPETUS FOR STUDY

Although the federal government has enacted many policies that support biofuel production since 1978, the rapid increase in production of the 2000s has brought biofuels under increased scrutiny. Economic and environmental concerns, such as effects on food prices, land use, and air pollution, have been raised. However, directly connecting repercussions from biofuel policy to measurable effects has many uncertainties because of complex interrelationships between effects and the paucity of information at appropriate scales. In part because of these concerns, the U.S. Congress requested the National Research Council to convene an expert committee to examine the economic and environmental effects of increasing biofuel production. The next two sections summarize the potential economic and environmental consequences of achieving RFS that have been raised since the enactment of RFS.

---

[4]Much of the biodiesel produced in the United States was exported before the market downturn. Biodiesel consumption in the United States was 316 million gallons in 2008 and 222 million gallons in 2010 (EIA, 2010b,c).

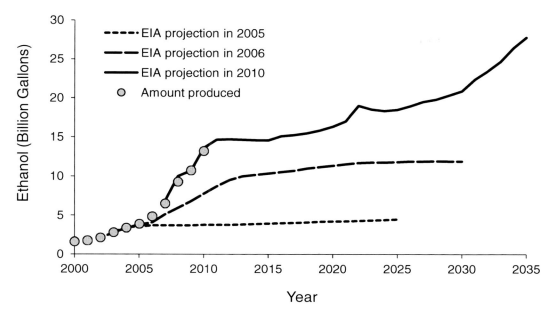

FIGURE 1-3 Amount of fuel ethanol produced projected by EIA in 2005 (before enactment of EPAct), 2006 (after enactment of EPAct), and 2010 (after enactment of EISA).
DATA SOURCES: EIA (2005, 2010a,b).

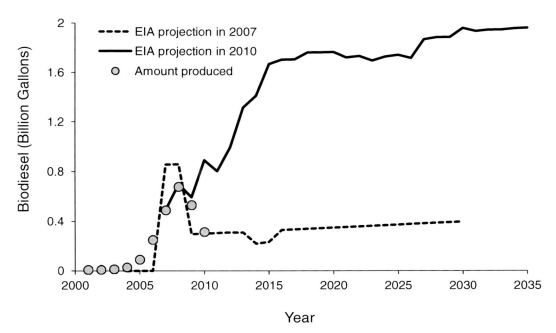

FIGURE 1-4 Amount of biodiesel produced projected by EIA in 2007 (before enactment of EISA) and 2010 (after enactment of EISA).
DATA SOURCES: EIA (2005, 2006, 2010a,b); NBB (2011).

Potential Economic Consequences of Achieving RFS2

The United States imports large quantities of oil from overseas each year. In 2008 when crude oil prices fluctuated widely, the United States spent $10-$38 billion each month on overseas oil imports. In 2010, oil product imports amounted to 17.4 percent of all U.S. imports. (Additional information on economics of petroleum-based fuel is available in Appendix G.) Domestic production of biofuels presents an opportunity to reduce oil importation. However, cellulosic biofuels are not a cost-competitive source of energy compared to petroleum-based fuels, even in most leading biofuel-producing countries (Kojima et al., 2007; Steenblik, 2007). Corn-grain ethanol could not compete with fossil fuels in the U.S. marketplace without mandates, subsidies, tax exemptions, and tariffs until the oil-price increase in the recent years. This lack of competitiveness raises questions about the use of government resources to support biofuels. Not only does the federal government expend resources in its subsidized loans and grants for development of the industry, but it also forgoes revenue in the form of tax credits for fuel blenders.

Because feedstocks for corn-grain ethanol and soybean biodiesel are feed and food crops, competition between the biofuel industry and food and feedstuff markets in the United States also has been raised as a concern. The diversion of corn to biofuel production was widely criticized in 2008 as a cause of the spike in food prices (Lipsky, 2008; Mitchell, 2008; Rosegrant, 2008; Timmer, 2009), even though U.S. corn production increased and exports remained constant (USDA-ERS, 2010). Several organizations representing livestock producers who purchase agricultural commodities for animal feed also suggested that price spikes in animal feed were linked to increasing biofuel production. They expressed concern that high feed prices would remain as biofuel production using food-based feedstocks increases and that the profitability of their livestock operations would be negatively affected (Brandenberger, 2010; Lobb, 2010).

Some livestock producer organizations contend that competition for feedstock between fuel and animal feed can be at least partially alleviated by using grass, forage, and coproducts from biofuel production, such as dried distiller grains and soybean meal as feed. However, other producers are concerned that those feedstuffs would affect the quality of their products (Stillman et al., 2009; Jonker, 2010). Pork and poultry producers cannot substitute forage or grass for corn, and distillers grains can only be used in small percentages with nonruminants (Brandenberger, 2010; Lobb, 2010; Spronk, 2010). Scholars contended that biofuel is a contributing, but not the only, factor that drives food and animal feed prices. Rather, food and animal feed prices are driven by a combination of supply and use, exchange rates and macroeconomic factors, and linkages between energy and agricultural markets (Abbott et al., 2009, 2011; Baffes and Haniotis, 2010; Trostle et al., 2011).

Cellulosic bioenergy feedstock—for example, crop and forest residues, perennial grasses, urban-derived waste materials, and other sources—is less likely to directly compete with food and animal feed. However, some cellulosic feedstocks have other uses. For example, representatives of the forestry and paper industries are concerned that the RFS2 mandate and other incentive policies distort wood-product markets (Noe, 2010). They worried that a growing demand for wood products for biofuel production would increase costs for their industries and that incentives favorable to biofuels will put them at a further disadvantage for purchasing timber. Reduced competitiveness would, in turn, cause job losses in the forestry and paper mill sectors (Noe, 2010). However, forest owners suggested that a developing biofuel industry can introduce new viable markets for forest biomass and encourage conservation of working forests (Tenny, 2010).

Potential Environmental Consequences of Achieving RFS2

Early interests in displacing fossil fuels with biofuels stem from desire to improve energy security. Potential GHG benefits compared to fossil fuels have been an additional motivation for federal support (EPA, 2002; Ribeiro et al., 2007). However, critics assert that the GHG benefits of biofuels might have been overstated. If all direct and indirect emissions associated with biofuel production are included in GHG accounting, their GHG benefits might not be as high as previously believed; some suggest that production and use of bio-fuels could result in higher GHG emissions than that of petroleum-based fuels (Fargione et al., 2008; Searchinger et al., 2008; Plevin et al., 2010).

Growing biomass for fuels will likely increase the demand on U.S. agricultural and for-estry output. The external costs or unintended consequences of agriculture, such as deple-tion of natural resources and environmental degradation, and the need to mitigate those external costs, have been recognized and documented (NRC, 2010b). For example, water scarcity is a key concern as many aquifers in the United States have been pumped exten-sively to provide water for agriculture and other competing uses (NRC, 2010b; USGS, 2010). Soil erosion from agricultural land and transport of dissolved nutrients from fertilizers that were not taken up by plants have contributed to reduced water quality (NRC, 2010b).

The question for estimating environmental consequences of achieving RFS2 is how much bioenergy feedstock production and processing will add to existing concerns, par-ticularly the nation's consumptive water use, water quality, and soil erosion (NRC, 2010b). Some critics assert that expanding corn-grain ethanol production would further degrade the environment (Donner and Kucharik, 2008). Others suggest growing a mix of bioenergy crops and using appropriate management practices could achieve the multiple goals of providing biomass for food, feed, and fuel and enhancing the natural resource base simul-taneously (Robertson et al., 2008).

Purpose of the Study

The purpose of this study is to provide an independent assessment of the economic and environmental benefits and costs associated with achieving RFS2 that have been raised. The committee distinguishes between the opportunities that biofuels can provide—which studies have suggested could provide many benefits if done right (Tilman et al., 2009)—and the effects of meeting the consumption mandate of 36 billion gallons of different types of biofuels by 2022; this study focuses on latter effects. Although the committee was asked to discuss "means by which the federal government could prevent or minimize adverse impacts of the RFS on the price and availability of animal feedstocks, food, and forest products, including options available under current law" (Box 1-1), it was not asked to discuss whether or how RFS2 could be modified to better achieve energy security and GHG reduction. The committee also was not asked to compare carbon mitigation potential of using biomass for fuels against using biomass for other purposes. Thus, alternative uses of biomass such as biopower for electric vehicles or combined coal and biomass for fuel and their effects on energy security and GHG emissions are not discussed.

This study relies on data from literature published up to the time of its preparation. Therefore, some topics are discussed in more detail than others depending on the amount of published literature on each subject. Fuel ethanol is discussed more frequently than other types of biofuels for two additional reasons:

- Under RFS2, over 40 percent of biofuels to be consumed to meet the mandate in 2022 will be conventional biofuel, most likely corn-grain ethanol. In addition, consumption mandates for advanced biofuels and cellulosic biofuels could be met, at least in part, by ethanol.
- Some issues, such as compatibility with existing infrastructure, are only applicable to fuel ethanol.

When possible, the report describes empirical evidence for the economic and environmental effects of increasing U.S. biofuel production. However, biofuel production is a developing industry, which has complex interactions with other sectors including agriculture, forestry, transportation, and energy. In cases in which the published literature provides diverging information or quantitative effects or model predictions that span a wide range, the committee sought to explain the sources of differences or areas of uncertainties.

The committee recognized that the effects of biofuels depend on the type of feedstock used, how and where it is grown, conditions prior to establishment of the bioenergy crops, logistics involved in feedstock transport and storage, conversion processes used to produce fuels from biomass, distribution of biofuels to the end users, and engine technology and performance given the specified blend of biofuels. Any assessment of environmental and economic benefits and costs is contingent upon details of each of these steps in a given system. Therefore, Chapter 2 provides background information on the biofuel supply chain. In addition, the economic and environmental effects of biofuels depend on site-specific conditions and the quantities of feedstock used. Because cellulosic biomass for fuels was not produced in large quantities as of 2011, its supply could only be estimated on the basis of experimental yields and assumed economic conditions. Therefore, Chapter 3 describes the projected supply of cellulosic biomass up to 2022 in the United States. The economics of biofuel production (including factors that influence the cost of biofuels such as feedstock prices and costs of conversion) and the economic effects of increasing biofuel production (including the extent to which biofuel production affects food prices, the linkage between biofuel production, animal feed prices, and coproducts, and the effect on the federal budget) are discussed in Chapter 4. Chapter 5 describes the projected or estimates of local, regional, and global effects of biofuel production on the environment. Chapter 6 presents potential economic, technical, environmental, social, and policy factors that could prevent the United States from meeting RFS2 mandates without passing judgment as to whether the mandate should be met.

REFERENCES

Abbott, P.C., C. Hurt, and W.E. Tyner. 2009. What's Driving Food Prices? March 2009 Update. Washington, DC: Farm Foundation.

Abbott, P.C., C. Hurt, and W.E. Tyner. 2011. What's Driving Food Prices in 2011? Oak Brook, IL: Farm Foundation.

Baffes, J., and T. Haniotis. 2010. Placing the 2006/08 Commodity Price Boom into Perspective. Washington, DC: World Bank.

Brandenberger, J. 2010. Input from the National Turkey Federation. Presentation to the Committee on Economic and Environmental Impacts of Increasing Biofuels Production, May 3.

Carolan, M.S. 2009. A sociological look at biofuels: Ethanol in the early decades of the twentieth century and lessons for today. Rural Sociology 74(1):86-112.

Dimitri, C., and A. Effland. 2007. Fueling the automobile: An economic exploration of early adoption of gasoline over ethanol. Journal of Agricultural and Food Industrial Organization 5(2):11.

Donner, S.D., and C.J. Kucharik. 2008. Corn-based ethanol production compromises goal of reducing nitrogen export by the Mississippi River. Proceedings of the National Academy of Sciences of the United States of America 105:4513-4518.

Duffield, J.A., I. Xiarchos, and S.A. Halbrook. 2008. Ethanol policy: Past, present, and future. South Dakota Law Review 53:101-128.

EIA (Energy Information Administration). 1995. Renewable Energy Annual 1995. Washington, DC: U.S. Department of Energy.

EIA (Energy Information Administration). 2005. Annual Energy Outlook 2005—With Projections to 2025. Washington, DC: U.S. Department of Energy.

EIA (Energy Information Administration). 2006. Annual Energy Outlook 2006—With Projections to 2030. Washington, DC: U.S. Department of Energy.

EIA (Energy Information Administration). 2008. Energy kids: Ethanol. Available online at http://www.eia.doe.gov/kids/energy.cfm?page=tl_ethanol. Accessed August 5, 2010.

EIA (Energy Information Administration). 2009a. Annual Energy Outlook 2009—With Projections to 2030. Washington, DC: U.S. Department of Energy.

EIA (Energy Information Administration). 2009b. Annual Energy Review 2008. Washington, DC: U.S. Department of Energy.

EIA (Energy Information Administration). 2010a. Annual Energy Outlook 2010—With Projections to 2035. Washington, DC: U.S. Department of Energy.

EIA (Energy Information Administration). 2010b. Annual Energy Review 2009. Washington, DC: U.S. Department of Energy.

EIA (Energy Information Administration). 2010c. Biodiesel overview. Available online at http://www.eia.gov/totalenergy/data/monthly/pdf/sec10_8.pdf. Accessed June 30, 2011.

EIA (Energy Information Administration). 2010d. Fuel ethanol overview. Available online at http://www.eia.gov/totalenergy/data/monthly/pdf/sec10_7.pdf. Accessed June 30, 2011.

EIA (Energy Information Administration). 2010e. International Energy Outlook 2010. Washington, DC: U.S. Department of Energy.

EIA (Energy Information Administration). 2010f. International Energy Outlook 2010. Washington, DC: U.S. Department of Energy.

EIA (Energy Information Administration). 2010g. International energy statistics. Available online at http://tonto.eia.doe.gov/cfapps/ipdbproject/IEDIndex3.cfm?tid=5&pid=54&aid=2. Accessed August 3, 2010.

EPA (U.S. Environmental Protection Agency). 2002. A Comprehensive Analysis of Biodiesel Impacts on Exhaust Emissions. Washington, DC: U.S. Environmental Protection Agency.

EPA (U.S. Environmental Protection Agency). 2010a. Inventory of U.S. Greenhouse Gas Emissions and Sinks: 1990–2008. U.S. Environmental Protection Agency.

EPA (U.S. Environmental Protection Agency). 2010b. Regulation of Fuels and Fuel Additives: Changes to Renewable Fuel Standard Program; Final Rule. Washington, DC: U.S. Environmental Protection Agency.

Fargione, J., J. Hill, D. Tilman, S. Polasky, and P. Hawthorne. 2008. Land clearing and the biofuel carbon debt. Science 319(5867):1235-1238.

Finlay, M.R. 2003. Old efforts at new uses: A brief history of chemurgy and the American search for biobased materials. Journal of Industrial Ecology 7(3-4):33-46.

Giebelhaus, A.W. 1980. Farming for fuel: The alcohol motor fuel movement of the 1930s. Agricultural History 54(1):173-184.

Jonker, J. 2010. Input from the National Milk Producers Federation. Presentation to the Committee on Economic and Environmental Impacts of Increasing Biofuels Production, March 5.

Knothe, G. 2001. Historical perspectives on vegetable oil-based diesel fuels. Inform 12(November):1103-1107.

Kojima, M., D. Mitchell, and W. Ward. 2007. Considering Trade Policies for Liquid Biofuels. Washington, DC: The World Bank.

Koplow, D. 2007. Biofuels—At What Cost? Government Support for Ethanol and Biodiesel in the United States: 2007 Update. Geneva: The Global Subsidies Initiative of the International Institute for Sustainable Development.

Lipsky, J. 2008. Commodity prices and global inflation. Available online at http://www.imf.org/external/np/speeches/2008/050808.htm. Accessed September 22, 2010.

Lobb, R. 2010. Input from the National Chicken Council. Presentation to the Committee on Economic and Environmental Impacts of Increasing Biofuels Production on May 3.

Melaina, M.W. 2007. Turn of the century refueling: A review of innovations in early gasoline refueling methods and analogies for hydrogen. Energy Policy 35:4919-4934.

Mitchell, D. 2008. A Note on Rising Food Prices. Washington, DC: The World Bank.

NBB (National Biodiesel Board). 2011. Biodiesel production estimates CY2005—current. Available online at http://www.biodiesel.org/resources/fuelfactsheets/. Accessed April 27, 2011.

Noe, P. 2010. Input from the American Forest and Paper Association. Presentation to the Committee on Economic and Environmental Impacts of Biofuels Production, May 3.

NRC (National Research Council). 2010a. Advancing the Science of Climate Change Washington, DC: National Academies Press.

NRC (National Research Council). 2010b. Toward Sustainable Agricultural Systems in the 21st Century. Washington, DC: National Academies Press.

Plevin, R.J., M. O'Hare, A.D. Jones, M.S. Torn, and H.K. Gibbs. 2010. Greenhouse gas emissions from biofuels' indirect land use change are uncertain but may be much greater than previously estimated. Environmental Science & Technology 44(21):8015-8021.

Ribeiro, N.M., A.C. Pinto, C.M. Quintella, G.O. da Rocha, L.S.G. Teixeira, L.L.N. Guarieiro, M.D. Rangel, M.C.C. Veloso, M.J.C. Rezende, R.S. da Cruz, A.M. de Oliveira, E.A. Torres, and J.B. de Andrade. 2007. The role of additives for diesel and diesel blended (ethanol or biodiesel) fuels: A review. Energy & Fuels 21(4):2433-2445.

Robertson, G.P., V.H. Dale, O.C. Doering, S.P. Hamburg, J.M. Melillo, M.M. Wander, W.J. Parton, P.R. Adler, J.N. Barney, R.M. Cruse, C.S. Duke, P.M. Fearnside, R.F. Follett, H.K. Gibbs, J. Goldemberg, D.J. Mladenoff, D. Ojima, M.W. Palmer, A. Sharpley, L. Wallace, K.C. Weathers, J.A. Wiens, and W.W. Wilhelm. 2008. Agriculture—Sustainable biofuels redux. Science 322(5898):49-50.

Rosegrant, M. 2008. Hearing on Biofuels and Grain Prices: Impacts and Policy Responses before the Senate Committee on Homeland Security and Governmental Affairs, May 7.

Searchinger, T., R. Heimlich, R.A. Houghton, F. Dong, A. Elobeid, J. Fabiosa, S. Tokgoz, D. Hayes, and T.-H. Yu. 2008. Use of U.S. croplands for biofuels increases greenhouse gases through emissions from land-use change. Science 319(5867):1238-1240.

Solomon, B.D., J.R. Barnes, and K.E. Halvorsen. 2007. Grain and cellulosic ethanol: History, economics, and energy policy. Biomass and Bioenergy 31:416-425.

Spronk, R. 2010. Input from the National Pork Producers Council. Presentation to the Committee on Economic and Environmental Impacts of Increasing Biofuels Production, May 3.

Steenblik, R. 2007. Government Support for Ethanol and Biodiesel in Selected OECD Countries. A Synthesis of Reports Addressing Subsidies for Biofuels in Australia, Canada, the European Union, Switzerland and the United States. Geneva: International Institute for Sustainable Development.

Stillman, R., M. Haley, and K. Mathews. 2009. Grain prices impact entire livestock production cycle. Amber Waves 7(1):24-27.

Tenny, D. 2010. Input from the National Alliance of Forest Owners. Presentation to the Committee on Economic and Environmental Impacts of Biofuels Production, May 3.

Tilman, D., R. Socolow, J.A. Foley, J. Hill, E. Larson, L. Lynd, S. Pacala, J. Reilly, T. Searchinger, C. Somerville, and R. Williams. 2009. Beneficial biofuels—The food, energy, and environment trilemma. Science 325(5938):270-271.

Timmer, C.P. 2009. Causes of High Food Prices. Manila, Philippines: Asian Development Bank.

Trostle, R., D. Marti, S. Rosen, and P. Westcott. 2011. Why Have Food Commodity Prices Risen Again? Washington, DC: U.S. Department of Agriculture - Economic Research Service.

Tyner, W.E. 2008. The US ethanol and biofuels boom: Its origins, current status, and future prospects. BioScience 58(7):646-653.

USDA-ERS (U.S. Department of Agriculture - Economic Research Service). 2010. Corn: Market outlook. USDA feed grain baseline, 2009-19. Available online at http://www.ers.usda.gov/briefing/corn/2010baseline.htm. Accessed March 14, 2011.

USDA-RD (U.S. Department of Agriculture - Rural Development). 2010. Bioenergy program for advanced biofuels payments to advanced biofuel producers. Available online at http://www.rurdev.usda.gov/rbs/busp/9005Biofuels.htm. Accessed August 4, 2010.

USGS (U.S. Geological Survey). 2010. Groundwater depletion. Available online at http://ga.water.usgs.gov/edu/gwdepletion.html. Accessed November 17, 2010.

2

Biofuel Supply Chain

The biofuel supply chain involves producing biomass feedstock; harvesting, collecting, storing, and transporting the feedstock to the biorefinery; converting the biomass to fuel at the biorefinery; distributing biofuels to end users; and, finally, using the fuel. Biomass is procured from diverse environments, each associated with different economic costs for production and collection. These differing conditions contribute to a range of economic costs for feedstocks and environmental effects. Each subsequent stage of biofuel production and use could incur positive or negative effects on the economics of producing biofuels, the economic effects on other sectors, and the environment. This chapter examines the supply chains of food-based biofuels that are produced and nonfood-based biofuels that are likely to be produced in the United States within the 2022 timeline as established by the Renewable Fuel Standard amended by the Energy Independence and Security Act of 2007 (RFS2). Other feedstocks and conversion technologies that are not likely to be deployed by 2022 also are discussed.

FOOD-BASED BIOFUELS

Corn-Grain Biofuels

Feedstock

As of 2010, the primary feedstock for biofuel produced and consumed in the United States was corn grain. The majority of corn acreage is found in the Midwest (Figure 2-1). Corn yield in the United States has been increasing over recent decades (Cassman and Liska, 2007). The national average reached 165 bushels per acre in 2009 and was 156 bushels per acre in 2010. An advantage of grains as biofuel feedstock is that they are relatively dense and efficient to store and transport; have well-established production, harvest, storage, and transport supply chains or systems; and are commodity crops with well-established grades and standards that facilitate marketing and trading.

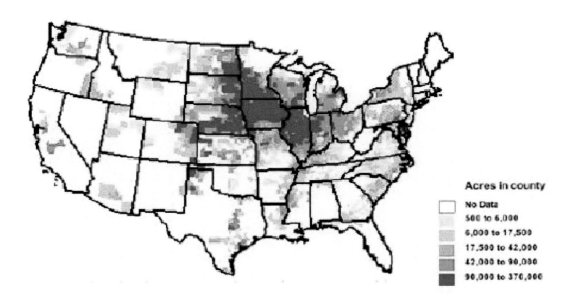

FIGURE 2-1 Distribution of planted corn acres in the United States in 2008.
SOURCE: USDA-ERS (2010).

Conversion

Starch from grains can be converted to ethanol by biochemical pathways. Most corn is dry milled—that is, the grain is ground to a meal, and then the starch from the grain is hydrolyzed by enzymes to glucose. The 6-carbon sugars are then fermented to ethanol by natural yeast and bacteria. The fermented mash is separated into ethanol and residue by distillation (Figure 2-2). The residue can be marketed wet as a dairy or cattle feedstuff or as dried distillers grain with solubles (DDGS) as a dairy, cattle, swine, and poultry feedstuff (Schwietzke et al., 2008). The theoretical yield of converting corn starch to ethanol is 112 gallons per dry ton (Patzek, 2006). A survey of U.S. ethanol plants conducted in 2008 reported average ethanol yield of 100 gallons per dry ton (Mueller, 2010).

Products and Coproducts

DDGS is a coproduct of grain ethanol production. Starch from grains is fermented to ethanol and the remaining protein, oil, yeast, minerals, and fiber form the coproduct DDGS, which is mostly used as an animal feedstuff (Nichols et al., 2006). For every bushel of corn grain used for ethanol production, about one-third comes out as DDGS, one-third as ethanol, and one-third as carbon dioxide (CO_2). The wet-mill process also produces corn oil and high-fructose corn syrup as coproducts.

Distribution and Use

Many biofuels, including ethanol, are more soluble in water than petroleum-based fuels, requiring biofuels to be stored and handled more carefully to avoid water contamination. If

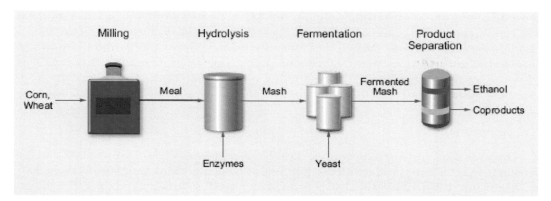

FIGURE 2-2 Processing steps for converting corn grain to ethanol.
SOURCE: Schwietzke et al. (2008). Reprinted with permission from IEA Bioenergy.

ethanol picks up water, it might not meet the fuel-ethanol specifications because the fraction of water exceeds the allowable amount. For that reason and because it is highly corrosive, ethanol cannot be transported in existing pipelines used for petroleum and is distributed by rail cars, barges, and trucks in the United States (USDA-AMS, 2007).

Ethanol has been used as a gasoline blending component or substitute for many years (see Chapter 1). It is almost exclusively produced outside a normal petroleum refinery and shipped to the refiner or distributor for blending into finished gasoline. The industry has developed stringent specifications for ethanol quality, such as ASTM D4806, so that all batches of ethanol can be treated equally and the final user is assured of getting a product that is fit for its desired purpose. All ethanol that meets this specification can be assumed to have the same performance as a fuel, regardless of its source. The refiner or blender does not have to be concerned about the source of the ethanol or how it was transported to the facility as long as it meets the specifications when it arrives.

Ethanol has a high octane value,[1] a beneficial characteristic, but it requires petroleum-refinery operational adjustments that reduce the value of the additional octane. It also has a high blending Reid vapor pressure (RVP)[2]; RVP is about 20 pounds per square inch (psi; or 136,895 Pascal) at 10-percent ethanol and even higher at lower concentrations. This high RVP can cause drivability problems for the fuel, namely vapor locking,[3] even if all other specifications are met. Therefore, the petroleum refiner has to reduce the amount of light hydrocarbons, such as butanes and hexanes, blended into gasoline.

If ethanol is used as a gasoline substitute, there will be a reduction in the amount of hydrogen produced by the naphtha reformer in a standard petroleum refinery. Hydrogen is a valuable coproduct of petroleum refining because it is used in upgrading hydrocarbons to more valuable products. This loss in hydrogen production would have to be compensated with either an increase in hydrogen production from within the refinery or an increase in purchased hydrogen produced via steam-methane reforming. Refiners that have access to a hydrogen pipeline system will usually just increase hydrogen purchases. Those facilities without access to merchant hydrogen would have

---

[1]Octane value is a measure of the maximum compression ratio at which a particular fuel can be used in an engine without some fuel and air mixture self-igniting or so-called "engine knocking."

[2]Reid vapor pressure is a measure of volatility.

[3]Vapor locking is the interruption of flow of fuel in an internal-combustion engine caused by vapor.

to modify their operations to maintain or increase hydrogen availability. This modification can sometimes be achieved by increasing reformer severity, but as petroleum-based gasoline demand declines and diesel demand increases, the refinery ultimately would need to build a hydrogen plant. Hydrogen production from methane releases CO_2 as a byproduct. This additional CO_2 production offsets some of the CO_2 reduction from using the biofuels.[4]

Status as of 2010

The amount of corn used for corn-grain ethanol has been increasing since 2000, and the percentage of U.S. corn production used for fuel ethanol increased dramatically from 2005 to 2009 (Figure 2-3). In 2010, about 40 percent of corn yield was used to produce 13.2 billion gallons of ethanol (USDA-NASS, 2010; RFA, 2011b). Given that RFS2 consumption mandate for conventional ethanol is 15 billion gallons per year from 2015 to 2022, the mandate can be achieved with a small increase in corn grain for ethanol.

The number of corn-grain ethanol biorefineries in the United States and the total capacity to produce ethanol has been increasing rapidly since 2002 (Figure 2-4). The average capacity build rate of ethanol biorefineries from 2001 to 2009 was about 25 percent with a substantial expansion in 2006 and 2007 (RFA, 2002, 2003, 2004, 2005, 2006, 2007, 2008, 2009, 2010). In January 2011, there were about 200 biorefineries (Figure 2-5) that converted corn starch into ethanol and had a combined installed (also known as nameplate) capacity[5] of 14.1 billion gallons of ethanol per year (RFA, 2011a). A list of food-based ethanol refineries is available in Appendix H.

Motor gasoline consumption in the United States was about 9 million barrels per day (or 138 billion gallons per year) in the same period (EIA, 2010). As of 2010, gasoline for light-duty vehicles in the United States was sold mostly as 90-percent gasoline blended with 10-percent ethanol by volume (E10). In October 2010, the U.S. Environmental Protection Agency (EPA) approved a waiver that allows the use of E15 in model year 2007 or newer light-duty vehicles. In January 2011, EPA extended the waiver to model year 2001 to 2006 light-duty vehicles (EPA, 2011). Despite the regulatory change, E15 had not been implemented at the time this report was written because gasoline retailers would have to install new tanks and pumps to accommodate the blend and no reliable system had been developed to prevent misfueling of older vehicles. Some ethanol is sold as E85 blend for use in flex-fuel vehicles. As of 2010, there were about 2,000 E85 stations in the United States.

Corn starch also can be converted to biobutanol via the acetone-butanol-ethanol (ABE) fermentation pathway (Ezeji et al., 2007). Coproducts include alcohols with lower molecular weight than butanol and acetone. However, only a small number of companies have pursued biobutanol from corn starch, and the development of that technology remains in the precommercial stage. Challenges to producing biobutanol include its toxicity to the microorganisms that ferment sugar for its production and reducing the yield of coproducts to maximize butanol yield. If corn grain is the source of the sugars for fermentation, a residue similar to DDG will also be produced. However, the DDG from conversion of corn grain to biobutanol might require additional processing to remove any toxic biobutanol and acetone residue before it could be used as an animal feedstuff. Biobutanol is significantly less soluble in water than ethanol. It could be a drop-in

---

[4]At constant diesel production, every gallon of ethanol increases diesel CO_2 emissions by about 1.4 lbs. This offsets the reduction in the CO_2 attributable to ethanol use by about 10 percent.

[5]The full-load continuous rating of the process plant as designed.

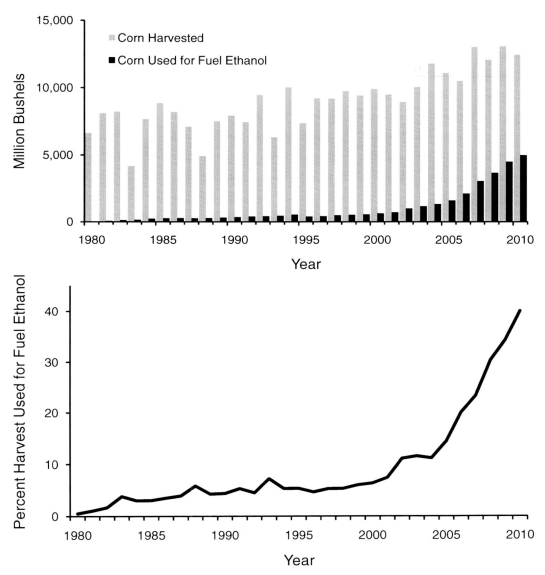

FIGURE 2-3 U.S. corn production and use as fuel ethanol from 1980 to 2009.
DATA SOURCES: USDA-ERS (2010), USDA-NASS (2010).

fuel: that is, a nonpetroleum fuel that is compatible with existing pipelines and delivery mechanisms for petroleum-based fuels. However, extensive testing would be required to confirm its compatibility with existing infrastructure. Its blending RVP is much lower than that of ethanol, and its octane is similar to regular-grade gasoline. As of 2010, there were no industry standard fuel-grade specifications and no accepted limits on the amount of biobutanol that can be safely blended into gasoline without damaging engine components. Because biobutanol properties are similar to regular gasoline, its major impact on the operation of the other refinery units would be a displacement of conventional petroleum-based, gasoline-blending components.

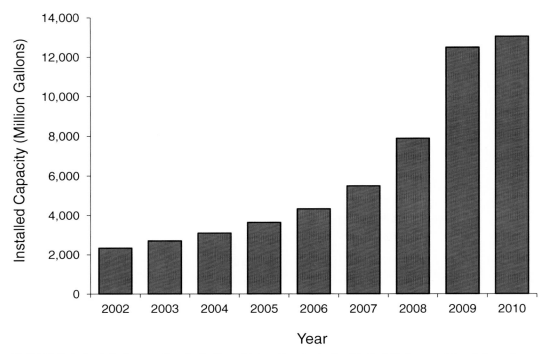

FIGURE 2-4 Installed capacity of all ethanol biorefineries in the United States combined, from January 2002 to January 2010.
DATA SOURCES: RFA (2002, 2003, 2004, 2005, 2006, 2007, 2008, 2009, 2010).

Biofuels from Vegetable Oils and Animal Fats

Feedstock

Several countries, mostly in Europe, produce biodiesel from a variety of feedstocks, including rapeseed oil, palm oil, and soybean oil (Reijnders, 2009; de Vries et al., 2010). In the United States, biodiesel is produced mostly from soybean oil. Other vegetable oils and animal fats constitute a small fraction of biodiesel feedstock. As in the case of corn, soybean is mostly grown in the Midwest (Figure 2-6) and has established markets and infrastructure for storage and delivery.

Conversion

The most widely available commercial chemical conversion technology is transesterification of triglycerides to produce biodiesel (Knothe, 2001). Soybean is the typical feedstock in the United States even though corn, canola, oil palm, camelina, jatropha, used yellow grease, and animal fats can also be used to produce biodiesel through this process. A typical biodiesel refinery will extract the oil from the feedstock and use acid or base catalysts in excess alcohol (methanol) to convert the triglycerides into fatty acid methyl esters (FAMEs) or biodiesel. The process flow of biodiesel production is shown in Figure 2-7.

Thermochemical processes use a combination of heat and chemical catalysis to alter the biomass and convert it into a hydrocarbon closer in composition to diesel and gasoline than

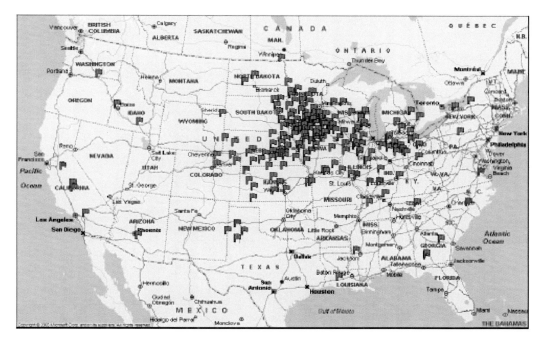

FIGURE 2-5 Location of ethanol biorefineries in the United States as of September 2010.
NOTE: Green flags indicate locations of operating biorefineries and red flags indicate locations of biorefineries under construction.
SOURCE: Urbanchuk (2010).

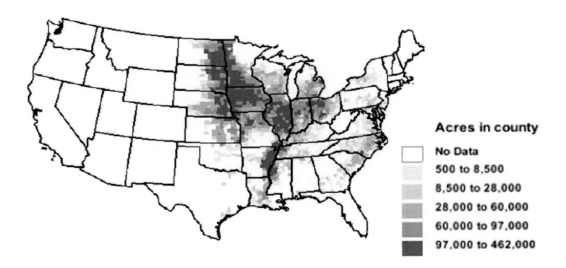

FIGURE 2-6 Distribution of planted soybean acres in the United States in 2008.
SOURCE: USDA-ERS (2010).

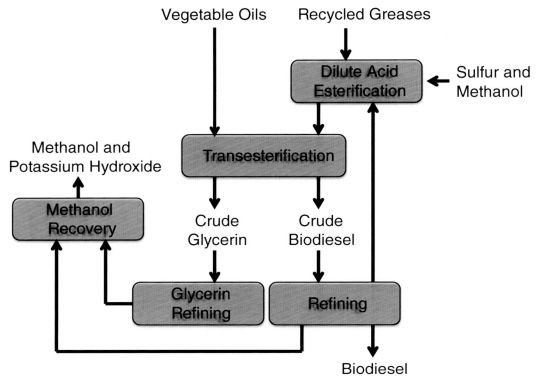

FIGURE 2-7 Process flow of biodiesel production.
SOURCE: Bain (2007).

conventional FAMEs. Neste Oil's NExBTL process (Neste Oil, 2011a), CANMET's "Super-Cetane™" (CETC, 2008), and "green diesel," developed and marketed by UOP, use large volumes of hydrogen and a catalyst to hydrogenate triglycerides recovered from animals or crop oils into a high-cetane diesel fuel (Kalnes et al., 2009). The severe hydrotreatment removes all of the oxygen from the triglyceride and saturates all of the olefinic bonds in the fatty acids. The primary products from this hydrogenation are water, CO_2, propane, and a mixture of normal paraffins. The normal paraffin mixture is called green diesel. This renewable diesel is fully compatible with petroleum-based diesel. It can even be produced by coprocessing triglycerides along with other petroleum streams in conventional refinery diesel hydrotreaters. In late 2010, Neste Oil announced the start-up of a dedicated 265 million gallons per year biodiesel NexBTL unit in Singapore using palm oil, waste fats, and greases as feedstock. At the time this report was written, a similar sized unit was scheduled to start up in Rotterdam in the first half of 2011 (Neste Oil, 2011a). As of early 2011, Neste Oil was operating two smaller units in Finland with a combined capacity of 125 million gallons of biodiesel per year (Neste Oil, 2011b).

Most large refining companies and technology vendors have performed laboratory studies and commercial trials that have demonstrated the feasibility of coprocessing triglycerides with existing diesel hydrodesulfurization (HDS) units (Renewable Diesel Subcommittee of the WSDA Technical Work Group, 2007; Melis, 2008). The amount of triglyceride that can be coprocessed is a function of the current limitations of the hydrotreater and the properties of the triglyceride. Conoco-Phillips and Tyson formed a joint venture to coprocess animal fats in Conoco-Phillips existing diesel hydrotreaters to produce 175 million gallons of renewable diesel each year. This coprocessing was discontinued when the U.S. Internal Revenue Service ruled that the tax credit for biodiesel did not apply to material coprocessed with petroleum.

Products and Coproducts

A typical transesterification plant can produce biodiesel from virgin oil and requires methanol, potassium hydroxide, and heat and electricity. The process results in the generation of glycerol and other impurities. Glycerol can be sold commercially for pharmaceutical formulation, soap production, and other uses before the market saturates. It also can be used as a feedstock to produce hydrogen, but technical improvements are needed to prove this pathway scalable and economically viable. Soybean seeds yield about 18-percent oil and the remaining meal, the primary product of soybean production, is sold as a highly nutritious animal feedstuff. Because of the high yield of the meal, this coproduct provides better monetary returns per ton of seed than the oil used in biofuel production.

Green diesel or renewable diesel generally has poor cold-flow properties (many products are solid at room temperature). Aside from the poor cold-flow properties, it is fully compatible with petroleum diesel and can use the existing distribution infrastructure. Production of green diesel does not result in any coproducts of significant volume. The coproducts from the production of green diesel are primarily water, CO_2, and propane.

Distribution and Use

Because of the flexibility in feedstocks that can be used to produce biodiesel, biodiesel refineries are more widely distributed geographically than refineries that produce corn-grain ethanol and include many that are processing waste oil as a feedstock rather than agricultural products (Figure 2-8). It is currently the largest volume class of biofuels after ethanol. FAME biodiesel is also mostly distributed by truck, barge, or rail. Although pipeline distribution would be the most economical option, it requires further experimentation. FAME has poor cold-flow properties, which could pose problems for delivery in colder climates. The fuel needs to be stored in special heated tanks to keep it fluid if it is to be used in northern tier states. The current specifications for FAMEs do not set limits on cold-flow properties, which are required to be reported to the customer only. The lack of limits makes refinery planning and operation more difficult. For many conventional petroleum crude oils, cold-flow properties limit the amount of diesel that can be produced. Operating a refinery capable of accepting FAMEs with different cold-flow properties would be difficult and costly. The base petroleum streams would have to be produced for the "worst case scenario" of poor cold-flow properties. Cold-flow properties do not always blend predictably, so the refineries have to produce conservative blends. It is costly for a refinery to reblend or reprocess product that is off specification for cold-flow properties.

Commercially relevant quantities of FAME have been blended into diesel fuel for several years. Quality control and economic and feedstock availability issues, however, have limited its growth as a petroleum-diesel replacement. FAME is more chemically active than petroleum-based diesel, and it degrades and forms corrosive acids during storage. Exposure to air and water accelerates this degradation. In addition, FAME can undergo biological degradation in contact with water. This process forms a "scum" at the oil-water interface that will plug downstream filters, including those owned by the final user, such as vehicle fuel filters or home heating system filters. Although ASTM specifications exist

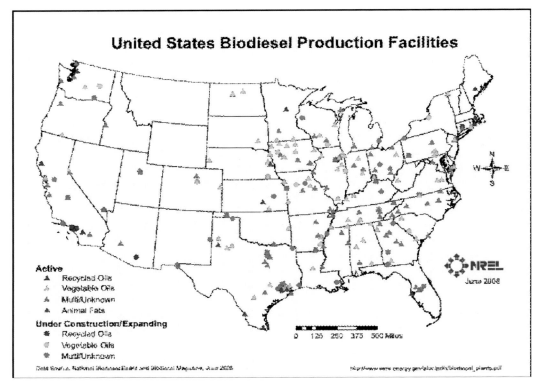

FIGURE 2-8 Biodiesel refineries in the United States (2008).
SOURCE: NREL (2008).

for FAME, they are frequently updated as new contaminants or problems are identified. FAME is produced via several processes from different feedstocks that lead to significant quality variations. It has good lubricity properties and can reduce the amount of lubricity additive required for ultra-low-sulfur diesel.

FAME can be blended with petroleum-based diesel at any percentage, and those blends are compatible with petroleum-diesel engines (DOE-EERE, 2010). The most popular bio-diesel blend in the United States is B20—that is 80-percent petroleum-based diesel and 20-percent biodiesel (NREL, 2005). As of 2010, there were about 650 B20 stations in the United States.

Although a few dedicated green diesel hydrotreater projects are in various stages of development, most large refining companies and technology vendors have performed laboratory studies and commercial trials that have demonstrated the feasibility of copro-cessing triglycerides on existing diesel HDS units. The amount of triglycerides that can be coprocessed is a function of the current limitations of the hydrotreater and the properties of the triglyceride.

Triglycerides consume roughly five times more hydrogen per barrel of feedstocks than typical diesel HDS feedstocks. The heat of reaction is also roughly five times higher than typical HDS feedstocks. These two factors usually limit the amount of triglyceride that can be coprocessed to about 10 percent of the HDS feedstocks. Unit modification to increase hy-drogen make-up capacity and to handle the additional heat of reaction would be required to coprocess significantly higher amounts of triglycerides.

Most HDS units can handle the incremental amount of water and propane that are produced from limited coprocessing, but again these facilities would have to be upgraded to process additional volumes. The oxygen removed via triglyceride hydrogenation produces some CO_2 in addition to water. Up to half of the total oxygen can be released in the form of CO_2. Most HDS units are not designed to handle CO_2 as a reaction product. If the HDS unit has a recycle-gas, amine-based hydrogen sulfide (H_2S) scrubber, then the CO_2 will be recovered together with the H_2S. This will minimize the effect of CO_2 on the HDS operation but will increase the load on the amine system and the sulfur plant, reducing the refinery's ability to process high-sulfur crudes.

Without a recycle-gas scrubber, CO_2 can build up in the recycle gas. This buildup has negative effects on unit operation. Processing triglycerides can also affect the HDS preheat system. All biobased triglycerides contain olefinic bonds as well as some free fatty acids. The olefinic bonds can interact and form gums either in storage or in the preheat train. These gums deposit in the preheat exchangers, the furnace, and the catalyst bed and degrade unit performance. The free fatty acids can also cause additional corrosion in the preheat exchangers. HDS preheat train metallurgy would most likely have to be upgraded to process significant volumes of triglycerides.

Because most refineries control diesel cold-flow properties by adjusting the back-end distillation cut point of the diesel components, including green diesel in the diesel pool requires an end-point reduction in the other blend components. Reducing the diesel end point decreases the amount of diesel that can be produced from a barrel of crude oil. This heavy stream that can no longer be included in diesel would have to be processed by fluidized catalytic cracking (FCC), the vacuum gas oil hydrocracker, or blended into heavy fuel oil. All of these alternate options are usually less profitable than including it in diesel fuel.

Status as of 2010

In 2008, an estimated 16 percent of soybean production in the United States was used to produce biodiesel (USDA-NASS, 2010). Biodiesel production increased from 9 million gallons per year in 2001 to 532 million gallons per year in 2009 (EIA, 2010), but members of the National Biodiesel Board reported a total production capacity of 2.7 billion gallons per year (NBB, 2010). A list of biodiesel refineries is provided in Appendix I. Nonfood oils produced from algae or dedicated bioenergy crops, such as camelina, are expected to be used as biomass feedstock in the future.

NONFOOD-BASED BIOFUELS

Cellulosic Feedstock

Cellulose,[6] hemicellulose,[7] and lignin[8] provide the structural components of plant cells. Those plant materials can be used to produce biofuels, commonly referred to as cellulosic biofuels. Potential feedstocks for cellulosic biofuels include agricultural residues, dedicated energy crops, forest resources, and municipal solid waste.

---

[6] A complex carbohydrate, $(C_6H_{10}O_5)n$, that forms cell walls of most plants.

[7] A matrix of polysaccharides present in almost all plant cell walls with cellulose.

[8] A complex polymer that occurs in certain plant cell walls. Lignin binds to cellulose fibers and hardens and strengthens the cell walls of plants.

Agricultural Residues

Crop residues include leaves, stalks, cobs (corn), and roots. Those being considered for cellulosic biofuel production include corn stover, corn cobs, sorghum stalks, wheat straw, cotton residue, and alfalfa stems. Crop residues are sometimes called crop wastes, but this is a misnomer as they help maintain soil quality (including fertility, structure, and other physical, chemical, and biochemical qualities) and reduce or mitigate soil erosion (Blanco-Canqui and Lal, 2009b). The amount of residues to be left to achieve those functions depends on soil conditions, crop yield, crop management practices (for example, tillage and crop rotation), prevailing climate conditions, and topography of the land.

Corn stover can be recovered from the millions of acres that annually produce corn grain in United States, primarily in the Midwest (Figure 2-1). Brechbill and Tyner (2008) summarized a range of estimates for corn stover yield as a proportion of grain yield. The range is usually around a 1:1 ratio between grain and stover measured in dry weight (Johnson et al., 2006). The amount of stover available for biofuel production is typically much less than the amount produced, however, in part because it is not mechanically possible or environmentally desirable to collect all corn stover (Petrolia, 2006; Graham et al., 2007; Wilhelm et al., 2007). Recovery of 30 to 50 percent of stover has been reported as achievable, with the lower amount considered an average for sustainable removal rates from many studies (Beach and McCarl, 2010; Miranowski and Rosburg, 2010; Schnepf, 2010). The amounts estimated as recoverable depend on the harvest techniques used (Brechbill and Tyner, 2008), seasonal factors associated with the weather, and the costs of harvest, which encapsulate other factors such as the price of oil and tax and other policies that affect a farmer's bottom line.

The amount of stover removed from the field affects erosion and maintenance of soil resources. The amount of stover retained on the field to minimize erosion and maintain soil resources varies with biophysical conditions and crop management practices. Some fields, or locations within fields, will not support crop residue removal without excess soil erosion or loss of soil organic matter, while other sites might benefit from partial residue removal (Blanco-Canqui and Lal, 2009a,b; NAS-NAE-NRC, 2009). Blanco-Canqui and Lal (2009a,b) suggested that the amounts of residue needed to maintain soil organic matter (SOM) levels are larger than those needed to prevent soil erosion. Clay et al. (2010) pointed out the importance of estimating below-ground or root residue carbon in estimating SOM maintenance levels. Although some long-term residue removal studies exist (Karlen et al., 1994), they are inadequate to serve the diversity of landscapes, residue types, and farming practices from which residues might be removed. Therefore, some authors suggested long-term research to create site-specific guidelines for residue harvest (Karlen et al., 1994; Andrews, 2006; Clay et al., 2010).

In addition to concerns about the environmental effects of crop residue use, there are other unresolved problems associated with the timing of residue harvest, transportation of residues to biorefineries, and deterioration during storage (Brechbill and Tyner, 2008; Schnepf, 2010). (See also the later section in this chapter, "Storage and Delivery of Cellulosic Feedstocks.") These same issues affect energy crops in general and, to a lesser degree, some forest residues.

The national average yield for corn in the years 2006-2010 was 155 bushels per acre (USDA-NASS, 2010). Assuming a 1:1 ratio of dried stover to grain by weight, and taking into account corn grain has 16-percent moisture, 3.7 dry tons of stover were available per acre. Most corn is produced in the Midwest,[9] where the average production in 2006-2010

---

[9]Illinois, Indiana, Iowa, Kansas, Kentucky, Michigan, Minnesota, Missouri, Nebraska, Ohio, South Dakota, and Wisconsin accounted for over 80 percent of the corn acres planted in 2006-2010.

was 162 bushels per acre or 3.8 dry tons of available stover per acre. Because of the environmental considerations cited above, the actual amount of stover removed would be much less. Harvestable corn stover has been estimated to be 0.7-3.8 dry tons per acre.[10]

The U.S. Department of Agriculture (USDA) projected that corn acreage will hover at 88-92 million acres through 2021 (USDA, 2011). However, it forecasted that corn yield per acre will increase from 154 bushels per acre in 2008-2009 to 180 bushels per acre in 2020-2021 (USDA, 2010, 2011). Others projected greater increases; for example, the National Corn Growers Association estimated yields will reach 205 bushels per acre by 2020 (NCGA, 2010), while Monsanto predicted yields closer to 225 bushels per acre in that timeframe (Fraley, 2010). EPA estimated that 22 percent, or 82 million dry tons, of corn stover in the Midwest could be harvested in 2022; however, EPA's estimate assumed that the density of corn plants would increase substantially by that time (EPA, 2010b).

Dedicated Bioenergy Crops

Dedicated bioenergy crops refer to nonfood perennial crops that are grown primarily for use as bioenergy feedstocks. Examples of dedicated bioenergy crops include switchgrass, *Miscanthus*, mixtures of native grasses, and short-rotation woody crops such as hybrid poplar and willow. The following section describes species characteristics and summarizes yield data from published literature. The research in these species is still in its infancy; therefore, many of the yield results are from trial plots and do not account for site quality or growing conditions, which heavily influence crop yields from production agriculture (Johnston et al., 2009; Lobell et al., 2009). Other studies examined the crops under systems designed for forage rather than bioenergy feedstock production.

Switchgrass

Switchgrass (*Panicum virgatum* L.) is native to the United States. It is a deep-rooted, warm-season perennial prairie grass and has traditionally been used for soil erosion control, forage, wildlife habitat, and landscaping. Switchgrass can tolerate a wide range of soil conditions, including soil too poor to support row-crop production (Bransby et al., 2010). It is a highly productive and nutrient-efficient plant that can grow on acidic or infertile soils with few inputs (Jung et al., 1988). It is endemic to the prairie and oak savanna ecosystems of the United States. Though most of its original habitat has been disrupted to accommodate farming, it can grow from Texas to the Dakotas and east to the Atlantic Coast.

Of the perennial native grasses under consideration as potential bioenergy crops, switchgrass has been the most extensively studied, particularly following its selection as the herbaceous model species for the U.S. Department of Energy's (DOE's) Feedstock Research Program (Wright and Turhollow, 2010). Propagated from seeds, the stems can grow up to 8 feet tall (El Bassam, 2010). Switchgrass is typically harvested in late October or early November in the United States, after its top growth has died back (Rinehart, 2006). Its low mineral content at harvest time lends itself to efficient combustion and low-exhaust gas emissions (El Bassam, 2010). Stands of switchgrass can be productive 10 or more years.

Potential switchgrass yields range between 0.89 and 17.8 dry tons per acre.[11] In a review of studies on switchgrass yields, Schnepf (2010) found that yield averages were be-

---

[10]A compilation of estimates for harvestable corn stover from the literature is used in the economic analyses in Chapter 4 and presented in an associated appendix.

[11]A compilation of switchgrass yields from the literature is used in the economic analyses in Chapter 4 and presented in an associated appendix.

tween 3 and 6 dry tons per acre. Yields vary by agronomic conditions; factors that affect productivity include the amount of precipitation and its timing, growing degree days, and temperature (Wullschleger et al., 2010). In the prairie regions west of the 100th meridian, lower precipitation and lower temperatures constrain biomass yields (Nelson et al., 2010).

Yields also vary because of nitrogen availability. At one extreme, researchers found no response of switchgrass yield to nitrogen fertilization in southern England (Christian et al., 2002). At the other extreme, researchers suggested 200 tons per acre of nitrogen as an optimal rate of nitrogen fertilization for Alamo switchgrass at a site in Texas (Muir et al., 2001). Optimal rates of nitrogen fertilization depend on the use for switchgrass,[12] number of harvests per year,[13] nitrogen mineralization in soil organic matter, and potential symbiotic relationship with mycorrhizae and other microorganisms. Because of its adaptation to a large geographic area, soil type and acidity do not appear to affect switchgrass productivity (Parrish and Fike, 2005). Wullschleger et al. (2010) concluded from their review of field trials that it is likely that switchgrass productivity benefits from moderate fertilizer application but that the response rates are uncertain. They found that biomass yield is not related to plot size or row spacing. However, the relative success of stand establishment and crop management practices, such as irrigation and the number of harvests per season, will likely influence overall yield.

Though switchgrass can be grown in most of the eastern United States, empirical studies and model estimates show that it is likely to be most productive across the middle of the eastern United States from eastern Kansas through Virginia (Jager et al., 2010; Wullschleger et al., 2010) and in the lower Mississippi Valley (Thomson et al., 2009). Thomson et al. (2009) adapted the Environmental Policy Integrated Climate (EPIC) model to simulate switchgrass yield. This model simulates agroecosystem processes and has been widely used to simulate crop yields under different agronomic conditions. They predicted that yields will be highest in the lower Mississippi Valley regions and parts of the Southeastern United States, especially Florida. Other regions with potentially high yields are the central Corn Belt, parts of the Ohio River Valley, and parts of Tennessee (Figure 2-9). The authors noted that the model predicted that the regions with the highest potential productivities for switchgrass are also the highest-producing agricultural lands in the country. Jager et al. (2010) used multiple regression analyses to derive a model for predicting lowland and upland switchgrass yields across the United States on the basis of variables associated with climate, soils, and management (for example, precipitation, temperature, nitrogen fertilization, stand age, and locations). Jager et al. (2010) predicted somewhat different locations for the highest switchgrass yields, particularly areas in Virginia, North Carolina, and Tennessee, and higher average yields (Figure 2-10).

Miscanthus

Miscanthus is a genus of perennial, warm-season grasses native to eastern Asia, northern India, and sub-Saharan Africa. Under favorable conditions, it is highly productive in terms of biomass compared to other grasses and agricultural crops (Beale and Long, 1995; Bransby et al., 2010). *Miscanthus* has been used in China and Japan for grazing, paper-making, and building materials and has been naturalized around the world as an ornamental plant in landscaping (Stewart et al., 2009). Research into its use as a bioenergy crop has been under way since the 1980s (Heaton et al., 2010). Most research in the United States has been conducted on Giant Miscanthus (*Miscanthus* × *giganteus*), a sterile, hybrid

---

[12]High nitrogen content is desirable in switchgrass for forage, but not in switchgrass for bioenergy.

[13]Switchgrass managed for forage is harvested more frequently than switchgrass managed for bioenergy.

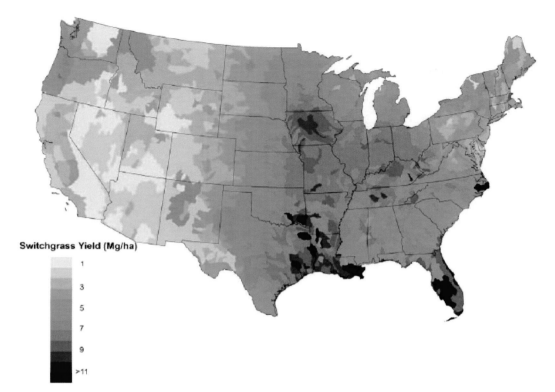

Switchgrass Yield (Mg/ha)

1
3
5
7
9
>11

FIGURE 2-9 Long-term (30-year average) switchgrass yield in the United States as simulated by the Environmental Policy Integrated Climate (EPIC) model.
SOURCE: Thomson et al. (2009).

variety propagated by rhizome division. Stands in Illinois can grow to more than 12 feet tall. The stems are usually one-half to three-quarters of an inch in diameter. Harvest takes place between early December and early March (Pyter et al., 2009), though harvesting is usually avoided the first year to allow time for the stand to establish itself and survive the first winter. Stands of *Miscanthus* can grow productively for more than 15 years.

Yield estimates from U.S. studies range from 6.3 to 17 dry tons per acre.[14] It is possible to harvest even higher yields from *Miscanthus* if it is harvested in the summer. However, even though there is a 30-50 percent loss of biomass, most harvesting is done during the winter when plant moisture and nutrient content is low and biomass is therefore more amenable to processing.

Miscanthus could be an efficient use of land in terms of producing biomass. However, there are still many unanswered questions about *Miscanthus* production at a large scale in the United States. First, optimum nitrogen application is uncertain. For example, yields on high-quality cropland in Illinois in one study of *M. × giganteus* averaged 13.5 dry tons per acre without irrigation and with 22.5 pounds per acre of applied nitrogen fertilizer (Heaton et al., 2008). However, Heaton et al. (2009) found that over the course of 3 years

---

[14]A compilation of *Miscanthus* yields from the literature is used in the economic analyses in Chapter 4 and presented in an associated appendix. Through 2010, more research on *Miscanthus* had taken place in Europe than in the United States, and many of these European studies are included to increase the robustness of the dataset.

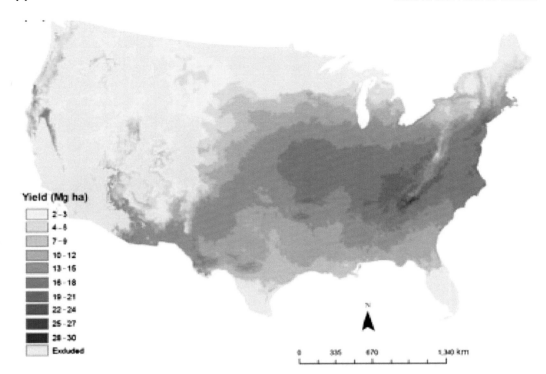

FIGURE 2-10 Map of potential switchgrass yield in low-land ecotype predicted by Jager et al.'s empirical model.
SOURCE: Jager et al. (2010). Reprinted with permission from Wiley Blackwell.

M. × *giganteus* removed 12 times more nitrogen from the soil than was added. It has been proposed that *Miscanthus* may have nitrogen-fixing properties (Davis et al., 2010); this needs to be determined to ensure that *Miscanthus* planted on a large scale does not deplete the soil of nutrients. Second, *M.* × *giganteus* uses water more efficiently and produces more biomass per unit water than corn (Heaton et al., 2010). However, *M.* × *giganteus* produces so much more biomass than corn that the total quantity of water consumed in a given plot could be higher if *M.* × *giganteus* is planted instead of corn. McIsaac et al. (2010) found that total evapotranspiration is greater in *M.* × *giganteus* than in switchgrass or corn under comparable conditions and therefore could reduce surface water flows and groundwater recharge if planted widely in place of corn. Finally, questions about the tolerance of *M.* × *giganteus* to temperature extremes, the conditions necessary for it to overwinter the first year, and its susceptibility to pest pressures have yet to be resolved (Heaton et al., 2010).

Cultivation of *M.* × *giganteus* may be limited in some locations because of the expense of asexual propagation and establishment. Sterility would limit inadvertent spread of a highly productive nonnative species. Research is under way to develop a seeded variety of *Miscanthus*, but its potential to become an invasive species would have to be evaluated before it is planted widely (Heaton et al., 2010).

In addition to the mechanics of its biology and its productivity on a large scale, more research is needed to determine the suitability of *Miscanthus* to different locations in the United States. While it has been demonstrated to yield well on marginal land in Ireland that tends to waterlog in the winter and dry out in the summer (Clifton-Brown et al., 2007), additional research needs to be conducted to determine its productivity on a wider range of land productivity classes and climate zones in the United States, especially those with moisture limitations or flooding. *Miscanthus* could compete for land currently used for corn production if prices were competitive because *Miscanthus* has been shown to yield well under conditions similar to corn production (Heaton et al., 2008). Furthermore, Miguez et al. (2008, 2009) developed a model predicting where and how well *Miscanthus* would yield. Somerville et al. (2010) applied the model to predict *Miscanthus* yield in the United States in the third year of planting assuming no competition from other crops or vegetation (Figure 2-11). The model projects that there is substantial overlap between acres currently used for corn and acres on which *Miscanthus* would be highly productive (Figure 2-1). High levels of productivity were predicted in some areas of the southern and eastern United States that are less suitable for corn.

Native Grasses

Native grasses are ecologically adapted for the environment that they occupy naturally; thus, the implicit assumption is that cultivation costs for native grass feedstock will be minimal, particularly if the agronomic practices emulate the species' environmental conditions. Experiments are being conducted on the suitability of native prairie grasses other

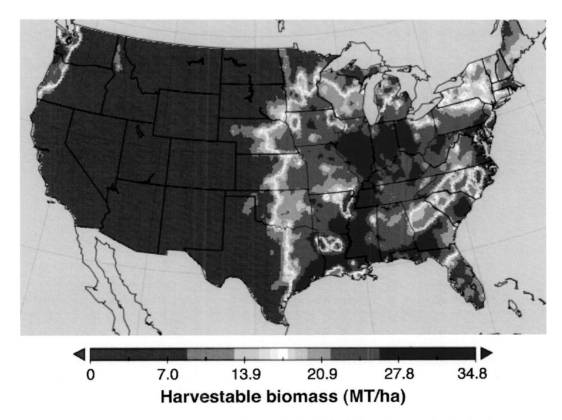

FIGURE 2-11 Projected annual average harvestable yield of *M. × giganteus* in the third year after planting.
SOURCE: Somerville et al. (2010). Reprinted with permission from G. Bollero, University of Illinois.

than switchgrass to serve as biofuel feedstocks in addition to or planted with switchgrass (Tilman et al., 2006; ANL, 2010). Warm-season candidates include big bluestem (*Andropogon gerardii* Vitman), a perennial, rhizomatous grass that grows up to 8 feet tall (Owsley, 2002); little bluestem (*Schizachyrium scoparium* Michx. Nash), a widely distributed native that tolerates a broad variety of soil conditions (USDA-NRCS, 2002); Indiangrass (*Sorghastrum nutans* L. Nash), which grows 3-7 feet tall under a variety of soil types; and eastern gamagrass (*Tripsacum dactyloides* L.), a long-lived, perennial relative of corn that is typically 2-3 feet tall but can reach 10 feet in height (Lady Bird Johnson Wildflower Center, 2011). Research has also been conducted on prairie cordgrass (*Spartina pectinata*) (Energy Biosciences Institute, 2011) and coastal panicgrass (*Panicum amarum* Ell. var. amarulum [A.S. Hitchc. & Chase] P.G. Palmer) (Henson and Fenchel, 2007; Mayton et al., 2011).

The warm-season grasses big bluestem, little bluestem, prairie cordgrass, and Indiangrass are native to most of the United States, while eastern gamagrass is native to the Great Plains and eastward. Weimer and Springer (2007) found that the yields from 3 years of harvests of three varieties of big bluestem planted in five different states averaged 2.04, 1.96, and 1.76 dry tons per acre. Owsley (2002) found that "Kaw" cultivar of big bluestem yielded 1.8 to over 3.5 dry tons per acre per year, depending on location and fertilizer inputs. Big bluestem can tolerate drought, and it has the genetic capacity to grow in a wide range of growing seasons, though breeding research has yet to exploit this capability. However, it is more difficult to establish than switchgrass (Anderson et al., 2008).

Eastern gamagrass is also difficult to establish. It requires more water and nitrogen and is less tolerant of poor soils than big bluestem, but it can yield 2.7-7.2 dry tons per acre per year (Anderson et al., 2008). However, big bluestem had better fermenting properties, making it more suitable for bioenergy processing than eastern gamagrass (Weimer and Springer, 2007). Prairie cordgrass occurs in most prairie grasslands and is tolerant to salinity and wet soils; therefore, it is likely to be productive on low, wet areas that are not conducive to switchgrass or big bluestem production (Gonzalez-Hernandez et al., 2009). Indiangrass yields vary with precipitation, with the better yields reportedly reaching 3.5 dry tons per acre per year (Tober et al., 2009).

For cooler and possibly less humid environments, DOE has started to investigate the cool-season perennial reed canarygrass (*Phalaris arundinacea* L.), a grass used for forage and hay, as a viable bioenergy feedstock. Reed canarygrass is native to all regions of the United States, with the exceptions of Alaska, Hawaii, and the South. It can grow over 8 feet tall and is tolerant of poorly drained soils (El Bassam, 2010). Though it is difficult to establish and does not yield well until the second or third year, once established it is long-lived and high-yielding (Anderson et al., 2008). In Europe, yields of about 1.5-8.6 dry tons per acre per year with liquid manure fertilizer have been demonstrated. In New York, Salon et al. (2010) reported an average yield of 4 dry tons per acre per year. More broadly, yields in the United States have typically ranged from 4.5 to 6.75 dry tons per acre (Anderson et al., 2008). However, the location of reed canarygrass is important because it is an invasive species in wetland areas.

Other cool-season possibilities include Canadian wildrye (*Elymus Canadensis* L.) (Lady Bird Johnson Wildflower Center, 2011), smooth bromegrass (*Bromus inermis* Leyss.), wildryes (for example, *Elymus glaucus*, *Elymus cinereus*, and *Leymus triticoides*), Sandberg bluegrass (*Poa secunda*), mountain brome (*Bromus marginatus*), and wheatgrasses (*Elymus lanceolatus* and *Elymus trachycaulus*) (El-Nashaar et al., 2009). Wildryes, *Leymus* spp., are found throughout the United States, while wheatgrasses (*Elymus* spp.) are more common west of the Great Plains, though they can be found in the Northeast as well. Many wildryes are well adapted to low precipitation and can tolerate high altitudes and salinity (Anderson et al.,

2008). Wildryes may be suited to the Mountain West because they require little additional precipitation or nutrient input. Anderson et al. (2008) reviewed studies that found yields of 1.35-9 dry tons per acre in that region. Tall wheatgrass yields are reported to be around 5 dry tons per acre per year (Salon et al., 2010).

Some authors have suggested that native stands of grasses, combined with other herbaceous species, can be sustainably harvested at low ecological cost and could have large enough yields to permit economic use (Tilman et al., 2006; Fargione et al., 2009). Native prairies are associated with many environmental services, especially those associated with wildlife and soil and nutrient conservation (Jarchow and Liebman, 2011). Combining these landscapes with economic production of biomass feedstocks is an appealing concept (Tilman et al., 2006; Jordan et al., 2007). Others have suggested that complex mixtures of plant species are at best similar to or lower yielding than simpler ones or monocultures under most economic production conditions (Sanderson et al., 2004; Adler et al., 2009; De Haan et al., 2010). However, native grass mixtures might sustain economic harvests of biomass without loss of environmental benefits in particular locations (Sanderson et al., 2004; De Haan et al., 2010).

Little research has been conducted into the suitability of native grasses (with the exception of switchgrass) as biomass in monoculture or mixed culture. Though some studies have looked at the possible yields of other native grasses, few have been conducted under conditions suited for bioenergy production, as opposed to forage, and there has been no control for site quality across studies that examine native grass yields. Furthermore, almost no research has been done to improve their germplasm for biomass production. Research and development could result in substantial increases in yield.

Short-Rotation Woody Crops

Short-rotation woody crops are intensively managed, fast-growing trees or woody shrubs. Several species have been evaluated for their biofuel utility and suitability for use in different agroecological regions in the United States. Trees and shrubs provide an advantage over residues or grasses because they can be stored while still growing until market demand warrants harvest. Because they are denser than herbaceous cellulosic feedstocks, they offer easier storage and transport logistics after harvest compared to agricultural residues or perennial grasses (Hinchee et al., 2009). Biomass from woody crops can also be mixed with other woody feedstock, such as forest-industry residues, forest-product residues, or urban wood residues (Johnson et al., 2007).

Species native to the United States that hold promise as bioenergy feedstocks include hybrid poplars (*Populus* spp.), willow (*Salix* spp.), black locust (*Robinia pseudoacadia* L.), sycamore (*Platanus* spp.), sweetgum (*Liquidambar styraciflua* L.), and loblolly pine (*Pinus taeda* L.) (Johnson et al., 2007; Hinchee et al., 2009; Kline and Coleman, 2010). Eucalyptus (*Eucalyptus* spp.), native to Australia, is also a potential feedstock in southern, frost-free locations (Hinchee et al., 2009; Kline and Coleman, 2010). Rotation lengths for these short-rotation tree crops can vary from 1 year to over 15 years, depending on the tree or shrub species, the productivity of the site, and the management of the plantation.

Most woody crops for bioenergy will likely be produced east of the Mississippi because rainfall levels are higher. Poplar and willow have great potential to provide woody biomass for bioenergy because they are highly productive and stand to benefit from extensive genetic research or harvesting improvements. Under the right conditions, *Populus* spp. plantings can be harvested 3-4 years after cutback and more than eight harvests are possible during the life of the plantation (El Bassam, 2010). Eastern cottonwood (*Populus deltoides* M.) can produce 5 dry tons per acre per year when managed intensively (Mercker, 2007). A survey of forest management practitioners conducted by Kline and Coleman (2010) in the Southeast reported that hybrid poplars achieved yields of 2.25 dry tons per acre per year on plantations, but 6.75 dry tons per acre per year on trial plots. Riemenschneider et al. (2001) found yields of 7 dry tons per acre per year on highly productive sites and sustained yields of 5 dry tons per acre per year in the North Central United States. The productivity of poplars has been improved through selection, clonal propagation, and hybridization (Heilman and Stettler, 1985), and advances in the future should be able to take advantage of the recent mapping of the poplar genome (Tuskan et al., 2006). Though hybrid poplars can grow in most of the eastern United States and in the Pacific Northwest, their productivity, range, and therefore usefulness as woody biomass feedstock is currently limited by their high demand for nutrients and water (Johnson et al., 2007; Hinchee et al., 2009; Kline and Coleman, 2010).

Willow (*Salix* spp.) is highly productive and can be harvested after 3 years (Walsh et al., 2003; El Bassam, 2010). It performs well under coppice management, in which new growth sprouts directly from cut stumps. At least eight harvests can be obtained from a plantation over 25 years (Volk et al., 2004; Johnson et al., 2007; El Bassam, 2010). It is a highly water-dependent crop and prefers moist soil in cold, temperate climates. Relative to its productivity, it is a nutrient-efficient plant when cultivated at the plantation scale (El Bassam, 1996). Insect susceptibility is not an extensive problem for willow, and the diversity of willow species offers rich potential for breeding disease resistance (El Bassam, 2010). Most willow trials have been conducted in the Northeast and Midwest (Ruark et al., 2006); willow grows in the Southeast, but extensive study has not yet been conducted (Kline and Coleman, 2010). Volk et al. (2006) found that fertilized and irrigated willow grown in 3-year rotations has yielded more than 10.8 dry tons per acre per year. Though irrigation may not be economically viable for willow plantations, the results demonstrate the tree's yield potential. Trials with irrigation in New York have produced yields of 3.4-4.6 dry tons per acre per year in the first rotation, and the best-producing trees produced 18-62 percent more in the second rotation (Ruark et al., 2006). Relatively little work has been done with willow in terms of breeding improvements (Ruark et al., 2006), but specialized planting and harvesting equipment has already been developed (Walsh et al., 2003). Research and development could improve biomass yields from willow.

Black locust (*Robinia pseudoacadia* L.), sycamore (*Platanus* spp.), sweetgum (*Liquidambar styraciflua* L.), loblolly pine (*Pinus taeda* L.), and eucalyptus (*Eucalyptus* spp.) are other potential woody biomass crops. Black locust is native to the Southeastern United States and has many characteristics amenable to feedstock production. It has rapid initial growth and responds favorably to coppice management. It also is leguminous, so it can fix nitrogen from the atmosphere, even if the soil is of poor quality. Because of its drought tolerance, black locust is often used to prevent erosion or remediate soils (El Bassam, 2010). One study in Kansas found yields averaged 4.2 dry tons per acre per year over an 8-year cycle (King et al., 1998), while the yields in another Kansas study averaged over 3.15 dry tons per acre per year over a 6-year cycle (Geyer, 2006). Bongarten et al. (1992) found black locust yield to average 1.4-3.6 dry tons per acre per year over 3 years of growth in upland Georgia Piedmont. Yield growth from genetic improvement of black locust could be possible, but research in this area has not been a high priority (King et al., 1998), perhaps in part because the tree is susceptible to a borer insect, its growth does not compete with poplar over time, and it presently lacks a market (Kline and Coleman, 2010).

American sycamore (*Platanus occidentalis* L.) is native to the eastern half of the United States, though its habitat does not extend north into northern Wisconsin and Maine or south into Florida and southern Texas. A study in South Carolina found yields for American sycamore to be 2.3 dry tons per acre per year (Davis and Trettin, 2006), and Kline and Coleman's survey of forest practitioners (2010) reported 4.1 dry tons per acre per year. Steinbeck (1999) found that sycamore produced 5.6 dry tons per acre per year of woody biomass when managed intensively. However, though the rapid early growth of sycamore is an advantage for establishment, it is susceptible to disease and pests (Mercker, 2007; Kline and Coleman, 2010). It is tolerant of wet soil but cannot withstand drought conditions (Kline and Coleman, 2010). Therefore, though there is potential, more breeding advances are needed before sycamore can contribute substantially to woody biomass production.

Over the course of 7 years, yields of sweetgum (*Liquidambar styraciflua* L.) were similar to sycamore, though sweetgum has slower initial growth (Davis and Trettin, 2006). It grows in the warm, temperate climate of the Southeast under a variety of soil conditions and is tolerant of nutrient and water shortages. Kline and Coleman (2010) found yields averaged 2.7 dry tons per acre per year, though yields reached 4.1 dry tons per acre in trials. Davis and Trettin (2006) found average productivity ranged from 1.8 to 2.7 dry tons per acre per year; however, they postulated that yields would be closer to those found for poplar when yields beyond the seventh year were calculated. Respondents to Kline and Coleman's survey (2010) suggested that sweetgum could be a promising candidate for biomass production because of its overall productivity and its tolerance of less than ideal growing conditions, even though it is hampered by its slow initial growth rate.

Though it is not a hardwood, some consider loblolly pine a candidate for bioenergy production (Kline and Coleman, 2010). Loblolly pine is fast growing and can tolerate a wide range of soils, though it prefers moderately acidic soil. Annual growth is substantially affected by the availability of water. A tree native to the Southeast, it is valued for its lumber and as a source of wood pulp and is already grown on more than 32 million acres of plantations. As reported by Kline and Coleman (2010), respondents suggested that loblolly pine could be a bioenergy crop because of its prevalence in the Southeast and the extensive research and management experience associated with it. They reported yields averaging 4.1 dry tons per acre per year. Other studies have found that commercial plantations yielded 0.45-2.25 dry tons per acre per year, but under intensive management, yields could be as high as 2.25-5.4 dry tons per acre per year (Jokela et al., 2004; Coyle et al., 2008). However, loblolly pine rotations are longer compared to hardwoods, and it is more likely to be a feedstock for thermochemical conversion than biochemical because chemicals in the pinewood impede biochemical conversion (Kline and Coleman, 2010).

Eucalyptus is not a native species to the United States, but its rapid growth makes it a strong contender for bioenergy cropping. With intensive management, Prine et al. (2000) found that *Eucalyptus amplifolia* has the potential to yield up to 11.2 dry tons per acre per year under ideal conditions in northeastern and perhaps northwestern Florida, while *Eucalyptus grandis* could yield as much as 16 dry tons per acre per year in central and southern Florida. *E. grandis* prefers sandy or organic soils (Stricker et al., 2000) and is productive on land previously mined for phosphate. It could be harvested after 3 years, with an additional five harvests every 3 years (Prine et al., 2000). Eucalyptus thrives under coppice management, and plantations may be productive from 10 to 25 years (Stricker et al., 2000). However, eucalyptus is not tolerant of frost, and therefore can only be grown in Florida and along the Gulf Coast (Kline and Coleman, 2010). A freeze-tolerant variety, which could expand the number of acres suitable for eucalyptus, has been developed through genetic engineering, but it has not been approved by USDA's regulatory process (Hinchee et al., 2009).

Table 2-1 lists the advantages and limitations of woody crops for biomass production. The table is restricted to species in the Southeastern United States to allow easy comparison

TABLE 2-1 Range, Advantages, and Limitations for Commercial Biomass Tree Species in the Southeastern United States

Species	Preferable Range	Advantages	Limitations
Sweetgum	Uplands throughout the Southeast	Native species Most adaptable hardwood across region Fairly well known/studied Improvement likely in medium term	Moderate productivity Limited commercial experience
Eucalyptus	Florida and South Coastal Plain (hardiness zone 8b or higher to reduce freeze damage)	Highest growth rates Adaptable to marginal sites Multiple products/markets Improvement likely in near term	Exotic species Frost vulnerable Water requirements
Cottonwood and Poplar Hybrids	Alluvial bottomlands and low river terraces	Potential for high growth rates under right conditions Extensive genetic research Existing commercial stands Improvements likely in medium term	Narrow site requirement Variable productivity Requires intensive management and inputs Not drought tolerant
Sycamore	Well drained bottomlands	Fast growing first 2 to 3 years (but not a recommended species with current cultivars)	Narrow site requirement Chronic disease problems Not drought tolerant
Loblolly Pine	Established pine plantations and native pine areas throughout region	Well developed operations (50+ years of intense research and development) with 13×10^6 hectares in the Southeastern United States Better productivity than most hardwoods on same sites Broad genetic potential Improvements ongoing	Undesired traits for biochemical conversion to ethanol

SOURCE: Kline and Coleman (2010). Reprinted with permission from Elsevier.

because the Southeast is a favorable location for the majority of candidate species. However, the use of woody crops is not necessarily limited to this region. Woody bioenergy crop development is likely to use different species in different regions, favoring the ones that grow the quickest with few inputs within a particular region. There is potential to obtain woody biomass from many kinds of tree and shrub species, but some technological challenges have to be resolved. Otherwise, conversion of woody biomass is costly and limited to a few species.

Forest Resources

Forests with the highest productivity in the United States are concentrated in the Pacific Northwest (Figure 2-12). However, many of the productive forests in the Pacific Northwest are federally owned, and the definitions of renewable biomass in the Energy Independence and Security Act (EISA) of 2007 exclude biomass harvested from federal lands (Chapter 1). Figure 2-13 shows forestland under private ownership. Because the Southeast has high productivity, mostly private ownership, and a large number of existing tree plantations, this region is expected to supply much of the woody biomass feedstock for biofuel production. Other regions could also contribute, but given lower growth or public ownership, they are likely to supply modest quantities.

EISA also limits which forest products can be used for cellulosic biofuel that qualify under the mandate. (See Chapter 1 for EISA's definition of renewable biomass.) EISA limits

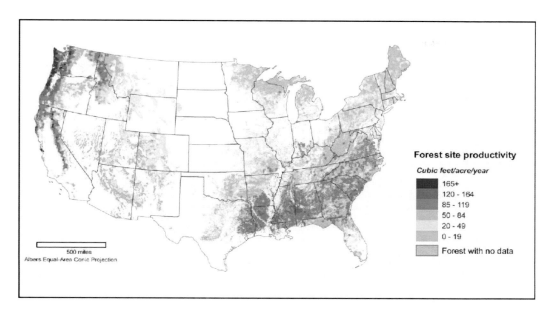

FIGURE 2-12 Net primary forest productivity in the conterminous United States.
SOURCE: Perry et al. (2007).

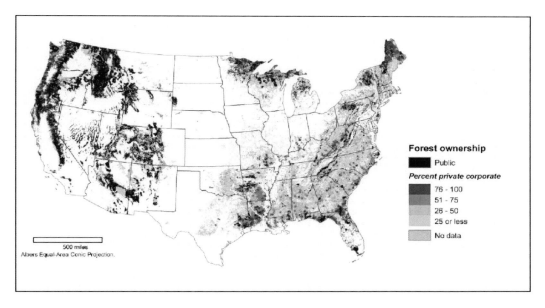

FIGURE 2-13 Forestland in the conterminous United States by ownership category.
SOURCE: Perry et al. (2007).

use from tree plantations to those established on or before December 19, 2007. According to Smith et al. (2009), there were approximately 50 million acres of planted timberland in the United States in 2007. About 80 percent of this planted land, however, is located in the southern United States. The definition of "planting" is restricted to include only activities where humans have physically planted a large proportion of the trees. One exception is made for coppice practices, whereby trees are regenerated from the sprouts of the previous generation of tree trunks. Use of natural regeneration techniques, such as the use of seed trees (as is the standard with hardwood forest management) or stand establishment with advanced regeneration in the understory, does not fit the EISA definition of "planting." The definition discourages the establishment of new plantations for biofuel purposes.

Although the law might have intended to limit the clearing of native forests for plantations to support biofuel production, it could increase competition for wood products because existing plantations are typically already dedicated for other wood products such as saw wood or pulp. This increase in competition for wood products would increase timber prices and spur new investments in plantations. These new softwood plantations in the Southeast, in turn, would displace natural hardwood stands (Sohngen and Brown, 2006). Thus, despite the intent of the law, there could be some negative consequences associated with increased biofuel production from wood.

Tree residues from logging operations on plantations fitting the definition are included in the definition of renewable biofuels. Also included, however, are residues from subsequent milling operations, as long as these residues are not mixed with other chemicals or materials during manufacturing. Slash from other logging operations and precommercial thinning can be used for biofuel production for meeting RFS2 as long as the material is obtained from nonfederal forestlands and as long as the material is not from old-growth forests. This definition effectively allows material from nonplanted forests that are harvested for traditional forestry purposes, including final harvests and thinning operations. Supply of residues might not be consistent throughout the year; for example, snowy regions often have "spring breakup" times when logging trucks cannot operate on many roads to prevent spring thaw-related damage.

Municipal Solid Waste

Municipal solid wastes (MSWs) contain paper, paperboard, textiles, wood, yard trimmings, and food scraps, which are biological material that could be used to generate bioenergy. In 2009, the United States generated 243 million tons of MSW, 165 million tons of which were the aforementioned biological material (EPA, 2010a). At the time this report was written, most municipalities paid for the collection and disposal of MSW. Many locations have recycling programs where most of the higher value portions of the waste stream, such as metals, paper, and plastic, are separated from the bulk trash and recycled. In 2009, the proportion of recovered biological materials were 62-percent paper and paperboard, 15-percent textiles, 14-percent wood, 2.5-percent food, and 60-percent yard trimmings (EPA, 2010a). The remainder of the waste stream is either landfilled or incinerated. Increasing recovery of biological waste stream can provide opportunities for using MSW as biofuel feedstocks (Milbrandt, 2005; Perlack et al., 2005; NAS-NAE-NRC, 2009). If MSW were to be used as biofuel feedstock, the quantity supplied would likely be highest close to urban centers with large populations.

There are two major barriers to the conversion of MSW to liquid transportation fuels. First, many large cities that have large volumes of MSW (to achieve economy of size) installed trash incinerators over the last 25 years to recover the energy in the MSW and to reduce the economic and environmental impact of their MSW. Any new biofuel facility for MSW would compete economically with the existing incineration facilities. Second, MSW is a mixed stream that is highly heterogeneous. Not all metals and plastics can be economically removed from the waste stream. There will always be some nonbiogenic material in MSW. It will also contain food wastes that have microorganisms that could contaminate a biochemical conversion plant. MSW would probably have to be sterilized before being used in a biochemical conversion process. It will also contain some level of toxic substances, such as mercury in batteries, pesticide residues, and paints.

MSW might be better suited for a thermochemical conversion process; however, this technology is still in development. Because of the potential contamination issues, thermochemical conversion technologies will most likely be developed using "cleaner," more uniform feedstocks such as wood or switchgrass. Once the technology is mature, MSW would become an attractive feedstock.

Storage and Delivery of Cellulosic Feedstocks

Most agricultural biomass production (with the exception of forest products) is seasonal and results in a large volume of feedstock material being generated, transported, and stored at the refining facility in a short timeframe. Yet, the refinery facility needs to produce biofuels throughout the year to maximize productivity and prevent degradation of its biofuel product. Corn stover, switchgrass, and other similar materials are harvested during a short period in the fall and would have to be stored for year-round supply to a biorefinery. They are usually baled when collected for transport and can be stored near the field or biorefinery. Based on the feedstock type and storage location, infrastructure or action might be needed to control moisture of any biomass feedstocks and ensure adequate long-term storage. Loss of dry matter is a primary concern of storing biomass feedstocks. It could vary from 1-percent loss each year in cool regions with low humidity and rainfall to 25-percent loss each year in warm regions with high humidity and rainfall (Hess et al., 2007). Moisture control in storage facilities has to be adjusted to accommodate varying volumes of feedstock. Research also suggested that dry-matter loss depends on how biomass is stored. For example, switchgrass stored as round bales was reported to have lower dry-matter loss than switchgrass stored as rectangular bales in the Southeast (Larson et al., 2010).

Transporting cellulosic feedstock over long distances to biorefineries might not be economically feasible because biomass is bulky and has about one-third of the energy density of crude oil. To avoid transporting bulky biomass, some researchers have suggested setting up regional preprocessing infrastructure to clean, sort, chop or grind, control moisture, densify, and package the feedstocks before transporting them to biorefineries. Pretreatment could be carried out at the preprocessing centers if biomass is to be converted to fuels via biochemical pathways (Carolan et al., 2007). Torrefaction and liquefaction using pyrolysis or other thermochemical techniques also have been proposed to reduce moisture and break down some of the biomass ultrastructure (Sadaka and Negi, 2009; Yan et al., 2009). In the case of torrefaction, a dense "bale" of solid biomass can be produced on-site. Torrefaction and pyrolysis are only suitable for feedstocks for thermochemical conversion to fuel. This biomass "bale" could enable more energy-dense materials to be transported, increasing the efficiency of the delivery system, while reducing the water content of the material—a key issue with conversion and processing costs. Liquefaction, on the other hand, involves producing pyrolytic or other thermochemically derived liquid from biomass in the field and transporting this intermediate material to the refinery either by tanker truck, rail, barge, or pipeline (Pootakham and Kumar, 2010). Only small trials of these technologies have been conducted. Large-scale, in-the-field processing has yet to be fully realized, and there could be significant energy, infrastructure, and economic barriers to deployment.

In contrast, MSW is available year-round and forest resources can be harvested year-round so that long-term storage might not be necessary. In addition, reducing woody biomass to wood chips on-site for volume reduction is necessary prior to transport to another location and is also a well-established technology.

Conversion Technologies

Two types of technology are likely to be used to convert cellulosic biomass to fuels to meet the Renewable Fuel Standard: biochemical and thermochemical conversion (NAS-NAE-NRC, 2009).

Biochemical Conversion

Biochemical pathways for converting cellulosic biomass into fuels follow the generalized process of pretreatment to release carbohydrates from the lignin shield, breaking down cellulose and hemicellulose to release sugars, fermentation of sugar to ethanol, distillation to separate the ethanol from the dilute aqueous solution, and conversion of the residue to electricity (Figure 2-14). The release of sugars from cellulose and hemicellulose is often incomplete because physical and chemical associations between the major components in biomass hinder the hydrolysis of cellulose into fermentable sugars (Alvira et al., 2010). To

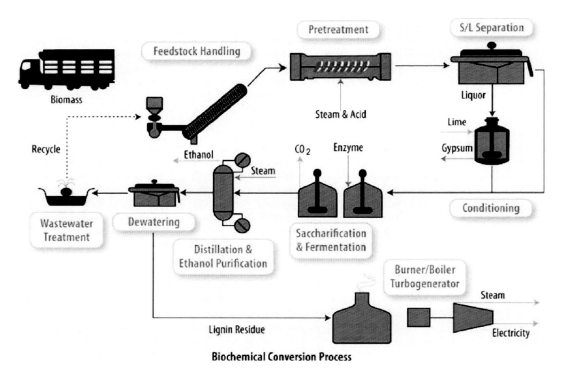

Biochemical Conversion Process

FIGURE 2-14 Model of a lignocellulosic-based ethanol biochemical refinery.
SOURCE: Foust et al. (2009). Reprinted with permission from Springer-Verlag.

address this issue, several pretreatment pathways have been investigated (Eggeman and Elander, 2005; Mosier et al., 2005; Yang and Wyman, 2007; Sendich et al., 2008). Given the variability in composition of different feedstocks, not all pretreatment pathways yield the desired results. For instance, alkaline-based pretreatment for the removal of lignin was shown effective in agricultural residues, such as stover, but not a good choice for softwood species (Chandra et al., 2007). Acid pretreatment has resulted in improvements in hydrolysis of nearly all biomass feedstocks into sugars, but the toxicity of the compounds derived from these reactions affects the fermentation steps (Oliva et al., 2003), increases the cost of processing, and increases the environmental impact by requiring neutralization and disposal of byproducts. Because of the severity of the pretreatment steps, the resulting sugar streams are not homogeneous, can undergo degradation, and result in enzyme and fermentation compatibility issues with the fermenting organisms (Alvira et al., 2010).

After pretreatment, cellulolytic enzymes are used to hydrolyze the cellulose polymers to 5-carbon and 6-carbon sugars (xylose and glucose). Unlike glucose, xylose is not readily fermented to ethanol. Yeasts or bacteria have been genetically modified or metabolically engineered to ferment both glucose and xylose to enhance yield of ethanol from cellulose (Aristidou and Penttila, 2000; Sonderegger et al., 2004; Nevoigt, 2008). The challenge is to develop glucose- and xylose-fermenting microorganisms that can withstand antimicrobial agents released during the pretreatment and hydrolysis steps and that are not inhibited by high alcohol concentrations.

Distillation in cellulosic ethanol production is the same as that in corn-grain ethanol production. Solids from different portions of the conversion process can be concentrated by centrifugation and burned in a boiler to generate steam and electricity for the biorefinery, particularly if the feedstock is rich in lignin.

Thermochemical Conversion

The mechanisms of thermochemical conversion include high temperature, pressure, chemicals, and catalysts to transform lignocellulosic biomass into many different products, including ethanol, butanol, green diesel, super diesel, Fischer-Tropsch (F-T) liquids, pyrolysis oils, and green gasoline (see Figure 2-15 for an example). The most advanced technologies include gasification to produce syngas followed by F-T synthesis or methanol synthesis, and pyrolysis or liquefaction to bio-oil followed by catalytic upgrading. Thermochemical conversion processes are not as feedstock-specific as biochemical conversion, allowing for a wide range of biomass feedstock to be used in its processing; this provides opportunities for refineries to be built in any location where adequate biomass can be produced to maintain their operations.

Gasification

Gasification is a process in which biomass (or other carbon-containing feedstock) reacts with limited oxygen (or air) and steam at high temperatures to produce syngas (Huber et al., 2006). The syngas is then conditioned to remove impurities and shifted to desired ratio of hydrogen to carbon monoxide (CO). The clean synthesis gas can be used to produce ethanol (Figure 2-16), whereby the anaerobic bacterium *Clostridium ljungdahlii* ferments syngas into ethanol (Huber et al., 2006). Alternatively, the synthesis gas can be used in F-T reactors to produce liquid hydrocarbon fuels using cobalt, iron, or ruthenium catalysts. The F-T technology has been used commercially for over 50 years and produces fuels that are compatible with conventional petroleum products. The gasification of biomass, however,

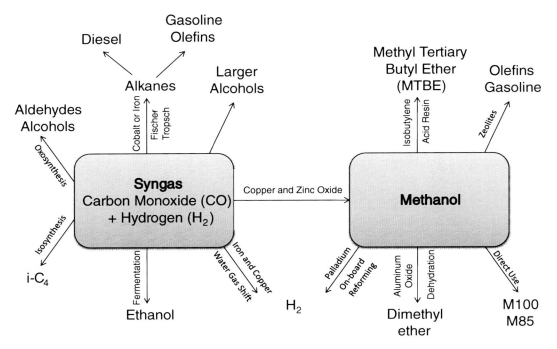

FIGURE 2-15 Thermochemical conversion pathways and products.
SOURCE: Adapted from Spath and Dayton (2003).

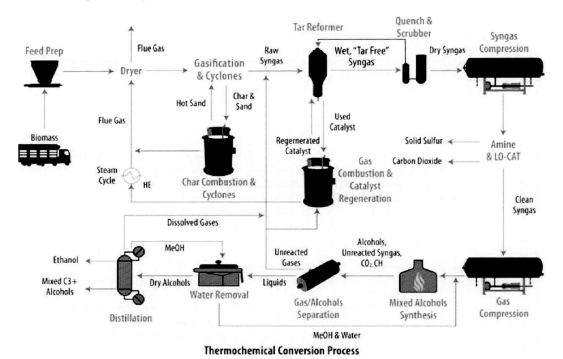

FIGURE 2-16 Schematic diagram of a thermochemical conversion refinery to produce ethanol.
SOURCE: Foust et al. (2009). Reprinted with permission from Springer-Verlag.

has proven to be more difficult than coal or natural gas gasification (Huber et al., 2006; NAS-NAE-NRC, 2009).

Gasification and upgrading of syngas into liquid "drop-in" fuels has been done successfully at the pilot scale using various feedstocks. Various manufacturers in Europe and North America have recently entered the market as equipment suppliers, including over 50 companies offering various gasification configuration options (Balat et al., 2009). Catalysts for F-T synthesis are well developed for coal gasification systems but have not demonstrated long-term performance when applied to biomass. Small quantities of contaminants in the syngas stream can render catalyst beds inactive, and scrubbing and cleaning gas processes are costly and yet to be fully deployed.

Fast Pyrolysis and Liquefaction

Pyrolysis, a process that uses high temperature under anaerobic conditions, breaks down biomass into a volatile mixture of hydrocarbons. This mixture of hot gases is condensed into a bio-oil. Several commercial-scale pyrolysis facilities are currently in operation; however, these are lower severity processes designed to produce specialty chemicals, not fuels. A number of public and privately funded organizations are actively developing fast pyrolysis technologies for fuel production. In fast pyrolysis, small biomass particles are mixed with hot solids to rapidly heat the biomass and thermally decompose its components. The solid can either be inert or have catalytic properties. The hot vapors are separated from the solids and cooled to condense a liquid phase (bio-oil), and the noncondensable gases along with the solid char are burned to supply heat for the process. The raw bio-oil is an emulsion, rendering it incompatible with conventional petroleum oils and requiring additional upgrading. The most frequently proposed upgrading technology is hydrotreating. Although hydrotreating is a well-established petroleum refining technology, little commercial experience exists for upgrading pyrolysis oils.

Liquefaction of biomass also produces bio-oil, but through controlled reaction rates and reaction mechanisms using pressure, gases, and catalysts. Catalysts used for upgrading liquefaction products include alkali, metals, and nickel and ruthenium heterogeneous catalysts. Opportunities in nascent technologies exist in the form of liquefaction of biomass for production of bio-oil, including the use of hydrothermal technologies. Hydrothermal liquefaction uses subcritical or supercritical water to liquefy biomass into a bio-oil. Elevated temperatures (200-600°C) are used in a pressurized vessel containing biomass (5-40 MegaPascals), de-polymerizing and converting cellulose, lignin, and hemicellulose into a soluble mixture that can be upgraded and processed in similar fashion as pyrolytic based bio-oil (Elliott et al., 1991; Demirbas, 2009). Liquefaction with other supercritical fluids has been shown to overcome successfully many of the pyrolytic bio-oil processing shortcomings for stover and wood. Methanol reaches supercritical conditions at milder temperatures (238°C) and pressures (8 MegaPascals) than water. Liquefaction with methanol is almost half as energy intensive as pyrolysis (500°C), resulting in conversion of solid biomass into a bio-oil in excess of 90 percent on a weight-by-weight basis (Demirbas, 2000; Balat, 2008; Soria et al., 2008). The use of a single step system is favored in the long run as it will require less operational and infrastructure complexity. The primary advantage of these liquefaction systems is that they do not require pretreatment and can work with high-moisture biomass feedstocks and MSW streams. Current technical barriers include residence times for high liquefaction to occur on certain feedstocks and batch or semi-batch processing, resulting in limited throughput and scalability issues.

Bio-oil has poor volatility, high viscosity, coking, corrosiveness, and poor cold-flow properties (Huber et al., 2006) and would have to be upgraded if it were to be used as transportation fuel. It can be upgraded to fuels in the gasoline and diesel range by hydrode-oxgenation or zeolite upgrading—processes similar to the ones used by current petroleum refining (Huber et al., 2006; Roberts et al., 2010). Catalytic upgrading reduces the oxygen level of the bio-oil and increases the hydrogen proportion, leading to the production of saturated C-C bonds that are fully compatible with petroleum infrastructure and use. Catalyst effectiveness and stability are areas of intense research and development; many companies are experimenting with proprietary technology that could prove to be commercially feasible and meet environmental standards. Catalysts can be easily rendered inactive by deposition of coke and tars, and represent an operational cost and environmental problem, as regeneration of catalysts might involve a combination of high temperature and use of organic solvents and produce a waste stream and evaporative emissions. Alternatively, the bio-oils can be used to produce syngas via steam reforming. The syngas can then be used to produce transportation fuels using F-T synthesis as discussed earlier (Huber et al., 2006).

Products and Coproducts

Because of the large scale of biofuel production, sizeable quantities of coproducts will be produced. Therefore, the market for coproducts would have to complement the quantities produced for the coproducts to maintain value. Coproducts of ethanol production from lignocellulosic feedstock through biochemical conversion are rich in lignin and not suitable as animal feedstuffs. Many proposed lignocellulosic ethanol plants include a boiler and turbogenerator system to convert residue to electricity. Other potential coproducts in early stages of development are bioplastic (Snell and Peoples, 2009) and a performance enhancer in asphalt binder (McCready and Williams, 2008).

As discussed above, the products and coproducts of thermochemical conversion of lignocellulosic feedstock depend on the processes used. F-T liquids include paraffinic diesel that can be refined to high-quality diesel or jet fuel and naphtha that requires substantial refining to produce high-octane gasoline. Coproducts of F-T liquids include ash from the gasifier, CO_2, and some light hydrocarbon gases. Some pyrolysis processes a coproduct that is a solid (biochar) that has been proposed as a potential carbon sequestration mechanism and soil amendment (Roberts et al., 2010).

Distribution and Use

Ethanol

Delivery and use of cellulosic ethanol is the same as corn-grain ethanol (discussed earlier). Although all ethanol is the same regardless of source, EISA puts an additional accounting burden on the final blenders of ethanol-containing gasoline. There are separate requirements for "advanced renewable fuels" that can be met by the use of cellulosic ethanol. Therefore, these new ethanol sources require separate accounting from the existing ethanol sources. The additional accounting might not require new physical equipment, but it will definitely require additional labor and accounting software.

F-T Liquids

F-T diesel and gasoline are compatible with existing fuel infrastructure, and they could be used in combustion engines in conventional vehicles. Introduction of biomass-sourced F-T product would have little impact on refinery operations. The relative amounts of naphtha, jet fuel, diesel, and heavier material in the stream would affect crude-unit yields and downstream unit rate. The F-T material itself, however, is high quality with essentially no heteroatoms. The naphtha is paraffinic and does not reform well but can be included as a component of reformer feedstock or sold as a petrochemical feedstock. With the exception of fluidity properties, distillate range F-T products are excellent jet fuel and diesel components. The heavy F-T product readily cracks in either an FCC unit or a hydrocracker.

F-T products have already been approved as a jet fuel component at levels up to 50 percent of the fuel. Therefore, coprocessing F-T products with conventional crude oil would not disqualify jet fuel produced from the blend. The poor cold-flow properties of both jet fuel and diesel fuel will reduce the maximum end point of both products and reduce the amounts that can be recovered from the conventional crude oil. Processing the F-T product in a separate crude unit would give the refinery additional operating and blending flexibility.

Pyrolysis Oil

Typical raw, biobased pyrolysis oils are a wide boiling-range mixture of hydrocarbons that contain a significant amount of heteroatoms such as oxygen and nitrogen. Most refineries could process a small amount of this material as an alternate crude source; however, processing significant quantities would require major refinery modifications. Including even small amounts of pyrolysis oils in a refinery crude mix could cause operational problems. These oils are unstable and easily form gums even without being exposed to air. These gums can cause fouling in the crude-unit preheat train and in downstream units. They also contain some acidic components that can increase equipment corrosion rates.

The yields of the various pyrolysis components depend strongly on the type of biomass fed to the unit as well as pyrolysis conditions. The refiner that receives these oils has to treat them as an alternate crude source. That is, the refiner has to know what yields to expect and the qualities of the various cuts that will exit the crude unit. The pyrolysis oils will increase the hydrogen consumption on the naphtha pretreater and the distillate hydrotreater and require an increase in severity to remove the additional heteroatoms. Thermally derived oils generally make poor FCC feedstocks, so the FCC yields from the heavy portion of the pyrolysis oil could be low.

Biobased pyrolysis oils are not yet approved for use as a constituent of jet fuel. Tests are currently in progress for several different oils, but until pyrolysis oils are approved, inclusion of these oils in the crude-unit feedstock or the feedstock to any other unit that produces jet fuel will disqualify all potential jet fuel from that unit. The refiner might install a separate small distillation unit to segregate the biobased oils from the rest of the refinery products to keep them out of jet fuel. The separate crude unit would also allow the refiner to use upgraded metallurgy to minimize corrosion issues. Preheat train fouling would also be limited to this smaller unit and would not impact overall refinery operation. Desalting of the pyrolysis oil could increase the load on the wastewater treatment facility.

The miscibility of pyrolysis oils is a function of the pyrolysis severity. Severities that give high yields of pyrolysis oils produce materials that are immiscible with conventional petroleum streams. They cannot be blended with any refiner stream, either finished products or intermediate streams. Therefore, their upgrading for blending with gasoline or diesel fuel would require a new dedicated upgrading unit either at the pyrolysis unit or in the refinery.

Status as of 2010

As of 2010, small quantities of cellulosic biofuels were marketed commercially in the United States. Several companies are operating demonstration facilities or constructing commercial demonstration plants to convert cellulosic biomass to ethanol via biochemical conversion (Biofuels Digest, 2010a).

In Canada, Iogen has been operating its cellulosic ethanol demonstration plant in Ottawa (Tolan, 2004). It has been using wheat straw to produce ethanol since 2004 and had not reached full capacity as of 2010 (Table 2-2). The facility has an annual capacity of 482,000-584,000 gallons and can receive up to 30 dry tons of feedstock per day (Iogen, 2010b). Iogen is developing a commercial-scale plant in Saskatchewan, Canada. As a comparison, most corn-grain ethanol biorefineries operating in 2010 have nameplate capacity of over 40 million gallons per year (RFA, 2010).

Several companies are investigating pilot-scale plants in the United States and hope to scale them up to commercial-size plants by 2011-2012, as shown in Table 2-3. Among those companies listed, about half of them plan to produce ethanol via biochemical conversion (Figure 2-17). Thirty-eight percent of the companies plan to produce drop-in fuels using algae or cellulosic biomass as feedstocks (Figure 2-18). Operating experience from demonstration plants will help to confidently design a commercial-scale facility. DOE provided $564 million from the American Recovery and Reinvestment Act (111 P.L. 5) to 19 selected advanced biofuel projects to support the construction and operation of pilot, demonstration, and commercial-scale facilities (Table 2-4).

For cellulosic biofuels to become cost-competitive with petroleum products, significant improvements in the associated technologies will have to be achieved. These technologies include improved ways to produce biomass and much more cost-effective conversion technologies. Conversion technologies include both thermochemical and biochemical approaches. Strong incentives such as RFS2 have led to significant investments in research, development, and deployment by both public and private sponsors in the United States and overseas. These investments are expected to bear fruit in the 5- to 10-year framework, but it is doubtful that they could make biofuels cost-competitive without subsidies, unless the price of oil rises substantially or a disruptive innovation[15] for biofuel production is developed.

Growing biomass and converting the feedstock is subject to well-understood, inherent physicochemical constraints. Technology advances can get closer to the theoretical limits but cannot change the limits imposed by physics and chemistry. For example, successful

TABLE 2-2 Ethanol Production from the Iogen Demonstration Facility in Ottawa, Canada, 2005-2010

Year	Ethanol (gallons)
2005	34,223
2006	4,441
2007	686
2008	54,558
2009	153,495
2010	134,406

DATA SOURCE: Iogen (2010a).

---

[15]A term used in business and technology that refers to innovation that improves the product (typically by lowering the price of product substantially) that the market does not expect.

TABLE 2-3 Companies That Have Secured Funding for Demonstration of Nonfood-Based Biofuels

Company	Location	Feedstock	Conversion Technology	Product	Estimated Production (million gal/yr)					
					2010	2011	2012	2013	2014	2015
Abengoa	Hugoton, KS	Corn stover, sorghum stubble, switchgrass	Cellulosic feedstock Biochemical	Ethanol	0	0	0	15	15	15
ADM	Decatur, IL	Corn stover	Cellulosic feedstock Biochemical	Ethanol	0	0	1	1	1	1
AE Biofuels	Butte, MT	Switchgrass, straw, corn stover, bagasse	Cellulosic feedstock Biochemical	Ethanol	0.02	0.05	10.04	10.5	10.5	10.5
AltAir	Anacortes, WA	Camelina	Thermochemical	Drop-in fuel	0	0	100	100	100	100
American Process	Alpena, MI	Woody biomass	Cellulosic feedstock Biochemical	Ethanol	0	0.89	0.89	0.89	0.89	0.89
Amyris	Emeryville, CA	Sweet sorghum	Cellulosic feedstock Biochemical	Drop-in fuel	0.01	0.02	1.02	27.02	27.02	27.02
Aquatic Energy	Not specified, LA	Algae	Oil extraction	Oil feedstock	0	0.03	0.03	0.03	0.03	0.03
Aurora Algae	Vero Beach, FL	Algae	FAME	Drop-in fuel	0.01	0.01	0.5	0.5	0.5	0.5
BioProcess Algae	Shenandoah, IA	Algae	FAME	Drop-in fuel	0	0.01	0.01	0.01	0.01	0.01
BlueFire Renewables	Lancaster, CA	Woody biomass	Cellulosic feedstock Biochemical	Ethanol	0.01	3.91	3.91	22.91	22.91	22.91
BP Biofuels	Jennings, LA	Wheat, switchgrass	Cellulosic feedstock Biochemical	Ethanol	1.4	1.4	37.4	37.4	37.4	37.4
Buckeye Technologies/UF	Perry, FL	Woody biomass, sugarcane	Cellulosic feedstock Biochemical	Ethanol	0	0.01	0.01	0.01	0.01	0.01
Clearfuels	Not specified, HI	Bagasse, woody biomass	Thermochemical, Fischer-Tropsch	Drop-in fuel	0	0	0	0	0	18

continued

TABLE 2-3 Continued

Company	Location	Feedstock	Conversion Technology	Product	Estimated Production (million gal/yr)					
					2010	2011	2012	2013	2014	2015
Clearfuels	Commerce City, CO	Woody biomass, corn stover, bagasse	Thermochemical, Fischer-Tropsch	Drop-in fuel	0	0.07	0.07	0.07	0.07	0.07
Clearfuels	Collinswood, TN	Woody biomass	Thermochemical, Fischer-Tropsch	Drop-in fuel	0	0	0	0	20	20
Cobalt	Sausalito, CA	Woody biomass	Cellulosic feedstock Biochemical	Drop-in fuel Biobutanol	0.01	2.01	102	102	102	102
Coskata	Green County, AL	Woody biomass	Cellulosic feedstock Biochemical	Ethanol	0.05	0.05	55.05	55.05	55.05	55.05
DDCE	Vonore, TN	Corn stover, switchgrass	Cellulosic feedstock Biochemical	Ethanol	0.25	0.25	0.25	0.25	0.25	0.25
Dynamic Fuels	Geismar, LA	Animal fats, tallow, and vegetable oils	Thermochemical	Drop-in fuel	75	75	75	75	75	75
Fiberight	Blairstown, IA	Organic-based MSW	Cellulosic feedstock Biochemical	Ethanol	0.01	6	6	6	6	6
Fulcrum	Reno, NV	Organic MSW	Thermochemical, gasification	Ethanol	0.01	0.01	10.51	10.51	10.51	10.51
Gevo	St Joseph, MO	Corn, sugar cane, sugar beets	Cellulosic feedstock Biochemical	Drop-in fuel Biobutanol	1	1	16	300	300	300
Green Star Products	To be determined, UT	Algae	FAME	Drop-in fuel	0	0	2	2	2	2
Haldor Topsoe	Des Plains, IL	Woody biomass	Cellulosic feedstock Thermochemical	Drop-in fuel	0	0	0.8	0.8	0.8	0.8
HCL Clean Tech	Durham, NC	Woody biomass	Cellulosic feedstock Biochemical	Ethanol	0	0.01	0.01	0.01	0.01	0.01

TABLE 2-3 Continued

Company	Location	Feedstock	Conversion Technology	Product	Estimated Production (million gal/yr)					
					2010	2011	2012	2013	2014	2015
IneosBIO	Vero Beach, FL	Organic-based MSW	Gasification-Fermentation	Ethanol	0	8	8	8	8	8
Joule	Leander, TX	Algae	FAME	Drop-in fuel	0.01	0.01	0.01	0.01	0.01	0.01
Kent BioEnergy	Mecca, CA	Algae	Oil extraction	Oil feedstock	0.01	0.01	0.01	0.01	0.01	0.01
KiOR	Columbus, MS	Woody biomass	Thermochemical, pyrolysis	Drop-in fuel	0.23	0.23	0.23	80	80	120
KL Energy	Upton, WY	Sugarcane, bagasse, woody biomass	Cellulosic feedstock Biochemical	Ethanol	0	1.3	1.3	1.3	1.3	1.3
LiveFuels	To be determined	Algae	FAME	Drop-in fuel	0.01	0.01	0.01	0.01	0.01	0.01
Logos Technologies	Visalia, CA	Switchgrass, corn stover, woody biomass	Cellulosic feedstock Biochemical	Ethanol	0	0	0.8	0.8	0.8	0.8
LS9	Okeechobee, FL	Sugar cane	Cellulosic feedstock Biochemical	Drop-in fuel	0.1	10.1	10.1	10.1	10.1	10.1
Mascoma	Rome, NY	Switchgrass, woody biomass, ag waste	Cellulosic feedstock Biochemical	Ethanol	0.2	0.2	0.2	20.2	20.2	20.2
Murphy Oil	Hereford, TX	Organic-based MSW	Cellulosic feedstock Biochemical	Ethanol	0	0	115	115	115	115
PetroAlgae	Fellsmere, FL	Algae	Biochemical and thermochemical	Drop-in fuel	0.12	0.12	70.12	140.12	210.12	210.12
Phycal	Central Oahu, HI	Algae	Oil extraction	Oil feedstock	0	0.01	0.01	0.01	0.01	0.01
POET	Emmetsburg, IA	Corn stover	Cellulosic feedstock Biochemical	Ethanol	0.02	0.02	25.02	25.02	25.02	25.02

continued

TABLE 2-3 Continued

Company	Location	Feedstock	Conversion Technology	Product	Estimated Production (million gal/yr)					
					2010	2011	2012	2013	2014	2015
Powers Energy	Lake County, IN	Organic-based MSW	Gasification-Fermentation	Ethanol	0	0	0	160	160	160
Range Fuels	Soperton, GA	Woody biomass	Thermochemical, gasification	Methanol and ethanol	4	4	4	20	20	20
REII	Toledo, OH	Woody biomass, rice hulls, corn stover, straw, switchgrass	Thermochemical, Fischer-Tropsch	Drop-in fuel	0.02	0.02	0.35	0.35	0.35	0.35
Rentech	Rialto, CA	Woody biomass, corn stover, straw, bagasse, organic MSW	Thermochemical, Fischer-Tropsch	Drop-in fuel	0.15	0.15	0.15	8.15	259	259
Sapphire Energy	Columbus, NM	Algae	Thermochemical	Drop-in fuel	0.02	0.02	0.02	0.02	0.02	0.02
Solazyme	South San Francisco, CA	Algae	Thermochemical	Drop-in fuel	0.1	0.15	100.1	100.1	100.1	100.1
Solix	Durango, CO	Algae	FAME	Drop-in fuel	0.01	0.01	0.01	0.01	0.01	0.01
Terrabon	Bryan, TX	Woody biomass, sweet sorghum	Cellulosic feedstock Biochemical	Drop-in fuel	0.1	0.1	1.25	5	25	25
ThermoChem Recovery (TRI)	Durham, NC	Woody biomass	Thermochemical, Fischer-Tropsch	Drop-in fuel	0.01	0.01	0.01	0.01	0.01	0.01
Trenton Fuel Works	Trenton, NJ	Organic-based MSW	Cellulosic feedstock Biochemical	Ethanol	0	0	0	0	3.87	3.87
UOP-Aquaflow Bionomic	Hopewell, VA	Algae	Thermochemical	Drop-in fuel	0	0.01	0.01	0.01	0.01	0.01

TABLE 2-3 Continued

Company	Location	Feedstock	Conversion Technology	Product	Estimated Production (million gal/yr)					
					2010	2011	2012	2013	2014	2015
Virent	Madison, WI	Sugar beets, corn stover, sugarcane, woody biomass, switchgrass	Thermochemical	Drop-in fuel	0.01	0.01	0.01	0.01	0.01	0.01
ZeaChem	Boardman, OR	Woody biomass	Gasification-Fermentation	Ethanol	0.25	0.25	0.25	0.25	0.25	0.25

DATA SOURCE: Individual companies and Biofuel Digest (2010b).

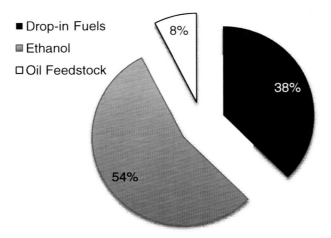

FIGURE 2-17 Percent companies with secured funding for demonstration of nonfood-based biofuels that plan to produce drop-in fuels, ethanol, and oil feedstock.
DATA SOURCE: Biofuels Digest (2010a).

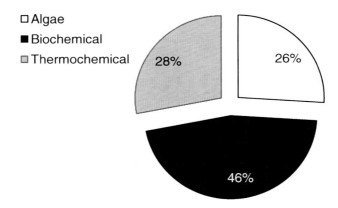

FIGURE 2-18 Percent companies with secured funding for demonstration of nonfood-based biofuels that plan to produce algal biofuels or cellulosic biofuels via biochemical or thermo-chemical pathways.
DATA SOURCE: Biofuels Digest (2010a).

application of fertilizers and pesticides has improved agricultural yields, but those improvements still operate within the same set of physicochemical constraints. Driven by high-technology tools from the physical sciences, including high-performance computing, genomics is producing a steady stream of results that have a huge potential to revolutionize many scientific fields, including those involved in the production of biofuels. Disruptive technologies work on processes subject to different constraints: that is, they operate under wider physicochemical boundaries, which may not turn out to be an advantage. Such a disruptive technology could happen in biotechnology because of the significant paradigm shift that biotechnology has undergone with the advances in genomics.

TABLE 2-4 Advanced Biofuel Projects Supported by the U.S. Department of Energy through the American Recovery and Reinvestment Act

Grantee	Location	DOE Grant Amount	Nonfederal Amount	Feedstock Source	Description
Algenol Biofuels Inc.	Fort Myers, FL	$25,000,000	$33,915,478	Industrial CO_2	This project will make ethanol directly from carbon dioxide (CO_2) and seawater using algae. The facility will have the capacity to produce 100,000 gallons of fuel-grade ethanol per year.
American Process Inc.	Alpena, MI	$17,944,902	$10,148,508	On-site mill-wood waste	This project will produce fuel and potassium acetate, a compound with many industrial applications, using processed wood generated by Decorative Panels International, an existing hardboard manufacturing facility in Alpena, MI. The pilot plant will have the capacity to produce up to 890,000 gallons of ethanol and 690,000 gallons of potassium acetate per year starting in 2011.
Amyris Biotechnologies, Inc.	Emerville, CA	$25,000,000	$10,489,763	Ceres engineered crops	This project will produce a diesel substitute through the fermentation of sweet sorghum. The pilot plant will also have the capacity to co-produce lubricants, polymers, and other petrochemical substitutes.
Archer Daniels Midland	Decatur, IL	$24,834,592	$10,946,609	Corn stover within a 50-mile radius of refinery	This project will use acid to break down biomass which can be converted to liquid fuels or energy. The facility will produce ethanol and ethyl acrylate, a compound used to make a variety of materials, and will also recover minerals and salts from the biomass that can then be returned to the soil.
Clearfuels Technology Inc.	Commerce City, CO	$23,000,000	$13,433,926	Unknown	This project will produce renewable diesel and jet fuel from woody biomass by integrating ClearFuels' and Rentech's conversion technologies. The facility will also evaluate the conversion of bagasse and biomass mixtures to fuels.
Elevance Renewable Sciences	Bolington, IL	$2,500,000	$625,000	Unknown	This project was selected to complete preliminary engineering design for a future facility producing jet fuel, renewable diesel substitutes, and high-value chemicals from plant oils and poultry fat.

continued

TABLE 2-4 Continued

Grantee	Location	DOE Grant Amount	Nonfederal Amount	Feedstock Source	Description
Gas Technology Institute	Des Plains, IL	$2,500,000	$625,000	Algae from PetroSun and Blue Marble; 1 T wood from Johnson Timber; and corn stover from Cargill	This project was selected to complete preliminary engineering design for a novel process to produce green gasoline and diesel from woody biomass, agricultural residues, and algae.
Haldor Topsoe, Inc.	Des Plains, IL,	$25,000,000	$9,701,468	Wood from UPM, MN	This project will convert wood to green gasoline by fully integrating and optimizing a multistep gasification process. The pilot plant will have the capacity to process 23 dry tons of feedstock per day.
ICM, Inc.	St. Joseph, MO	$25,000,000	$6,268,136	Corn stover from LifeLine Foods; other feedstocks from unknown sources	This project will modify an existing corn-ethanol facility to produce cellulosic ethanol from switchgrass and energy sorghum using biochemical conversion processes.
Logos Technologies	Visalia, CA	$20,445,849	$5,113,962	Corn stover from Next Step Biofuels Inc.; switchgrass from Ceres; wood from central California facilities	This project will convert switchgrass and woody biomass into ethanol using a biochemical conversion processes.
Renewable Energy Institute International	Toledo, OH	$19,980,930	$5,116,072	Unknown	This project will produce high-quality green diesel from agriculture and forest residues using advanced pyrolysis and steam reforming. The pilot plant will have the capacity to process 25 dry tons of feedstock per day.
Solazyme, Inc.	Riverside, PA	$21,765,738	$3,857,111	Sucrose, MSW organics, or switchgrass from unknown sources	This project will validate the projected economics of a commercial scale biorefinery producing multiple advanced biofuels. This project will produce algal oil that can be converted to oil-based fuels.
UOP LLC	Kapolei, HI	$25,000,000	$6,685,340	Unknown	This project will integrate existing technology from Ensyn and UOP to produce green gasoline, diesel, and jet fuel from agricultural residue, woody biomass, dedicated energy crops, and algae.

TABLE 2-4 Continued

Grantee	Location	DOE Grant Amount	Nonfederal Amount	Feedstock Source	Description
ZeaChem Inc.	Boardman, OR	$25,000,000	$48,400,000	Wood from Greenwood Resources	This project will use purpose-grown hybrid poplar trees to produce fuel-grade ethanol using hybrid technology. Additional feedstocks such as agricultural residues and energy crops will also be evaluated in the pilot plant.
Demonstration-Scale Projects (in alphabetical order)					
BioEnergy International, LLC	Clearfield County, PA	$50,000,000	$89,589,188	Corn from Lansing Trade Group, LLC	This project will biologically produce succinic acid from sorghum. The process being developed displaces petroleum-based feedstocks and uses less energy per ton of succinic acid produced than its petroleum counterpart.
Enerkem Corporation	Pontoc, MS	$50,000,000	$90,470,217	Plant located in landfill site	This project will construct a facility that produces ethanol fuel from woody biomass, mill residue, and sorted municipal solid waste. The facility will have the capacity to produce 19 million gallons of ethanol per year.
INEOS New Planet BioEnergy, LLC	Vero Beach, FL	$50,000,000	$50,000,000	Locally sourced organic MSW	This project will produce ethanol and electricity from wood and vegetative residues and construction and demolition materials. The facility will combine biomass gasification and fermentation and will have the capacity to produce 8 million gallons of ethanol and 2 megawatts of electricity per year by the end of 2011.
Sapphire Energy, Inc.	Columbus, NM	$50,000,000	$85,064,206	Onsite algae production	This project will cultivate algae in ponds that will ultimately be converted into green fuels, such as jet fuel and diesel, using the Dynamic Fuels refining process.
Increased Funding to Existing Biorefinery Projects					
Bluefire LLC	Fulton, MS	$81,134,686	$223,227,314	Unmerchantable timber and logging wastes from sites 75 to 100 mile radius from plant	This project will construct a facility that produces ethanol fuel from woody biomass, mill residue, and sorted municipal solid waste. The facility will have the capacity to produce 19 million gallons of ethanol per year.

One potential feedstock source for biofuels that continues to be pursued is algae. The concept of converting algae to fuels has been investigated for over 30 years. For example, DOE operated the Aquatic Species Program (ASP) that was terminated in 1995 because its product was shown to be uncompetitive with petroleum. However, 1995 was also the year when the first genome of a free-living organism was sequenced, that of *Haemophilus influenzae* Rd (Fleischmann et al., 1995), a date considered a milestone in the advancement of genomics. Application of genomics or synthetic biology to algal biofuel production could lead to a technology that will be able to produce large quantities of biofuels at competitive costs.

The past 15 years have seen a significant amount of research in genomics that has created a foundation on which a whole new set of technologies have been based. This foundation has led to another foray into algal biofuel research and development. Significant public and private investments have been committed, such as $600 million from ExxonMobil.

OTHER FEEDSTOCKS AND PROCESSING TECHNOLOGIES IN DEVELOPMENT

The technologies for producing biofuels from algae and cyanobacteria and hydrocarbon fuels[16] from biomass are discussed in the report *Liquid Transportation Fuels from Coal and Biomass: Technological Status, Costs, and Environmental Impacts* (NAS-NAE-NRC, 2009). Technologies for those biofuels are further from commercial deployment than cellulosic biofuels. Some examples of those biofuel refineries are included in Table 2-3.

Algal biofuels are attractive for many reasons. Algae and cyanobacteria can be cultivated in mass quantities in open ponds or in photoreactors. They would not necessarily compete with food for agricultural land or fresh water, and they would use CO_2 as a feedstock. Several types of fuel potentially can be produced from algae and cyanobacteria, including biodiesel, ethanol, and hydrocarbons from various conversion pathways. The spectrum of research and development into algal biofuels is broad and covers both the use of naturally occurring and genetically modified (including entirely synthetic) organisms as well as engineering systems that involve open ponds, closed bioreactors, and combinations of both. Companies are building or operating pilot- and demonstration-scale facilities[17] for cultivating algae and cyanobacteria and for converting them to biofuels (EMO, 2009). Algal and cyanobacterial biofuels have the potential to contribute to biofuels consumed in the United States in 2022, but the quantity and cost of production are highly uncertain and depend largely on the development of those industries in the next 5-10 years. The environmental effects of algal biofuel production are discussed in another NRC study on Sustainable Development of Algal Biofuels to be completed in 2012.

Production of hydrocarbon fuels directly from biomass is mostly in the research and development phase (Huber et al., 2005; Roman-Leshkov et al., 2007; Kunkes et al., 2008; Gürbüz et al., 2010). Some technological innovations might never reach the demonstration phase. Pilot and demonstration facilities need to be coupled with research and development programs to resolve issues identified during demonstration and to reduce costs (NAS-NAE-NRC, 2009). Commercial demonstrations are critical to proving and improving technologies, improving proficiency in technology operation, and quantifying the economics

---

[16]Fuels that are organic compounds that contains only carbon and hydrogen. Hydrocarbon fuels include petroleum and alkanes.

[17]A pilot facility is a small processing facility that is operated to gather information. The National Renewable Energy Laboratory defines a pilot biofuel facility as one that processes 1-10 dry tons of feedstock per day.

at large commercial scales (Sagar and van der Zwaan, 2006; Katzer, 2010). Demonstration, scaling up, and learning-by-doing takes time and, therefore, technologies for converting biomass to fuels that are not in pilot demonstration as of 2010 would need "fast-track" development to make any significant contribution to meeting RFS2.

CONCLUSION

The U.S. Congress has mandated that 36 billion gallons per year of biofuels in four categories—renewable fuels, advanced biofuels, cellulosic biofuels, and biomass-based diesel[18]—be consumed by U.S. transportation by 2022. The capacity of existing corn-grain ethanol facilities was estimated to be 14.1 billion gallons per year as of January 2011 and predicted to reach almost 15 billion gallons per year by 2012, essentially ensuring that production capacity is available to meet the legislated mandate for conventional biofuel. Similarly, existing biodiesel facilities are expected to meet that target for biomass-based diesel in the next several years. In contrast, whether and how the mandates for advanced biofuels and cellulosic biofuels will be met is uncertain.

At the time this report was written, the technologies for producing advanced and cellulosic biofuels were being developed and demonstrated at pilot scale. Many potential feedstocks, including crop residues, dedicated bioenergy crops, forest residues, and MSW, have been proposed. In the near term, crop and forest residues might be the most likely feedstocks because those resources are available and only investments in harvesting, storage, and transport are needed for delivery to biorefineries. Although MSW is available in large quantities, recovery rate of biological fraction is low. Many species of herbaceous perennial grasses and short-rotation woody crops have been suggested as potential dedicated bioenergy crops, and appropriate species would have to be selected on the basis of agronomic conditions to minimize the need for inputs (for example, irrigation and fertilizers) and maximize yield.

At the time this report was written, no proven commercial-scale technologies were available for converting lignocellulosic biomass to fuels. Another report estimated that the 16 billion gallons of ethanol-equivalent cellulosic biofuels could be produced by 2022 under an aggressive deployment scenario in which the capacity build rate doubles the historic capacity build rate of corn-grain ethanol (NAS-NAE-NRC, 2009). That scenario assumes that biorefineries with a collective capacity of 1 billion gallons per year will be available by 2015. The industry will expand at an average build rate of 50 percent per year over 6 years. At the maximum build rate, 2-4 billion gallons of cellulosic biofuels will be added each year to achieve 16 billion gallons of ethanol-equivalent cellulosic biofuels by 2022. Successful commercial-scale demonstration in the next few years and aggressive deployment thereafter will be key determinants of whether RFS2 could be met. The success of demonstration and the rate of deployment depend on many other factors, including research and development to improve the economics of biomass and biofuel production and to resolve issues that arise during demonstration, economic conditions, and investors' confidence.

---

[18]See section on "Renewable Fuel Standard" in Chapter 1 for definition of each category.

REFERENCES

Adler, P.R., M.A. Sanderson, P.J. Weimer, and K.P. Vogel. 2009. Plant species composition and biofuel yields of conservation grasslands. Ecological Applications 19(8):2202-2209.

Alvira, P., E. Tomas-Pejo, M. Ballesteros, and M.J. Negro. 2010. Pretreatment technologies for an efficient bioethanol production process based on enzymatic hydrolysis: A review. Bioresource Technology 101(13):4851-4861.

Anderson, W., M. Casler, and B. Baldwin. 2008. Improvement of perennial forage species as feedstock for bioenergy. Pp. 309-345 in Genetic Improvement of Bioenergy Crops, W. Vermerris, ed. New York: Springer.

Andrews, S. 2006. Crop Residue Removal for Biomass Energy Production: Effects on Soils and Recommendations. Washington, DC: U.S. Department of Agriculture - Natural Resources Conservation Service.

ANL (Argonne National Laboratory). 2010. Sustainable Bioenergy Crop Production Research Facility. Available online at http://www.bio.anl.gov/environmental_biology/terrestrial_ecology/bioenergy.html. Accessed February 7, 2011.

Aristidou, A., and M. Penttila. 2000. Metabolic engineering applications to renewable resource utilization. Current Opinion in Biotechnology 11(2):187-198.

Bain, R.L. 2007. World Biofuels Assessment. Worldwide Biomass Potential: Technology Characterizations. Golden, CO: National Renewable Energy Laboratory.

Balat, M. 2008. Mechanisms of thermochemical biomass conversion processes. Part 3: Reactions of liquefaction. Energy Sources Part A–Recovery Utilization and Environmental Effects 30(7):649-659.

Balat, M., M. Balat, E. Kirtay, and H. Balat. 2009. Main routes for the thermo-conversion of biomass into fuels and chemicals. Part 2: Gasification systems. Energy Conversion and Management 50(12):3158-3168.

Beach, R.H., and B.A. McCarl. 2010. U.S. Agricultural and Forestry Impacts of the Energy Independence and Security Act: FASOM Results and Model Description. Final Report. Research Triangle Park, NC: RTI International.

Beale, C.V., and S.P. Long. 1995. Can perennial C4 grasses attain high efficiencies of radiant energy-conversion in cool climates. Plant, Cell & Environment 18:641-650.

Biofuels Digest. 2010a. Advanced biofuels tracking database release 1.3. Available online at http://www.ascension-publishing.com/BIZ/ABTD13.xls. Accessed August 3, 2010.

Biofuels Digest. 2010b. Industry Data. Available online at http://biofuelsdigest.com/bdigest/free-industry-data/. Accessed November 12, 2010.

Blanco-Canqui, H., and R. Lal. 2009a. Corn stover removal for expanded uses reduced soil fertility and structural stability. Soil Science Society of America Journal 73:418-426.

Blanco-Canqui, H., and R. Lal. 2009b. Crop residue removal impacts on soil productivity and environmental quality. Critical Review in Plant Sciences 28:139-163.

Bongarten, B.C., D.A. Huber, and D.K. Apsley. 1992. Environmental and genetic influences on short-rotation biomass production of black locust (Robinia pseudoacacia L.) in the Georgia Piedmont. Forest Ecology and Management 55(1-4):315-331.

Bransby, D.I., D.J. Allen, N. Gutterson, G. Ikonen, E. Richard, W. Rooney, and E. van Santen. 2010. Engineering advantages, challenges and status of grass energy crops. Biotechnology in Agriculture and Forestry 66(2):125-154.

Brechbill, S.C., and W.E. Tyner. 2008. The Economics of Biomass Collection, Transportation, and Supply to Indiana Cellulosic and Electrical Utility Facilities. West Lafayette, IN: Purdue University.

Carolan, J.E., S.V. Joshi, and B.E. Dale. 2007. Technical and financial feasibility analysis of distributed bioprocessing using regional biomass pre-processing centers. Journal of Agricultural & Food Industrial Organization 5:Article 10.

Cassman, K.G., and A.J. Liska. 2007. Food and fuel for all: Realistic or foolish? Biofuels Bioproducts and Biorefining 1(1):18-23.

CETC (Canmet Energy Technology Centre). 2008. SUPERCETANE™ Technology at CanmetENERGY. Ottawa: Natural Resources Canada - Canmet Energy Technology Centre.

Chandra, R.P., R. Bura, W.E. Mabee, A. Berlin, X. Pan, and J.N. Saddler. 2007. Substrate pretreatment: The key to effective enzymatic hydrolysis of lignocellulosics? Advances in Biochemical Engineering/Biotechnology 108:67-93.

Christian, D.G., A.B. Riche, and N.E. Yates. 2002. The yield and composition of switchgrass and coastal panic grass grown as a biofuel in southern England. Bioresource Technology 83(2):115-124.

Clay, D., G. Carlson, T. Schumacher, V. Owens, and F. Mamani-Pati. 2010. Biomass estimation approach impacts on calculated soil organic carbon maintenance requirements and associated mineralization rate constants. Journal of Environmental Quality 39(3):784-790.

Clifton-Brown, J.C., J. Breuer, and M.B. Jones. 2007. Carbon mitigation by the energy crop, Miscanthus. Global Change Biology 13(11):2296-2307.

Coyle, D.R., M.D. Coleman, and D.P. Aubrey. 2008. Above- and below-ground biomass accumulation, production, and distribution of sweetgum and loblolly pine grown with irrigation and fertilization. Canadian Journal of Forest Research-Revue Canadienne De Recherche Forestiere 38(6):1335-1348.

Davis, A.A., and C.C. Trettin. 2006. Sycamore and sweetgum plantation productivity on former agricultural land in South Carolina. Biomass & Bioenergy 30(8-9):769-777.

Davis, S.C., W.J. Parton, F.G. Dohleman, C.M. Smith, S. Del Grosso, A.D. Kent, and E.H. DeLucia. 2010. Comparative biogeochemical cycles of bioenergy crops reveal nitrogen-fixation and low greenhouse gas emissions in a *Miscanthus* x *giganteus* agro-ecosystem. Ecosystems 13(1):144-156.

De Haan, L.R., S. Weisberg, D. Tilman, and D. Fornara. 2010. Agricultural and biofuel implications of a species diversity experiment with native perennial grassland plants. Agriculture, Ecosystems & Environment 1-2(137):33-38.

de Vries, S.C., G.W.J. van de Ven, M.K. van Ittersum, and K.E. Giller. 2010. Resource use efficiency and environmental performance of nine major biofuel crops, processed by first-generation conversion techniques. Biomass & Bioenergy 34(5):588-601.

Demirbas, A. 2000. Mechanisms of liquefaction and pyrolysis reactions of biomass. Energy Conversion and Management 41(6):633-646.

Demirbas, A. 2009. Pyrolysis mechanisms of biomass materials. Energy Sources Part A-Recovery Utilization and Environmental Effects 31(13):1186-1193.

DOE-EERE (U.S. Department of Energy - Energy Efficiency and Renewable Energy). 2010. Alternative and advanced fuels: Biodiesel. Available online at http://www.afdc.energy.gov/afdc/fuels/biodiesel.html. Accessed August 5, 2010.

Eggeman, T., and R.T. Elander. 2005. Process and economic analysis of pretreatment technologies. Bioresource Technology 96(18):2019-2025.

EIA (Energy Information Administration). 2010. Annual Energy Review 2009. Washington, DC: U.S. Department of Energy.

El-Nashaar, H.M., S.M. Griffith, J.J. Steiner, and G.M. Banowetz. 2009. Mineral concentration in selected native temperate grasses with potential use as biofuel feedstock. Bioresource Technology 100(14):3526-3531.

El Bassam, N. 1996. Renewable Energy: Potential Energy Crops for Europe and the Mediterranean Region. Rome: Food and Agricultural Organization.

El Bassam, N. 2010. Handbook of Bioenergy Crops: A Complete Reference to Species, Development and Applications. London: Earthscan.

Elliott, D.C., D. Beckman, A.V. Bridgwater, J.P. Diebold, S.B. Gevert, and Y. Solantausta. 1991. Developments in direct thermochemical liquefaction of biomass: 1983-1990. Energy & Fuels 5(3):399-410.

EMO (Emerging Markets Online). 2009. Algae 2020: Biofuels Markets and Commercialization Outlook. Houston, TX: Emerging Markets Online.

Energy Biosciences Institute. 2011. Developing prairie cordgrass as a cellulosic bioenergy crop. Available online at http://www.energybiosciencesinstitute.org/index.php?option=com_content&task=view&id=312. Accessed February 7, 2011.

EPA (U.S. Environmental Protection Agency). 2010a. Municipal Solid Wastes in the United States. 2009 Facts and Figures. Washington, DC: U.S. Environmental Protection Agency.

EPA (U.S. Environmental Protection Agency). 2010b. Renewable Fuel Standard Program (RFS2) Regulatory Impact Analysis. Washington, DC: U.S. Environmental Protection Agency.

EPA (U.S. Environmental Protection Agency). 2011. EPA grants E15 fuel waiver for model years 2001-2006 cars and light trucks/Agency continues review of public comments for an E15 pump label to help ensure consumers use the correct fuel. Available online at http://yosemite.epa.gov/opa/admpress.nsf/0/8206AB91F87CEC0 88525781F0059E65C. Accessed February 9, 2011.

Ezeji, T.C., N. Qureshi, and H.P. Blaschek. 2007. Bioproduction of butanol from biomass: From genes to bioreactors. Current Opinion in Biotechnology 18(3):220-227.

Fargione, J.E., T.R. Cooper, D.J. Flaspohler, J. Hill, C. Lehman, T. McCoy, S. McLeod, E.J. Nelson, K.S. Oberhauser, and D. Tilman. 2009. Bioenergy and wildlife: Threats and opportunities for grassland conservation. BioScience 59(9):767-777.

Fleischmann, R.D., M.D. Adams, O. White, R.A. Clayton, E.F. Kirkness, A.R. Kerlavage, C.J. Bult, J.F. Tomb, B.A. Dougherty, J.M. Merrick, K. McKenney, G. Sutton, W. Fitzhugh, C. Fields, J.D. Gocayne, J. Scott, R. Shirley, L.I. Liu, A. Glodek, J.M. Kelley, J.F. Weidman, C.A. Phillips, T. Spriggs, E. Hedblom, M.D. Cotton, T.R. Utterback, M.C. Hanna, D.T. Nguyen, D.M. Saudek, R.C. Brandon, L.D. Fine, J.L. Fritchman, J.L. Fuhrmann, N.S.M. Geoghagen, C.L. Gnehm, L.A. McDonald, K.V. Small, C.M. Fraser, H.O. Smith, and J.C. Venter. 1995. Whole-genome random sequencing and assembly of Haemophilus influenzae Rd. Science 269(5223):496-512.

Foust, T., A. Aden, A. Dutta, and S. Phillips. 2009. An economic and environmental comparison of a biochemical and a thermochemical lignocellulosic ethanol conversion processes. Cellulose 16(4):547-565.

Fraley, R. 2010. Crop yield gains: Understanding the past, envisioning the future. Presentation to the Committee on Economic and Environmental Impacts of Increasing Biofuels Production, July 14.

Geyer, W.A. 2006. Biomass production in the Central Great Plains USA under various coppice regimes. Biomass & Bioenergy 30(8-9):778-783.

Gonzalez-Hernandez, J.L., G. Sarath, J.M. Stein, V. Owens, K. Gedye, and A. Boe. 2009. A multiple species approach to biomass production from native herbaceous perennial feedstocks. In Vitro Cellular & Developmental Biology-Plant 45(3):267-281.

Graham, R.L., R. Nelson, J. Sheehan, R.D. Perlack, and L.L. Wright. 2007. Current and potential US corn stover supplies. Agronomy Journal 99(1):1-11.

Gürbüz, E.I., E.L. Kunkes, and J.A. Dumesic. 2010. Dual-bed catalyst system for C-C coupling of biomass-derived oxygenated hydrocarbons to fuel-grade compounds. Green Chemistry 12(2):223-227.

Heaton, E.A., F.G. Dohleman, and S.P. Long. 2008. Meeting US biofuel goals with less land: The potential of *Miscanthus*. Global Change Biology 14(9):2000-2014.

Heaton, E.A., F.G. Dohleman, and S.P. Long. 2009. Seasonal nitrogen dynamics of *Miscanthus x giganteus* and *Panicum virgatum*. Global Change Biology Bioenergy 1(4):297-307.

Heaton, E.A., F.G. Dohleman, and A.F. Miguez. 2010. *Miscanthus*. A promising biomass crop. Advances in Botanical Research 56:75-137.

Heilman, P.E., and R.F. Stettler. 1985. Genetic variation and productivity of *Populus trichocarpa* and its hybrids. II. Biomass production in a 4-year plantation. Canadian Journal of Forest Research 15(2):384-388.

Henson, J.F., and G.A. Fenchel. 2007. Plant Guide: Eastern Gamagrass. Baton Rouge, LA: U.S. Department of Agriculture - Natural Resources Conservation Service - National Plant Materials Center.

Hess, J.R., C.T. Wright, and K.L. Kenney. 2007. Cellulosic biomass feedstocks and logistics for ethanol production. Biofuels Bioproducts & Biorefining 1(3):181-190.

Hinchee, M., W. Rottmann, L. Mullinax, C.S. Zhang, S.J. Chang, M. Cunningham, L. Pearson, and N. Nehra. 2009. Short-rotation woody crops for bioenergy and biofuels applications. In Vitro Cellular & Developmental Biology-Plant 45(6):619-629.

Huber, G.W., J.N. Chheda, C.J. Barrett, and J.A. Dumesic. 2005. Production of liquid alkanes by aqueous-phase processing of biomass-derived carbohydrates. Science 308(5727):1446-1450.

Huber, G.W., S. Iborra, and A. Corma. 2006. Synthesis of transportation fuels from biomass: Chemistry, catalysts, and engineering. Chemical Reviews 106(9):4044-4098.

Iogen. 2010a. Demo plant fuel production. Available online at http://www.iogen.ca/. Accessed February 16, 2011.

Iogen. 2010b. Iogen's cellulosic ethanol demonstration plant. Available online at http://www.iogen.ca/company/demo_plant/index.html. Accessed October 19, 2010.

Jager, H., L.M. Baskaran, C.C. Brandt, E.B. Davis, C.A. Gunderson, and S.D. Wullschleger. 2010. Empirical geographic modeling of switchgrass yields in the United States. Global Change Biology Bioenergy 2(5):248-257.

Jarchow, M.E. and M. Liebman. 2011. Incorporating Prairies into Multifunctional Landscapes Establishing and Managing Prairies for Enhanced Environmental Quality, Livestock Grazing and Hay Production, Bioenergy Production, and Carbon Sequestration. Ames: Iowa State University.

Johnson, J.M.-F., M.D. Coleman, R. Gesch, A. Jaradat, R. Mitchell, D. Reicosky, and W.W. Wilhelm. 2007. Biomass-bioenergy crops in the United States: A changing paradigm. The American Journal of Plant Science and Biotechnology 1(1):1-28.

Johnson, J.M.F., R.R. Allmaras, and D.C. Reicosky. 2006. Estimating source carbon from crop residues, roots and rhizodeposits using the national grain-yield database. Agronomy Journal 98(3):622-636.

Johnston, M., J.A. Foley, T. Holloway, C. Kucharik, and C. Monfreda. 2009. Resetting global expectations from agricultural biofuels. Environmental Research Letters 4(1) doi:10.1088/1748-9326/4/1/014004.

Jokela, E.J., P.M. Dougherty, and T.A. Martin. 2004. Production dynamics of intensively managed loblolly pine stands in the southern United States: A synthesis of seven long-term experiments. Forest Ecology and Management 192(1):117-130.

Jordan, N., G. Boody, W. Broussard, J.D. Glover, D. Keeney, B.H. McCown, G. McIsaac, M. Muller, H. Murray, J. Neal, C. Pansing, R.E. Turner, K. Warner, and D. Wyse. 2007. Environment—Sustainable development of the agricultural bio-economy. Science 316(5831):1570-1571.

Jung, G.A., J.A. Shaffer, and W.L. Stout. 1988. Switchgrass and big bluestem responses to amendments on strongly acid soil. Agronomy Journal 80(4):669-676.

Kalnes, T.N., K.P. Koers, T. Marker, and D.R. Shonnard. 2009. A technoeconomic and environmental life cycle comparison of green diesel to biodiesel and syndiesel. Environmental Progress & Sustainable Energy 28(1):111-120.

Karlen, D.L., N.C. Wollenhaupt, D.C. Erbach, E.C. Berry, J.B. Swan, N.S. Eash, and J.L. Jordahl. 1994. Crop residue effects on soil quality following 10-years of no-till corn. Soil & Tillage Research 31(2-3):149-167.

Katzer, J.R. 2010. Sustainable research, development, and demonstration (RD&D). Industrial & Engineering Chemistry Research 49(21):10154-10158.

King, J.E., J.M. Hannifan, and R. Nelson. 1998. An Assessment of the Feasibility of Electric Power Derived from Biomass and Waste Feedstocks. Topeka: Kansas Corporation Commission.

Kline, K.L., and M.D. Coleman. 2010. Woody energy crops in the southeastern United States: Two centuries of practitioner experience. Biomass & Bioenergy 34(12):1655-1666.

Knothe, G. 2001. Historical perspectives on vegetable oil-based diesel fuels. Inform 12(November):1103-1107.

Kunkes, E.L., D.A. Simonetti, R.M. West, J.C. Serrano-Ruiz, C.A. Gartner, and J.A. Dumesic. 2008. Catalytic conversion of biomass to monofunctional hydrocarbons and targeted liquid-fuel classes. Science 322(5900):417-421.

Lady Bird Johnson Wildflower Center. 2011. Native plant database. Available online at http://www.wildflower.org/plants/. Accessed February 7, 2011.

Larson, J.A., D.F. Mooney, B.C. English, and D.D. Tyler. 2010. Cost analysis of alternative harvest and storage methods for switchgrass in the southeastern U.S. Paper read at the 2010 Annual Meeting of Southern Agricultural Economics Association, February 6-9, Orlando, FL.

Lobell, D.B., K.G. Cassman, and C.B. Field. 2009. Crop yield gaps: Their importance, magnitudes, and causes. Annual Review of Environment and Resources 34:179-204.

Mayton, H., J. Hansen, J. Neally, and D. Viands. 2011. Accelerated evaluation of perennial grass and legume feedstocks for biofuel production in New York state. Available online at http://nesungrant.cornell.edu/cals/sungrant/difference/upload/Cornell-Biofuels-Crop-Development.pdf. Accessed February 7, 2011.

McCready, N.S., and R.C. Williams. 2008. Utilization of biofuel coproducts as performance enhancers in asphalt binder. Transportation Research Record (2051):8-14.

McIsaac, G.F., M.B. David, and C.A. Mitchell. 2010. *Miscanthus* and switchgrass production in Central Illinois: Impacts on hydrology and inorganic nitrogen leaching. Journal of Environmental Quality 39(5):1790-1799.

Melis, S. 2008. Albemarle catalytic solutions for the co-processing of vegetable oil in conventional hydrotreaters. Catalysts Courrier 73:6-9.

Mercker, D. 2007. Short rotation woody crops for biofuels. Knoxville: University of Tennessee Agricultural Experiment Station.

Miguez, F.E., M.B. Villamil, S.P. Long, and G.A. Bollero. 2008. Meta-analysis of the effects of management factors on *Miscanthus x giganteus* growth and biomass production. Agricultural and Forest Meteorology 148(8-9):1280-1292.

Miguez, F.E., X.G. Zhu, S. Humphries, G.A. Bollero, and S.P. Long. 2009. A semimechanistic model predicting the growth and production of the bioenergy crop *Miscanthus × giganteus*: Description, parameterization and validation. Global Change Biology Bioenergy 1(4):282-296.

Milbrandt, A. 2005. A Geographic Perspective on the Current Biomass Resource Availability in the United States. Golden, CO: National Renewable Energy Laboratory.

Miranowski, J., and A. Rosburg. 2010. An Economic Breakeven Model of Cellulosic Feedstock Production and Ethanol Conversion with Implied Carbon Pricing. Ames: Iowa State University.

Mosier, N., C. Wyman, B. Dale, R. Elander, Y.Y. Lee, M. Holtzapple, and M. Ladisch. 2005. Features of promising technologies for pretreatment of lignocellulosic biomass. Bioresource Technology 96(6):673-686.

Mueller, S. 2010. 2008 National dry mill corn ethanol survey. Biotechnology Letters 32(9):1261-1264.

Muir, J.P., M.A. Sanderson, W.R. Ocumpaugh, R.M. Jones, and R.L. Reed. 2001. Biomass production of "Alamo" switchgrass in response to nitrogen, phosphorus, and row spacing. Agronomy Journal 93(4):896-901.

NAS-NAE-NRC (National Academy of Sciences, National Academy of Engineering, National Research Council). 2009. Liquid Transportation Fuels from Coal and Biomass: Technological Status, Costs, and Environmental Impacts. Washington, DC: National Academies Press.

NBB (National Biodiesel Board). 2010. NBB member plant locations. Available online at http://www.biodiesel.org/buyingbiodiesel/plants/default.aspx?AspxAutoDetectCookieSupport=1. Accessed November 9, 2011.

NCGA (National Corn Growers Association). 2010. National Corn Growers Association: Strategic Plan. Chesterfield, MO: National Corn Growers Association.

Nelson, R., M. Langemeier, J. Williams, C. Rice, S. Staggenborg, P. Pfomm, D. Rodgers, D. Wang, and J. Nippert. 2010. Kansas Biomass Resource Assessment: Assessment and Supply of Select Biomass-based Resources. Manhattan: Kansas State University.

Neste Oil. 2011a. Neste Oil starts up its new renewable diesel plant in Singapore. Available online at http://nesteoil.com/default.asp?path=1,41;540;1259;1261;13291;16384. Accessed January 10, 2011.

Neste Oil. 2011b. NExBTL diesel. Available online at http://www.nesteoil.com/default.asp?path=1,41,535,547,3716,3884. Accessed January 10, 2011.

Nevoigt, E. 2008. Progress in metabolic engineering of *Saccharomyces cerevisiae*. Microbiology and Molecular Biology Reviews 72(3):379-412.

Nichols, N.N., B.S. Dien, R.J. Bothast, and M.A. Cotta. 2006. The corn ethanol industry. Pp. 59-77 in Alcholic Fuels, S. Minteer, ed. Boca Raton, FL: CRC Press.

NREL (National Renewable Energy Laboratory). 2005. Biodiesel Blends. Golden, CO: National Renewable Energy Laboratory.

NREL (National Renewable Energy Laboratory). 2008. United States Biodiesel Production Facilities. Map Illustration. Golden, CO: National Renewable Energy Laboratory.

Oliva, J.M., F. Saez, I. Ballesteros, A. Gonzalez, M.J. Negro, P. Manzanares, and M. Ballesteros. 2003. Effect of lignocellulosic degradation compounds from steam explosion pretreatment on ethanol fermentation by thermotolerant yeast *Kluyveromyces marxianus*. Applied Biochemistry and Biotechnology 105:141-153.

Owsley, C.M. 2002. Review of Big Bluestem. Washington, DC: U.S. Department of Agriculture - Natural Resources Conservation Service.

Parrish, D.J., and J.H. Fike. 2005. The biology and agronomy of switchgrass for biofuels. Critical Reviews in Plant Sciences 24(5-6):423-459.

Patzek, T.W. 2006. A statistical analysis of the theoretical yield of ethanol from corn starch. Natural Resources Research 15(3):205-212.

Perlack, R.D., L.L. Wright, A.F. Turhollow, R.L. Graham, B.J. Stokes, and D.C. Erbach. 2005. Biomass as Feedstock for a Bioenergy and Bioproducts Industry: The Technical Feasibility of a Billion-Ton Annual Supply. Oak Ridge, TN: U.S. Department of Energy, U.S. Department of Agriculture.

Perry, C.H., M.D. Nelson, J.C. Toney, T.S. Frescino, and M.L. Hoppus. 2007. Mapping forest resources of the United States. In Forest Resources of the United States, 2007, W.B. Smith, P.D. Miles, C.H. Perry, and S.A. Pugh, eds. Washington, DC: U.S. Department of Agriculture - Forest Service.

Petrolia, D.R. 2006. The Economics of Harvesting and Transporting Corn Stover for Conversion to Fuel Ethanol: A Case Study for Minnesota. St. Paul: University of Minnesota.

Pootakham, T., and A. Kumar. 2010. A comparison of pipeline versus truck transport of bio-oil. Bioresource Technology 101(1):414-421.

Prine, G.M., D.L. Rockwood, and J.A. Stricker. 2000. Many short rotation trees and herbaceous plants available as energy crops in humid lower south. Proceedings of the Bioenergy 2000, Northeast Regional Bioenergy Program.

Pyter, R., E. Heaton, F. Dohleman, T. Voigt, and S. Long. 2009. Agronomic Experiences with *Miscanthus x giganteus* in Illinois, USA. Biofuels: Methods and Protocols:41-52.

Reijnders, L. 2009. Microalgal and terrestrial transport biofuels to displace fossil fuels. Energies 2(1):48-56.

Renewable Diesel Subcommittee of the WSDA Technical Work Group. 2007. Renewable Diesel Technology. Olympia: Washington State Department of Agriculture.

RFA (Renewable Fuels Association). 2002. Growing Homeland Energy Security. 2002 Ethanol Industry Outlook. Washington, DC: Renewable Fuels Association.

RFA (Renewable Fuels Association). 2003. Building a Secure Energy Future. 2003 Ethanol Industry Outlook. Washington, DC: Renewable Fuels Association.

RFA (Renewable Fuels Association). 2004. Synergy in Energy. 2004 Ethanol Industry Outlook. Washington, DC: Renewable Fuels Association.

RFA (Renewable Fuels Association). 2005. Home Grown for the Homeland. 2005 Ethanol Industry Outlook. Washington, DC: Renewable Fuels Association.

RFA (Renewable Fuels Association). 2006. From Niche to Nation. 2006 Ethanol Industry Outlook. Washington, DC: Renewable Fuels Association.

RFA (Renewable Fuels Association). 2007. Building New Horizons. 2007 Ethanol Industry Outlook. Washington, DC: Renewable Fuels Association.

RFA (Renewable Fuels Association). 2008. Changing the Climate. 2008 Ethanol Industry Outlook. Washington, DC: Renewable Fuels Association.

RFA (Renewable Fuels Association). 2009. Growing Innovation. America's Energy Future Starts at Home. 2009 Ethanol Industry Outlook. Washington, DC: Renewable Fuels Association.

RFA (Renewable Fuels Association). 2010. Climate of Opportunity. 2010 Ethanol Industry Outlook. Washington, DC: Renewable Fuels Association.

RFA (Renewable Fuels Association). 2011a. Building Bridges to a More Sustainable Future. 2011 Ethanol Industry Outlook. Washington, DC: Renewable Fuels Association.

RFA (Renewable Fuels Association). 2011b. Statistics. Available online at http://www.ethanolrfa.org/pages/statistics. Accessed April 27, 2011.

Riemenschneider, D.E., W.E. Berguson, D.I. Dickmann, R.B. Hall, J.G. Isebrands, C.A. Mohn, G.R. Stanosz, and G.A. Tuskan. 2001. Poplar breeding and testing strategies in the north-central U.S.: Demonstration of potential yield and consideration of future research needs. The Forestry Chronicle 77(2):245-253.

Rinehart, L. 2006. Switchgrass as a Bioenergy Crop. Butte, MT: National Center for Appropriate Technology.

Roberts, K.G., B.A. Gloy, S. Joseph, N.R. Scott, and J. Lehmann. 2010. Life cycle assessment of biochar systems: Estimating the energetic, economic, and climate change potential. Environmental Science & Technology 44(2):827-833.

Roman-Leshkov, Y., C.J. Barrett, Z.Y. Liu, and J.A. Dumesic. 2007. Production of dimethylfuran for liquid fuels from biomass-derived carbohydrates. Nature 447(7147):982-985.

Ruark, G., S. Josiah, D. Riemenschneider, and T. Volk. 2006. Perennial crops for bio-fuels and conservation. Paper read at the USDA Agricultural Outlook Forum—Prospering in Rural America, February 1-17, Arlington, VA.

Sadaka, S., and S. Negi. 2009. Improvements of biomass physical and thermochemical characteristics via torrefaction process. Environmental Progress & Sustainable Energy 28(3):427-434.

Sagar, A.D., and B. van der Zwaan. 2006. Technological innovation in the energy sector: R&D, deployment, and learning-by-doing. Energy Policy 34(17):2601-2608.

Salon, P.R., T. Horvath, M. van der Grinten, and H. Mayton. 2010. Yield and time of harvest of tall wheatgrass for biomass energy in New York. Paper read at the 2009 American Society of Agronomy/Crop Science Society of America/Soil Science Society of America Annual Meetings, November 1-5, Pittsburgh, PA.

Sanderson, M.A., R.H. Skinner, D.J. Barker, G.R. Edwards, B.F. Tracy, and D.A. Wedin. 2004. Plant species diversity and management of temperate forage and grazing land ecosystems. Crop Science 44(4):1132-1144.

Schnepf, R. 2010. Cellulosic Ethanol: Feedstocks, Conversion Technologies, Economics, and Policy Options. Washington, DC: Congressional Research Service.

Schwietzke, S., M. Ladisch, L. Russo, K. Kwant, T. Mäkinen, B. Kavalov, K. Maniatis, R. Zwart, G. Shahanan, K. Sipila, P. Grabowski, B. Telenius, M. White, and A. Brown. 2008. Gaps in the research of 2nd generation transportation biofuels. IEA Bioenergy T41(2):2008:01.

Sendich, E., M. Laser, S. Kim, H. Alizadeh, L. Laureano-Perez, B. Dale, and L. Lynd. 2008. Recent process improvements for the ammonia fiber expansion (AFEX) process and resulting reductions in minimum ethanol selling price. Bioresource Technology 99(17):8429-8435.

Smith, W.B., P.D. Miles, C.H. Perry, and S.A. Pugh. 2009. Forest Resources of the United States, 2007. Washington, DC: U.S. Department of Agriculture - Forest Service.

Snell, K.D., and O.P. Peoples. 2009. PHA bioplastic: A value-added coproduct for biomass biorefineries. Biofuels Bioproducts and Biorefining 3(4):456-467.

Sohngen, B., and S. Brown. 2006. The influence of conversion of forest types on carbon sequestration and other ecosystem services in the South Central United States. Ecological Economics 57(4):698-708.

Somerville, C., H. Youngs, C. Taylor, S.C. Davis, and S.P. Long. 2010. Feedstocks for lignocellulosic biofuels. Science 329(5993):790-792.

Sonderegger, M., M. Jeppsson, C. Larsson, M.F. Gorwa-Grauslund, E. Boles, L. Olsson, I. Spencer-Martins, B. Hahn-Hagerdal, and U. Sauer. 2004. Fermentation performance of engineered and evolved xylose-fermenting *Saccharomyces cerevisiae* strains. Biotechnology and Bioengineering 87(1):90-98.

Soria, A.J., A.G. McDonald, and B.B. He. 2008. Wood solubilization and depolymerization by supercritical methanol. Part 2: Analysis of methanol soluble compounds. Holzforschung 62(4):409-416.

Spath, P.L., and D.C. Dayton. 2003. Preliminary Screening—Technical and Economic Assessment of Synthesis Gas to Fuels and Chemicals with Emphasis on the Potential for Biomass-Derived Syngas. Golden, CO: National Renewable Energy Laboratory.

Steinbeck, K. 1999. Thirty years of short-rotation hardwoods research. Pp. 63-65 in Proceedings of the Tenth Biennial Southern Silvicultural Research Conference, J.D. Haywood, ed. Asheville, NC.

Stewart, J.R., Y. Toma, F.G. Fernandez, A. Nishiwaki, T. Yamada, and G. Bollero. 2009. The ecology and agronomy of *Miscanthus sinensis*, a species important to bioenergy crop development, in its native range in Japan: A review. Global Change Biology Bioenergy 1(2):126-153.

Stricker, J.A., D.L. Rockwood, S.A. Segrest, G.R. Alker, G.M. Prine, and D.R. Carter. 2000. Short Rotation Woody Crops for Florida. Available online at http://www.treepower.org/papers/strickerny.doc. Accessed February 8, 2011.

Thomson, A.M., R.C. Izarrualde, T.O. West, D.J. Parrish, D.D. Tyler, and J.R. Williams. 2009. Simulating Potential Switchgrass Production in the United States. Richland, WA: Pacific Northwest National Laboratory.

Tilman, D., J. Hill, and C. Lehman. 2006. Carbon-negative biofuels from low-input high-diversity grassland biomass. Science 314(5805):1598-1600.

Tober, D., W. Duckwitz, and M. Knudson. 2009. Indiangrass (*Sorghastrum nutans*) Biomass Trials. North Dakota, South Dakota, and Minnesota. Bismarck, ND: U.S. Department of Agriculture - Natural Resources Conservation Service.

Tolan, J.S. 2004. Iogen's demonstration process for producing ethanol from cellulosic biomass. Pp. 193-208 in Biorefineries—Industrial Processes and Products, B. Kamm, P.R. Gruber and M. Kamm, eds. Weinheim, Germany: Wiley-VCH.

Tuskan, G.A., S. DiFazio, S. Jansson, J. Bohlmann, I. Grigoriev, U. Hellsten, N. Putnam, S. Ralph, S. Rombauts, A. Salamov, J. Schein, L. Sterck, A. Aerts, R.R. Bhalerao, R.P. Bhalerao, D. Blaudez, W. Boerjan, A. Brun, A. Brunner, V. Busov, M. Campbell, J. Carlson, M. Chalot, J. Chapman, G.L. Chen, D. Cooper, P.M. Coutinho, J. Couturier, S. Covert, Q. Cronk, R. Cunningham, J. Davis, S. Degroeve, A. Dejardin, C. Depamphilis, J. Detter, B. Dirks, I. Dubchak, S. Duplessis, J. Ehlting, B. Ellis, K. Gendler, D. Goodstein, M. Gribskov, J. Grimwood, A. Groover, L. Gunter, B. Hamberger, B. Heinze, Y. Helariutta, B. Henrissat, D. Holligan, R. Holt, W. Huang, N. Islam-Faridi, S. Jones, M. Jones-Rhoades, R. Jorgensen, C. Joshi, J. Kangasjarvi, J. Karlsson, C. Kelleher, R. Kirkpatrick, M. Kirst, A. Kohler, U. Kalluri, F. Larimer, J. Leebens-Mack, J.C. Leple, P. Locascio, Y. Lou, S. Lucas, F. Martin, B. Montanini, C. Napoli, D.R. Nelson, C. Nelson, K. Nieminen, O. Nilsson, V. Pereda, G. Peter, R. Philippe, G. Pilate, A. Poliakov, J. Razumovskaya, P. Richardson, C. Rinaldi, K. Ritland, P. Rouze, D. Ryaboy, J. Schmutz, J. Schrader, B. Segerman, H. Shin, A. Siddiqui, F. Sterky, A. Terry, C.J. Tsai, E. Uberbacher, P. Unneberg, J. Vahala, K. Wall, S. Wessler, G. Yang, T. Yin, C. Douglas, M. Marra, G. Sandberg, Y. Van de Peer, and D. Rokhsar. 2006. The genome of black cottonwood, *Populus trichocarpa* (Torr. & Gray). Science 313(5793):1596-1604.

Urbanchuk, J. 2010. Current State of the U.S. Ethanol Industry. Washington, DC: U.S. Department of Energy.

USDA (U.S. Department of Agriculture). 2010. USDA Agricultural Projections to 2019. Washington, DC: U.S. Department of Agriculture.

USDA (U.S. Department of Agriculture). 2011. USDA Agricultural Projections to 2020. Washington, DC: U.S. Department of Agriculture.

USDA-AMS (U.S. Department of Agriculture - Agricultural Marketing Service). 2007. Ethanol Transportation Backgrounder: Expansion of U.S. Corn-Based Ethanol from the Agricultural Transportation Perspective. Washington, DC: U.S. Department of Agriculture.

USDA-ERS (U.S. Department of Agriculture - Economic Research Service). 2010. Farm program acres. Available online at http://maps.ers.usda.gov/BaseAcres/index.aspx. Accessed October 15, 2010.

USDA-NASS (U.S. Department of Agriculture - National Agricultural Statistics Service). 2010. Data and statistics: Quick stats. Available online at http://www.nass.usda.gov/Data_and_Statistics/Quick_Stats/index.asp. Accessed June 24, 2011.

USDA-NRCS (U.S. Department of Agriculture - Natural Resources Conservation Service). 2002. Plant Factsheet: Little Bluestem. Beltsville, MD: U.S. Department of Agriculture - Natural Resources Conservation Service - National Plant Materials Center.

Volk, T.A., T. Verwijst, P.J. Tharakan, L.P. Abrahamson, and E.H. White. 2004. Growing fuel: A sustainability assessment of willow biomass crops. Frontiers in Ecology and the Environment 2(8):411-418.

Volk, T.A., L.P. Abrahamson, C.A. Nowak, L.B. Smart, P.J. Tharakan, and E.H. White. 2006. The development of short-rotation willow in the northeastern United States for bioenergy and bioproducts, agroforestry and phytoremediation. Biomass & Bioenergy 30(8-9):715-727.

Walsh, M., D.G. De La Torre Uguarte, H. Shapouri, and S.P. Slinsky. 2003. Bioenergy crop production in the United States. Environmental and Resource Economics 24(3):313-333.

Weimer, P.J., and T.L. Springer. 2007. Fermentability of eastern gamagrass, big bluestem and sand bluestem grown across a wide variety of environments. Bioresource Technology 98(8):1615-1621.

Wilhelm, W.W., J.M.E. Johnson, D.L. Karlen, and D.T. Lightle. 2007. Corn stover to sustain soil organic carbon further constrains biomass supply. Agronomy Journal 99(6):1665-1667.

Wright, L., and A. Turhollow. 2010. Switchgrass selection as a "model" bioenergy crop: A history of the process. Biomass & Bioenergy 34(6):851-868.

Wullschleger, S.D., E.B. Davis, M.E. Borsuk, C.A. Gunderson, and L.R. Lynd. 2010. Biomass production in switchgrass across the United States: Database description and determinants of yield. Agronomy Journal 102(4):1158-1168.

Yan, W., T.C. Acharjee, C.J. Coronella, and V.R. Vasquez. 2009. Thermal pretreatment of lignocellulosic biomass. Environmental Progress & Sustainable Energy 28(3):435-440.

Yang, B., and C. Wyman. 2007. Pretreatment: The key to unlocking low-cost cellulosic ethanol. Biofuels Bioproducts & Biorefining 2:26-40.

3

Projected Supply of
Cellulosic Biomass

The consumption mandate for two of the four categories of biofuels listed in the Renewable Fuel Standard as amended by the Energy Independence and Security Act (EISA) of 2007 (RFS2) will likely be met by corn-grain ethanol and biodiesel, as discussed in Chapter 2. The remaining 20 billion gallons per year of mandated consumption is to be met with cellulosic biofuels or advanced fuels, which could include cellulosic biofuels, other types of biofuels derived from sugar or any starch other than corn starch, and imports of ethanol from sugarcane facilities in Brazil and elsewhere. Based on anticipated advances in conversion technologies, earlier studies suggested that over 550 million dry tons per year of nonfood-based resources, including agricultural residues, dedicated bioenergy crops, forest resources, and municipal solid wastes (MSWs), can potentially be produced in the United States (Perlack et al., 2005; NAS-NAE-NRC, 2009). However, the potentially available feedstock that would be supplied to biofuel refineries in the future depends on multiple factors: where the feedstock is grown and collected; expected crop, residue, or forest yields; competition for biomass from other uses (for example, electricity generation versus biofuel production); markets; technology development; public policies; and other unanticipated factors. Potential availability refers to the amount of cellulosic biomass that could be grown and harvested in the United States based on assumptions of recoverable yields from diverse farm and forest landscapes but without specific consideration of the costs of producing, harvesting, and delivering the biomass to a biorefinery. The study *Biomass as Feedstock for a Bioenergy and Bioproducts Industry: The Technical Feasibility of a Billion-Ton Annual Supply*[1] (Perlack et al., 2005) provided one of the first estimates of potential availability of cellulosic feedstocks in the United States. Supply refers to a schedule of amounts that would be delivered to biorefineries at different costs, taking accessibility of biomass, infrastructure, and other

---

[1]The report *U.S. Billion-Ton Update: Biomass Supply for a Bioenergy and Bioproducts Industry* (Perlack and Stokes, 2011) was released while the committee was preparing this report for public release. The committee did not have an opportunity to review the Perlack and Stokes (2011) report.

economic conditions into consideration. Taking all those factors into consideration, realized supply is likely to be much lower than potential availability.

This chapter describes the estimated supply of cellulosic biomass made by different groups, including the researchers at the University of California, Davis, the U.S. Environmental Protection Agency (EPA), the Biomass Research and Development Initiative, and researchers at the University of Tennessee. Other factors, such as biotechnology, competition for biomass with other sectors, weather-related losses, and pests and diseases, which are typically not considered in projecting biomass supply, contribute to uncertainties in feedstock supply and are discussed at the end of this chapter.

POTENTIAL SUPPLY OF BIOFUEL FEEDSTOCK AND LOCATION OF BIOREFINERIES

Several studies attempted to predict the most likely locations for biomass production and corresponding siting of the biofuel biorefineries for regulatory and other planning purposes (BRDB, 2008; English et al., 2010; EPA, 2010; Parker et al., 2010b; USDA, 2010). Some studies principally identify the regional availability of bioenergy feedstocks that could be used for biofuel production, while others also identify likely biorefinery locations. The following sections describe some of the approaches and assumptions used in the modeling of potential feedstock supply and biorefinery locations and compare projected locations for biorefineries among studies and with some of the proposed locations of cellulosic biofuel refineries. A comparison of the assumptions related to the types and amounts of feedstocks and the conversion rate to energy is provided in Table 3-1.

National Biorefinery Siting Model

Approach and Assumptions

The National Biorefinery Siting Model (NBSM) was developed by researchers at the University of California, Davis (Parker et al., 2010a; Tittmann et al., 2010; Parker, 2011). It integrates geographically explicit biomass resource assessments, engineering and economic models of the conversion technologies, models for multimodal transportation of feedstock and fuels based on existing transportation networks, and a supply chain optimization model that locates and supplies a biorefinery based on inputs from the other models (Parker et al., 2010a). To identify the location of biorefineries, the model first maximizes the profitability of the entire national biofuel industry. The profit maximized is the sum of the profits for each individual feedstock supplier and fuel producer. Costs minimized in the model are those associated with feedstock procurement, transportation, conversion to fuel, and fuel transmission to distribution terminals. Fuel production and selling price determine industry revenue. Coproduct revenues are included.

NBSM used data from the U.S. Department of Agriculture (USDA) National Agricultural Statistics Service (NASS) and Forest Service (USFS) provided by Skog et al. (2006, 2008) to project crop and woody biomass location and abundance and create spatially explicit estimates of biomass availability. NBSM constrained estimates for the supply of corn to be equal to the quantity needed to meet the RFS2 mandate of 15 billion gallons per year for conventional ethanol. Soybean and canola were assumed to be grown and used for biofuels. To limit the proportion of soybean dedicated to fuel production in the model, the use of soybean oil for biodiesel is limited to not more than 38 percent of all soy oil produced.

TABLE 3-1 Comparison of Assumptions in Biomass Supply Analyses

	Cellulosic Feedstocks	Other Feedstocks	Biomass Supply in 2022	Energy Conversion Ratio
National Biorefinery Siting Model	Corn stover Switchgrass Woody biomass	Greases MSW	500 million dry tons of all feedstock types	Varies by technology and feedstock type
EPA	Corn stover Switchgrass Other dedicated bioenergy crops Bagasse Sweet sorghum pulp Woody biomass	MSW	82 million dry tons of corn stover in 2022 10 million dry tons of bagasse 44 million dry tons of woody biomass	Over 90 gallons per dry ton, but varies by feedstock 94 gallons per dry ton for corn stover in 2022
USDA	Logging residue Dedicated bioenergy grasses Soybean Energy cane Sweet sorghum Canola Corn stover Straw	NA	42.5 million dry tons of logging residue	70 gallons per dry ton of logging residue Conversion ratios for other feedstocks not included in source
Biomass Research and Development Initiative	Corn stover Wheat straw Dedicated energy crops, including switchgrass Woody biomass including hybrid poplar and willow Sweet sorghum	NA	75-79 million dry tons of dedicated energy crops and annual energy crops like sweet sorghum in 2022 45 million dry tons of woody biomass in 2022 51-84 million dry tons of corn stover in 2022 20-32 million dry tons of wheat straw in 2022	80-90 gallons of ethanol per dry ton of switchgrass Conversion ratios for other feedstocks not included in source

NOTE: All analyses assumed that the 15-billion gallon mandate for conventional biofuel would be met by corn-grain ethanol.

NBSM also constrains cellulosic feedstock acquisition from all sources to an area within a 100-mile radius of the biorefinery site for the most part. Crop residue removal was constrained to levels that were estimated to prevent erosion. Those levels were estimated using local soil and landscape data and methods to estimate effects on soil quality and erosion (Nelson, 2002; Nelson et al., 2004, 2006). A combination of soil erosion (by wind and water) models were used to estimate the upper limit of crop residue removal. An amount of residue lower than the upper limit is considered to be removable without detrimental effects on the environment and resource base. The methods used combine detailed field-scale data on soil type, capability class, and slope from the USDA Natural Resources Conservation Service (NRCS) Soil Survey Geographic (SSURGO) database (USDA-NRCS, 2008) and an estimation of maximum rate of soil erosion not affecting productivity (the T value calculated using the Universal Soil Loss Equation; Renard et al., 1997). Residue amounts come from crop yields derived from the NASS database cited above. Wind erosion limits are also calculated using methods described by Nelson (2002). A limitation of these methods is that they do not account for some aspects of soil management, such as soil organic matter (SOM) maintenance. They do not include estimates of technical limits to stover recovery. Removal rates could be overestimated if the amount of stover left on the field is less than the amount needed to conserve SOM.

Switchgrass is modeled based on yields estimated at the Oak Ridge National Laboratory (Jager et al., 2010) using results from a survey of the most current agronomic literature and predictions from a switchgrass model developed by the Pacific Northwest National Laboratory (Thomson et al., 2009). Harvest costs are estimated using a model from the Idaho National Lab (Hess et al., 2009). Residue and cellulosic yield and cost estimates are resolved to the county level and calculated as an edge of field price. Costs and supply estimates are derived from the Policy Analysis System (POLYSYS) modeling framework (Walsh et al., 2003). POLYSYS estimated switchgrass to be available at $50 to $85 per dry ton, at the farm gate.

Estimates of forest biomass to the county level were derived from several sources. Accessible biomass estimates were guided by sustainability principles. However, the sustainability guidelines are site specific or region specific and could vary by ownership, with federal rules and guidelines, state guidelines, or professional standards used to guide management and harvest. In all cases, forest biomass generally included is the secondary output of other commercial forestry operations. Therefore, significant variability and uncertainty about resource access exists. Forest biomass available for biofuel production was estimated for the thinning of timberland with high fire hazard, logging residue left behind after anticipated logging operations for conventional products, treatment of Pinyon Juniper woodland and rangeland, normal thinning of private timberland, precommercial thinning on National Forest land in western Oregon and Washington, and unused mill residue.

EISA excludes credit for wood removed from federal lands, so NBSM provides separate estimates of forest biomass availability with and without federal lands. Forest resources were estimated to be available starting at $20-$30 per dry ton at the roadside, with the majority available at $45-$65 per dry ton, all depending on location, at the time of simulation. Pulpwood is available to biorefineries at suitable locations at up to $100 per dry ton. In addition to USFS data sources already noted, various models were used to estimate amounts available and costs for biomass harvest and removal (Biesecker and Fight, 2006).

Biorefineries are sited in or near cities in NBSM. No water constraints on biorefinery operation are assumed in the model for this reason. Water availability could limit the number of new refineries using cellulosic biomass in some regions, and this might be true for some existing corn-grain ethanol refineries as well (NRC, 2008). Corn-grain ethanol production is modeled using current information on ethanol refinery location, size, and cost. Optimistically, biorefineries are considered to be able to use mixed feedstocks for the most part. Where mixed feedstocks are available, corn-grain ethanol is produced up to the limit imposed by RFS2, and then crop residues, dedicated bioenergy crops, fats, oils and greases, and MSW all contribute to biofuel supply, with the mixtures varying locally.

Two conversion technologies are represented in the model: biochemical fermentation of ethanol from grain and cellulosic feedstocks and thermochemical production of biofuels from mostly cellulosic feedstocks. The use and costs of dilute acid hydrolysis followed by ethanol production from fermentation is modeled for lignocellulosic feedstocks. Although several thermochemical pathways could be used to convert cellulosic biomass to fuels (see Chapter 2), gasification followed by Fischer-Tropsch synthesis was used in NBSM to represent a larger class of thermochemical processes, including pyrolysis and other gasification technologies, that create biomass-derived diesel. Optimal combinations of feedstocks and technologies vary regionally. However, each simulation provides results of national or industry-wide fuel production at a given fuel price and identifies optimal locations and size of biorefineries and types of biomass resources used at each biorefinery. The selling prices of the product fuels are input parameters that are varied to create a supply curve through multiple model iterations across a range of prices ($1.00-$5.50 per gallon of gasoline equivalent) (Figure 3-1).

Results

The model predicted that the RFS2 consumption mandate of 36 billion gallons of biofuels by 2022 can be met at $2.90 per gallon of gasoline equivalent (or $1.91 per gallon of ethanol) at the time the most recent simulations were conducted (Figure 3-1). At this price, about 500 million dry tons of different types of biomass (including corn grain, fats, and oil) would be converted to biofuels nationally. Of the 500 million dry tons of biomass, 360 million are cellulosic biomass. The committee cautions that the estimated prices for various cellulosic feedstocks in NBSM are lower than the more recent estimates presented in

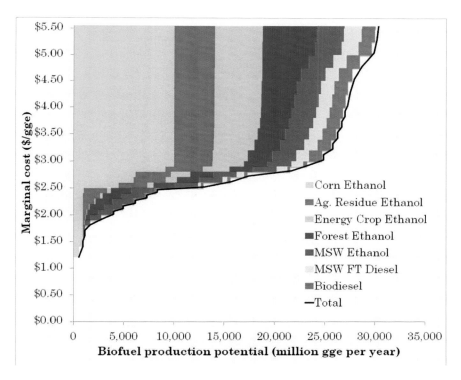

FIGURE 3-1 Biofuel supply and fuel pathways estimated from teh National Biorefinery Siting Model.
NOTE: About 500-600 million dry tons per year of biomass are considered available and recoverable at prices needed to meet RFS2 in 2022. Feedstock use reflects availability and price.
SOURCE: Jenkins (2010). Reprinted with permission from N.C. Parker, University of California, Davis.

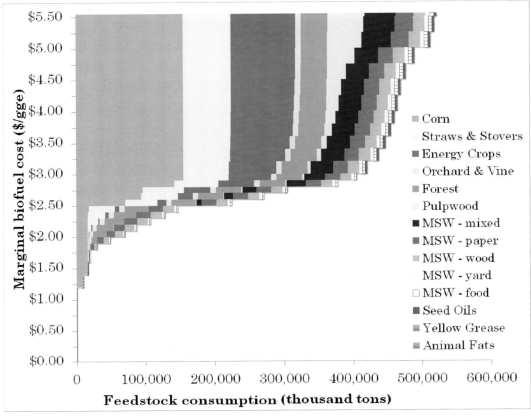

FIGURE 3-2 Biomass supply curves estimated by the National Biorefinery Siting Model.
SOURCE: Jenkins (2010). Reprinted with permission from N.C. Parker, University of California, Davis.

Chapter 4 of this report. The estimates of feedstock costs used in NBSM are at the farm gate and do not include the opportunity cost of cropland.[2] Although delivery costs were not included as part of feedstock costs, NBSM accounts for the cost of feedstock transport to biorefinery in its analysis.

Types and amounts of biomass vary with biofuel cost (Figure 3-2 and Table 3-2). The range and quantities of potential feedstocks increase with higher biofuel costs. As can be expected, agricultural residues are concentrated in the Corn Belt (Figure 3-3A). The areas with the highest yield per acre for perennial herbaceous grasses are not the areas with the largest supply (Figure 3-3B). The area with the highest quantity of forest residue is the Pacific Northwest followed by the Southeast (Figure 3-3C). As discussed in Chapter 2, models projecting yield on the basis of agronomic conditions suggest that *Miscanthus* and switchgrass are most productive on existing cropland. Because NBSM maximizes profitability of land use and commodity crops have higher value than perennial

---

[2]Opportunity cost is the net returns forgone by the producer for not using cropland to produce the next-best (or next most profitable) crop or product.

TABLE 3-2 Biomass Feedstocks for Integrated Biorefineries Projected by the National Biorefinery Siting Model in Selected Regions of the United States, at Prices Sufficient to Meet RFS2 Mandates

Region	Biorefinery Locations	Feedstock Type (in thousands of dry tons per year)						
		MSW	Forest	Pulpwood	Corn Grain	Crop Residues	Dedicated Bioenergy Crops	Orchard and Vineyard Wastes
North Central and Northeast	Grayling, MI	541	338	8.1				17.5
	Warren, MI		123					
	Port Huron, MI				500			
	Saginaw, MI				500	636	268	
	Marquette, WI		194	2.3				
	Rhinelander, WI		502					
	Bemidji, MN		289					
	Syracuse, NY	20.8	268		1,000	30.8	595	60.9
Mid-south	Fayetteville, AR	29	301	23.4		13	600	23.6
	Poplar Bluff, MO					282	527	
	Paducah, KY					147	865	
	Jackson, TN	12,800				63.1	1,240	
	Memphis, TN					881	479	
	Morristown, TN						1,120	
	Murfreesboro, TN					4.8	1,050	
Southeast	Huntsville, AL					4.8	811	
	Greenville, MS					836	306	
	Vicksburg, MS				1,000			
	Columbus, MS	28.2	538	58.3	540			1.3
	Waycross, GA		730	275				42.6
	Greenwood, SC	52.6	630	29.7		4.1	797	17.2
	Asheville, NC							
	Fayetteville, NC		646		600			
	Lumberton, NC	36.2	496	104			1,300	
	Danville, VA	75.8		34.7		22.1		

continued

TABLE 3-2 Continued

Region	Biorefinery Locations	MSW	Forest	Pulpwood	Feedstock Type (in thousands of dry tons per year)		Dedicated Bioenergy Crops	Orchard and Vineyard Wastes
					Corn Grain	Crop Residues		
Northern Plains	Nebraska City, NE					444	344	
	Norfolk, NE				2,070	11,100	241	
	Howard, SD				1,000	526	706	
	Pierre, SD					80.8	790	
	Sioux Falls, SD				1,210	1,260	95.8	
	Watertown, SD				1,000	989	371	
	Jamestown, SD				1,000	132	379	
Southern Plains	Garden City, KS				684	18.6	910	
	Guymon, OK					94.8	770	
	Keys, OK						1,030	
	Dumas, TX				1,000	139	1,120	
	Hereford, TX				115	113	931	

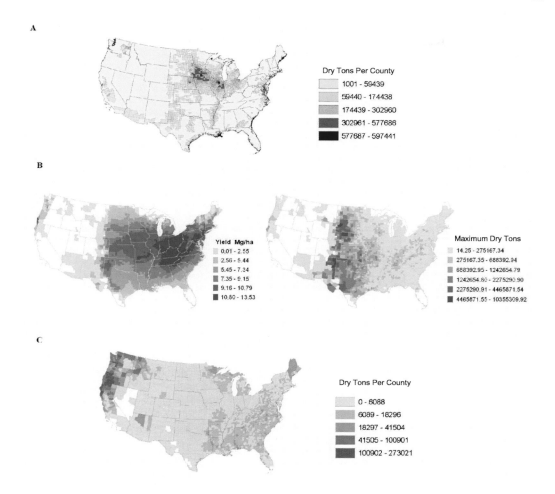

FIGURE 3-3 Principal amounts and locations of landscape-derived biomass feedstocks for biofuels from the National Biorefinery Siting Model.
A. Agricultural residue availability (dry tons per county).
B. Energy crop yields (Mg per hectare) and projections of maximum available supplies (maximum dry tons).
C. Estimated forest residue on public and private land (dry tons per county).
NOTE: Figure 3-3C also includes forest residues from federal forestland that do not qualify as renewable biomass under EISA's definition.
SOURCE: Jenkins (2010). Reprinted with permission from N.C. Parker, University of California, Davis.

herbaceous energy crops, the model estimated the perennial herbaceous energy crops would be planted on less productive lands. The projected distribution of crop residue, switchgrass, and *Miscanthus* supplies are consistent with another study that projects national cellulosic biomass supply using a multimarket equilibrium nonlinear mathematical programming model called the Biofuel and Environmental Policy Analysis Model (Khanna et al., 2011) and a study that projects regional cellulosic biomass supply in Michigan (Egbendewe-Mondzozo et al., 2010). Most of the supplies of woody biomass are projected to come from the north central and southeastern parts of United States. Biorefinery sites and the associated feedstock sheds identified by the model nationally are presented in Figure 3-4.

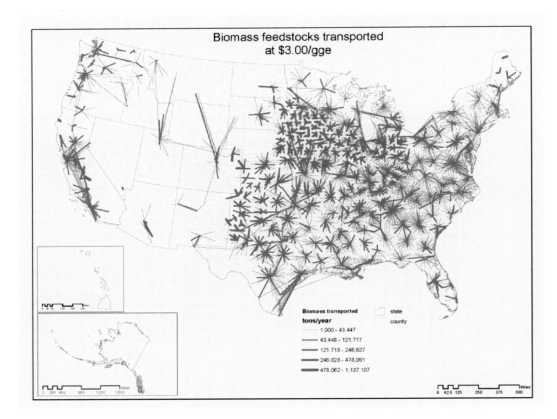

FIGURE 3-4 Biomass deliveries to the biorefineries needed to meet the RFS2 consumption mandate in 2022 projected by the National Biorefinery Siting Model.
SOURCE: Jenkins (2010). Reprinted with permission from N.C. Parker, University of California, Davis.

EPA

Approach and Assumptions

In its Regulatory Impact Analysis (EPA, 2010) for RFS2, EPA describes a transport tool that estimates the location of cellulosic biorefineries to be built to produce 16 billion gallons of cellulosic biofuels by 2022. Biomass data were derived from a number of sources, including NASS for agricultural residues, Elliot Campbell from Stanford University for bioenergy crops, and the U.S. Forest Service for forestry residue. MSW also was included as a potential feedstock for biofuels.

For each U.S. county, feedstock availability is estimated from the Forest and Agricultural Sector Optimization Model (FASOM, as discussed in Appendix K). FASOM was modified to reflect the current RFS2 program to include updated values for herbaceous energy crop yields, cellulosic ethanol conversions, and modifications to the accounting procedures for rangeland (Beach and McCarl, 2010). Switchgrass yields used in FASOM were derived from Thompson et al. (2009). Crop yields were projected to increase at current rates. The conversion yield from biomass to ethanol was assumed to be 90-94 gallons per dry ton depending on the feedstock (EPA, 2010).

The EPA siting tool also estimated cost information for each feedstock available. These costs included the roadside cost of production, transportation to move the feedstock to the centroid of its own county or from a neighboring county, and the secondary storage of that feedstock. Each county data point contains a list detailing the total cost of each feedstock available in that county. This list of feedstock availability and costs is used to choose feedstocks that a centrally located biorefinery would process. A biorefinery is assumed to process multiple feedstocks if they are available in the area. For each county, the cheapest feedstocks are selected for the biorefinery, and the volume of feedstock available at this price was converted to gallons, based on feedstock conversion modeled by FASOM (Tao and Aden, 2008), and added to a running count of the total volume of feedstock processed by that county, up to a maximum processing volume. Feedstock prices and conversion are reported by Beach and McCarl (2010). Biorefineries are assumed to be 100 million gallons per year in capacity, based on assumptions from FASOM and Carolan et al. (2007).

Capital costs associated with the increased volume were added to the total cost of the feedstock processing for that county. The model selected feedstock sources using a cost minimization algorithm that selects progressively more expensive feedstocks until the county either reaches a set 90-percent maximum processing volume or if adding another feedstock would produce a more expensive result on a price per gallon basis. At the end of this step, each potential county biorefinery location contained information regarding the cheapest total cost to produce cellulosic ethanol at that location. The most competitive locations were identified by comparing feedstock and capital costs. These locations resulted in a list of estimated least-cost biorefinery locations needed to meet 16 billion gallons per year mandated cellulosic biofuels. (The 4 billion gallons per year of advanced biofuels, which could be met by imports, were not included in this analysis.) The EPA Transport Tool does not take availability of water, permits, or human resources into account.

Results

The EPA Transport Tool estimated the majority of cellulosic biorefineries to be located in the upper Midwest and the Southeastern states (Figure 3-5). Because the tool projected

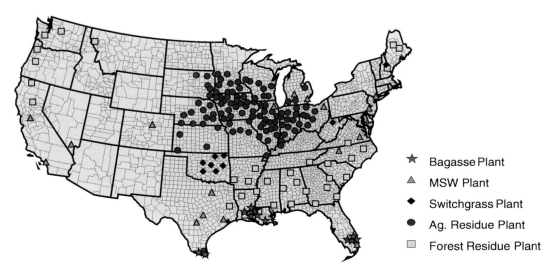

FIGURE 3-5 Locations of cellulosic facilities projected by the EPA Transport Tool.
SOURCE: EPA (2010).

biorefinery locations on the basis of price and the ease of recovery, corn stover would be the primary feedstock for cellulosic ethanol in 2022 (EPA, 2010; Table 1.8-13, p, 275). About 380 million dry tons of corn stover[3] were estimated to be available, of which 82 million dry tons could be supplied for producing 8 billion gallons of ethanol. Additional cellulosic biofuels were predicted to come from other crop residues in Florida, Texas, and Louisiana (primarily 10 million dry tons of bagasse associated with sugarcane harvest) and from dedicated bioenergy crops (primarily switchgrass) in the Southeastern states. Straw from wheat harvest in the Midwest or elsewhere was not included in EPA's model. Forestry residues also were considered likely feedstocks, primarily in the Southeastern United States and in a few locations in the Northwest and the Northeast (EPA, 2010). EPA estimated 44 million dry tons of woody biomass would be converted to biofuel in 2022.

USDA

Approach and Assumptions

USDA published the report *A USDA Regional Roadmap to Meeting the Biofuels Goals of the Renewable Fuels Standard by 2020* (USDA, 2010) to provide an overview of that agency's estimates for biofuel production by macro-region and biomass types. The report assumes that corn-grain ethanol production will meet the 15 billion gallons per year target for conventional fuels in 2020. The analysis in the report focuses on one scenario in which U.S. agriculture could produce enough biomass feedstock for biofuels to meet the remaining 21 billion gallons of biomass-based diesel, advanced biofuels, and cellulosic biofuels in the consumption mandate. USDA's analysis included corn stover, straw, soybean, canola, sweet sorghum, sugarcane for energy production, perennial grasses for energy production, and logging residue. It did not include MSW, algae, or manure as feedstocks. Energy conversion yields were calculated for each feedstock, but only the ethanol yield of logging residue was reported. This yield was assumed to be 70 gallons per dry ton, and residue availability was limited to recoverable amounts from permitted timber harvests on federal, state, and private lands as estimated by the U.S. Forest Service. The report is based on a synthesis of the professional judgment of scientists from the USDA Agricultural Research Service. Many details of assumptions used were not provided.

Results

The USDA report concluded that 27 million acres of cropland, representing 6.5 percent of all U.S. cropland, would be needed to produce enough biomass feedstock to meet RFS2. The amount of biomass supplied from that land was not reported, with one exception. The report estimated that 42.5 million dry tons of logging residues that are not used for any other purposes would be available for biofuel production in 2022. That estimate was based on actual data from the 2001-2005 period.

The Southeastern and Middle Atlantic states were estimated to provide the largest amount (10.5 billion gallons per year) of biofuels needed to meet the RFS2 mandate from dedicated bioenergy crops, soybean, energy cane, sweet sorghum, and logging residues. USDA attributed the high potential for biofuel production in this region to its robust growing season and estimated yield advantages of energy crops compared to corn. The Corn Belt

---

[3]In its calculation, EPA assumed a 1:1 of stover to grain. However, it did not account for the fact that corn grain contains 15 percent moisture. Thus, 1 dry ton of corn grain would yield about 0.85 dry ton of stover.

and surrounding states (central east) were estimated as the second largest source of feedstock for biofuels (9.09 billion gallons per year) with dedicated bioenergy crops, canola, soybean, sweet sorghum, corn stover, and logging residues as the principal feedstock sources. For the most part, USDA estimated little crop substitution in the most productive regions of the Corn Belt. Other regions of the United States are estimated to provide only small portions of the mandated fuels (Table 3-3).

Biomass Research and Development Initiative

Approach and Assessment

The Biomass Research and Development Board (BRDB) was established to coordinate federal research and development activities associated with biofuels, biopower, and bioproducts. The Biomass Research and Development Initiative was legislatively directed to address "feedstocks development, biofuels and biobased products development, and biofuels development analysis" (BRDB, 2010). BRDB released the report *Increasing Feedstock Production for Biofuels: Economic Drivers, Environmental Implication, and the Role of Research* (BRDB, 2008) that included modeling feedstock supply and distribution on the basis of the Regional Environment and Agriculture Programming (REAP) model (Johansson et al., 2007) linked to POLYSYS (described below and in Appendix K). The study focuses on the biofuel feedstocks needed to meet the EISA 2007 mandates. It used the REAP model to predict corn-grain ethanol production under a range of assumptions about technology development, crop yield, and oil supply. However, POLYSYS was used to simulate production of cellulosic feedstock because REAP does not have the capability to assess dedicated energy crops or the collection of crop residues. Three scenarios with varying contributions were presented. First, 20 billion gallons of advanced and cellulosic biofuels would be produced from cellulosic feedstock (corn stover, wheat straw, dedicated energy crops including switchgrass, and sweet sorghum) from U.S. cropland only. The committee judged the first scenario as unlikely because of the extent of land changes that would be incurred. The second scenario assumes that 16 billion gallons of biofuels would be produced from cellulosic feedstock from U.S. cropland and 4 billion gallons of biofuels would be from forest resources (such as hybrid poplar and willow). The third scenario assumed that all 4 billion gallons of advanced biofuels would be met by imports and the remaining 16 billion gallons of cellulosic biofuels would be produced from cropland and forest resources. As in the case of POLYSYS (discussed later in this chapter), that BRDB report used published data on yields and prices as a baseline and predicts changes from that baseline.

Results

Under the two scenarios, the BRDB estimated that 51-84 million dry tons of corn stover, 20-32 million dry tons of wheat straw, 45 million dry tons of woody biomass, and 75-79 million dry tons of dedicated perennial (such as switchgrass) and annual (such as sweet sorghum) energy crops could be supplied to meet the mandate in 2022. Switchgrass was estimated to yield 80-90 gallons of ethanol per ton. Energy conversion yields for other feedstocks were not provided.

Results from the BRDB study emphasized the importance of crop yield assumptions for land use and biomass feedstock production. Under various scenarios, the BRDB report predicted that most dedicated bioenergy crop production will be derived from the lower Mississippi Valley and Delta region, the Southeastern coastal plains and Gulf slope regions, and

TABLE 3-3 Potential Production of Biomass-Based Diesel, Advanced Biofuel, and Cellulosic Biofuel to Meet RFS2 in Different Regions of the United States as Projected by USDA

Region	States Within Region	Feedstock Type Available (In Order of Importance)	Volume of Ethanol Produced from Feedstock (billon gallons per year)	Volume of Biodiesel Produced from Feedstock (billion gallons per year)	Total Volume (billion gallons per year of ethanol equivalent)
Southeast and Hawaii	Alabama Arkansas Florida Georgia Hawaii Kentucky Louisiana Mississippi North Carolina South Carolina Tennessee Texas	Perennial grasses, soy oil, energy cane, biomass (sweet sorghum), and logging residues	10	0.01	10
Central East	Delaware Iowa Illinois Indiana Kansas Missouri Ohio Oklahoma Maryland Minnesota Nebraska North Dakota Pennsylvania South Dakota Wisconsin Virginia	Perennial grasses, canola, soy oil, biomass (sweet sorghum), corn stover, logging residues	8.8	0.26	9.2
Northeast	Connecticut Massachusetts Maine Michigan New Hampshire New Jersey New York Rhode Island Vermont West Virginia	Perennial grasses, soy oil, biomass (sweet sorghum), corn stover, logging residues	0.42	0.01	0.43
Northwest	Alaska Idaho Montana Oregon Washington	Canola, straw, logging residues	0.79	0.18	1.05

TABLE 3-3 Continued

Region	States Within Region	Feedstock Type Available (In Order of Importance)	Volume of Ethanol Produced from Feedstock (billon gallons per year)	Volume of Biodiesel Produced from Feedstock (billion gallons per year)	Total Volume (billion gallons per year of ethanol equivalent)
West	Arizona California Colorado New Mexico Nevada Utah Wyoming	Biomass (sweet sorghum, logging residues	0.06	0.00	0.06
Total			21	0.45	21

SOURCE: USDA (2010).

the Corn Belt (Figure 3-6). Crop residues are predicted to be derived most intensively from the Corn Belt region, based on the favored use of corn stover. Malcolm et al. (2009) added estimates of potential environmental effects to the projections in the BRDB analysis. Both analyses projected cellulosic biorefineries to be sited based on crop residue use (primarily corn stover) in the Corn Belt and perennial grass production in the Southeastern states. The BRDB report also predicted that forest resources from biofuels would be mostly derived from the Pacific Northwest and the northeastern tip of the United States (Figure 3-6).

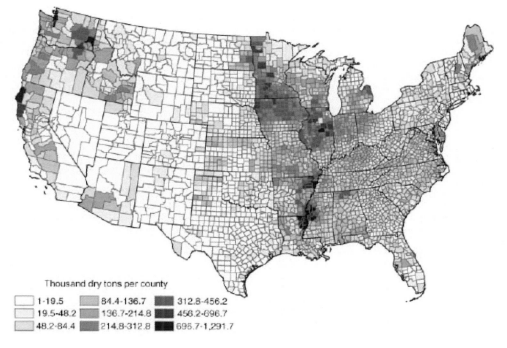

Thousand dry tons per county

☐ 1-19.5	▨ 84.4-136.7	■ 312.8-456.2
☐ 19.5-48.2	▨ 136.7-214.8	■ 456.2-696.7
▨ 48.2-84.4	▨ 214.8-312.8	■ 696.7-1,291.7

FIGURE 3-6 Projected locations and quantities of cropland and forest resources for producing 20 billion gallons of cellulosic biofuel based on REAP and POLYSYS.
SOURCE: BRDB (2008).

Estimates Based on POLYSYS and Related Models

POLYSYS, created at the University of Tennessee, has been widely used to estimate diverse effects of biofuel production from crops, crop residues, and perennial grasses (De La Torre Ugarte and Ray, 2000; Walsh et al., 2003, 2007; BRDB, 2008; Dicks et al., 2009; English et al., 2010; Larson et al., 2010; Perlack and Stokes, 2010). This model is described in Appendix K, but a brief summary focused on land use and crop choice is synthesized here from the studies cited. POLYSYS estimates livestock supply and demand, planted and harvested acres, crop yields, total production, exports, variable costs, demand by type of use, farm prices, cash receipts, government payments, and net realized income and agricultural income. Cattle are linked to land use by their consumption of pasture, hay, and grain. Hay and pasture activities can shift in the model in response to prices. Crops included in POLYSYS are the eight major U.S. crops (corn, grain sorghum, oats, barley, wheat, soybeans, cotton, and rice) and the livestock sector (beef, dairy, pork, lamb and mutton, broilers, turkeys, and eggs). Hay and edible oils and meals are not estimated in the model, but values are supplied externally instead. In large parts of the United States, the eight major crops predominate. In other areas, such as parts of the West Coast, Texas, and Florida, crops not included are more important, and the model would estimate any changes in these regions less reliably. POLYSYS incorporates data for 305 agricultural statistical districts (ASDs) based on NASS data and averages soil data from NRCS for dominant soil types within each district (USDA-NRCS, 2006). The State Soil Geographic (STATSGO) database aggregates soil information at a larger scale compared to the SSURGO database, which accounts for soil variation at the field scale. Independent regional linear programming (LP) models are used to model these ASDs, which are assumed to have homogeneous production characteristics (for example, rainfall, soil, crops, and climate characteristics). Proportions of tillage practices used were estimated at the county level (Larson et al., 2010). Data for crops and costs of production were derived from the University of Tennessee's Agriculture Budget System, which was developed from state agricultural extension budgets starting in 1995 and have been updated repeatedly since then (Larson et al., 2010).

POLYSYS includes all cropland, cropland used as pastures, hay land, and permanent pasture. To estimate land use in an ASD or larger aggregated regions, the crop supply model first determines the land area in each ASD available to (1) enter crop production, (2) shift production to a different crop, or (3) move out of crop production. Changes in land use are estimated based on expected crop productivity, cost of production, expected profit, and market conditions (Hellwinckel et al., 2010). However, the model's developers specified some portion of farmland as committed to crop production to reflect the inelastic nature of agricultural land supply, including the resistance to change that is part of most farming decisions (Walsh et al., 2003). Specific cropping systems are not modeled, but crop choice is constrained based on expert judgment, primarily provided by NRCS scientists. Once the land area that can be shifted is determined, the LP models allocate available acres among competing crops based on maximizing returns above costs.

POLYSYS also estimates the production of (nonirrigated) switchgrass, hybrid poplar, and willow. Commercial production of these crops is limited on the farms, so yields and costs are estimated in other ways (English et al., 2010). Switchgrass yield estimates from the Pacific Northwest National Laboratory switchgrass simulation model (Thomson et al., 2009; see also Chapter 2) and other sources of data for perennial grasses and short-rotation tree plantations (Gunderson et al., 2008; Jager et al., 2010; Perlack and Stokes, 2010) have been added to some of the 305 ASDs. Comparative advantage with respect to yields and prices for switchgrass produced are predicted for the Corn Belt and Southeastern regions of the United States (Walsh et al., 2007; Dicks et al., 2009).

Land shifts are estimated in more than 3,000 counties in the United States. To be included in the model's optimum solution, the net present value of bioenergy crops would have to be greater than the regional rental rate for pasture or for conventional crops. If some pastures are converted to bioenergy crop production, the remaining pasture would likely be managed more intensively, and more hay would be fed instead of grazing (English et al., 2010). Limits are placed on the rate (5 percent per year) and total amount of pasture that can be converted (20 percent). The amounts produced are determined by prices sufficient to meet mandated demands. These prices are supplied externally (Hellwinckel et al., 2010).

Locations of Department of Energy-Funded and Industry-Funded Advanced Biofuel Projects

As discussed in Chapter 2, the U.S. Department of Energy (DOE) funded a number advanced biofuel projects. Those planned facilities that have secured funding for demonstration of converting agricultural feedstocks or forestry materials to cellulosic or advanced biofuel are listed in Table 2-2 in Chapter 2 and shown in Figure 3-7.

FIGURE 3-7 Locations of advanced biofuel projects including pilot-scale, demonstration-scale, and commercial-scale projects funded by DOE (red dots) and proposed by industry (blue dots).
DATA SOURCES: Biofuels Digest (2010) and DOE-EERE (2011).

A number of companies that have DOE funding have proposed cellulosic biofuel facilities or begun construction of pilot-scale[4] or commercial-scale[5] facilities (Biofuels Digest, 2010). In each case, a combination of factors has motivated site selection, including feedstock supply, infrastructure, and federal, state, or local financial support for development at that location. In many instances, lists of proposed facilities or those under construction or in operation do not conform to those locations predicted by the siting models or analyses discussed. The site-specific nature of biofuel feedstock production, fuel demand and use, and other factors not easily reducible to general model formulations might influence industry decisions. For example, the models estimated that few biorefineries will be sited in the western and northeastern United States, other than the ones that rely on MSW as feedstock. Innovative businesses may recognize feedstock supplies and other advantages overlooked in larger, aggregated analyses.

Comparing Estimated Supplies from Different Studies

The different studies described above all concluded that the United States is capable of producing a sufficient quantity of cellulosic biomass to achieve the RFS2 mandate. The USDA and BDRB reports only estimated potential locations of feedstock supply. NBSM and the EPA Transport Tool also project locations of biorefineries on the basis of where feedstocks could be produced or harvested, feedstock costs, and other factors. In all cases, potential locations of feedstock supplies are based on agroecological classification. (See Chapter 2 for agroecological regions suitable for various biomass types.)

Although the different approaches were independent efforts to assess future feedstock production and related biofuel supplies, they have some commonalities (Table 3-4). All approaches account for the need to leave some residue in the field to prevent soil erosion, but none of them explicitly include water availability as a constraint. Many studies rely on similar data, on the use of a few critical models, or on the work of the same scientists or models. The federal government is the source of much data used by modelers in estimating feedstock supply and future biorefinery locations. For cropland use and other agricultural data nationwide, the USDA-NASS reports result from hundreds of surveys they conduct annually (USDA-NASS, 2011). Those reports cover most aspects of U.S. agriculture, including production and supplies of food and fiber, prices paid and received by farmers, farm labor and wages, farm finances, chemical use, and changes in the demographics of U.S. producers. USDA's Economic Research Service collects the annual Agricultural Resource Management Survey (ARMS). These data are USDA's primary source of information on the financial condition, production practices, resource use, and economic well-being of American farm households. ARMS provides observations on field-level and livestock management practices, the economics of farm businesses, and the characteristics of American farm households (for example, age, education, occupation, farm and off-farm work, types of employment, and family living expenses). The National Resources Inventory maintained by NRCS has been used to define farm structure (USDA-NRCS, 2010). These surveys form a series from 1982 and provide updated information on the status, condition, and trends of land cover and land use, land capability classes, soil and soil erosion, water and irrigation,

---

[4]A pilot demonstration for biofuel refinery is a facility that has the capacity to process 1-10 dry tons of feedstock per day.

[5]A commercial demonstration for biofuel refinery is a facility that has the capacity to process 700 dry tons of feedstock per day.

TABLE 3-4 Comparison of Models Used to Estimate Biomass Production by Region and Biorefinery Locations

Source	National Biorefinery Siting Model	U.S. Environmental Protection Agency (EPA)	Biomass Research and Development Initiative	University of Tennessee and Oak Ridge National Laboratory (ORNL)
Data or model sources for feedstocks — Crops	USDA-NASS (U.S. Department of Agriculture National Agricultural Statistics Service)	USDA-NASS USDA-ARMS (U.S. Department of Agriculture Agricultural Resource Management Survey) FASOM (Forest and agricultural sector optimization model)	USDA-NASS USDA-ARMS	USDA-NASS USDA-NRI (U.S. Department of Agriculture National Research Initiative) Data from ORNL
Crop residue	USDA-NASS SURGO (Soil survey geographic database) RUSLE (Revised Universal Soil Loss Equation)		USDA-NASS	USDA-NASS SURGO RUSLE
Dedicated bioenergy crops	PNNL (Pacific Northwest National Laboratory)	PNNL CRP (Conservation Reserve Program) land excluded	POLYSYS (Policy Analysis System Model) CRP land excluded	PNNL Jager et al. (2010)
Forest residue	FIA (Forest Inventory and Analysis National Program) U.S. Department of Agriculture Forest Service (USFS) (2010)	FIA FASOM	FIA	FIA USFS (2010)
Municipal Solid Wastes	EPA Arsova et al. (2008)	EPA		
Livestock		FASOM		USDA-NASS
Infrastructure	Bureau of Transportation Statistics			
Models	Original optimization model RUSLE FRCS (Fuel Reduction Cost Simulator)	FASOM Siting tool	REAP (Rural Energy for America Program) POLYSYS EPIC (Environment Policy Integrated Climate Model)	POLYSYS IMPLAN (Impact analysis for planning) economic modeling FRCS

continued

TABLE 3-4 Continued

Source	National Biorefinery Siting Model	U.S. Environmental Protection Agency (EPA)	Biomass Research and Development Initiative	University of Tennessee and Oak Ridge National Laboratory (ORNL)
Description/scenarios	Minimizes the cost of meeting biofuel targets on a national level, subject to local biomass supplies and infrastructure availability. Diverse price scenarios modeled.	Minimizes the cost of meeting biofuel targets on a national level, subject to local biomass supplies and local capital costs.	Uses projected crop yields, land availability, and targeted biomass requirements to meet biofuel demand. Diverse yield and demand scenarios modeled.	Uses projected crop yields, land availability, land use transformation, and targeted biomass requirements to meet biofuel demand. Diverse scenarios modeled, including different levels of demand for biomass and fuels, yield levels, and carbon storage payments.
Environmental restrictions on feedstock production — Erosion and maintenance of soil organic matter	Yes	Yes	Yes	Yes
Nutrients		Yes	Yes	Yes
Water		Yes		
Greenhouse gas		Yes	Yes	Yes
Biomass power considered	Yes			
Biorefinery capital costs	Yes	Yes		
Specific locations of biorefineries	Yes	Yes		
References	Parker (2011) and Parker et al. (2010a)	EPA (2010)	BRDB (2008)	De La Torre Ugarte and Ray (2000), Walsh et al. (2003, 2007), Dicks et al., (2009), English et al. (2010), Jager et al. (2010)

and related resources on the nation's nonfederal lands.[6] The USFS data form the basis for many assessments of biomass availability from those sectors (USFS, 2010).

Comparisons of projected biorefinery locations and feedstock supplies identified from different studies indicate that there are similarities among studies (Figures 3-4, 3-5, 3-6, and 3-7)—a large amount of crop residues can be derived from the Corn Belt; herbaceous perennial crops will likely be planted in the Southeast; the Pacific Northwest, North Central, and Northeast regions can supply forest residues; and large quantities of MSW, if included, can be supplied from larger urban areas (primarily near urban areas in the Northeast and western states). Both EPA and BRDB estimated that 10 billion gallons of ethanol-equivalent biofuel would be derived from crop residues or dedicated bioenergy crops and 4 billion gallons of ethanol-equivalent biofuels would be derived from forest resources.

Some differences in the outcome of the feedstock supply and biorefinery siting studies were observed. NBSM projected more perennial grass crop production along the western edge of the Corn Belt in the southeastern and northern prairie regions than EPA, English et al. (2010), or BRDB (2008). Compared to the estimations of biorefinery locations provided by NBSM and the feedstock locations provided by USDA and BRDB, EPA projected fewer facilities located in the Corn Belt region, more in the Southeastern United States, and more in California (presumably associated with MSW conversion). The influence of capital costs associated with biorefinery permitting and construction was estimated and used in EPA's model but not in other siting models. These costs are estimated by EPA to be lower in the Southeast and Midwest than elsewhere. EPA's projected biorefinery locations were sited in locations with the lowest capital costs. In EPA's analysis, minimizing capital costs was more important than maximizing the yield of perennial grasses or the availability of forest residues. Nonetheless, many biorefineries are projected to be built in similar regions to those derived from other modeling efforts.

UNCERTAINTIES ABOUT CELLULOSIC FEEDSTOCK PRODUCTION AND SUPPLY

Although NBSM and other studies estimated that 500-600 million dry tons of biomass could be supplied to biorefineries for fuel production, several factors could alter that supply: competition for biomass, potential for pests and diseases, and yield increase as a result of research. Farmers' willingness to grow or harvest feedstocks also can affect supply, which is discussed in Chapter 6 in the context of social barriers to achieving RFS2.

Cellulosic bioenergy crops can be grown for markets other than biofuels. For example, bioenergy crops can be used for power generation (electricity or combined heat and power) or as forage or bedding for animals. Most states (36 out of 50) have set standards that require the electricity sector to generate a portion of the electricity from renewable or alternative sources. Although NBSM accounted for biomass allocated for electricity generation, competition for feedstock between the two sectors could drive up the price of feedstock. The technology for producing fiberboard from sawdust and other residues has improved (Ye et al., 2007; Yousefi, 2009), and crop and wood residues can be used for that purpose and further increase the competition for feedstock.

In the case of agricultural residue, it becomes a commodity with value instead of being a residue that incurs an additional cost of its removal when new market opportunities become available. As discussed in Chapter 2, leaving a portion of crop residue can protect

---

[6]Nonfederal lands include privately owned lands, tribal and trust lands, and lands controlled by state and local governments.

land from soil erosion and maintain soil carbon. The crop residues removed can be used for animal bedding (Tarkalson et al., 2009).

The price and supply of bioenergy feedstocks likely to be available to biorefineries depend partly on competition with other uses, in addition to other factors including production, harvesting, and transportation costs. Whether biorefineries can compete for biomass with other sectors will depend on the prices that other sectors are willing to pay for the feedstocks. For example, the likely value of crop residues as bedding to animal producers is the cost of replacing the residues with a substitute.

Competition for feedstock could intensify during periods of weather extremes (for example, drought or flood) if crops are lost to pests and diseases. Fungal diseases that could affect switchgrass have been reported (Gustafson et al., 2003; Crouch et al., 2009). Insect pests could affect the establishment of switchgrass stands; for example, grass seedlings were reported to be susceptible to grasshoppers, crickets, corn flea beetle, and cinch bug (Landis and Werling, 2010). A preliminary study suggested the yellow sugarcane aphid and the corn leaf aphid as potential pests of *Miscanthus × giganteus* (Crouch et al., 2009). Although severe pest and disease outbreaks have not been observed for herbaceous perennial crops outside the tropics (Karp and Shield, 2008), the pest and disease dynamics could change if cultivation of these crops increases and become more intensive.

Short-rotation woody crops are susceptible to diseases and pests. Rust diseases can affect poplar and willow severely (Royle and Ostry, 1995). In addition to diseases, insect pests such as cottonwood leaf beetle and defoliators, sap feeders, and stem borers can attack poplar and willow (Landis and Werling, 2010).

Cultivar selection, breeding, and genomic approaches can result in bioenergy crops that are resistant to pests and diseases, suitable for their specific agronomic conditions, and have other desirable characteristics as biofuel feedstock (Bouton, 2007; Nelson and Johnsen, 2008). Increase in yield per acre as a result of agronomic and genetic research (Mitchell et al., 2008; Jakob et al., 2009; Wrobel et al., 2009) could alleviate competition for feedstocks among different sectors.

CONCLUSION

Several studies estimated that the United States has the capability to produce adequate biomass feedstock for production of 16-20 billion gallons of cellulosic biofuels to meet RFS2. Different types of feedstocks predominate in different regions. In the North Central and Northeast regions, forestry residues are most important. In the southeastern United States, forest residues and perennial grasses are most important. In the prairie regions of the United States, crop residues, corn grain, and perennial grasses are predicted to be produced. Some studies constrain the feedstock supply by price with the intent to simulate feedstock supply at a reasonable cost to biorefineries. However, the studies discussed above do not address the gap between the price that farmers are willing to sell their biomass feedstock and the price that biorefineries are willing to pay. The next chapter assesses the economics of feedstock supply in detail. Most studies also constrain the feedstock supply by limiting the amount of crop residues that could be harvested with the intent of minimizing soil erosion. However, soil erosion is only one of many environmental factors that have to be considered in large-scale production of bioenergy feedstock. Chapter 5 discusses various environmental effects to be considered. Knowing feedstock supply and biorefinery locations, local or biorefinery-specific environmental consequences of biofuel production also can be estimated or anticipated. Potential harvestable biomass feedstock is unlikely to be the limiting factor in meeting RFS2. At the same time, limits associated with the diverse economic and environmental effects of achieving the RFS2 mandate by 2022 could reduce the amount of biomass feedstock projected to be available in the United States for cellulosic biofuels by independent studies.

REFERENCES

Arsova, L., R. van Haaren, N. Goldstein, S. Kaufman and N. Themelis. 2008. The state of garbage in America. BioCycle 49(12):22.

Beach, R.H., and B.A. McCarl. 2010. U.S. Agricultural and Forestry Impacts of the Energy Independence and Security Act: FASOM Results and Model Description. Final Report. Research Triangle Park, NC: RTI International.

Biesecker, R.L., and R.D. Fight. 2006. My Fuel Treatment Planner: A User Guide. Portland, OR: U.S. Department of Agriculture - Forest Service.

Biofuels Digest. 2010. Industry Data. Available online at http://biofuelsdigest.com/bdigest/free-industry-data/. Accessed November 12, 2010.

Bouton, J.H. 2007. Molecular breeding of switchgrass for use as a biofuel crop. Current Opinion in Genetics & Development 17(6):553-558.

BRDB (Biomass Research and Development Board). 2008. Increasing Feedstock Production for Biofuels: Economic Drivers, Environmental Implication, and the Role of Research. Washington, DC: U.S. Department of Agriculture.

BRDB (Biomass Research and Development Board). 2010. BR&D: Biomass Research and Development. Available online at http://www.usbiomassboard.gov/. Accessed November 4, 2010.

Carolan, J., S. Joshi, and B. Dale. 2007. Technical and financial feasibility analysis of distributed bioprocessing using regional biomass pre-processing centers. Journal of Agricultural & Food Industrial Organization 5(2):Article 10.

Crouch, J.A., L.A. Beirn, L.M. Cortese, S.A. Bonos, and B.B. Clarke. 2009. Anthracnose disease of switchgrass caused by the novel fungal species *Colletotrichum navitas*. Mycological Research 113:1411-1421.

De La Torre Ugarte, D.G., and D.E. Ray. 2000. Biomass and bioenergy applications of the POLYSYS modeling framework. Biomass and Bioenergy 18(4):291-308.

Dicks, M.R., J. Campiche, D. De La Torre Ugarte, C. Hellwinckel, H.L. Bryant, and J.W. Richardson. 2009. Land use implications of expanding biofuel demand. Journal of Agricultural and Applied Economics 41(2):435-453.

DOE-EERE (U.S. Department of Energy-Energy Efficiency and Renewable Energy). 2011. Integrated biorefineries. Available online at http://www1.eere.energy.gov/biomass/integrated_biorefineries.html. Accessed May 12, 2011.

Egbendewe-Mondzozo, A., S.M. Swinton, R.C. Izaurralde, D.H. Manowitz, and X. Zhang 2010. Biomass Supply from Alternative Cellulosic Crops and Crop Residues: A Preliminary Spatial Bioeconomic Modeling Approach. Staff Paper No. 2010-07. East Lansing: Michigan State University.

English, B.C., D.G. De La Torre Ugarte, C. Hellwinckel, K.L. Jensen, R.J. Menard, T.O. West, and C.D. Clark. 2010. Implications of Energy and Carbon Policies for the Agriculture and Forestry Sectors. Knoxville: The University of Tennessee.

EPA (U.S. Environmental Protection Agency). 2010. Renewable Fuel Standard Program (RFS2) Regulatory Impact Analysis. Washington, DC: U.S. Environmental Protection Agency.

Gunderson, C.A., E.B. Davis, H.I. Jager, T.O. West, R.D. Perlack, C.C. Brandt, S.D. Wullschleger, L.M. Baskaran, E.G. Wilkerson, and M.E. Downing. 2008. Exploring Potential U.S. Switchgrass Production for Lignocellulosic Ethanol. Oak Ridge, TN: Oak Ridge National Laboratory.

Gustafson, D.M., A. Boe, and Y. Jin. 2003. Genetic variation for *Puccinia emaculata* infection in switchgrass. Crop Science 43(3):755-759.

Hellwinckel, C.M., T.O. West, D.G.D. Ugarte, and R.D. Perlack. 2010. Evaluating possible cap and trade legislation on cellulosic feedstock availability. Global Change Biology Bioenergy 2(5):278-287.

Hess, J.R., K.L. Kenney, L.P. Ovard, E.M. Searcy, and C.T. Wright. 2009. Commodity-Scale Production of an Infrastructure-Compatible Bulk Solid From Herbaceous Lignocellulosic Bioamss. Volume A: Uniform-Format Bioenergy Feedstock Supply Design System. Idaho Falls: Idaho National Laboratory.

Jager, H., L.M. Baskaran, C.C. Brandt, E.B. Davis, C.A. Gunderson, and S.D. Wullschleger. 2010. Empirical geographic modeling of switchgrass yields in the United States. Global Change Biology Bioenergy 2(5):248-257.

Jakob, K., F.S. Zhou, and A. Paterson. 2009. Genetic improvement of C4 grasses as cellulosic biofuel feedstocks. In Vitro Cellular & Developmental Biology-Plant 45(3):291-305.

Jenkins, B. 2010. National Biorefinery Siting Model: Optimizing Bioenergy Development in the U.S. Presentation. Presentation to the Committee on Economic and Environmental Impacts of Increasing Biofuels Production, March 5.

Johansson, R., M. Peters, and R. House. 2007. Regional Environment and Agriculture Programming Model. Washington, DC: U.S. Department of Agriculture - Economic Research Service.

Karp, A., and I. Shield. 2008. Bioenergy from plants and the sustainable yield challenge. New Phytologist 179(1):15-32.

Khanna, M., X. Chen, H. Huang, and H. Onal. 2011. Supply of cellulosic biofuel feedstocks and regional production patterns. American Journal of Agricultural Economics 93(2):473-480.

Landis, D.A., and B.P. Werling. 2010. Arthropods and biofuel production systems in North America. Insect Science 17(3):220-236.

Larson, J.A., B.C. English, D.G.D. Ugarte, R.J. Menard, C.M. Hellwinckel, and T.O. West. 2010. Economic and environmental impacts of the corn grain ethanol industry on the United States agricultural sector. Journal of Soil and Water Conservation 65(5):267-279.

Malcolm, S.A., M. Aillery, and M. Weinberg. 2009. Ethanol and a Changing Agricultural Landscape. Washington, DC: U.S. Department of Agriculture.

Mitchell, R., K.P. Vogel, and G. Sarath. 2008. Managing and enhancing switchgrass as a bioenergy feedstock. Biofuels Bioproducts & Biorefining-Biofpr 2(6):530-539.

NAS-NAE-NRC (National Academy of Sciences, National Academy of Engineering, National Research Council). 2009. Liquid Transportation Fuels from Coal and Biomass: Technological Status, Costs, and Environmental Impacts. Washington, DC: National Academies Press.

Nelson, C.D., and K.H. Johnsen. 2008. Genomic and physiological approaches to advancing forest tree improvement. Tree Physiology 28(7):1135-1143.

Nelson, R.G. 2002. Resource assessment and removal analysis for corn stover and wheat straw in the eastern and midwestern United States: Rainfall and wind-induced soil erosion methodology. Biomass & Bioenergy 22:349-363.

Nelson, R.G., M. Walsh, J.J. Sheehan, and R. Graham. 2004. Methodology for estimating removable quantities of agricultural residues for bioenergy and bioproduct use. Applied Biochemistry and Biotechnology 113:13-26.

Nelson, R.G., J.C. Ascough II, and M.R. Langemeier. 2006. Environmental and economic analysis of switchgrass production for water quality improvement in northeast Kansas. Journal of Environmental Management 79(4):336-347.

NRC (National Research Council). 2008. Water Implications of Biofuels Production in the United States; Committee on Water Implications of Biofuels Production in the United States. Washington, DC: National Academies Press.

Parker, N., Q. Hart, P. Tittman, M. Murphy, R. Nelson, K. Skog, E. Gray, A. Schmidt, and B. Jenkins. 2010a. Development of a biorefinery optimized biofuel supply curve for the western United States. Biomass & Bioenergy 34(11):1597-1607.

Parker, N., Q. Hart, P. Tittman, M. Murphy, R. Nelson, K. Skog, E. Gray, A. Schmidt, and B. Jenkins. 2010b. National Biorefinery Siting Model: Spatial Analysis and Supply Curve Development. Washington, DC: Western Governors Association.

Parker, N.C. 2011. Modeling Future Biofuel Supply Chains Using Spatially Explicit Infrastructure Optimization. Ph.D. Graduate Group in Transportation Technology and Policy, University of California, Davis.

Perlack, R.D., and B.J. Stokes. 2010. Update of the "billion ton" study. Presentation to the Committee on Economic and Environmental Impacts of Increasing Biofuels Production, May 3.

Perlack, R.D., and B.J. Stokes (Leads). 2011. U.S. Billion-Ton Update: Biomass Supply for a Bioenergy and Bioproducts Industry. Oak Ridge, TN: Oak Ridge National Laboratory.

Perlack, R.D., L.L. Wright, A.F. Turhollow, R.L. Graham, B.J. Stokes, and D.C. Erbach. 2005. Biomass as Feedstock for a Bioenergy and Bioproducts Industry: The Technical Feasibility of a Billion-Ton Annual Supply. Oak Ridge, TN: Oak Ridge National Laboratory.

Renard, K.G., G.R. Foster, G.A. Weesies, D.K. McCool, and D.C. Yoder. 1997. Predicting Soil Erosion by Water: A Guide to Conservation Planning with the Revised Soil Loss Equation (RUSLE). Washington, DC: U.S. Department of Agriculture.

Royle, D.J., and M.E. Ostry. 1995. Disease and pest control in the bioenergy crops poplar and willow. Biomass & Bioenergy 9(1-5):69-79.

Skog, K.E., R.J. Barbour, K.L. Abt, E.M. Bilek, F.Burch, R.D. Fight, R.J. Hugget, P.D. Miles, E.D. Reinhardt, and W.D. Sheppard. 2006. Evaluation of Silvicultural Treatments and Biomass Use for Reducing Fire Hazard in Western States. Madison, WI: U.S. Department of Agriculture - Forest Service.

Skog, K.E., R. Rummer, B. Jenkins, N. Parker, P. Tittmann, Q. Hart, R. Nelson, E. Gray, A. Schmidt, M. Patton-Mallory, and G. Gordon. 2008. A Strategic Assessment of Biofuels Development in the Western States. Proceedings of the Forest Inventory and Analysis (FIA) Symposium.

Tao, L., and A. Aden. 2008. Technoeconomic Modeling to Support the EPA Notice of Proposed Rulemaking (NOPR). Golden, CO: National Renewable Energy Laboratory.

Tarkalson, D.D., B. Brown, H. Kok, and D.L. Bjorneberg. 2009. Irrigated small-grain residue management effects on soil chemical and physical properties and nutrient cycling. Soil Science 174(6):303-311.

Thomson, A.M., R.C. Izarrualde, T.O. West, D.J. Parrish, D.D. Tyler, and J.R. Williams. 2009. Simulating Potential Switchgrass Production in the United States. Richland, WA: Pacific Northwest National Laboratory.

Tittmann, P., N. Parker, Q. Hart, and B. Jenkins. 2010. A spatially explicit techno-economic model of bioenergy and biofuels production in California. Journal of Transport Geography 18(6):715-728.

USDA (U.S. Department of Agriculture). 2010. A USDA Regional Roadmap to Meeting the Biofuels Goals of the Renewable Fuels Standard by 2022. Washington, DC: U.S. Department of Agriculture.

USDA-NASS (U.S. Department of Agriculture - National Agricultural Statistics Service). 2011. Agency overview. Available online at http://www.nass.usda.gov/About_NASS/index.asp Accessed February 3, 2011.

USDA-NRCS (U.S. Department of Agriculture - Natural Resources Conservation Service). 2006. U.S. general soil map (STATSGO2). Available online at http://soils.usda.gov/survey/geography/statsgo/. Accessed November 30, 2010.

USDA-NRCS (U.S. Department of Agriculture - Natural Resources Conservation Service). 2008. Soil Survey Geographic (SSURGO) Database. Available online at http://soils.usda.gov/survey/geography/ssurgo/. Accessed November 30, 2010.

USDA-NRCS (U.S. Department of Agriculture - Natural Resources Conservation Service). 2010. National Resources Inventory. Available online at http://www.nrcs.usda.gov/technical/NRI/. Accessed September 23, 2010.

USFS (U.S. Department of Agriculture - Forest Service). 2010. Forest Inventory and Analysis National Program. Available online at http://fia.fs.fed.us/tools-data/ Accessed November 18, 2010.

Walsh, M., D.G. De La Torre Uguarte, H. Shapouri, and S.P. Slinsky. 2003. Bioenergy crop production in the United States. Environmental and Resource Economics 24(3):313-333.

Walsh, M.E., D.G. De La Torre Ugarte, B.C. English, K. Jensen, C. Hellwinckel, R.J. Menard, and R.G. Nelson. 2007. Agricultural impacts of biofuels production. Journal of Agricultural and Applied Economics 39(2):365-372.

Wrobel, C., B.E. Coulman, and D.L. Smith. 2009. The potential use of reed canarygrass (*Phalaris arundinacea* L.) as a biofuel crop. Acta Agriculturae Scandinavica Section B-Soil and Plant Science 59(1):1-18.

Ye, X.P., J. Julson, M.L. Kuo, A. Womac, and D. Myers. 2007. Properties of medium density fiberboards made from renewable biomass. Bioresource Technology 98(5):1077-1084.

Yousefi, H. 2009. Canola straw as a bio-waste resource for medium density fiberboard (MDF) manufacture. Waste Management 29(10):2644-2648.

4

The Economics and Economic Effects of Biofuel Production

The supply of biofuels depends on the availability and price of feedstocks. As discussed in Chapter 3, a sufficient quantity of cellulosic biomass could be produced in the United States to meet the Renewable Fuel Standard, as amended in the Energy Independence and Security Act (EISA) of 2007 (RFS2) mandate. However, buyers of biomass would have to offer a price that incentivizes suppliers to provide the requisite amount. For a cellulosic biomass market to be feasible, the price offered by suppliers would have to be equal to or lower than what buyers would be willing to pay and still make a profit. The first part of this chapter describes an economic analysis that estimates what the price of different types of biomass would need to be for producers to supply the bioenergy market and the cost of converting the biomass to fuel.

After this examination of the economics of producing biofuels from cellulosic biomass, the chapter turns to look at the effects of biofuel production on related sectors of the U.S. economy. The newly emergent biofuel market intersects with established markets in agriculture, forestry, and energy. The competition for feedstock created by increased production of biofuels could have substantial economic impacts on the prices of agricultural commodities, food, feedstuffs, forest products, fossil fuel energy, and land values. Therefore, the second part of this chapter examines the price effects that biofuel policy can have on competing markets.

Along with the prices of commodities, biofuel production will likely alter the availability of these products, which may change where they are produced and where they are demanded. The third part of the chapter therefore examines the effects of biofuel production in the United States on the balance of trade. Effects on the imports and exports of grains, livestock, wood products and woody biomass, and petroleum are discussed.

In addition to its interaction with commodity markets and trade, the biofuel industry also has economic effects related to federal spending. To make biofuels competitive in the energy market, the federal government supports biofuels through the RFS2 mandate and additional policy instruments discussed in Chapter 1. Tax credits and a tariff influence government revenue and expenditures. Support policies for biofuels also affect other government programs tailored to agricultural production, conservation, and human nutrition. The fourth section of this chapter reviews the federal and state policies that are related to biofuels or affected by biofuel policies and the observed and anticipated economic effects of biofuel-support policies on other government initiatives. The rationale for public support for these policies is also examined.

Because of the costs that biofuel policies incur, alternative options have been proposed to achieve similar policy goals. The final section provides an overview of alternatives that could possibly reduce or mitigate these costs while still encouraging biofuel production. It also examines how biofuel policy may interact with federal policy to reduce carbon emissions. Both policies have or would have reduction of greenhouse-gas (GHG) emissions as an objective.

ESTIMATING THE POTENTIAL PRICE OF CELLULOSIC BIOMASS

As of 2011, a functioning market for cellulosic biomass does not exist. Therefore, the committee chose to model possible prices based on production results found in published literature. This section explains the model, along with its assumptions and results, and estimates the cost of converting biomass to liquid fuel. It was not feasible for the committee to model every possible conversion pathway and biofuel product in the duration of this study. Thus, biochemical conversion of biomass to ethanol was used as an illustration in this analysis. The first part evaluates the production costs of various potential biorefinery feedstocks, assuming constant biorefinery processing costs. The second part analyzes the costs for various biorefining technologies, assuming constant feedstock costs.

Crop Residues and Dedicated Bioenergy Crops

If a cellulosic feedstock market were in existence, the data on market outcomes would be collectable. For instance, the purchase price for feedstocks could be obtained by surveying biorefineries, and the marginal costs of producing and delivering biomass feedstocks to a biorefinery could be calculated based on observed production practices. Presumably, if the market is operating, the price the biorefinery pays would be equal to or above the marginal cost of production and delivery. However, at the time this report was written, a commercial-scale cellulosic biorefinery and feedstock supply system did not exist in the United States. As a consequence, industry values were not available to estimate or otherwise assess the biomass supplier's marginal cost or supply curve and the biorefinery's derived demand for biomass.

The <u>Bio</u>fuel <u>Break</u>even model (BioBreak) was used to evaluate the costs and feasibility of a local or regional cellulosic biomass market for a variety of potential feedstocks.[1] BioBreak is a simple and flexible long-run, breakeven model that represents the local or regional feedstock supply system and biofuel refining process or biorefinery. BioBreak calculates the maximum amount that a biorefinery would be willing to pay for a dry ton of biomass delivered to the biorefinery gate. This value, or willingness to pay (WTP), is a function of the price of ethanol, the conversion yield (gallons per dry ton of biomass) the

---

[1] The BioBreak model was originally developed as a research tool to estimate the biorefinery's long-run, breakeven price for sufficient biomass feedstock to supply a commercial-scale biorefinery and the biomass supplier's long-run, breakeven price for supplying sufficient feedstock to operate such a biorefinery at capacity. An earlier version of the model was used in the NAS-NAE-NRC report *Liquid Transportation Fuels from Coal and Biomass: Technological Status, Costs, and Environmental Impacts* (2009b).

biorefinery can expect with current technology, and the costs of processing the feedstock (Box 4-1).

BioBreak also calculates the minimum value that a biomass feedstock producer would be willing to accept for a dry ton of biomass delivered to the biorefinery. This value, or willingness to accept (WTA), depends on the biomass feedstock producer's supply cost (that is, opportunity cost, production cost, and delivery cost) of supplying biomass in the long run (Box 4-2). A local or regional biofuel market for a specific feedstock will only exist or be sustained if the biofuel processor can obtain sufficient feedstock and the feedstock producers can deliver sufficient feedstock at a market price that allows both parties to break even in the long run. For the analysis, BioBreak calculated the difference or "price gap" between the supplier WTA and the processor WTP for each feedstock scenario. If the price gap is zero or negative, a biomass market is feasible (Jiang and Swinton, 2009). If the price gap is positive, a biofuel market cannot be sustained under the assumed feedstock production and conversion technology.

The BioBreak model is based on a number of assumptions. First, it assumes that the typical biomass feedstock producer minimizes costs and produces at the minimum point on the long-run average cost curve. Second, it assumes a yield distribution for biomass crops based on the expected mean yield and variation in yield within a region. Third, it assumes a transportation cost based on the average hauling distance for a circular capture region (that is, the biomass supply area) with a square road grid.[2] Fourth, the model assumes that the biorefinery has a 50-million gallon annual capacity. The model is flexible and can be rescaled to consider other facility sizes. This scale is chosen because it is assumed to be the minimum scale necessary to be competitive in the ethanol market. A smaller scale will imply lower WTP. Fifth, the model assumes that each biorefinery uses a single feedstock, and this feedstock is available without causing market disruptions (for example, changes in land rental prices) within the biomass capture region. Most biorefineries will likely be built to use locally sourced material for input (Babcock et al., 2011; Miranowski et al., 2011), but to the extent that they source material from outside the capture region, the actual WTA will be higher than is estimated in the results presented in this chapter. Sixth, beyond solving for alternative oil price scenarios, the impact of energy price uncertainty on biofuel investment is not considered. If potential investors require a higher return because of future energy market uncertainty (that is, a risk premium), actual WTP will be lower and the price gap will be higher than the price gap estimates presented in this chapter. With energy market uncertainty, a price gap estimate below zero will satisfy the necessary condition for development of a feedstock market (that is, both biomass supplier and biomass processor will break even in the long run), but it may not be sufficient to induce investment.[3]

---

[2]Due to heterogeneity in nontransportation production costs within the capture region, BioBreak uses the average distance rather than the capture region distance. Although the transportation cost per unit of biomass will be higher at the edge of the capture region, the supplier's minimum willingness to accept will not necessarily be strictly increasing with distance due to heterogeneity in production and opportunity costs. Even with higher transportation costs, a biomass supplier at the edge of the capture region with low production costs may be willing to supply biomass at a lower price than a biomass supplier with relatively high production costs located close to the biorefinery. BioBreak assumes that the average hauling distance within the capture region is representative of the location of the last unit of biomass purchased by the biorefinery to meet the biorefinery feedstock demand. Using the capture region distance would provide the correct estimate of the supplier's willingness to accept if the last unit of biomass purchased by the biorefinery is located at the edge of the capture region but would overestimate the supplier's willingness to accept in all other cases.

[3]For additional information on BioBreak model assumptions and limitations, refer to Appendix K and to Miranowski and Rosburg (2010).

BOX 4-1
Calculating Willingness to Pay (WTP)

Equation (1) details the processor's WTP, or the derived demand, for 1 dry ton of cellulosic material delivered to a biorefinery.

$$WTP = \{P_{gas} * E_V + T + V_{CP} + V_O - C_I - C_O\} * Y_E \tag{1}$$

The market price of ethanol (or revenue per unit of output) is calculated as the energy equivalent price of gasoline, where P_{gas} denotes per gallon price of gasoline and E_V denotes the energy equivalent factor of gasoline to ethanol. Based on weekly historical data for conventional gasoline and crude oil, the following relationship between the price of gasoline and oil is assumed: $P_{gas} = 0.13087 + 0.023917*P_{oil}$. Beyond direct ethanol sales, the ethanol processor also receives revenues from tax credits (T), coproduct production (V_{CP}), and octane benefits (V_O) per gallon of processed ethanol. Biorefinery costs are separated into two components: investment costs (C_I) and operating (C_O) costs per gallon. The calculation within brackets in Equation (1) provides the net returns per gallon of ethanol above all nonfeedstock costs. To determine the processor's maximum WTP per dry ton of feedstock, a conversion ratio is used for gallons of ethanol produced per dry ton of biomass (Y_E). Therefore, Equation (1) provides the maximum amount the processor can pay per dry ton of biomass delivered to the biorefinery and still break even. The values of the variables in Equation (1) are based on the following assumptions.

Price of Oil (P_{oil})

The processor's breakeven price of the price of oil per barrel is a critical parameter. Based on Cushing Crude Spot Prices (EIA, 2010c), oil briefly increased to $145 per barrel in July 2008 but decreased to $30 per barrel the last week of 2008. It increased to $48 per barrel the first week of 2009 and ended 2010 at $90. Given the high volatility in crude oil spot prices, rather than simulating or specifying a single price for oil, the difference between the WTP and WTA was calculated for three oil price levels: $52, $111, and $191, which are the low, reference, and high price projections for 2022 from the EIA Annual Energy Outlook (2010a) in 2008$.

Energy Equivalent Factor (E_V) and Octane Benefits (V_O)

Per unit, ethanol provides a lower energy value than gasoline. The energy equivalent ratio (E_V) for ethanol to gasoline was fixed at 0.667. While ethanol has a lower energy value than pure gasoline, ethanol is an octane enhancer. Blending gasoline with ethanol, even at low levels, increases the fuel's octane value. For simplicity, the octane enhancement value (V_O) was fixed at $0.10 per gallon.

Coproduct Value (V_{CP})

For coproduct value (V_{CP}), the estimation is simplified by assuming that excess energy is the only coproduct from the proposed biorefinery.[1] Aden et al. (2002) estimated that cellulosic ethanol production yields excess energy valued at approximately $0.14-$0.21 per gallon of ethanol, after updating to 2007 energy costs (EIA, 2008a). Without specifying the source of coproduct value, Khanna and Dhungana (2007) used an estimate of around $0.16 per gallon for cellulosic ethanol. Huang et al. (2009) found that switchgrass conversion yields the largest amount of excess electricity followed by corn stover and aspen wood. The model assumed a fixed coproduct value of $0.18 per gallon for switchgrass, Miscanthus, wheat straw, and alfalfa, while corn stover and woody biomass coproduct values were fixed at $0.16 and $0.14 per gallon.[2]

Conversion Ratio (Y_E)

The conversion ratio of ethanol from biomass (Y_E) is expected to vary based on feedstock type (because of variations in cellulose, hemicellulose, and lignin content), conversion process, and biorefinery efficiency. Research estimates for the conversion ratio have ranged from as low as 60 gallons per dry ton to theoretical values as high as 140 gallons dry per ton (see Appendix M, Table M-1). Eliminating theoretical values and outliers on either end, the reported range for the conversion ratio is approximately 65 to 100 gallons per dry ton. Based on the large variation within the research estimates, the model assumed a conversion ratio with a mean value of 70 gallons per dry ton as representative of current and near future technology (baseline scenario) and a mean of 80 gallons per dry ton as representative of the long-run conversion ratio in the sensitivity analysis.

Nonfeedstock Investment Costs (C_I)

Investment or capital costs for a cellulosic biorefinery have been estimated to be four to five times higher than a starch-based ethanol biorefinery of similar size (Wright and Brown, 2007). The biorefinery cost estimates used in this application of the model were based on research estimates and numbers provided by Aden et al. (2002), with cost adjustments to ensure consistency with the conversion rate and storage assumptions. Given cost adjustments and updating to 2007 values, the model assumed a mean (likeliest) value of $0.94 ($0.85) per gallon for biorefinery capital investment cost in the baseline scenario.[3]

Operating Costs (C_O)

Operating costs were separated into two components: enzyme costs and nonenzyme operating costs. Nonenzyme operating costs, including salaries, maintenance, overhead, insurance, taxes, and other conversion costs, were fixed at $0.36 per gallon. Aden et al. (2002) assumed that enzymes were purchased and set enzyme costs at $0.10 per gallon, and these enzyme cost estimates were used in the NAS-NAE-NRC (2009b) report on liquid transportation fuels from coal and biomass. Other (nonupdated) published estimates for enzymes have ranged between $0.07 and $0.25 per gallon. Discussions with industry sources indicate that enzyme costs may run between $0.40 and $1.00 per gallon given current yields and technology. The decrease in enzyme costs anticipated by Aden et al. (2002) and used in the NAS-NAE-NRC (2009b) report has not materialized. For the simulation in this report, the assumption was that the enzyme cost has a mean (likeliest) value of $0.46 ($0.50) per gallon but is skewed to allow for cost reductions in the near future.

Biofuel Production Incentives and Tax Credits (T)

To account for potential tax credits for cellulosic ethanol producers, the tax credit (T) for cellulosic ethanol producers designated by the Food, Conservation, and Energy Act of 2008 of $1.01 per gallon was considered in the sensitivity analysis and was denoted as the "producer's tax credit."[4]

---

[1] The coproduction of higher value specialty chemicals may reduce production costs; however, the committee could not find any economic evaluations of such options

[2] The coproduct value is fixed based on the percentage of lignin, cellulose, and hemicellulose reported by Huang et al. (2009) for each feedstock type. In the studies, the only biorefinery products are ethanol and electricity. All biomass that is not converted to ethanol is burned to produce energy. Energy that is not consumed by the biorefinery is exported to the electricity grid. There are some small differences in the assumed biorefinery energy requirements. Ignoring these small differences, any biomass that is not converted to ethanol will be burned to produce electricity. Thus, the coproduct value would decrease as ethanol yield increases. There are also small differences in the composition (energy content) of the biomass feedstocks. Overall, the coproduct values are a small fraction of the overall cost to produce biofuels, so these small variations in composition and yield have only a minor effect on overall economics.

[3] For parameters with an assumed skewed distribution in Monte Carlo analysis, the "likeliest" value denotes the value with the highest probability density.

[4] The processor's tax credit was only considered in the sensitivity analysis and not included in the baseline scenario results.

BOX 4-2
Calculating Willingness to Accept (WTA)

The biomass supplier's WTA per unit of feedstock delivered to the biorefinery is detailed in Equation (2).

$$WTA = \{(C_{ES} + C_{Opp}) / Y_B + C_{HM} + SF + C_{NR} + C_S + DFC + DVC * D\} - G \qquad (2)$$

The supplier's WTA for 1 dry ton of delivered cellulosic material is equal to the total economic costs the supplier incurs to deliver 1 unit of biomass to the biorefinery less the government incentives received (G) (for example, tax credits and production subsidies). Depending on the type of biomass feedstock, costs include establishment and seeding (C_{ES}), land and biomass opportunity costs (C_{Opp}), harvest and maintenance (C_{HM}), stumpage fees (SF), nutrient replacement (C_{NR}), biomass storage (C_S), transportation fixed costs (DFC), and variable transportation costs calculated as the variable cost per mile (DVC) multiplied by the average hauling distance to the biorefinery (D). Establishment and seeding cost and land and biomass opportunity cost are most commonly reported on a per acre scale. Therefore, the biomass yield per acre (Y_B) is used to convert the per acre costs into per dry ton costs, and Equation (2) provides the minimum amount the supplier can accept for the last dry ton of biomass delivered to the biorefinery and still break even. The values of the variables in Equation (2) are based on the following assumptions.[1]

Nutrient Replacement (C_{NR})

Uncollected cellulosic material adds value to the soil through enrichment and protection against rain, wind, and radiation, thereby limiting erosion that would cause the loss of vital soil nutrients such as nitrogen, phosphorus, and potassium. Biomass suppliers will incorporate the costs of soil damage and nutrient loss from biomass collection into the minimum price they are willing to accept. After adjusting for 2007 costs, estimates for nutrient replacement costs range from $5 to $21 per dry ton. Based on the model's baseline oil price ($111 per barrel) and research estimates, nutrient replacement was assumed to have a mean (likeliest) value of $14.20 ($15.20) per dry ton for stover, $16.20 ($17.20) per ton for switchgrass, $9 per ton for Miscanthus, and $6.20 per dry ton for wheat straw. At the high oil price ($191 per barrel), nutrient replacement costs increase by about $1.35 per dry ton. At the low oil price ($52 per barrel), nutrient replacement costs decrease by about $1.00 per dry ton.

Harvest and Maintenance Costs (C_{HM}) and Stumpage Fees (SF)

Harvest and crop maintenance cost (C_{HM}) estimates for cellulosic material have varied based on harvest technique and feedstock. Estimates of harvest costs range from $14 to $84 per dry ton for corn stover, $16 to $58 per dry ton for switchgrass, and $19 to $54 per dry ton for Miscanthus, after adjusting for 2007 costs.[2] Estimates for nonspecific biomass range between $15 and $38 per dry ton. Costs for woody biomass collection up to roadside range between $17 and $50 per dry ton. Spelter and Toth (2009) find total delivered costs (including transportation) about $58, $66, $75, and $86 per dry ton[3] for woody residue in the Northeast, South, North, and West regions, respectively.[4] Using the timber harvesting cost simulator outlined in Fight et al. (2006), Sohngen et al. (2010) found costs for harvest up to roadside to be about $25 per dry ton, with a high cost scenario of $34 per dry ton. Depending on the feedstock, the model assumed a mean value of $27-$46 per dry ton for harvest and maintenance with an additional stumpage fee with a mean value of $20 per dry ton for short-rotation woody crops (SRWC).

Transportation Costs (DVC, DFC, and D)

Previous research on transportation of biomass has provided two distinct types of cost estimates: (1) total transportation cost; and (2) breakdown of variable and fixed transportation costs. Research estimates for total corn stover transportation costs range between $3 per dry ton and $32 dry per ton. Total switchgrass and Miscanthus transportation costs have been estimated between $14 and $36 per dry ton, adjusted to 2007 costs.[5] Woody biomass transportation costs are expected to range between $11 and $30 per dry ton. Based on the second method, distance variable cost (DVC) estimates range between $0.09 and $0.60 per dry ton per mile, while distance fixed cost (DFC) estimates range between $4.80 and $9.80 per dry ton, depending on feedstock type. The BioBreak model used the latter method of separating fixed and variable transportation costs. One-way transportation distance (D) has been evaluated up to around 140 miles for woody biomass and between 5 and 75 miles for all other feedstocks. BioBreak calculates the average hauling distance (D) as a function of annual biorefinery biomass demand, annual biomass yield, and biomass density using the formulation by French (1960) for a circular area with a square road grid. The average hauling distance ranges between 13 and 53 miles.

Storage Costs (C_S)

Due to the low density of biomass compared to traditional cash crops such as corn and soybean, biomass storage costs (C_S) can vary greatly depending on the feedstock type, harvest technique, and type of storage area. Adjusted for 2007 costs, biomass storage estimates ranged between $2 and $23 per dry ton. The mean (likeliest) cost for woody biomass storage was $11.50 ($12) per dry ton, while corn stover, switchgrass, Miscanthus, wheat straw, and alfalfa storage costs were assumed to have mean (likeliest) values of $10.50 ($11) per dry ton.

Establishment and Seeding Costs (C_{ES})

Corn stover, wheat straw, and forest residue suppliers were assumed to not incur establishment and seeding costs (C_{ES}), whereas all other feedstock suppliers would have to be compensated for their establishment and seeding costs. Costs vary by initial cost, stand length, years to maturity, and interest rate. Stand length for switchgrass ranges between 10 and 20 years with full yield maturity by the third year. Miscanthus stand length ranges from 10 to 25 years with full maturity between the second and fifth year. Interest rates used for amortization of establishment costs range between 4 and 8 percent. Amortized cost estimates for switchgrass establishment and seeding, adjusted to 2007 costs, are between $30 and $200 per acre. Miscanthus establishment and seeding cost estimates vary widely, based on the assumed level of technology and rhizome costs. Establishment costs for wood also vary by species and location. Cubbage et al. (2010) reported establishment costs of $386-$430 and $520 per acre for yellow pine and Douglas Fir, respectively (2008$). The model assumed a mean established cost value of $40 per acre per year for switchgrass, $150 per acre per year for Miscanthus, $52 per acre per year for SRWC, and a fixed $165 establishment and fertilizer cost for alfalfa.

Opportunity Costs (C_{Opp})

To provide a complete economic model, the opportunity costs of using biomass for ethanol production were included in BioBreak. Research estimates for the opportunity cost of switchgrass and Miscanthus ranged between $70 and $318 per acre while estimates for nonspecific biomass opportunity cost ranged between $10 and $76 per acre, depending on the harvest restrictions under Conservation Reserve Program (CRP) contracts. Opportunity cost of woody biomass was estimated to range between $0 and $30 per dry ton. Depending on the region, the model assumed a mean opportunity cost of $50-$150 per acre for switchgrass and $75-$150 per acre for Miscanthus.[6]

Biomass Yield (Y_B)

Biomass yield is variable in the near and distant future due to technological advancements and environmental uncertainties. For simulation, the mean yield of corn stover was approximately 2 dry tons per acre. Switchgrass grown in the Midwest was assumed to have a distribution with a mean (likeliest) value around 4 (3.4) dry tons per acre on high-quality land and 3.1 dry tons per acre on low-quality land.[7] Miscanthus grown in the Midwest was assumed to have a mean (likeliest) value of 8.6 (8) dry tons per acre on high-quality land and 7.1 (6) dry tons per acre on low-quality land.[8] Switchgrass grown in the South-Central region has a higher mean yield of around 5.7 dry tons per acre. For the regions analyzed, the Appalachian region provides the best climatic conditions for switchgrass and Miscanthus with assumed mean (likeliest) yields of 6 (5) and 8.8 (8) dry tons per acre, respectively. Wheat straw, forest residues, and SRWC were assumed to be normally distributed with mean yields of 1, 0.5, and 5 dry tons per acre. First-year alfalfa yield was fixed at 1.25 dry tons per acre

continued

Box 4-2 Continued

(sold for hay value), while second-year yield was fixed at 4 dry tons per acre (50-percent leaf mass sold for protein value), resulting in 2 dry tons per acre of alfalfa for biomass feedstock during the second year.

Biomass Supplier Government Incentives (G)

For biomass supplier government incentives (G), the dollar for dollar matching payments provided in the Food, Conservation, and Energy Act of 2008 up to $45 per dry ton of feedstock for collection, harvest, storage and transportation is used, and it is denoted as "CHST." The CHST payment was considered in the sensitivity analysis rather than the baseline scenario because the payment is a temporary (2-year) program and might not be considered in the supplier's long-run analysis. Although the BioBreak model is flexible enough to account for any additional biomass supply incentives, the establishment assistance program outlined in the 2008 farm bill is not considered because implementation details were not finalized at the time the model was run.

[1]Further detail and references for the parameters can be found in Appendix K.

[2]Harvest and maintenance costs were updated using USDA-NASS agricultural fuel, machinery, and labor prices from 1999-2007 (USDA-NASS, 2007a,b).

[3]Based on a conversion rate of 0.59 dry tons per green tons.

[4]Northeast includes Pennsylvania, New Jersey, New York, Connecticut, Massachusetts, Rhode Island, Vermont, New Hampshire, and Maine. South refers to Delaware, Maryland, West Virginia, Virginia, North Carolina, South Carolina, Kentucky, Tennessee, Florida, Georgia, Alabama, Mississippi, Louisiana, Arkansas, Texas, and Oklahoma. States in the North region are Minnesota, Wisconsin, Michigan, Iowa, Missouri, Illinois, Indiana, and Ohio. West includes South Dakota, Wyoming, Colorado, New Mexico, Arizona, Utah, Montana, Idaho, Washington, Oregon, Nevada, and California.

[5]Transportation costs were updated using USDA-NASS agricultural fuel prices from 1999-2007 (USDA-NASS, 2007a,b).

[6]The corn stover harvest activity was developed for a corn-soybean rotation alternative and has no opportunity cost beyond the nutrient replacement cost. A continuous corn alternative, used by 10-20 percent of Corn Belt producers, was developed for corn stover harvest but not included in the BioBreak results presented in this report. The continuous corn production budgets, developed by state extension specialists, are always less profitable than corn-soybean rotation budgets with or without stover harvest. Continuous corn has an associated yield penalty or forgone profit (opportunity costs) relative to the corn-soybean rotation that occurs irrespective of stover harvest. Thus, a comparative analysis of stover harvest with a corn-soybean rotation and with continuous corn may be misinterpreted.

From the rotation calculator provided by the Iowa State University extension services with a corn price of $4 per bushel, a soybean price of $10 per bushel, and a yield penalty of 7 bushels per acre, the lost net returns to switching from a corn-soybean rotation to continuous corn equal around $62 per acre (ISUE, 2010).

[7]Plot trials were evaluated at 80 percent of their estimated yield.

[8]This is a significantly lower assumed yield than previous research has assumed or simulated (Heaton et al., 2004; Khanna and Dhungana, 2007; Khanna, 2008; Khanna et al., 2008).

For this report, the BioBreak model was used to evaluate the cost and feasibility of seven different feedstocks: corn stover, alfalfa, switchgrass, *Miscanthus*, wheat straw, short-rotation woody crops, and forest residue.[4] Corn stover was considered from a corn-soybean

---

[4]Although similar economic costs of biofuel were used in the NAS-NAE-NRC reports *America's Energy Future: Technology and Transformation* (2009a) and *Liquid Transportation Fuels from Coal and Biomass: Technological Status, Costs, and Environmental Impacts* (2009b), the values differ for a number of reasons. First, the current biofuel cost estimates and biomass yield assumptions included several studies published since the earlier reports were completed. Second, the gasoline equivalent price of ethanol was revised based on improved statistical information. Third, the enzyme price assumptions used for hydrolyzing biomass in 2008 were no longer valid in 2010, and these prices were updated based on current estimates. Finally, the BioBreak model was improved with the addition of a Monte Carlo process to better reflect the distribution of observations from published studies underlying the parameters of the model.

rotation (CS).[5] A 4-year corn stover-alfalfa rotation with 2 years of each crop (that is, CCAA) also was included. To account for regional variation in climate and agronomic characteristics, the WTP and WTA for switchgrass were evaluated in three regions: Midwest (MW), South-Central (SC), and Appalachia (App).[6] *Miscanthus* was also evaluated in the Midwest and Appalachian regions, while corn stover and wheat straw were assumed to be produced on cropland used for production in the Midwest and Pacific Northwest[7] regions, respectively. To account for the heterogeneity in Midwest land quality, perennial grasses (switchgrass and *Miscanthus*) on high quality (HQ) and low quality (LQ) Midwest cropland were also considered. This is not an exhaustive list of potential feedstocks or of the potential variation in productivity across the United States, but it provides information on 13 combinations of the most widely discussed feedstocks in regions where they are likely to be produced. The 13 combinations evaluated were: corn stover (CS), stover-alfalfa, alfalfa, Midwest switchgrass (HQ), Midwest switchgrass (LQ), Appalachian switchgrass, South-Central switchgrass, Midwest *Miscanthus* (LQ), Midwest *Miscanthus* (HQ), Appalachian *Miscanthus*, wheat straw, short-rotation woody crops (SRWC), and forest residues.

BioBreak derives a point estimate of WTA, WTP, and the price gap for a biorefinery with a fixed capacity and a local feedstock supply area. The point estimates are based on a number of assumptions and a number of parameter inputs. Since many of these parameter inputs are uncertain, BioBreak uses Monte Carlo simulation to assess the implications of this uncertainty on the results.[8] Monte Carlo simulation permits parameter variability, parameter correlation, and sensitivity testing not available in fixed parameter analysis.[9] For this analysis, distributional assumptions for each parameter were based on empirical data updated to 2007 values and verified with industry information when available.[10] If appropriate data were insufficient or not available, a distribution was constructed to fit available data or a range of industry values was obtained. A sensitivity analysis was then performed to determine importance. Monte Carlo simulation with parameter distributional assumptions captures the range of variability found in the estimates in the literature, which were used in this analysis. Boxes 4-1 and 4-2 summarize the equations used to calculate the biorefinery's WTP and the biomass feedstock supplier's WTA and the assumptions used in this committee's analysis for the BioBreak model parameters. Appendix K provides further details about the assumptions for the feedstock supply costs. Summary tables of parameter assumptions used in the analysis are available in Appendix L, while Appendix M provides a review of the literature used to construct the parameter assumptions.

---

[5]Compared to a corn-soybean rotation, corn from continuous corn production has a yield penalty but produces more stover over the course of the rotation. If the price of stover were sufficiently high, a farmer could find it more profitable to switch to continuous corn production because the additional stover revenue would more than offset the yield penalty (that is, opportunity cost). Whether this would occur in practice is in dispute.

[6]Midwest includes North Dakota, South Dakota, Nebraska, Kansas, Iowa, Illinois, and Indiana. South-Central applies to Oklahoma, Texas, Arkansas, and Louisiana. Appalachian refers to Tennessee, Kentucky, North Carolina, Virginia, West Virginia, and Pennsylvania.

[7]Washington, Idaho, and Oregon.

[8]For the Monte Carlo simulations, BioBreak uses Oracle's spreadsheet-based program Crystal Ball®.

[9]See NAS-NAE-NRC (2009b) for an example of BioBreak applied in a fixed parameter analysis.

[10]Costs were updated using USDA-NASS agricultural prices from 1999-2007 (USDA-NASS, 2007a,b).

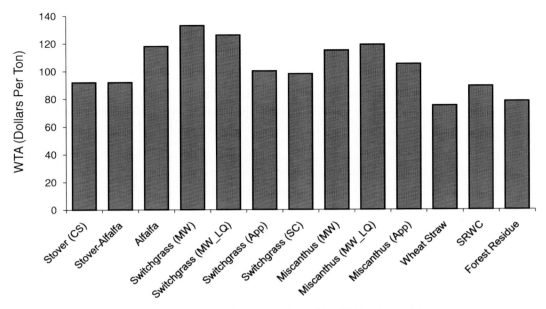

FIGURE 4-1 Biomass supplier WTA per dry ton projected by BioBreak model.
NOTE: Baseline scenario (no policy incentives, $111/barrel oil, 70 gallons per dry ton).

WTA

Given the parameter assumptions and an oil price of $111 per barrel, the biomass supplier's average cost or WTA per ton of biomass delivered to the biorefinery ranges between $75 per dry ton for wheat straw in the Pacific Northwest to $133 per dry ton for switchgrass grown on high-quality land in the Midwest. Figure 4-1 provides the supply cost per dry ton for all 13 feedstock-rotation combinations in the analysis.[11] Regional characteristics play a significant role. Switchgrass and *Miscanthus* grown on high-quality Midwest cropland have relatively high costs because of high land opportunity costs and lower yields relative to the Appalachian and South Central regions.

BioBreak derives the price gap between the biomass producer's supply cost and the processor's derived demand for biomass delivered to the biorefinery. Table 4-1 provides the biofuel processor's WTP, biomass supplier's WTA, and the price gap given the parameter assumptions and no policy incentives (for example, no blender's tax credit or supplier payment).

This analysis ignores that RFS2, which requires that any cellulosic biofuel produced up to the mandated quantity be consumed, could influence feedstock producers and investors' decision-making. Indeed, suppliers might be willing to invest in biofuel facilities irrespective of the economics described here if the consumption mandate of RFS2 is perceived as being rigid because the mandate provides a market for the biofuel. If the mandate is not perceived as being rigid, it will be difficult to induce private-sector investment. The complexities in the mechanisms for renewable identification numbers (RINs) for cellulosic

---

[11]The parameter draws and calculations were repeated 10,000 times resulting in 10,000 values for WTP, WTA, and the difference value (WTP-WTA) for each scenario. The value provided is the mean over the 10,000 calculations for each feedstock.

TABLE 4-1 BioBreak Simulated Mean WTP, WTA, and Difference per Dry Ton Without Policy Incentives

	WTA	WTP	WTA-WTP (per dry ton)	Price Gap in Dollars per Gallon of Ethanol	Price Gap in Dollars per Gallon of Gasoline Equivalent
Stover (CS)	$92	$25	$67	$0.96	$1.43
Stover-Alfalfa	$92	$26	$66	$0.94	$1.42
Alfalfa	$118	$26	$92	$1.31	$1.97
Switchgrass (MW)	$133	$26	$106	$1.51	$2.28
Switchgrass (MW LQ)	$126	$27	$99	$1.41	$2.13
Switchgrass (App)	$100	$26	$74	$1.06	$1.59
Switchgrass (SC)	$98	$26	$72	$1.03	$1.53
Miscanthus (MW)	$115	$26	$89	$1.27	$1.90
Miscanthus (MW LQ)	$119	$27	$93	$1.33	$1.98
Miscanthus (App)	$105	$27	$79	$1.13	$1.69
Wheat Straw	$75	$27	$49	$0.70	$1.04
SRWC	$89	$24	$65	$0.93	$1.39
Forest Residues	$78	$24	$54	$0.77	$1.16

NOTE: Oil price is assumed to be $111 per barrel and conversion efficiency of biomass to fuel is assumed to be 70 gallons per dry ton.

biofuels could lead investors to conclude the cellulosic mandate is not rigid (see Chapter 6 for further discussion of RINs).

Without policy intervention, no feedstock market is feasible in economic terms in the baseline scenario. The price gap that would need to be closed to sustain a feedstock market ranges between $49 per dry ton for wheat straw to $106 per dry ton for switchgrass grown on high-quality land in the Midwest. Figure 4-2 provides a graphical depiction of the price gap for all 13 feedstock-rotation combinations (see also Box 4-3).

The breakeven values and resulting price gaps depicted in Figure 4-2 are sensitive to assumptions and parameters used in the analysis. One key parameter in the BioBreak model is the price of oil (see Box 4-1). The price of oil drives the processor's derived demand for feedstock given biomass conversion cost and influences biomass supply cost through production costs. An increase (decrease) in the price of oil increases (decreases) what the processor can pay per dry ton of each feedstock and break even in the long run. At the same time, an increase (decrease) in the oil price increases (decreases) harvest and transportation costs resulting in a higher (lower) biomass supplier long-run breakeven cost. Given the assumptions, the effect on the processor's derived demand price from an oil price change dominates the effect on the biomass supply cost. Therefore, the price gap (WTA − WTP) decreases with higher oil prices and vice versa.

The results in Table 4-1 and Figures 4-1 and 4-2 assume an oil price of $111 per barrel. At an oil price of $191 per barrel, the price gap is eliminated for several feedstocks, including stover (CS), switchgrass (App, SC), *Miscanthus* (App), wheat straw, SRWC, forest residue, and stover-alfalfa. Remaining feedstocks have a price gap between $5 and $23 per dry ton. Correspondingly, the price gap increases to between $110 and $168 per dry ton of biomass with an oil price of $52 per barrel. The breakeven price is also sensitive to the conversion rate of biomass to ethanol. The baseline results assume a conversion rate of 70 gallons per dry ton

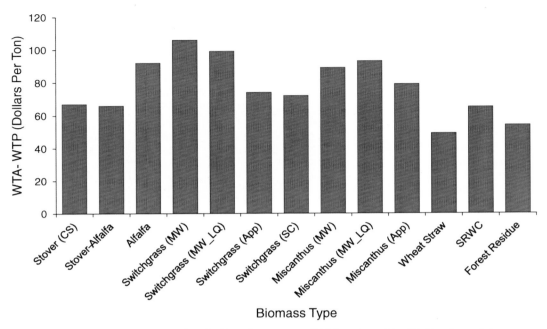

FIGURE 4-2 Gap between supplier WTA and processor WTP projected by BioBreak model.
NOTE: No policy incentives (WTA – WTP, $111 per barrel oil, 70 gallons per dry ton).

of biomass for all types of feedstocks (see Box 4-1), but potential advances in the conversion process may increase this rate. An increase in the biomass conversion rate increases the biorefinery returns per unit of feedstock converted and therefore reduces the price gap. Figure 4-4 provides sensitivity results of the processor WTP for South-Central switchgrass to the price of oil and conversion rate. Sensitivity results for other feedstocks are similar.

The results presented above assume no policy incentives. Any policy incentives for either the processor or supplier will decrease the price gap needed for market viability. The 2008 farm bill provides a $1.01 per gallon tax credit to cellulosic biofuel blenders. Figure 4-5 displays the price gap when the blender's credit is included. Given the blender's tax credit, the price gap drops significantly, resulting in viable feedstock markets for stover (CS), stover-alfalfa, wheat straw, SRWC, and forest residues (that is, WTP > WTA or WTA – WTP < 0). The remaining feedstocks have a gap between $1 and $35 per dry ton. Similarly, any policy incentive to suppliers, such as the U.S. Department of Agriculture's (USDA's) Biomass Crop Assistance Program in the 2008 farm bill, which provides payments for establishing bioenergy crops and collecting biomass, would further decrease the price gap and, given BioBreak's baseline assumptions, result in viable feedstock markets for all feedstocks in the analysis (for more on the Biomass Crop Assistance Program, see Box 4-4 in section "Potential Changes Caused by Biofuel Policy"). Policy incentives for carbon emissions could also affect the price gap (as discussed later in the section "Interaction of Biofuel Policy with Possible Carbon Policies").

One benefit of using Monte Carlo simulation to derive the breakeven values is the ability to capture the variability found in the literature for each parameter in the model. For the BioBreak application presented here, the Monte Carlo simulation was conducted using

BOX 4-3
Gap in Forest Residue Demand and Supply

The market for forest residue exemplifies the gap between WTP and WTA for cellulosic feedstock. Many existing studies assume that a large proportion of wood will be available for cellulosic ethanol production through harvesting of residues. However, these studies often ignore the likely costs of extracting residues. Figure 4-3 shows that, in most cases, forest residues are not collected because the costs of extracting additional residues are likely to be high relative to the value. Figure 4-3 also shows the marketable components of a typical tree. The bole of the tree is the main marketable log from the stump at the bottom up to a diameter of 7 inches or so. This material typically is cut into lumber of some sort, depending on the form of the tree. From 7 inches or so up to around 4 inches, the main log of the tree is likely used as pulpwood. The additional stems at the top of the tree, the branches, and the leaves have traditionally been left as slash in the forest. The reason these components have been left as slash is largely economic—the cost of extracting this additional material is greater than the value of selling it. As shown in the right hand side of Figure 4-3, WTP for sawtimber is typically much higher than the marginal cost of extracting the large stems. WTP for extracting pulpwood, however, is close to, or equal to, the marginal cost of extracting the pulp component of timber, and WTP for biomass material is less than the marginal cost of extracting the additional material.

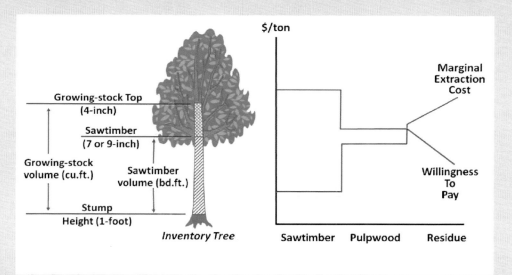

FIGURE 4-3 Components of trees and their use and value in markets.
NOTE: Sawtimber represents the largest part of the stump, up to about 9 inches in diameter. Pulpwood represents the rest of the stump up to 4 inches. The remainder is residue (often referred to as "slash").
SOURCE: Adapted from Figure 8 in Perlack et al. (2005).

10,000 draws from the assumed distribution for each parameter. From each draw, WTA, WTP, and the price gap were calculated for each feedstock. The results presented so far have been the mean values over all 10,000 calculations. Using the distributional assumptions outlined in Appendix L, which are based on literature summarized in Appendix M, Table 4-2 provides the estimated WTA value for each feedstock at select percentiles over the 10,000 Monte Carlo simulations at an oil price of $111 per barrel. The values in Table 4-2 provide a sensitivity range for the breakeven feedstock supply cost based on the parameter variation found in the literature.

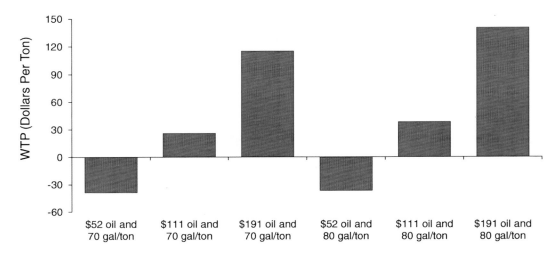

FIGURE 4-4 Sensitivity of WTP for switchgrass (SC) to the price of oil and ethanol conversion rate without policy incentives.

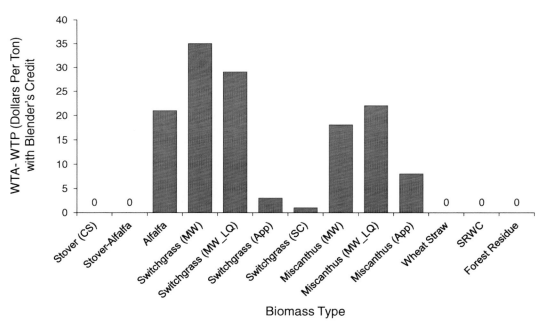

FIGURE 4-5 Gap between supplier WTA and processor WTP with blender's credit only projected by BioBreak model.

NOTE: WTA – WTP under the assumptions of $111 per barrel of oil and a biomass to fuel conversion efficiency of 70 gallons per dry ton.

TABLE 4-2 BioBreak Simulated WTA Value Without Policy Incentives by Percentile

	10%	25%	40%	MEAN	60%	75%	90%
Stover (CS)	$81	$87	$91	$92	$95	$97	$101
Stover-Alfalfa	$87	$89	$91	$92	$93	$95	$97
Alfalfa	$114	$116	$118	$118	$119	$120	$122
SG (MW)	$109	$118	$125	$133	$135	$144	$159
SG (MW_LQ)	$115	$121	$124	$126	$128	$132	$136
SG (App)	$87	$93	$97	$100	$102	$107	$115
SG (SC)	$85	$90	$93	$98	$98	$103	$112
Misc (MW)	$102	$109	$113	$115	$118	$122	$127
Misc (MW_LQ)	$99	$107	$113	$119	$121	$129	$142
Misc (App)	$91	$98	$103	$105	$108	$113	$119
Wheat Straw	$65	$70	$73	$75	$77	$80	$86
Farmed Trees	$78	$83	$86	$89	$91	$94	$100
Forest Residue	$68	$73	$76	$78	$80	$83	$88

NOTE: Oil price is assumed to be $111 per barrel and conversion efficiency of biomass to fuel is assumed to be 70 gallons per dry ton.

Comparing Feedstock Cost Estimates of the BioBreak Model with Other Studies

The cost estimates generated by the model are highly dependent on the assumptions used and the parameters considered. The way costs are treated and the comprehensiveness of which economic costs are included in the biomass supply chain and in ethanol processing varies by study. For example, the *U.S. Billion-Ton Update: Biomass Supply for a Bioenergy and Bioproducts Industry* (Perlack and Stokes, 2011) relies on the University of Tennessee's POLYSYS modeling system to estimate the marginal cost for supplying biomass to range from $40-$60 and average about $50 per dry ton of harvested biomass at the farm gate. BioBreak costs for wheat straw to the farm gate average about $40 per dry ton, corn stover about $55-$60 per dry ton, and switchgrass in the Appalachian and South Central Regions about $65 per dry ton without land opportunity costs included and $80 with land opportunity costs. Preliminary results indicate that much of the switchgrass would be produced on converted pasturelands that would have low opportunity costs. Handling, possibly drying, storing, and transporting low-density dry biomass to the biorefinery is a logistical challenge and costly (see Chapter 6).

Another study by Khanna et al. (2010) developed costs of production for corn stover, wheat straw, *Miscanthus*, and switchgrass for a number of potential producing states and then used these costs of production to develop biomass supply curves. Again, these were comprehensive costs at the farm gate that included land opportunity costs and were developed for the low-cost scenario assuming the availability of CRP land on which to produce switchgrass and *Miscanthus*. Farm-gate low-cost scenario estimates ranged from $44 to over $110 per dry ton for *Miscanthus* and $55 to $105 per dry ton for switchgrass. The high-cost scenarios were higher than those reported for BioBreak above. Corn stover estimates ranged from a low of $63 for no-till CS rotation to a low of $99 per dry ton for conventional till with median values of over $110 per dry ton for both CS tillage options. Again, the costs reported from the Khanna et al. (2010) did not include costs beyond the farm gate for transportation and storage.

The biomass cost estimates derived using the BioBreak model are typically higher than most similar studies because the model is inclusive of all economic costs (including opportunity costs of land) involved in producing, harvesting, storing, and delivering the last dry ton of biomass to the biofuel processing facility through the biomass supply chain.

Likewise, the biomass conversion costs account for all long-run costs in processing biomass to ethanol and include coproduct returns from a biorefinery of given capacity.

Finally, most studies assume biomass production costs are independent of crude oil prices; however, there are two factors that cause biomass production costs to increase as crude oil prices increase. First, part of the variability in crude price is due to the value of the dollar relative to other currencies. This same effect has been shown to influence crop prices (Abbott et al., 2009). Any increase in crude price caused by a devalued dollar would also increase opportunity costs for the land and fuel-based biomass production costs. It would also raise the demand for biofuels. Second, a portion of the cost of harvest, transportation, and nutrient replacement is related to the cost of fossil fuels. This concurrent increase in biomass cost would increase the apparent crude price at which biofuels would become cost competitive. The BioBreak model has attempted to incorporate these price effects on WTA.

Cost of Converting Cellulosic Biomass to Liquid Fuels

As mentioned earlier, along with the cost of harvesting and transporting biomass to a biorefinery is the cost of converting it into fuel. No commercial-scale facilities currently exist for the production of liquid fuels from cellulosic biomass. The conversion cost data used in the BioBreak analysis are based on laboratory or pilot-scale performance information and estimated investment and operating cost data for an optimized nth biorefinery that uses biochemical conversion of corn stover to ethanol (Aden et al., 2002). A recent report by Anex et al. (2010) compared the cost to produce liquid biofuels biochemically and thermochemically. The report examined the costs of fermentation to produce ethanol, fast pyrolysis to produce a gasoline or diesel "drop-in" fuel, and gasification and Fischer-Tropsch (F-T) to produce a gasoline or diesel "drop-in" fuel. Because the three technologies produce fuels with different energy contents, the results are presented in terms of gallons of gasoline equivalent. RFS2 is written in terms of gallons of ethanol equivalent, so a lesser volume of "drop-in" fuels from pyrolysis or F-T based technologies are required to satisfy RFS2. (See Table 1-1 in Chapter 1.) The study was based on a consistent biorefinery size of 2,205 dry tons per day of corn stover. All capital and operating costs are referenced to 2007. Corn stover is priced at $75 per dry ton, delivered to the biorefinery on a year-round basis. The only significant products from the biorefinery are the liquid fuel and electricity or gaseous fuel generated from the unconverted biomass. A required selling price for the liquid fuel is calculated to give a 10-percent discounted cash flow rate of return on a fully equity-financed project with a 20-year life.

The cellulosic ethanol costs in the paper were based on the equipment in a 2002 study by the National Renewable Energy Laboratory (NREL) (Aden et al., 2002), updated to 2007 construction and cellulase costs. The Anex et al. study (2010) used a nominal ethanol yield of 68 gallons per dry ton of biomass, which was the maximum demonstrated yield at the time the study was conducted and is close to the yield that this committee uses in the BioBreak analysis.

The gasification and F-T economics in Anex et al. (2010) are based on an NREL report prepared by Swanson et al. (2010). Two cases were evaluated: A high-temperature (HT), entrained flow, slagging gasification system and a lower temperature, fluidized bed, non-slagging gasification system. The HT system produces more fuel per ton of biomass, but its capital cost is higher. Overall cost to produce is slightly lower for the HT case because of the higher liquid yield. Biomass gasification has been attempted by several groups at the pilot scale. Operational difficulties have been encountered, but the gasification and F-T technology are well established for coal. Therefore, the cost data and yields for the gasification and F-T scheme can be considered reasonably reliable once the operational difficulties are overcome.

The fast pyrolysis economics are based on Wright et al. (2010). Fast pyrolysis of biomass for fuel production is a relatively new technology with little published information on yields, potential operational problems, or required equipment. The process uses equipment that is common in the petroleum refining industry, such as hydroprocessing, hydrocracking, hydrogen production, and high-temperature solids circulation similar to the fluid catalytic cracking process.

Kior, a privately funded company that is developing catalytic pyrolysis technology, submitted a Form S-1 to the U.S. Securities and Exchanges Commission (Kior, 2011) that contained additional information on capital requirements and overall yields. The technology and equipment proposed by Kior are similar to that used in the Wright et al. (2010) study, except that Kior has included a boiler and turbogenerator system to convert the off-gas and excess char into electricity. The capital costs included in Wright et al. were much lower than those reported by Kior. Although the boiler and turbogenerator represent a large capital investment, they are required to recover the energy contained in the nonliquid products as in the case of ethanol biorefineries. The Kior capital estimate is for a first-of-its-kind facility and its current usage is closer to an nth plant than a pioneer plant, but it is based on a fully developed cost estimate prepared by a major engineering company. In contrast, Wright et al.'s cost estimate is a "scoping quality" estimate for a fully developed technology. When adjusted to the same feed rate using a 0.6 scaling factor, capital cost estimated by Wright et al. was 43 percent of the capital cost estimated by Kior. Wright et al. (2010) acknowledged that some aspects of technology, such as solids removal from the pyrolysis oil, have yet to be developed and demonstrated.

The cases reported by Wright et al. (2010) and Kior (2011) were evaluated. The raw pyrolysis oil has to be hydrotreated before it can be used as a fuel. The two cases in Wright et al. (2010) differ in the source of hydrogen used to hydrotreat the pyrolysis oil. In the first case, part of the pyrolysis oil is used as feedstock to an on-site hydrogen plant to produce hydrogen. In the second case, hydrogen is purchased from an off-site plant that uses natural gas to produce the hydrogen. Producing hydrogen on site from bio-oil product lowers the liquid yield and increases the capital cost for the project. Kior's case is similar to the hydrogen purchase case by Wright et al. (2010), with the exception of the capital costs (as discussed earlier) and yield estimates.

The pertinent information from the published studies is summarized in Table 4-3 along with a calculation of the number of biorefineries and capital investment required, the number of acres of land necessary to produce the biomass (assuming all biomass for bioenergy comes from dedicated bioenergy crops), and the annual subsidies that would be required to support the industry at various crude oil prices. Table 4-3 demonstrates that catalytic pyrolysis and fast pyrolysis are promising technologies; they can produce "drop-in" products that are compatible with the existing petroleum distribution system. However, pyrolysis still requires substantial research and development before it is economically viable without subsidies.

The three crude prices used in Table 4-3 to calculate subsidies are from the three crude price scenarios for 2022 listed in the 2010 Annual Energy Outlook (EIA, 2010a). Only the high crude price scenario eliminates the need for subsidies to support a biofuel industry. All other price scenarios require either subsidies for the biofuel industry or additional taxes on petroleum products to narrow the price gap between petroleum fuels and biofuel. Without

TABLE 4-3 Summary of Economics of Biofuel Conversion

	Ethanol		Gasification and F-T		Pyrolysis, Hydrogen Purchase	
	90 Gallons Per Dry Ton	70 Gallons Per Dry Ton	High Temp	Low Temp	High Yield	Kior
Single Plant Capital, Million Dollars	380	380	606	498	200	463
Fuel Produced, Million Gallons Per Year						
Million Gallons Per Year	69.5	52.4	41.7	32.3	58.2	43.1
Million Gallons of Gasoline Equivalent Per Year	46.3	34.9	41.7	32.3	58.2	48.9
Cost to Produce						
Nth Plant	375	500	430	480	210	324
Pioneer Plant	650	850	800	750	350	N/A
Number of Plants to Meet 16 billion gallons of ethanol-equivalent biofuels in 2022	230	305	256	331	183	218
Capital Costs Required to Meet RFS2, Billion Dollars	88	116	155	165	37	101
Price Gap, Billion Dollars Per Year						
At $52 Per Barrel	25	39	31	37	8	20
At $111 Per Barrel	10	24	16	21	–7	5
At $191 Per Barrel	–10	3	–4	1	–28	–16
Biomass Feed Requirements						
Million Dry Tons Per Year	178	236	175	226	133	159
Million Acres at 5 Tons Per Acre	36	47	35	45	27	32

SOURCES: Aden et al. (2002); Anex et al. (2010); R. Anex (University of Wisconsin, Madison, personal communication on August 23, 2011); Swanson et al. (2010); Wright et al. (2010).

these subsidies or taxes, the biofuel industry would not expand to meet RFS2 requirements. An increase of $25 per dry ton in the price of biomass increases the annual subsidies required by $5 billion to $10 billion per year. Figure 4-6 shows a graphical breakdown of the production costs.

The capital-related costs in Figure 4-6 include the average depreciation and the assumed 10-percent return on investment for the 20-year life of the project. In the discounted cash flow analysis used to develop these costs, the capital charges are higher in the early years of the project and decline throughout the life of the project. The per-gallon, capital-related operating costs are determined by dividing this average annual effective cost of capital (depreciation plus return on investment) by the annual fuel production. The annual effective cost of capital varies from 12 to 14 percent of the total capital investment for the various projects. Another way of defining these costs is to assume they are an effective capital recovery factor for the capital investment. This range of capital recovery factors would give an effective rate of return of about 12 percent for a 20-year project.

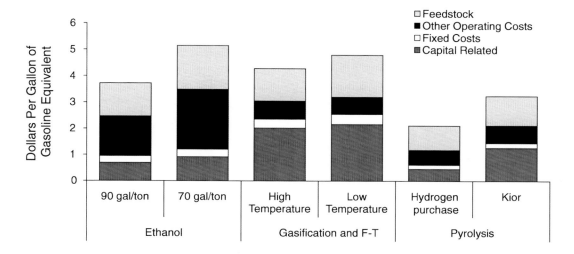

FIGURE 4-6 Breakdown of biomass conversion costs.

The 10-percent after-tax rate of return used in these studies is probably on the low side of returns that would be required to attract capital for a new, high-risk project. The economics also assume that the project is fully equity financed. None of these projects has yet to be demonstrated commercially, implying that they are high-risk investments. High-risk investments usually require higher returns or leveraging (borrowing) of capital to reduce the risk. Either of these would increase the effective cost of capital for at least the early projects, so the total production cost numbers are probably low.

The costs in Table 4-3 and Figure 4-6 are pre-tax wholesale costs at the biorefinery gate. "Drop-in" fuels, such as those produced by pyrolysis and gasification and F-T, can use the existing petroleum infrastructure for delivery to the final consumer. Transportation and distribution costs for drop-in fuels would be similar to current petroleum products transportation costs of $0.02-$0.05 per gallon. Cellulosic ethanol would continue to be shipped by rail, barge, and truck for blending at the final distribution point with costs of $0.10-$0.50 per gallon. Construction of an ethanol pipeline system would reduce transportation costs but would require additional capital investment. Nominal pipeline construction costs typically exceed $1 million per mile (Smith, 2010).

Producing enough biomass to meet RFS2 could require 30-60 million acres of land, excluding the high yield, hydrogen-purchase pyrolysis case in Table 4-3. If all biomass for cellulosic biofuels is produced from dedicated energy crops, the amount of land needed would be at the high end of the estimate. The use of corn stover, wheat straw, other crop residues, and forest residues would reduce the amount of acres needed.

PRIMARY MARKET AND PRODUCTION EFFECTS OF U.S. BIOFUEL POLICY

Because RFS2 creates another market for crops, particularly for corn, and a possible incentive to shift land from food crops to biomass feedstocks, the mandate has repercussions for related commodity markets. The prices of grain and oilseed crops, food, animal feed, and wood products have all experienced upward pressure coinciding with the rapid expansion of the biofuel market. Coproducts from biofuel have also introduced competition in feed

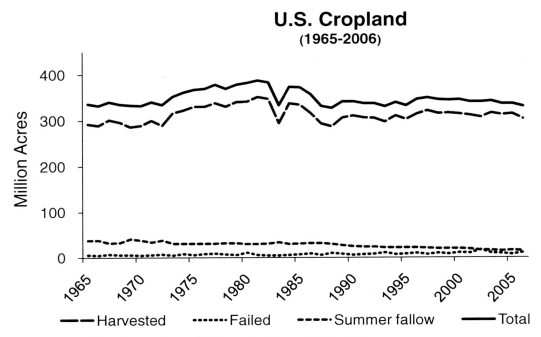

FIGURE 4-7 Allocation and use of U.S. cropland from 1965 to 2006.
DATA SOURCE: USDA-ERS (2007).

markets. Increasing biofuel in the transportation fuel market could affect domestic gasoline and diesel prices. Demand for feedstocks to meet traditional needs and those of the biofuel market increases competition for land. Although several attempts have been made to tie price and resource use effects to biofuel expansion, there is little agreement in the economic literature about the effects that can be attributed to biofuel expansion. Therefore, this section presents what has happened recently to resource prices and use relative to biofuel expansion rather than a cause-and-effect empirical analysis of biofuel expansion.

Agricultural Commodities and Resources

This section reviews the primary feedstuff and food crops, market series, and cropland resource base published by the U.S. Department of Agriculture (USDA) National Agricultural Statistics Service (NASS). Figure 4-7 provides an indication of what has happened over time to total cropland and total harvested cropland. Both series peaked in 1981 and have slowly trended downward through 2006 to about 300 million acres, with reportedly a slight increase since 2006. There has been concern over land-use change associated with the expansion of the biofuel industry. The continuous reallocation of existing cropland along with productivity growth has supported increased output even though overall cropland acres are decreasing in the United States. However, that may not be the case in other parts of the world.

Changes in production levels follow changes in demand for and net returns to certain crops. The major field crops in terms of harvested acreage are corn, soybean, wheat, and hay (Figure 4-8). As Figure 4-8 indicates, domestic acreage for corn and soybean has been increasing, hay acreage has been relatively constant, and wheat acreage has been declining.

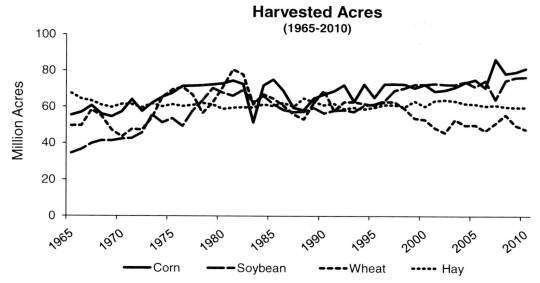

FIGURE 4-8 Harvested acres of corn for grain, soybean (all), wheat, and hay (all) from 1965 to 2009. DATA SOURCE: USDA-NASS (2010).

These adjustments are in response to differences in relative commodity prices and differences in yield and productivity growth affecting net returns on these crops over time.

When considering agricultural commodity prices over time, media sources frequently refer to nominal prices, abstracting from the temporal impacts of inflation on prices. Although substantial nominal price shocks occurred in the mid-1970s and again in the beginning of 2006, it is important to remove inflationary impacts on prices or to convert from nominal to real prices by using an appropriate deflator as presented in Figure 4-9. Despite short-run price shocks, real commodity prices have been decreasing over the long run as a result of total factor productivity growth in the agricultural sector. Real price shocks for corn, soybean, and wheat were concurrent with higher oil real prices in the mid-1970s. Beginning in 2006, real commodity prices have tended to demonstrate increased fluctuation but modest overall increases.

Several attempts have been made to link real price changes to increased feedstock demand for biofuels, but results vary significantly between studies (see also the section "Food Prices" later in this chapter). U.S. demand for corn as an ethanol feedstock accounts for more than 40 percent of crop use (even though one-third of the grain weight is returned as a feedstuff source in dried distillers grains with solubles [DDGS]). All other things equal, corn prices will increase if there is increased corn demand for ethanol production. However, the magnitude of the price effect is not clearly established (see also "Effect of Short-term Price Spikes on Livestock Producers" in this chapter) and depends on several factors, such as biofuel expansion, drought, flooding, crop failures, exchange rate shifts, government price supports, and trade restrictions. In the short run, increased corn feedstock demand may cause a substantial corn price shock, but, in the long run, production resources will shift to increase corn supply and moderate price increases.

Figures 4-10 and 4-11 provide an indication of how production has shifted in the United States in the three major commodity crops from the 1965 to 2010 crop years. The production

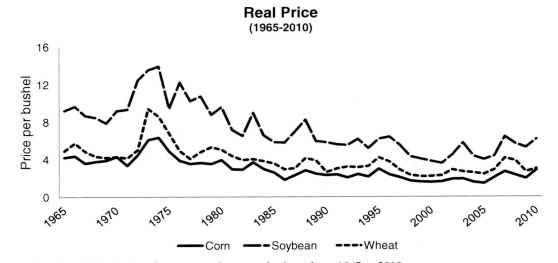

FIGURE 4-9 Real prices for corn, soybean, and wheat from 1965 to 2010 crop year.
NOTE: 1990-1992=100.
DATA SOURCE: Price index for Producer Prices Paid Index from USDA-NASS and USDA-ERS.

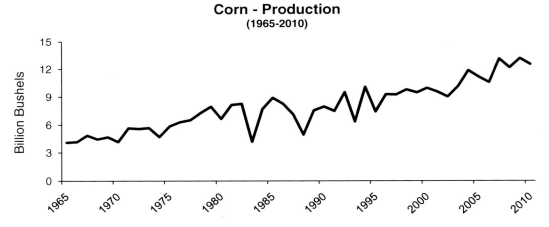

FIGURE 4-10 U.S. corn-grain production from 1965 to 2010 crop year.
DATA SOURCE: USDA-NASS (2010).

of corn and soybean has been increasing significantly over time, while wheat production peaked in 1981 and has been slowly declining, although with significant annual fluctuation up to the 2010 crop year.

At the same time, Figure 4-12 indicates that domestic consumption of corn has increased significantly. The increase in corn production began in 1975 (well before the demand for biofuel). Thus, in addition to the demand for biofuel, the increase in domestic consumption of corn can be attributed to increases in feed and residual use (USDA-NASS, 2010). Domestic consumption of soybean, possibly affected by the demand for soybean oil as a biodiesel feedstock, has held steady. Domestic wheat consumption has remained relatively flat in recent years.

Another perspective on how biofuel production is affecting the domestic and global markets is to view patterns in U.S. net exports of corn, soybean, and wheat. It could be argued that net exports would decline, especially with increasing domestic consumption of corn, and to a lesser extent of soybean oil, for biofuel feedstock. Figure 4-13 indicates that

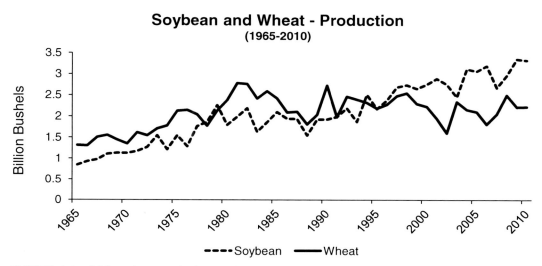

FIGURE 4-11 U.S. soybean and wheat (all) production from 1965 to 2010 crop year.
DATA SOURCE: USDA-NASS (2010).

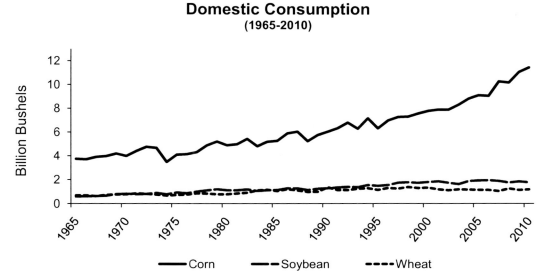

FIGURE 4-12 U.S. domestic consumption of corn, soybean, and wheat from 1965 to 2010 crop year.
DATA SOURCE: USDA-NASS (2010).

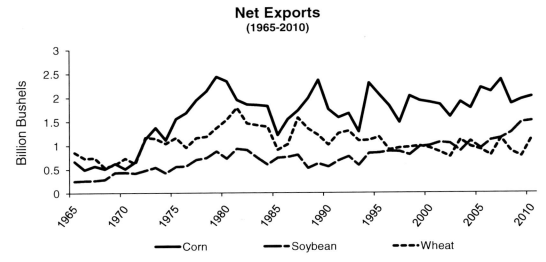

FIGURE 4-13 U.S. net exports of corn, soybean, and wheat from 1965 to 2010 crop year.
NOTE: Total year exports minus total year imports.
DATA SOURCE: USDA-FAS (2010).

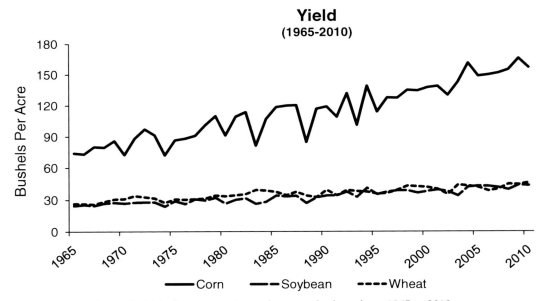

FIGURE 4-14 Annual yields for corn grain, soybean, and wheat from 1965 to 2010 crop year.
DATA SOURCE: USDA-NASS (2010).

net corn exports have held fairly steady, net soybean exports have actually increased, and wheat exports have declined, while yields have been steadily increasing for all three crops, as shown in Figure 4-14.

What accounts for the increase in corn and soybean production and the gradual decline in wheat production? Essentially, market forces determine the allocation of resources.

Producers select the most profitable combination of crops to produce on the cropland acres they farm. The prices, yields, seed technology, and government programs make corn and soybean more profitable than wheat. In addition, income improvement and diet adjustments in developing countries create growing demand for feed grains and oilseeds to produce animal products as well as biofuels.

In summary, the committee made the following observations on what is happening in U.S. agricultural commodity markets and resource use. First, cropland acreage has been declining slowly over time. Second, shifts in acreage for grain and oilseed crops were under way before the advent of biofuel expansion even though biofuel expansion and growing export demand have encouraged the shift in recent years.

Food Prices

The diversion of land to corn production and a greater demand for corn from the biofuel industry discussed in the previous section coincided with an aberrant rise in food prices in the mid-2000s. Between 2004 and 2008, the price of the staple commodities (wheat, corn, soybean, and rice) grew an average of 102 percent (Figure 4-15) (IMF, 2010). Even though real prices had been at an all-time low in the late 1990s and early 2000s (Babcock et al., 2010), the rapid nature of the increase was disruptive to food processors and to households. The Food Consumer Price Index (CPI), calculated by USDA's Economic Research Service, increased from 2.4 percent in 2006 to 4 percent in 2007 and grew a further 5.5 percent in 2008 (USDA-ERS, 2011b). Food banks and international development organizations expressed

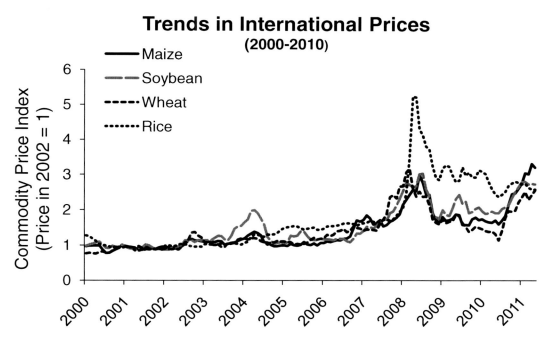

FIGURE 4-15 Trends in real international prices of key cereals: 1960 to May 2008.
SOURCE: Adapted from Abbott et al. (2011).

particular concern for households that allocated a large percentage of income to food (Crompton, 2008; Lustig, 2008; Reinbold, 2008; von Braun, 2008; Wiggins and Levy, 2008).

During this commodity price spike, which peaked between 2007 and 2009, controversy ensued over the role of increased ethanol production in increased food prices. However, much of the debate used the term "food prices" in an imprecise and often contradictory manner. Specifically, some analysis of that period focused on the effect of ethanol production on raw agricultural commodity prices at either the farm level or the international market level. Other analysis focused on the effect of ethanol production on prices of processed food products at the consumer retail level. Frequently, both types of analysis were reported to the public under the label of "the effect on food prices."

As discussed below, the nature of the U.S. food marketing system implies that changes in agricultural commodity prices and changes in retail food product prices do not correlate on a 1:1 basis. Much of the confusion in that debate, and the wide range of estimated effects of ethanol production on "food prices" during that period, was due to these uses of imprecise terminology. Consequently, the remainder of this section uses specific terminology to discuss the potential price effects of expanded biofuel production. First, the term "agricultural commodity prices" refers to the prices of raw agricultural products at the farm or international market level. Second, the term "retail food prices" refers to the prices of consumer food products at the grocery retail level.

Effects on Agricultural Commodity and Retail Food Prices: Lessons from 2007-2009

Estimates of ethanol's influence on global agricultural commodity prices during the 2007-2009 period were as high as 70 percent (Table 4-4). Determining the extent to which biofuel production affected agricultural commodity and retail food prices is difficult because most prices at the time were also influenced by the high price of oil, greater speculation activity in commodity markets, the changing value of the dollar relative to other currencies, drought in some major production regions, export restrictions imposed by some countries, and more demand for food from the growth in population and incomes in developing countries (Trostle, 2008; Baffes and Haniotis, 2010). Some combination of these events is likely to continue to influence prices. Though the increase in the Food CPI dropped to historically low levels (1.8 percent for 2009 and 0.8 percent for 2010, the lowest rate since 1962 [USDA-ERS, 2011b]), food prices are still much higher than they were at the beginning of the decade (IMF, 2010), and food inflation in 2011 is projected to return to the historic average of between 2 and 3 percent (USDA-ERS, 2011b).

Furthermore, because of the interrelationships of agricultural commodity markets and the competition for production resources among agricultural commodities, a price change in one agricultural commodity can affect prices in other agricultural commodity markets (see "Agricultural Commodities and Resources" above). Thus, any secondary price effect needs to be taken into account in the analysis of the effects of ethanol on commodity or retail prices. Also, the magnitude of price changes at the farm, international market, or retail level resulting from increased ethanol production is determined by the complex nature of the food marketing system and the transmittal of price changes through that system. Thus, though price changes at each level of the food system are jointly determined, the size of the price changes at each level may differ.

The range of agricultural commodity price increases assigned to increased ethanol production tended to decrease with the passage of time as additional data became available and more accurate analysis could be conducted (Abbott et al., 2009; Baffes and Haniotis,

TABLE 4-4 Estimates of Effect of Biofuel Production on Agricultural Commodity Prices, 2007-2009

Author	Coverage and Key Assumptions	Key Effects of Biofuels on Agricultural Commodity Prices
Banse et al. (2008)	2001-2010; Reference scenario without mandatory biofuel blending, 5.75% mandatory blending scenario (in EU member states), 11.5% mandatory blending scenario (in EU member states)	Price change under reference scenario, 5.75% blending and 11.5% blending, respectively: Cereals: –4.5%, –1.75%, +2.5% Oilseeds: –1.5%, +2%, +8.5% Sugar: –4%, –1.5%, +5.75%
Baier et al. (2009)	24 months ending June 2008; historical crop price elasticities from academic literature; bivariate regression estimates of indirect effects	Global biofuel production growth responsible for 17%, 14%, and 100% of the rises in corn, soybean, and sugar prices, respectively, and 12% of the rise in the IMF's agricultural commodity price index.
Lazear (2008)	12 months ending March 2008	U.S. ethanol production increase accounted for 33% of the rise in corn prices. U.S. corn-grain ethanol production increased global food prices by 3%.
IMF (2008)	Estimated range covers the plausible values for the price elasticity of demand	Range of 25-45% for the share of the rise in corn prices attributable to ethanol production increase in the United States.
Collins (2008)	2006/07-2008/09; Two scenarios considered: (1) normal and (2) restricted, with price inelastic market demand and supply	Under the normal scenario, the increase in ethanol production accounted for 30% of the rise in corn price. Under the restricted scenario, ethanol could account for 60% of the expected increase in corn prices.
Glauber (2008)	12 months ending April 2008	Increase in U.S. biofuels accounted for about 25% of the rise in corn prices; U.S. biofuel production accounted for about 10% of the rise in IMF global agricultural commodity price index.
Lipsky (2008) and Johnson (2008)	2005-2007	Increased demand for world biofuels accounts for 70% of the increase in corn prices.
Mitchell (2008)	2002–mid-2008; ad hoc methodology: effect of movement in dollar and energy prices on food prices estimated, residual allocated to the effect of biofuels	70-75% of the increase in agricultural commodities prices was due to world biofuels and the related consequences of low grain stocks, land use shifts, speculative activity, and export bans.
Abbott et al. (2009)	Rise in corn price from about $2 to $6 per bushel accompanying the rise in oil price from $40 in 2004 to $120 in 2008	$1 of the $4 increase in corn price (25%) due to the fixed subsidy of $0.51 per gallon of ethanol.
Rosegrant (2008)	2000-2007; Scenario with actual increased biofuel demand compared to baseline scenario where biofuel demand grows according to historical rate	Increased biofuel demand is found to have accounted for 30% of the increase in weighted average grain prices, 39% of the increase in real maize prices, 21% of the increase in rice prices, and 22% of the rise in wheat prices.
Fischer et al. (2009)	(1) Scenario based on the IEA's WEO 2008 projections; (2) variation of WEO 2008 scenario with delayed 2nd generation biofuel deployment; (3) aggressive biofuel production target scenario; (4) variation of target scenario with accelerated 2nd generation deployment	Increase in prices of wheat, rice, coarse grains, protein feed, other food, and nonfood, respectively, compared to reference scenario: (1) +11%, +4%, +11%, –19%, +11%, +2% (2) +13%, +5%, +18%, –21%, +12%, +2% (3) +33%, +14%, +51%, –38%, +32%, +6% (4) +17%, +8%, +18%, –29%, +22%, +4%

SOURCE: Timilsina and Shrestha (2010).

2010; Timilsina and Shrestha, 2010).[12] In addition, given the wide range of scenarios used by the authors summarized in Table 4-4 and the level of uncertainty about future scenarios, a precise estimate of the effect of expanded biofuel production on agricultural commodity prices is likely to be impossible. Instead, a range of possible price effects during 2007-2009 is probably most instructive in understanding the potential effect of biofuel production on agricultural commodity prices. For purposes of this analysis, a range of 20 to 40 percent increase in agricultural commodity prices is used.

The next step in analyzing the effect of expanded biofuel production on retail food prices is to convert the change in agricultural commodity prices (20-40 percent) into a change in retail grocery prices paid by consumers. This conversion depends critically on the size and nature of the marketing process between the farm level and the retail level. The cost of this process, typically defined as the "marketing margin" of food products, includes the cost of all processing, transportation, and distribution activities that occur between the sale of agricultural commodities at the farm level and the purchase of a consumer food product at the retail level. The marketing margin on an agricultural commodity is typically measured as "the difference between the price paid by consumers and that obtained by [farm-level] producers" for that quantity of the agricultural commodity contained in the consumer food product (Tomek and Robinson, 1983, pp. 120-122). To the extent that an agricultural commodity undergoes a greater degree of processing before reaching the final consumer or is costlier to transport and distribute, that agricultural commodity would have a greater marketing margin when measured as the share of the price paid by the consumer. Only if the marketing margin approaches zero would the full effect of an agricultural commodity price change be transmitted to the final consumer on a one-to-one basis (Gardner, 1975, 1987).

Those agricultural commodities most affected by the production of biofuels in 2007-2009 (primarily corn, soybean, and wheat) typically undergo a high degree of processing before reaching grocery consumers in the United States. Thus, the marketing margin on those agricultural commodities tends to be relatively high. Many corn-based consumer products (for example, corn flakes or corn syrup) had a marketing margin of 95 percent, while wheat-based consumer products (for example, bread or bakery products) had a marketing margin of approximately 90 percent, and soybean-based products (for example, shortening or margarine) had a marketing margin of 84 percent in 2006. On average, including all other food products, the marketing margin for all agricultural products has been approximately 81 percent in recent years (USDA-ERS, 2011c).

Consequently, in determining the increase in the consumer retail price of a food product that results from an increase in the price of the agricultural commodity contained in that food product, both the increase in the price of the agricultural commodity and the marketing margin of the consumer food product have to be considered. In the case of corn, for example, if the range of a 20 to 40 percent price increase for corn is used with a marketing margin of 95 percent, then the retail price of grocery food products containing corn would

---

[12]This section is based on the recent economic literature as of 2011 on the effect of ethanol production on agricultural commodity prices and retail food prices during 2007-2009. Much of the analysis conducted at that time suggested that the price effects of increased ethanol production were larger than the subsequent analysis. This is likely the result of the additional or improved data available to researchers during 2009-2010 and is not a reflection on the quality of the earlier analysis.

be a 1 to 2 percent increase in price at the retail level.[13] In the case of animal products, the marketing margin is lower relative to other foods. In 2010, the marketing margins for chickens (broilers), pork, and beef averaged 58, 69, and 54 percent (USDA-ERS, 2011c). Feed is the dominant cost in producing animal products. For broilers, feed costs are 69 percent for the cost of meat production (see "Feed Prices and Animal Production" below). Given that a broiler diet is predominantly corn and that the price of the major ingredients shadows that of corn (Donohue and Cunningham, 2009), the impact of biofuel production on broiler feed prices is likely to be in the range of the 20-40 percent for agricultural commodities summarized in Table 4-4. Considering the marketing margin and the contribution of feed costs to animal production costs, the impact of biofuels on the retail price of broiler meat during 2007-2009 is likely to have been in the range of 5.8 to 11.6 percent.[14] The actual increase will vary with the time span and local market conditions.

Livestock product prices also are complicated by the fact that each livestock sector (beef, pork, chicken, eggs, and dairy) has a different cycle (Abbott et al., 2011). Beef has the longest cycle (up to 2 years) followed by pork and poultry. The retail prices of beef and pork rose by 11 and 14 percent, respectively, from 2008 to 2011. The price increases in 2011 began in 2008 to 2009 when livestock producers began reducing herd sizes as a result of lower profitability at the high commodity prices reached in 2008. Reduction in herd sizes takes time and so does the consequent rise in retail prices.

Another measure of the effect of ethanol on retail food prices is the effect on the CPI. As before, such a measure would be determined by the change in the agricultural commodity price at the farm level and the marketing margin of a retail food product containing that commodity. In addition, however, the effect of a change in agricultural commodity price on the CPI also depends on the weight of that retail food product in the "representative market basket" used to measure changes in retail food prices paid by consumers (the sum of all weights in the food basket has to be 100 percent). For example, the weight on purchases of "cereals and products" is 4.5 percent, indicating that consumers spend 4.5 percent of their total food expenditures on those products.[15] Thus, the 20 to 40 percent increase in the price of an agricultural product such as corn would result in a 2 to 4 percent increase in the prices of corn-based food products at the retail level. This would result in an increase of 0.045 to 0.090 percentage points in the Food-Consumed-At-Home CPI (Capehart and Richardson, 2008).[16]

---

[13]Calculated as the percent increase in the price of the agricultural commodity * (1.0 – the marketing margin on corn-based food products) (Gardner, 1975, 1987). Thus, if the increase in the price of the agricultural commodity corn is 20 percent and the marketing margin on corn-based food products is 95 percent, then 20 * (1 – .95) = 1.0 percent increase in retail level prices of corn-based food products. Similarly, using the 20 to 40 percent increase in agricultural commodity prices for wheat and soybean would yield a retail food price increase of 2.0 to 4.0 percent for wheat-based retail food products and 3.2 to 6.4 percent for soybean-based retail food products. Other researchers have come to similar conclusions using different assumptions about agricultural commodity price increases or food product marketing margins or by using different estimation methods (Jensen and Babcock, 2007; Leibtag, 2008; Perrin, 2008; Trostle, 2008; CBO, 2009). It should also be noted that this is a measure of the effect on the prices of food consumed at home. The marketing margin for food consumed away from home is likely to be larger than the marketing margin for food consumed at home. Thus, an increase in a given agricultural commodity price would be expected to result in a smaller percentage increase in the price of food consumed away from home.

[14]Calculated as (1 – the marketing margin of 0.58) * 69 percent of production costs due to feed * 20-40 percent increase in the feed costs due to biofuel production.

[15]This category includes flour and prepared flour mixes, breakfast cereals, rice, pasta, and cornmeal (Capehart and Richardson, 2008).

[16]Calculated as 20 percent * (1 – .95) * .045 = 0.00045. This would be the effect of a change in corn prices on the CPI-Food-At-Home Index. If wheat prices, soybean prices, and meat and poultry prices are affected by the increase in corn prices, then the total effect of an increase in corn prices would be the sum of all of these individual price changes.

This overview of the price effects on U.S. retail food prices from the 2007-2009 increase in agricultural commodity prices provides important lessons in considering the economic consequences between 2011 and 2022. First, at any point in time, a multitude of factors affects movements in agricultural commodity prices. The expansion of ethanol production was not the only cause of the agricultural commodity price increase in 2007-2009 and is not likely to be a major factor determining price movements. For example, substantial increases in demand for livestock products in developing countries witnessing rapid economic growth, such as China, are a major driver of feed grain and meat exports (Roland-Holst, 2010). As incomes increase, the income elasticity of demand for high-valued food products is high, driving feed, livestock, and other food prices higher. Second, the effects on agricultural (farm-level) commodity prices and consumer retail food prices need to be defined clearly in examining the price effects on expanded biofuel production under RFS2. Inaccurate definitions of the effect of biofuel production on "food prices" yield inaccurate and misleading information about economic consequences and an increased likelihood of mistaken policy reactions.

Feed Prices and Animal Production

The price of feed dominates the cost of the production of animal products. For example, in 2008 feed ingredient costs were 69 percent of live-production costs of broiler chickens (Donohue and Cunningham, 2009). Grains are the primary energy feedstuffs, and oilseed meals are the primary protein feedstuffs in concentrates fed to pigs, poultry, dairy cows, and feedlot cattle. Corn accounted for about 94 percent of grains fed to animals in 2009-2010 with sorghum, wheat, oats, and barley making up the remainder (USDA-ERS, 2011a). The reason that corn dominates the energy component of animal feeds is that the yield of usable energy (that is, calories of metabolizable energy) per acre of land is more than double that from other grains.

The relationship between the cost of a feed ingredient, such as corn, and the final cost of a nutritionally complete feed depends on many factors, such as switching to cheaper substitutes, adjusting the nutrient density of the feed, and decreasing product quality (for example, marbling of beef). Animal nutritionists use least-cost feed formulation programs to optimize these adjustments to maximize profits. When actual least-cost complete ration compositions and commodity costs in the Northeastern United States were used to capture the adjustments in feedstuff choices that livestock producers made to maximize returns during the commodity price spike in 2007, it was found that each $1 per ton increase in the price of corn increases feed costs by $0.59, $0.50, $0.67, and $0.45 per ton for dairy, hogs, broilers, and layers, respectively (Schmit et al., 2009).

Actual data from U.S. broiler producers during the period of time that ethanol underwent rapid expansion illustrate the extent to which feed costs increased as a result of increased commodity prices. In May 2005, broiler feed averaged $156 per ton. It increased to $284 per ton in May 2008 and $335 per ton in May 2011 (Collett and Villega, 2005, 2008, 2011). A key question is how much of this increase was due to government mandates and blender tax credits. A recent study (Babcock and Fabiosa, 2011) partitioned the cause of the increase in corn prices during the price spike between 2005 and 2009 into three causes: those due to nonethanol factors, those due to the increase in ethanol production from all other market causes, and those due to the increase in ethanol production caused by mandates and tax credits. That study found that the increase in ethanol production contributed 36 percent to the average increase in corn prices (which is toward the high end of the 20-40 percent bracket used in the report) but found that government policies that resulted from

EISA and RFS2 contributed only 8 percent to the increase. The remaining 28 percent of the increase was due to increases in ethanol production caused by other market forces, including inflated demand from the ban on methyl tertiary butyl ether and from periods of high oil prices that markedly increased ethanol prices. This analysis suggests that only around 8 percent of the increase in livestock feed prices between 2005 and 2009 can be attributed to EISA and RFS2, while the remaining 92 percent was caused by other market forces.

Effect of Corn-Grain Ethanol and Soybean Biodiesel

As of 2010, about 40 percent of the corn used in the United States is fed to animals, and about 40 percent is fermented for fuel ethanol production. One-third of the mass of corn grain used for ethanol comes out as DDGS, which also is fed to animals (USDA-ERS, 2011a). The price of other grains closely tracks the price of corn, and the proportions of various grains fed to livestock have not changed appreciably during the rapid rise in corn use for ethanol. This is largely because corn acreage and yields have increased in concert with higher total demand, resulting in no net decrease in the amount of corn used for animal feeding (see Figures 2-3 and 2-5). When coproducts from ethanol production are included, the total supply of corn-based feedstuffs continues to increase even as greater quantities of corn are diverted to biofuels.

Until 2015, most of the increase in U.S. biofuel production will be conventional, corn-grain ethanol. Between 2006 and 2015, full implementation of the RFS2 mandate, simultaneous with implementation of European Union (EU) biofuel mandates,[17] is expected to increase coarse grain prices in the United States by 12.6 percent (Taheripour et al., 2010). Biofuel production also raises returns to cropland, which may in turn encourage conversion of some pastureland to crops or, alternatively, lead to more intensive use of existing cropland coupled with high-yielding varieties to enhance coarse grain production. If pastureland is the primary source of land that supports the increased feedstock production, one study projected 7.55 million hectares (18.7 million acres) of land would be converted from pasture to crop production in the United States between 2006 and 2015 (Taheripour et al., 2010). Loss of pastureland would increase the cost of production of cattle and sheep and likely cause a shift to more intensive production systems, increased fertilization and other input use on remaining pasturelands, and increased time in feedlots.

The demand for feed-grade vegetable and animal fats has increased as a result of their use as biodiesel feedstock and as an energy substitute for corn. This has resulted in price changes for feed fats that mirror that of corn (Donohue and Cunningham, 2009).

Effect of Short-Term Price Spikes on Livestock Producers

As has been discussed, biofuels are only one of many factors influencing commodity prices. However, the biofuel market competes directly with the livestock market for feedstuffs. Because purchasing and investment decisions are based on the production cycle of the animal, short-term spikes in grain prices caused by the competition between markets can place financial stress on livestock producers, particularly those of animals with longer production cycles or less flexible diets.

---

[17]The EU has approved a mandate requiring 10 percent of transportation fuels to be derived from biofuels by 2020. Hertel et al. (2010) estimate that, by 2015, ethanol will be 1 percent of EU transportation and biodiesel will be 5.25 percent.

Demand for corn-grain ethanol is driven by oil prices (especially when oil prices are high), government mandates and environmental regulations, and biofuel incentives. These factors make the demand for corn-grain ethanol production more inelastic and may increase the sensitivity of corn prices to supply side shocks, such as those due to weather or disease, by as much as 50 percent (Hertel and Beckman, 2010). Livestock producers have some capacity to accommodate short-term price spikes by switching to other energy feedstuffs or choosing to feed lower energy diets, but these options vary among livestock species. As costs of grains and other high-energy feedstuffs increase, cattle may be fed an increasing portion of pasture grass, hay, silages, and waste products from human food production. Such diet changes could diminish the grade of meat and fat content of milk. Nonruminants like poultry, swine, catfish, and tilapia are more reliant on grains and have a much more limited ability to use other energy sources than ruminants. Consequently, nonruminants will continue to rely on grains, especially in the short run, even as grain prices increase. For example, a typical U.S. broiler diet is composed of about 60-percent corn and 25-percent soybean meal. This ratio held steady even during the 2 years from October 2006 to October 2008 when feed costs increased by about two-thirds (Donohue and Cunningham, 2009). During this 2-year period, the feed ingredient component of broiler production costs increased from $0.13 per pound to $0.31 per pound live weight produced. This translates into an 80-percent increase in total live-production cost. The cumulative effect of the increased feed costs to the broiler industry exceeded $7.8 billion during those 2 years (Donohue and Cunningham, 2009).

The animal producer's ability to pass increased production costs in the short run on to consumers is limited because increased prices of animal products decrease the quantity demanded. Furthermore, the reproductive pipeline makes it difficult for producers to quickly respond to increased feed costs by reducing animal numbers. The time between breeding parent stock to retail sales of fresh product from the resulting offspring ranges from 10 weeks for broiler meat to about 10 months for milk and pork to about 30 months for beef. This production lag means that beef products consumed today are based on production decisions made more than 2 years ago, and spikes in the price of corn in the interim markedly affect producers' profits.

Ethanol Coproducts

Fermentation of a bushel of corn (56 pounds) using the dry-mill process yields about 2.7 gallons of ethanol and about 17.5 lbs of DDGS that contains 10-percent moisture. This coproduct is richer in protein, fat, minerals, and fiber relative to corn. Because of the high fiber content, ruminants can use higher amounts in their diets than poultry or swine. Sales of DDGS account for about 16 percent of the industry's revenues (Taheripour et al., 2010), and the added-to profit has become critical in maintaining biorefineries' economic viability during times when ethanol prices are low. The wet-mill process yields 11-13 lbs of corn gluten feed, 2.6 lbs of corn gluten meal, and 1.6 lbs of corn oil, all of which can be used as feed ingredients.

The coproducts from corn fermentation decrease the impact of diverting corn from the livestock feed market to ethanol production by almost one third. The proportion of the concentrate component of livestock diets contributed by DDGS increased from 1.3 to 10.3 percent from 2001 to 2008 (Taheripour et al., 2010). Currently, use in animal feeds is able to absorb all high-quality coproducts produced in the United States, although exports of DDGS have increased as well. Assuming corn-grain ethanol production using dry milling increases to the maximum amount permitted by RFS2 (15 billion gallons in 2015), about 98 billion lbs of DDGS would be produced from 5.6 billion bushels of corn. When priced appropriately, this amount of DDGS can be easily absorbed by the animal feeding industry in the United States using existing technology. In addition, there is growing demand internationally for DDGS, and the export market has considerable room to expand.

The use of DDGS in livestock feeds has important potential regional economic effects. DDGS has to be dried before it can be transported long distances, adding to feed costs. Feeding DDGS in a wet form to cattle and hogs eliminates this additional expense and improves the economics of using DDGS. Relocation of large-scale livestock producers to the proximity of corn-grain ethanol producers has occurred, and this trend will likely continue.

Biodiesel produced from oilseeds, such as soybean or sunflower, leaves behind a protein-rich meal that is an excellent feedstuff for poultry, pigs, and dairy cattle. The supply, and consequently price, of this coproduct will likely be affected by the combined effect of RFS2 and EU mandates. Oilseed meal prices are also likely to be depressed by rapidly expanding DDGS availability. DDGS averages up to 30-percent protein content and can substitute for oilseed meals in dairy and swine feeds. However, as of 2010, DDGS was priced more closely with corn than oilseed meals.

Effect of Cellulosic Biofuels

The extent to which cellulosic and other second-generation biofuels raise the cost of feedstuffs fed to animals depends greatly on the mix of feedstocks used. Many potential feedstocks (for example, perennial grasses and short-rotation woody crops) will likely be grown on existing pasturelands. The yields of perennial grasses grown specifically for bioenergy feedstocks are considerably higher on the prime agricultural land that is currently used to grow grains and oilseeds. Although a policy goal of RFS2 is to prevent production of cellulosic feedstock from interfering with feedstuffs and food crops, land conversion or land-cover change could happen. Animal feed costs will likely increase in proportion to the extent that the production of second-generation feedstocks occurs on lands that produce feed grains or were previously in pasture. A USDA study (Gehlhar et al., 2010) was undertaken to examine the effect of full implementation of RFS2 on a variety of key economic components using the U.S. applied general equilibrium (USAGE) model modified to capture conventional ethanol production, second-generation ethanol production from dedicated energy crops, other advanced biofuels, and land allocation for feedstock production. This study assumed feedstock production would occur on land that was previously in crops when such a change would be economically advantageous. Although the methodology of this study is not completely documented, the results suggest that full implementation of RFS2 would result in an additional 3-5 percent increase in corn prices by 2022 and would have only a minor impact on the price of concentrate feedstuffs (Gehlhar et al., 2010).

Cellulosic biofuel production does not result in appreciable amounts of coproducts that have feeding value for livestock (Chapter 2). Thus, the significant mitigating effect of coproducts on livestock feed prices observed with DDGS in corn-grain ethanol and soybean biodiesel will not be applicable to these new fuel sources. However, cellulosic dedicated energy crops may take land that previously was used to graze cattle, thereby limiting the availability of that resource for the livestock sector.

Wood Products

Timber prices have risen 2.7-3 percent per year since the early part of the last century (Figure 4-16) (Haynes, 2008). This long-term rise in prices preceded recent policies that support biofuels and suggests continuing scarcity in wood resources over time, although the price increase may have been exacerbated by the depletion of old-growth stocks first

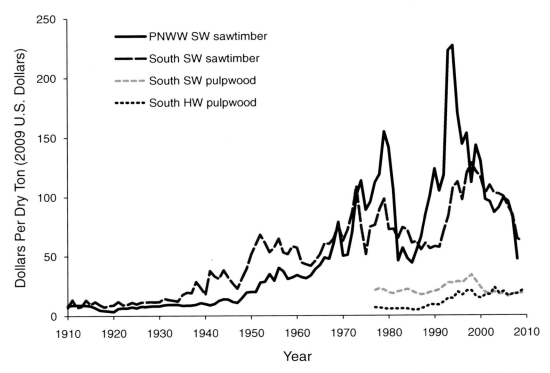

FIGURE 4-16 Historical U.S. timber stumpage prices for the Pacific Northwest west-side of the Cascades softwood sawtimber (PNWW), Southern softwood (SW) sawtimber, Southern softwood pulpwood, and Southern hardwood (HW) pulpwood.
NOTE: Prices are real prices deflated by the all-product producer price index (1982=100). The stumpage price is the value of wood standing in the forest before it is cut.
DATA SOURCES: PNWW SW sawtimber prices from Sohngen and Haynes (1994), Haynes (2008), and Warren (2010); Southern SW sawtimber from Haynes (2008) and Timber-Mart South (2010); Southern SW and HW pulpwood prices from Timber-Mart South (2010).

on private and then on public land in the Pacific Northwest. Nevertheless, given the long time lags between planting trees and harvesting them, continuing changes in the types of products demanded, technological change, and competition with international supplies, it has proven difficult to fully balance supply (for example, investments) and demand in wood resources in the United States over time. Despite the long-term trend upward, timber prices have fallen substantially since their highs in the 1990s and early 2000s.

Current Market for Wood as Energy

In recent years, up to 132 million dry tons (250 million m³) of roundwood equivalent[18] have been used to produce energy, though as electricity, not transportation fuel (Figure 4-17). This includes industrial roundwood used directly to produce energy as well as residues, black liquor from the pulping process, and fuelwood harvested from the forest. Industrial energy users in Figure 4-17 are mostly pulp mills and other large integrated wood

---

[18]For definitions of wood products, please see Appendix D.

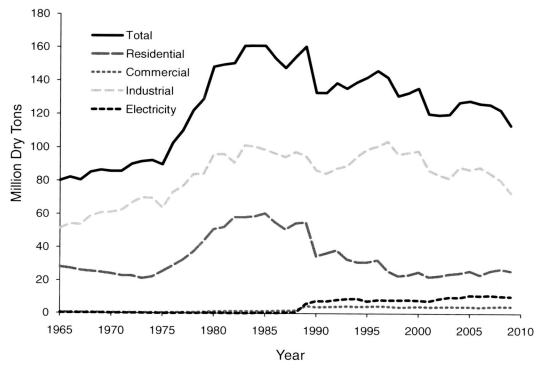

FIGURE 4-17 Wood used in energy.
NOTE: Conversion from Btus assumes 1 Quadrillion Btus = 4 billion ft³, following Howard (2007).
SOURCE: EIA (2010b).

producers that have boilers installed to use residues from the milling or pulping process. These sources of demand for wood were fairly steady from the 1980s to recently. They have declined as overall output in the wood products industry has declined. Wood used in electricity has increased in recent years as energy companies have cofired wood with coal to reduce emissions for air-quality regulations and as additional states have implemented Renewable Portfolio Standards that allow wood to be used for renewable electricity.

Residential wood use for energy initially declined after peaking in the early 1980s but has recently increased. For the most part, residential wood use is in the form of firewood for home heating. Recently, homeowners have increasingly been using wood pellets in wood-burning stoves, and pellets have been used in some industrial processes. Spelter and Toth (2009) estimated that there are about 850,000 heating stoves in the United States and that these and other sources demand more than 2.3 million dry tons of heating pellets. At the time of the publication of this report, pellets were made mostly from sawdust and other residue from milling of lumber and production of plywood. Between 1998 and 2006, the total quantity of sawdust and other residues used in industrial wood production was about 13 million dry tons, and the total supply was about 20 million dry tons. This gap, however, has narrowed in recent years as the supply of residues in the United States has fallen with the decline in wood manufacturing.

Potential Changes Caused by Biofuel Policy

Though many factors beyond biofuel policy affect prices, increased demand for woody biomass due to changes in competition for resources, economic incentives, or technology could have substantial effects on local woody biomass markets and harvesting decisions. Under current conditions, sawtimber is about 3 times more valuable than pulpwood (see Figure 4-16) because of the higher value end-uses it can provide. With any timber harvest, there is a distribution of log sizes. Sawtimber logs typically are the larger logs (more than 7 inches in diameter), and pulpwood logs are the smaller, lower-value logs. Sawtimber products tend to be higher value than pulpwood products and thus are expected to carry higher value. Historically, hardwood pulpwood has been of even lower value than softwood pulpwood, but since the early 2000s, hardwood pulpwood prices have achieved parity with softwood pulpwood as a result of improved technologies for making pulp with hardwoods. Estimates of the value of fuelwood are more difficult to obtain because most of the value lies in the cost of the delivery itself. For example, delivery costs for pulpwood or sawtimber in the southern United States are approximately $35 per dry ton, and fuelwood delivered prices are about $35-$37 per dry ton, according to Delivered Price Benchmark Service from Forest 2 Market, Inc. The difference between the delivery value and the cost of deliver is the stumpage value, which would be about $0-$2 per dry ton. Prices are similar in other parts of the country: for example, New Hampshire reports recent fuel chip stumpage prices of $1-$3 per dry ton (NHDRA, 2010). Pulpwood is the closest marketable commodity that could enter woody biomass markets, and delivered softwood pulpwood prices in the South are substantially higher than fuelwood prices, in the range of $45-$55 per dry ton (Forest2Market, 2010; Timber-Mart South, 2010). These prices represent current conditions. If a commercial woody biomass refinery is built, it would require 1,000-2,000 dry tons of biomass per day to operate efficiently. The competition for resources created by a biorefinery entering the woody biomass market would have profound effects on the local price for woody feedstock.

Economic incentives could raise the profitability of fuelwood and spur additional extraction of residues for market uses, such as cellulosic ethanol production. One way this could be done is by explicitly subsidizing the extraction of residue material (Figure 4-18A). To get a sense for the size of the subsidies necessary, studies in California (Jenkins et al., 2009; Sohngen et al., 2010) found that the cost of removing residues through whole-tree harvesting systems could be as high as $50 per dry ton at the roadside. The costs to deliver 80 or 100 miles (a typical distance for the study site in California by Sohngen et al., 2010) would add $18-$22 per dry ton, suggesting total delivery costs of $72 per dry ton. WTP for this material in California currently is about $30 per dry ton for delivery, suggesting that subsidies would need to be up to $42 per dry ton to induce removals, depending on the region of the country.

A second way in which additional residues may be extracted would be if markets for cellulosic materials began to grow as a result of technological advancement. For example, if the technology for producing cellulosic ethanol improved, then demand for forest-based cellulose would increase from the current low levels, driving up the demand for wood materials in general (assuming that technology does not improve so much that demand for the raw material input falls). This increase in demand for woody biomass in general would influence not only residue recovery but also sawtimber and pulpwood markets. The effect of this increase in demand can be seen in Figure 4-18B, as a shift in the entire demand function. Rising demands for cellulosic materials would have ripple effects through the entire market. Higher values for residues would increase residue collection but would also compete with some lower end pulpwood away from the pulp market. This in turn would

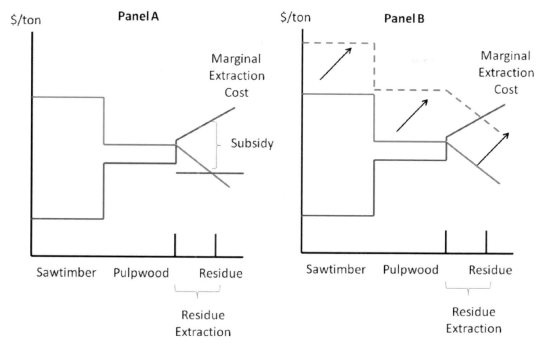

FIGURE 4-18 Effect of subsidy (panel A) and increase in demand (panel B) on extraction of residues from the forest floor for biomass energy markets.

increase pulpwood prices, causing some additional sawtimber at the lower end to be used as pulpwood, which results in higher sawtimber prices. Therefore, though current wood biofuel prices are low compared to pulpwood and sawtimber prices, improving technologies for cellulosic biofuels could raise prices for wood inputs.

The restrictive definition of woody biomass eligible for RFS2 was discussed in Chapters 1 and 2. If technological changes made cellulosic biofuels economically competitive, woody biomass prices could increase because of supply limitations. However, if technologies for cellulosic biofuels do not fully develop and there is no strong demand for cellulosic material, the effects in markets would be modest. In this case, subsidies would be needed to sustain the market (Sedjo, 2010). Figure 4-18, for instance, shows that a subsidy potentially raises the value of the biomass material above the value of pulpwood. Regardless of whether the subsidy is paid to consumers of biomass feedstocks or producers of feedstocks, higher prices for biomass material caused by subsidies could spur conversion of traditional pulpwood supplies to biofuel feedstocks. There is speculation that the initial subsidies in USDA's Biomass Crop Assistance Program raised the value of pulpwood by subsidizing the use of a wide range of forest materials as bioenergy feedstocks (Box 4-4). Thus, if subsidies are large enough, and not well targeted, they can have unintended consequences and strong implications in markets.

The implications of a demand increase due to RFS2 are large. Sedjo and Sohngen (2009) and Sohngen et al. (2010) used a global timber model to examine a case in which all material needed to produce the cellulosic biofuel mandate would be derived from forests by 2022. Rather than assuming that the material would be derived from residues alone, they assumed that the entire mandate would be derived from existing forest resources within the United

BOX 4-4
Biomass Crop Assistance Program

The Biomass Crop Assistance Program (BCAP) was written into the 2008 farm bill (Section 9001 of the Food, Conservation and Energy Act of 2008 [110 P.L. 234]) as a way to support the establishment of bioenergy crops in selected project areas. The program has two parts. One part provides a subsidy to agricultural producers, individuals, or companies who collect and deliver biomass material to production facilities. The law requires that the biomass is certified to have been collected or harvested with an approved conservation plan. The subsidy amount is $1 per dry ton for every $1 per dry ton paid by production facilities for biomass material, up to a maximum of $45 per dry ton. The payment is limited to 2 years. In an effort to avoid shifting material that already has productive uses into this new "market," the law defines eligible material for the subsidy restrictively. Eligible material can be forest material from public land that is taken lawfully and does not have a different market use (for example, material from precommercial thinning or invasive plant treatments) or any renewable organic matter from nonfederal land that is not eligible to receive payments from Title I of the farm bill. Animal waste, food waste, yard waste, and algae are not permitted to obtain payments.

The other part of BCAP provides a subsidy for the establishment of biomass energy crops within specific "BCAP Project Areas." Project areas will be specified by the USDA Farm Services Agency. The payments are for up to 75 percent of the establishment cost and annual payments for up to 5 years for perennial biomass crops and annual payments up to 15 years for woody biomass crops. Most private land qualifies, although land currently in the Conservation Reserve Program (or Wetland or Grassland Reserve Programs) is not permitted to participate.

Between June 2009 and February 2010, enrollment for the part of BCAP that provides payments for collecting, harvesting, storing, and transporting biomass was open and active on a preliminary basis with around $235 million in funding. As of August 2010, these preliminary funds, announced through a Notice of Funds Available, had been spent on around 7.1 million dry tons of material. USDA issued a final rule for BCAP in October 2010.

Many large timber mills registered as BCAP facilities during the enrollment period. Therefore, the bulk of funds has been spent on woody biomass delivered to these facilities. Of the total payments so far, $184 million were spent on woody biomass, $37 million on waste materials, and only about $250,000 on agricultural biomass. Maine, California, Alabama, Georgia, and South Carolina are the top five states, followed closely by Michigan and New Hampshire. With the exception of California, these states tend to be large timber producers and not large crop producers with the types of crops or residues that can be readily harvested for biomass energy. New Hampshire is not a particularly large timber producer, but numerous pellet producers and biomass energy production facilities that use forest inputs for fuel are located there.

To a large extent, the emphasis on the current BCAP payments on forestry activities makes perfect sense, given that forest materials are the most widely available cellulosic bioenergy feedstock. While the technology to convert trees to liquid fuels is costly and not yet commercialized, the technology to convert trees to electricity is available and substantially less costly. Many timber mills already produce electricity with residues from milling operations (including sawdust and black liquor). A number of states now have Renewable Portfolio Standards in place that provide incentives for biomass electricity production. The current data point to the fact that existing boilers using wood for energy have been the largest beneficiary of the subsidies in BCAP so far.

However, there was concern that the subsidy payment for collecting and delivering biomass was creating competition and increasing the price of biomass for other users, such as manufacturers and nurseries. Therefore, it was specified in the final rule that BCAP payments can only be applied to biomass material that cannot be used in higher-value products, such as particle board or composite panels. At the time this report was written, it remained to be seen whether this modification to the payments was implemented effectively (Stubbs, 2010).

Nationally, BCAP subsidies do not appear to have affected timber prices substantially to date. This is largely due to the economic slowdown that has reduced prices for timber in general. Although BCAP is not expected to have substantial effects nationally, some regions could experience important economic effects from increases in fuelwood production. For example, fuelwood stumpage prices have doubled since 2005 in New Hampshire, with the bulk of the increase occurring in 2008 and 2009 as the BCAP program picked up steam (Manomet, 2010).

States (that is, they shift the demand for wood products outward). This is an unlikely scenario in the long run, but it illustrates the potential effects of using only forests to meet the cellulosic biomass mandates. Initially, the cellulosic mandate would use a modest 7.4 million dry tons (14 million m^3) of national timber harvest. Given that the United States consumes about 217 million dry tons (410 million m^3) per year, this is a small increase. By 2022, however, the cellulosic mandate would use 169 million dry tons (320 million m^3) or over 75 percent of the total national timber harvest. Their results indicate that timber prices rise on average by around 5 percent, that U.S. timber production rises by about 7 percent, and that consumption of industrial wood in the United States declines by about 10 percent.

This increase in timber production by 7 percent amounts to only about an additional 15.9 million dry tons (30 million m^3) in timber harvests, not nearly enough to match the 169.3 million dry tons (320 million m^3) requirement for cellulosic markets. Even though industrial wood consumption declines by 10 percent, or about 21.2 million dry tons (40 million m^3), because of higher prices for timber, the United States would have to import additional wood to make up the difference. The model projects that industrial roundwood imports rise 10-fold to meet the timber shortage in the United States caused by the cellulosic biofuel standard.

The Sub-Regional Timber Supply model (SRTS) has also been applied to examine the effects of expanding demand for biomass energy from forest resources in the South (for example, Galik et al., 2009). The SRTS model has substantially more detail than the study by Sedjo and Sohngen (2009) described above and clearly delineates forest residues from other materials; however, the model has not been used explicitly to model cellulosic biofuels. The results for expansion in demand for biomass energy illustrate the implications of rising demand for forest resources.

The specific example considered with the SRTS model is the renewable portfolio standard in North Carolina, which requires 12.5 percent of electricity to be produced with renewable sources, including forests, by 2022 (see Galik et al., 2009). SRTS was used to examine the potential for forest materials, including residues from logging and milling, pulpwood substitution, and new investments in timber resources to be used to meet the renewable portfolio standards.

The authors modeled timber demand and supply in a three-state region, given that supply for the North Carolina market is generated from forests in Virginia, North Carolina, and South Carolina. The wood requirements for this renewable portfolio standard would be as much as 9.4 million dry tons per year from this region, which is nearly as much as the 11.8 million dry tons per year harvested for traditional timber uses (for example, sawtimber and pulpwood). The SRTS model finds that only about 3.5 million dry tons per year (or 38 percent of the total) of this supply could be met with residuals. The rest would have to be met with substitution from other products, higher prices for other products, and increased timber harvesting in the region.

Market Effects Beyond Biofuels

As with food prices, biofuel policy is a complicating factor but not the only force influencing the price of wood products. One example to this effect is the large impact of the economic slowdown in the late 2000s on timber prices in the United States. Although forest output rose globally, timber output fell substantially within the United States (Figure 4-19). The decline in output largely resulted from the precipitous slowdown in housing starts that began in 2006 (U.S. Census Bureau, 2010). After hitting over 2 million in 2005, housing starts fell below 600,000 in 2009. Pulpwood and plywood production also fell, but reductions in these outputs were not as dramatic as lumber output.

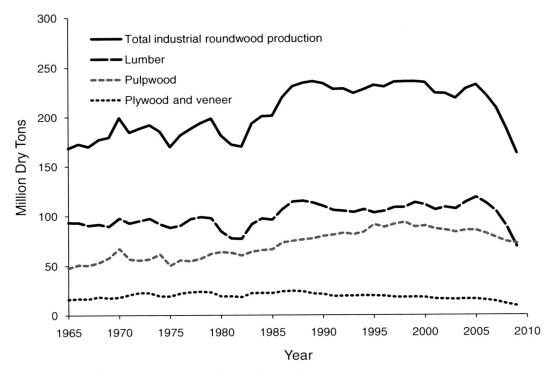

FIGURE 4-19 Industrial wood output in the United States.
SOURCE: Howard (2007); updated data to 2009 as a result of personal communication.

Historically, the United States has been a net importer of industrial wood (Figure 4-20), with about 30 percent of total industrial wood consumption being satisfied by imports in recent years (Howard, 2007). By far, the largest trading partner for the United States is Canada, which provides about 85 percent of total wood imports (Howard, 2007). Imports from Canada grew dramatically after 1990 as output from federal forests in the Pacific Northwest declined, and the closest substitutes in construction were found in Canadian wood (Haynes, 2003). Although Canada is the largest wood products trading partner, the United States is increasingly importing wood from South America. Since 1989, pulpwood imports from South America have risen by about 7 percent per year (USITC, 2010b).

Fossil Fuels

EPA (2010) projected that achieving RFS2 in 2022 will reduce oil imports by 0.9 million barrels per day, a 9.5-percent reduction. Because the United States is the largest consumer of oil, this reduction in demand will lower the world price of oil. According to EPA's model, meeting RFS2 could decrease the price of oil by $1.05 per barrel in 2022 (EPA, 2010).

USDA's Economic Research Service (ERS) has also analyzed the effects of introducing 36 billion gallons of biofuels into the transportation economy. It modeled scenarios in 2022 with $80 per barrel oil and $101 per barrel oil. ERS's results found that, for either price, achieving RFS2 in 2022 would reduce crude oil import prices by about 4 percent, gasoline

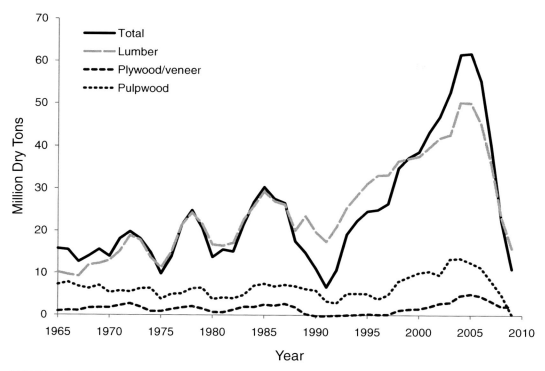

FIGURE 4-20 Net imports of wood into the United States (imports – exports).
SOURCE: Howard (2007); updated data to 2009 as a result of personal communication.

prices by about 8 percent, and the price of motor fuels (gasoline blended with ethanol) by about 12 percent (Gehlhar et al., 2010).

Land Prices

What does the RFS2 mandate mean for land prices? Nearly all the implications of the mandate indicate that land prices will be driven upward. One direct demand factor results from a potential increase in land used for dedicated biofuel crops. For example, if the United States produces 16 billion gallons of cellulosic biofuels by 2022, 30-60 million acres of land might be required for cellulosic biomass feedstock production from forests, pastures, croplands, land from the Conservation Reserve Program, and cropland pasture (land that was once in crops but is not in crops currently) (see Table 4-3). An indirect effect would result from biofuels produced from crop residues because new demand for surplus residue would increase the overall value of land. Thus, although the use of crop residues could reduce that amount of land needed directly for cellulosic feedstock production, the RFS2 mandate still would increase the overall demand for land.

Although it is clear that RFS2 will increase the demand for land and will raise land prices, the exact extent of the effect has not yet been estimated. The use of marginal agricultural land for dedicated bioenergy crops has been proposed as a mechanism to alleviate competition for cropland. Cai et al. (2011) defined marginal agricultural land as land that has "low inherent productivity for agriculture, is susceptible to degradation, and is high risk for agricultural production." Based on this definition, they estimated that about 168 million acres of marginal land are available in the United States. However, the availability of marginal land does not imply that bioenergy crops would be grown on those lands. Swinton et al. (2011) analyzed the increase in crop-planted area during the 2006 to 2009 spike in field crop prices. They found that despite an attendant gain in typical profitability of 64 percent, the area of crop-planted area only increased by 2 percent. Even if the profitability doubles, the area of crop-planted area was projected to increase by 3.2 percent, which is about 7.4-10 million acres. Swinton et al. (2011) reasoned that if farmers are reluctant to expand crop-planted area with familiar crops in the short run, they will even be less likely to expand crop-planted area with less familiar perennial dedicated bioenergy crops, which require longer time for crops to establish than annuals. Given the price gap between WTA and WTP for cellulosic biomass, it seems even less likely that farmers would be willing to expand crop-planted area to grow dedicated bioenergy crops. (See also section "Social Barriers" in Chapter 6.) The size of the increase will differ depending on land and crop productivity, other land uses in the region, and the growth of the biofuel sector locally.

EFFECTS OF BIOFUEL PRODUCTION ON THE BALANCE OF TRADE

Effect on Import and Export of Grains

To the extent that biofuel production leads to or has led to increases in the price of corn and other agricultural commodities, the quantity of these commodities exported could be expected to decrease to the extent that export quantity demanded responded to the higher price. The United States is a major exporter of corn, wheat, and soybeans, as well as some animal products. Higher crop prices eventually would lead to higher livestock product prices and reductions in exports of those products. Since 2002, however, while crop commodity prices were rising, exports of many of these commodities held steady or even increased (see Figure 4-13). The main reason for this occurrence is the huge depreciation in the U.S. dollar between 2002 and 2008. With a lower value for the U.S. dollar, commodity prices did not increase nearly so much in other currencies such as the euro or yen. Therefore, exports were not as affected as would have been expected in the absence of the depreciation in the U.S. dollar.

Livestock Production and Trade

Increased animal product costs as a result of the simultaneous implementation of RFS2 and EU biofuel mandates are expected to decrease the global value of livestock industries by about $3.7 billion (2006$) cumulatively between 2006 and 2015 (Taheripour et al., 2010). Most of this decrease would occur outside the United States, which would observe only a minor reduction ($0.9 billion) in its livestock and processed livestock products. The effect in the United States is buffered by the increasing availability of coproducts from corn-grain ethanol production, especially DDGS (see earlier section "Ethanol Coproducts"). Changes in livestock production are not predicted to occur evenly across species. Ruminants are better adapted to use DDGS than nonruminants and would be affected less. This is reflected in an expected increase in the trade balance for ruminant products of $135 million but a decrease for nonruminant products of $40 million (in 2006$) (Taheripour et al., 2010).

Effect on Import and Export of Wood Products and Woody Biomass

As discussed earlier (see section "Wood Products"), the United States imports a large quantity (over 30 percent) of its wood resources from outside the country (Howard, 2007). Most imports come from Canada, but in recent years, other countries have increased their exports of wood products to the United States. Although the United States is a net importer of wood, it exports some products. One example is wood pellets, which are currently made most often from sawdust obtained from milling operations. This market is relatively small at present but is growing and could continue expanding if demand remains strong in other countries that use pellets for industrial heating (Spelter and Toth, 2009).

Current estimates suggest that the RFS2 mandate would likely increase wood imports into the United States. Wood is the most widely available cellulosic bioenergy feedstock in the United States at present, and it will be an important source of supply for cellulosic biofuel refineries as RFS2 is implemented in the next 11 years. Sedjo and Sohngen (2009) suggested that up to 75 percent of wood currently used by wood producers could shift into biofuel production if the RFS2 mandate pushes supply prices high enough. A shift in industrial wood from traditional uses to biofuels in turn would cause the United States to import more industrial wood from elsewhere. The scale of this effect, however, cannot be precisely estimated at this time.

Effect on Import and Export of Petroleum

Between 2010 and 2022, imports of crude oil are projected to decline slightly, due in part to increased fuel efficiency standards in vehicles and to the RFS2 mandate. As mentioned earlier, EPA (2010) projected that achieving RFS2 in 2022 will reduce oil imports by 0.9 million barrels per day, a 9.5-percent reduction that would save $41.5 billion that year. EPA also estimated that 2 billion gallons of ethanol will be imported to meet RFS2; therefore, the estimated net savings would be $37.2 billion (2010).

BUDGET, WELFARE, AND SOCIAL VALUE EFFECTS OF RFS2

Government policies that support biofuels interact with each other and other federal subsidy programs, particularly those involving farming, conservation, and nutrition. Because they involve tax credits and tariffs, these policies also affect federal government revenue. Government programs are shaped, in part, by public opinion of the value of biofuels, particularly as they pertain to the environment.

Distribution of Benefits and Costs

An economic analysis of the U.S. biofuel policy must consider the three elements that support that policy: (a) the consumption mandate requiring the use of biofuels as an input in the production of transportation fuels, (b) the federal tax credit for biofuels used in the production of transportation fuels, and (c) the import tariff on ethanol used in the production of transportation fuels. This section considers the likely welfare consequences of each of the three elements both in isolation and in combination with the other two.

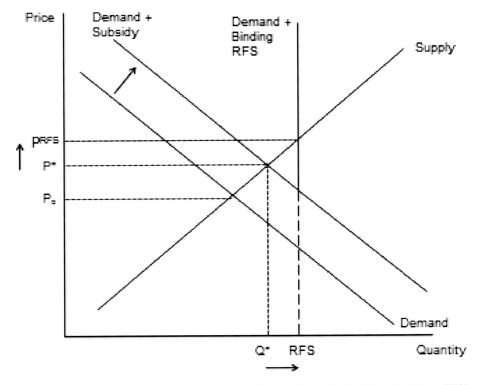

FIGURE 4-21 Price and quantity of biofuel demanded and supplied with and without RFS2 mandate and tax credit.
NOTE: P_O = price of biofuel without tax credit; P^* = price of biofuel with tax credit; Q^* = quantity of biofuel with tax credit; RFS = mandate quantity; P^{RFS} = price with tax credit and RFS2 mandate.

Consumption Mandate

The consumption mandate for RFS2 is described in Chapter 1 (see section on "Renewable Fuel Standard"). Under a consumption mandate, if the price of renewable biofuel is greater than the price of gasoline, the mandate would have the effect of raising the price of transportation fuels. Such an increased input cost would cause an increase in the cost of transportation fuels and a kink in the biofuel demand curve as shown in Figure 4-21. In essence, the biofuel demand curve becomes totally insensitive (inelastic) to the biofuel price because that level of consumption is required by the mandate. The higher price of biofuel as a result of the mandate would lead to a higher price of blended fuel and, consequently, reduced blended fuel consumption because of this higher price. Because the price elasticity of demand for transportation fuels is inelastic, the reduction in consumption would be small. Thus, the consumption mandate would result in the following effects, assuming no changes in technology, vehicle fuel-use efficiency, use of flex-fuel vehicles, and transportation fuel infrastructure:

1. With an increase in the price and a decrease in the quantity of transportation fuels consumed, the welfare of fuel consumers will decrease;

2. With an increase in the price and a decrease in the quantity of transportation fuels consumed, the welfare of biofuel producers will increase;[19]
3. With an increase in the quantity of biofuel demanded, the demand for biofuel feedstock will increase;
4. With an increase in the price and quantity of biofuel feedstock demanded, the welfare of feedstock producers will increase;
5. With an increase in the price and quantity of feedstock demanded, the demand for resources used to produce feedstock (for example, land, labor) will increase;
6. With the increase in the demand for resources used to produce feedstock, an increase in the price and quantity of resources used will increase the welfare of owners of resources used in the production of feedstock (for example, landowners);
7. With the increase in the price of resources used to produce feedstock, competing users of those resources will pay a higher price to retain those resources and the quantity of those resources used for production of other products will decrease;
8. As the quantity of resources used in the production of feedstock increases, the quantity of those resources used in the production of other goods (for example, food, livestock feed) decreases;
9. As the quantity of resources used in the production of other goods decreases, the quantity supplied of these other goods will decrease and their prices will increase;
10. As the price of these other goods increases and the quantity decreases, the welfare of consumers of these goods will decrease and the change in the welfare of producers of these goods will be determined by the price elasticity of demand for these goods; and
11. Though the mandate has no direct effect on federal government expenditures, the indirect effects could include:
    a. To the extent that RFS2 increases the prices of agricultural commodities that receive commodity program support payments, the size and cost of those payments will decrease;
    b. To the extent that RFS2 increases the prices of agricultural commodities and, therefore, the price of food and expenditures for those federal programs whose payments are related to the food price level (that is, expenditures that are increased to reflect the Consumer Price Index such as the Supplemental Nutrition Assurance Program [SNAP], the Special Supplemental Assistance Program for Women, Infants, and Children [WIC], or Social Security programs) will increase.

Tax Credit for Blended Biofuels

A second policy tool to support biofuels is the tax credit provided to blenders for using biofuels. As discussed in Chapter 1, tax credits exist to encourage the blending of corn-grain ethanol, biodiesel, and cellulosic biofuel into transportation fuel. Because the credit for corn-grain ethanol has been in place the longest and much more corn-grain ethanol has been consumed in the U.S. transportation market, this section will focus on the economic effect of the tax credit for corn-grain ethanol.

---

[19]The effect on the welfare of transportation fuel producers will depend upon the price elasticity of the demand for fuel. If the price elasticity of demand for fuel is inelastic, the percentage increase in price will be greater than the percentage decrease in quantity. Thus, the welfare of fuel producers will increase. Most studies of the price elasticity of demand for fuel have concluded that the demand for fuel is inelastic with regard to price (Espey, 1996, 1998; Graham and Glaister, 2002; Goodwin et al., 2004).

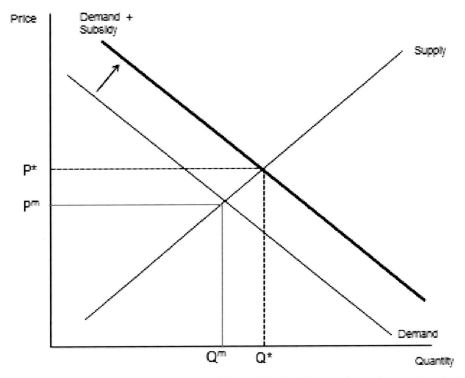

FIGURE 4-22 Effect of the Volumetric Ethanol Excise Tax Credit on price and quantity of ethanol.
NOTE: P^m = market price of ethanol without tax credit; P^* = market price of ethanol with tax credit; Q^m = quantity
of ethanol without tax credit; Q^* = quantity of ethanol with tax credit.

Under the legislation in action when this report was written, fuel blenders received a tax credit of $0.45 per gallon of corn-grain ethanol blended with gasoline, known as the Volumetric Ethanol Excise Tax Credit (VEETC). As illustrated in Figure 4-22, the effect of VEETC is to shift the demand for biofuel up and to the right because the blender is willing to pay more for every gallon of ethanol that receives the tax credit. In Figure 4-22, P^m and Q^m are the market price and quantity without the tax credit and P^* and Q^* those values with the tax credit.

The demand curve for ethanol is derived from the demand for gasoline transportation fuel. Therefore, the impact of the tax credit is to increase the price of ethanol and the quantity of ethanol produced relative to the absence of the tax credit. The impact on consumers is difficult to predict as the tax credit would be shifted among the blender, biofuel producer, and consumer depending on market supply and demand conditions.

This increased use of ethanol would result in an increase in the demand for biofuel feedstock, as noted in item 3 in the list of effects in the previous section. Following this change in the market for biofuel feedstock, the welfare consequences of a tax exemption would be identical to those noted in items 4 to 10 of the previous section, given no other changes.[20] For item 11 (the effect on the budget of the federal government), however, the

---

[20]The size of these welfare changes, however, would depend on the relative size of the change in biofuel consumption caused by the mandate versus the tax exemption.

tax credit policy would also result in the loss of federal tax revenue equal to the volume of gasoline displaced by biofuel multiplied by the VEETC ($0.45 per gallon) (see "Federal Fuel Tax Revenue" below). Thus, this results in either less funding available for government projects or higher taxes elsewhere to compensate for the shortfall in revenue.

Import Tariff

The third policy instrument to support the production and use of domestic biofuels is a tariff on imported ethanol. The current tariff is $0.54 per gallon plus 2.5 percent of import value; at recent ethanol prices, the total tariff equals about $0.59 per gallon. However, imported ethanol receives the same blender's tax credit as domestic ethanol ($0.45 per gallon), so the net tariff is about $0.14 per gallon. An import tariff on ethanol (considered in isolation from the consumption mandate and the tax credit) normally causes a decrease in the quantity of ethanol imports. Reduced imports would lead to increased domestic production. Given that imported sugarcane ethanol can be used to fulfill the other advanced biofuels category once that category of RFS2 becomes binding, it is not clear that the tariff at that point would reduce imports substantially.

Policy Interactions

Though the analysis above is indicative of each policy's effect, the use of these policies in combination can result in consequences that can be similar or, in particular circumstances, offsetting in nature (de Gorter and Just, 2010). If, for example, the consumption mandate policy is used in combination with the import tariff policy, the results of the mandate policy would be identical to those outlined above for the quantity of biofuel used to fulfill the consumption mandate (regardless of the domestic or foreign source of the biofuel).[21] However, RFS2 has four categories of biofuels, each with separate GHG rules. If the imports are based on sugarcane, they can be used for the "advanced biofuel" category, which has its own criteria, so the markets would be segregated. Beyond that point, any additional consumption of biofuel would be determined by the price of biofuel (including the import tariff) relative to the price of gasoline. If the price of biofuel (mileage-adjusted) is less than the price of gasoline, then a quantity of biofuel beyond the mandate would be consumed (up to any technical limit or "blend wall"). If the price of biofuel is greater than the price of gasoline, then no biofuel beyond the mandated level would be consumed.

If the import tariff policy is used in combination with the tax credit policy, then the welfare effects identified above would be the same, but the size of the anticipated welfare effects would be determined by the relative sizes (on a per gallon basis) of the import tariff and the tax credit. The net effect of these two policies would then determine the price of biofuel and its use as a substitute for gasoline in the production of transportation fuels. As before, if the price of biofuel (on a mileage-adjusted basis) resulting from the net effect of the combined policy is less than the price of gasoline, then the quantity of biofuel used would increase (until reaching any technical limitation).

---

[21] Though the direction of any welfare changes of a combined mandate and tariff policy would be identical to those identified for the mandate policy, the magnitude of the combined policies would likely be greater than those of a consumption mandate policy used in isolation. This result would occur because the combined policy is likely to increase the magnitude of any price changes that occur (i.e., the price of biofuel is likely to be greater under the combined policy). In addition, if the tariff applied to the price of imported biofuel is prohibitive (i.e., the cost of imported biofuel is greater than the cost of domestic biofuel), then a zero quantity of biofuel would be imported and the welfare consequences would be the same as if the consumption mandate had been used in isolation.

A combination of the consumption mandate policy and the tax credit policy can result in a wider variety of consequences. Once the consumption mandate is filled (that is, the quantity of biofuel consumed equals the quantity of biofuel required by the consumption mandate), any additional use of biofuel in the production of transportation fuel will be determined by the relative prices of biofuel and gasoline. If the price of biofuel is less than the price of gasoline, then the quantity of biofuel consumed would be greater than the mandated level. In this case, the welfare effects of a combined policy would be the same as indicated above, but the magnitude of these changes would be larger for a combined policy than for a mandate policy alone. In addition to the welfare effects noted above, a combined policy would also result in a loss of federal fuel tax revenue equal to the quantity of gasoline consumption displaced by biofuel multiplied by the level of the tax credit. In the case of corn-grain ethanol, the Government Accountability Office (2011) found the existence of both policies to be redundant because the infrastructure for the ethanol industry has been developed and no longer needs the additional incentive of the VEETC to create capacity to meet the RFS2 mandate. If, on the other hand, the price of biofuel is greater than the price of gasoline, then no additional biofuel would be consumed beyond the mandated quantity. However, if the cost difference is less that the level of the tax credit, then the blender's credit could still induce additional biofuel production.

Finally, if all three policies are used in combination, then the welfare effects of the combined policy would be determined by (a) the mandated consumption quantity and (b) the net price effect of the tax credit and the import tariff policy. The import tariff would increase the price of biofuel. The net effect of these two policies would then determine the price of biofuel and its use as a substitute for gasoline in the production of transportation fuel. If the price of biofuel (on a mileage-adjusted basis) resulting from the net effect of the combined tax credit and tariff policy is greater than the price of gasoline, then the quantity of biofuel consumed would equal the quantity of biofuel required by the mandate. If the net price of biofuel is less than the taxed price of gasoline, then the quantity consumed would exceed the mandated quantity. All of these projections are based on no other changes in the biofuel system.

State-Level Policies

In addition to the federal-level policies affecting biofuel consumption and production, state government policies also affect the biofuel market. These policies include a varying combination of incentives (construction grants, capital cost subsidies, tax incentives, loans and leases, rebates, exemptions) and regulations that influence the consumption of biofuel (mandates, air quality, carbon intensity, climate change initiatives). Because such policies vary widely across states, it is difficult to determine the level of subsidy or mandate that would be created by these state-level policies when considering their effects on a nationwide basis (Tyner, 2008). Comparing the magnitudes of federal and state policies, however, suggests that the welfare effects of federal policies are of greater magnitude than state-level policies (Box 4-5) (Steenblik, 2007; Koplow, 2009). For example, state tax credits often are in the range of $0.20 per gallon but can be as high as $1.00 per gallon (Kentucky). At the same time, state subsidies "add substantially to the profitability of production from existing facilities [but] they are often provided up to an annual limit" (Steenblik, 2007, p. 25). In some states, mandates or tax credits are contingent on the use of feedstock produced in the same state (Steenblik, 2007). Local levels of government sometimes provide production subsidies in the form of property tax abatements, economic development loans, infrastructure subsidies, or free use of land (Steenblik, 2007). In addition, such state-level policies are likely to

---

BOX 4-5
State and Federal Subsidy Expenditures on Energy in Texas

In 2008, the Comptroller of the State of Texas produced an extensive report examining the existing and potential resources Texas can employ to meet its energy demands (TCPA, 2008). This report included a chapter on the subsidies that various energy sources received in 2006, the last year with complete data when the report was being prepared.

According to the report, governments provide subsidies in the form of tax incentives (such as depletion allowances, accelerated depreciation, and reductions in excise taxes), direct spending for government services, the assumption of certain types of liability or risk by the government, government ownership of energy production, access to resources on government land, and tariffs.

The report breaks down the subsidies into two categories, those coming from the federal government and those coming from the State of Texas. Texas is a large consumer and producer of energy of many kinds. As such, its support of various energy industries through incentives for increased production is probably higher than most states. The state subsidies are for the energy used in Texas alone. Table 4-5 shows federal and state subsidies for various energy types as a percent of total consumer spending for the sources.

TABLE 4-5 Energy Subsidies as Percentage of Consumer Spending on That Source

Energy Source	Federal Subsidies	Texas Subsidies
Oil and Gas	0.5	1.5
Coal	6.9	0
Nuclear	20.9	0
Ethanol	26.5	0
Biodiesel	9.9	3.1
Wind	11.6	0.2
Solar	12.3	9.2
Other Renewable	0.5	<0.1

SOURCE: TCPA (2008).

Total federal subsidies for corn-grain ethanol and biodiesel sold in 2006 were $4.71 billion and $92 million on 4.8 billion gallons of ethanol and 250 million gallons of biodiesel. These subsidies totaled $0.98 per gallon for ethanol and $0.38 per gallon for biodiesel.

Substituting corn-grain ethanol or biodiesel, the only current alternatives for liquid transportation fuel for gasoline and diesel, will have a substantial effect on government tax revenues. RFS2 requires the inclusion of an additional approximately 20 to 25 billion gallons per year of biofuels in the transportation fuels sold in the United States by 2022. With the current tax structure, this would reduce state and federal excise tax revenues by over $10 billion per year.

---

have their greatest effect on the location of biofuel refineries among states rather than the total national biofuel production capacity (Cotti and Skidmore, 2010).

Federal Budget Effects

The net effect of RFS2 on the federal budget would be determined by changes in the following:

- The cost of farm commodity program payments;
- The cost of other USDA programs, including conservation programs;

- The cost of nutrition and income transfer programs that are affected by changes in the price of food through the CPI;
- The cost of biofuel production subsidy programs;
- The federal fuel tax revenue forgone due to tax credits, in particular the VEETC for corn-grain ethanol; and
- The tariff revenue generated or forgone by the tariff on imported ethanol.

Though these budget changes can be difficult to estimate with precision, the provisions of RFS2, the tax credits, and the import tariff and past experience with biofuel production can suggest the direction and general magnitude of the changes that would occur for each budget component.

Agricultural Commodity Programs

An increase in biofuel production encouraged by the RFS2 mandate can indirectly produce savings in federal payment programs that support agricultural commodities; however, the circumstances under which such savings are realized is rather specific and limited. Under the Food, Conservation, and Energy Act of 2008, commodity support programs consist of a direct payment program, a countercyclical payment program, and a marketing assistance loan or equivalent loan deficiency program.[22] To determine the effect of RFS2 on the budget cost of these programs, each program is considered. The first type of payments—direct payments—are fixed payments provided to crop producers regardless of the market price received by crop producers. As noted earlier, if an increase in the production of biofuel feedstocks results in an increase in competition for those resources (for example, land) that produce other crops, the prices of those other crops would be expected to increase. This increase in crop prices would not affect the budget cost of the direct payment program, however, because direct payments are paid to crop producers regardless of the price level. Thus, under no circumstances would meeting the RFS2 mandate generate savings in the budget cost of the direct payment program.

The second type of payments—countercyclical payments—are paid when the market price for a crop is less than the effective target price of that crop. The effective target price of a crop is calculated as the legislated target price of that crop minus the direct payment of that crop. For example, under the Food, Conservation, and Energy Act of 2008, the target price for corn is $2.63 per bushel and the direct payment for corn is $0.28 per bushel in the 2012 crop year. The effective target price for the 2012 crop year will be $2.35. If the market price, without the implementation of RFS2, is $2.35 per bushel or more, the budget savings of RFS2 will be zero (that is, no countercyclical payment would be made because the market price is greater than the effective target price even without the price effect of RFS2). Thus, only if the market price would be less than $2.35 without the implementation of RFS2, and then increases to above $2.35 with RFS2, can any budget savings in the countercyclical payment program be realized. Similarly, the marketing assistance loan program establishes a marketing loan rate of $1.95 per bushel in 2012. If the market price of corn is less than $1.95, corn producers would be eligible for a loan deficiency payment equal to the difference between the marketing loan rate and the market price. Thus, only if the market price would

---

[22]Crops included in these programs include corn, wheat, soybean, rice, cotton, peanuts, grain sorghum, barley, oats, and other oilseeds. The Food, Conservation, and Energy Act also established the Average Crop Revenue Election program (ACRE) as an alternative to the Direct and Countercyclical Payment program (DCP). Since a large majority of commodity producers have chosen to remain enrolled in the DCP program, this analysis will examine the budget consequences of biofuel production on the budget cost of the DCP program.

be less than $1.95 without the implementation of RFS2, and then increases to above $1.95 with RFS2, can any budget savings in the marketing loan assistance program be realized.[23]

Long-term projections of U.S. agricultural commodity prices from 2011 to 2021 suggest that market prices will exceed effective target prices during that period. For example, USDA projections of corn (ranging from $4.10 to $5.20), soybean ($10.25 to $11.45), and wheat prices ($5.45 to $6.50) would exceed the existing effective target prices during the period (USDA, 2011). If such projections hold true, then the change in the budget cost of commodity programs attributable to an expansion of biofuel production (which presumably would result in price levels higher than these levels) would be zero because market prices would exceed effective target prices (thus, no countercyclical payments would be made) and direct payments would be unchanged (paid regardless of market price levels). Similarly, long-term price projections from the Food and Agricultural Policy Research Institute (FAPRI, 2010) also suggest that market prices of these three agricultural commodities would exceed effective target prices during the period, again supporting a conclusion that expanded biofuel production would likely result in no change in the budget cost of commodity programs.[24]

Conservation Programs

The Conservation Reserve Program (CRP)—a conservation program in which farmers sign contracts with the federal government to take land out of crop production for a period of time for which they receive payment—is the largest federal conservation program directed at agricultural land. Under the 2008 farm bill, CRP is limited to 32 million acres, down from 39.2 million acres under the previous farm bill. Only land that was planted to an agricultural commodity in four of the previous 6 years from 1996 to 2001 or land that is suitable as a riparian buffer is eligible for CRP. Land is typically enrolled for 10-15 years, and, at the time this report was written, participants received an average payment of $44 per acre. In the summer of 2010, 31.3 million acres were enrolled. The program was estimated to cost $1.7 billion in fiscal year 2010, $250 million less than fiscal year 2009 (Cowan, 2010).

The effects of biofuel policy regarding expenditures on CRP are uncertain. Higher commodity prices, to which biofuel may be a contributing factor, could entice growers to remove acres from the program. CRP costs would decrease if further acres were not enrolled to replace acres leaving the program. The likelihood of declining acres is unknown: In the previous two general sign-ups for CRP (2006 and 2010), the number of acres bidding for the program exceeded the number of acres accepted (Cowan, 2010). However, if acres left the program and were not replaced by other enrollments and if keeping maximum enrollment is an important objective of CRP, then payment rates would have to increase to compete with commodity prices. Likewise, competitive CRP payment rates could incentivize producers to keep the most sensitive land in the program. This action could increase

---

[23]Thus, if the market price is below $1.95, a corn producer would be eligible for a target price of $2.63 that would consist of the market price (below the loan rate of $1.95) plus the loan deficiency payment (equal to $1.95 minus the market price) plus the countercyclical payment (equal to $2.35 minus the loan rate of $1.95) plus the direct payment ($0.28), thereby providing the target price of $2.63.

[24]The assumptions regarding biofuel production policies used in each of these studies should be noted. USDA assumes that all tariff and tax credit policies for ethanol will remain in place for the entire 2011-2021 period. Similarly, the FAPRI projections incorporate the mandates contained in EISA but assume the mandate regarding use of cellulosic ethanol is waived (i.e., only the EISA mandate of 15 billion gallons of conventional ethanol is continued) and that all tariffs and tax credits for biofuels are extended for the entire period.

the cost of the program, depending on the new payment rate and the number of acres that remain in CRP.

Harvest and grazing are allowed on CRP land under certain conditions. For example, if the government has defined a drought as a disaster, the harvesting of hay or grazing of cattle may be permitted under the 2008 farm bill. Routine harvesting may also be allowed to manage invasive species (Cowan, 2010). It has been suggested that bioenergy feedstocks could be cultivated on CRP land as long as certain criteria were met, such as requiring the harvest to occur after the bird-nesting season. CRP payments would be reduced in accordance with the revenue generated from harvesting biomass. With appropriate environmental restrictions, such compromise would allow conservation and biomass production to coexist while also reducing government CRP payments. However, an attempt in 2008 to classify some CRP land as eligible for haying and grazing for animal feed purposes was suspended by a lawsuit requesting further environmental review, and similar authorization was not made in 2009 or 2010 (Cowan, 2010). At the time this report was written, it appeared unlikely that cultivating biomass would be permissible on CRP land.

Nutrition and Income Transfer Programs

As noted earlier (see "Food Prices"), assessment of any change in food prices resulting from expanded biofuel production would have to consider the effect on agricultural commodity prices and the transmission of that commodity price effect through the food system to the retail level. Similarly, those two factors have to be considered in assessing the effect of an expansion of biofuel production on the budget cost of nutrition and income transfer programs. The experience of the 2007-2009 period provides useful examples for understanding the likely impact of expanded biofuel production under RFS2.

Nutrition and other income assistance programs are often adjusted for changes in the general price level as a means of protecting the real purchasing power of program recipients. This adjustment is based on the annual change in the CPI. If an increase in the production of biofuel feedstocks results in increased competition for resources (such as land) used to produce agricultural commodities for food, crop prices would be expected to increase, thereby increasing food prices and the food component of the CPI. In turn, the budget cost of those programs tied to the CPI would increase. For example, two programs would be affected by changes in food prices—SNAP and WIC. As discussed earlier (see section "Primary Market and Production Effects of U.S. Biofuel Policy" in this chapter), the role of biofuels in the increase in commodity prices is extensively debated in the literature. The Congressional Budget Office (2009) conducted a study that examined the effect of corn-grain ethanol use on food retail prices and GHG emissions between April 2007 and April 2008. It assumed that 10-15 percent of the increase resulted from biofuels. This translated to an increase in annual expenditures on SNAP of $500 to $800 million and less than $75 million on WIC in the 2009 fiscal year.

To the extent that ethanol production is now a permanent part of the commodity market through the RFS2 mandate, these increases could become a permanent increase in the annual cost of these programs. At the same time, because other programs are also adjusted according to changes in the CPI, these estimates are likely to be lower-end estimates of past and future budget costs for such programs. For example, programs such as Social Security, military or civilian retirement programs, and Supplemental Security Income are adjusted for changes in the CPI, while such items as food purchases for military personnel are affected directly by the prices of food purchased. The budget cost of such programs, and thus the increase in the budget cost of these programs attributed to expanded biofuel production, would likely dwarf the estimated budget costs of the SNAP or WIC programs. For example, the SNAP program had a projected total budget cost of $54.4 billion in 2009. This compares to an annual budget cost of Social Security and Supplemental Security Income payments of $709.6 billion in 2009 (OMB, 2010). These programs are also adjusted by changes in the CPI to protect the real purchasing power of program recipients. Thus, any calculation of the effect of expanded biofuel production on budget cost of Social Security and other federal programs would likely be much larger than the estimated cost for the SNAP program or other similar programs. However, an increase in raw commodity prices such as corn translates into a much smaller increase in food prices (see earlier section "Food Prices" in this chapter).

These estimates suggest the likely direction and magnitude of changes in budget costs that would be observed with an increase in biofuel production under RFS2. The primary effect on budget costs for nutrition and income transfer programs would be through the increase in retail food prices that would be related to a possible increase in competition for production resources (such as land) resulting from increased production of biofuel feedstock. In particular, the production of crops dedicated to cellulosic biofuel production could result in increased agricultural commodity prices if the production of those crops displaces production of crops devoted to food. For those crops that could be used for both food (for example, corn for grain) and for cellulosic feedstock (for example, corn stover), such joint products could compete more favorably for production resources at the farm level. For example, in the case of corn, producers might find it profitable to continue corn production for food, corn-grain ethanol, and cellulosic biofuel, rather than producing a crop devoted only to biofuel production. In such a case, the shift of acreage from food crops to crops devoted only to biofuel might be limited. Therefore, the effect of expanded biofuel production on agricultural commodity production and on retail food prices could be small.

Federal Fuel Tax Revenue

Another effect of expanded biofuel production on the federal budget is through the federal tax credits for biofuels blended with motor fuel. An estimate of the revenue forgone by the federal government is determined by the size of the federal tax credit provided to blenders of biofuels, the energy equivalence of biofuels and gasoline, and the quantity of biofuel receiving the tax credit (CBO, 2010). The federal tax credit for corn-grain ethanol, the VEETC, has had a nominal value of $0.45 per gallon since 2005. Considering all of the factors noted above, however, this tax credit by some calculations has had a gasoline-equivalent value of $1.78 per gallon in forgone revenue for the federal gasoline excise tax (CBO, 2010). The federal tax credit for ethanol is estimated to be a tax expenditure (that is, forgone revenue for the federal gasoline excise tax) of about $2.9 billion in the 2007 fiscal year, compared to $921 million in 1999 (EIA, 2008b). GAO (2011) found that the VEETC resulted in $5.4 billion in forgone revenue in 2010, which would grow to $6.75 billion in 2015 with increased ethanol production (GAO, 2011).

Because the monetary cost of cellulosic biofuel and other advanced biofuel is expected to be substantially greater than the cost of corn-grain ethanol, the nominal value of the federal tax credit is $1.01 for cellulosic biofuel. When all of the factors noted above are included, however, the cost of this tax credit, in terms of federal gasoline excise tax revenue forgone, could be $3.00 per gallon of cellulosic biofuel (CBO, 2010).

Import Tariff Revenue

The impact of expanded biofuel production on changes in tariff revenue also would affect the federal budget. The potential budget effects of a tariff on ethanol imports depends on the tariff rate charged on imports of ethanol and the quantity of ethanol imports. The effect of these trade restrictions will be determined by the "restrictiveness" of the trade restraint (or "size" of the import tariff) and the responsiveness of market participants (producers and consumers of ethanol in both domestic and foreign markets) to changes in market prices (USITC, 2009).[25]

As noted above, the tariff on imported ethanol has two parts—the 2.5 percent tariff on all ethanol, and the $0.54 per gallon duty on all ethanol imported for use as transportation fuel. Since the $0.54 per gallon duty is the only portion of the restriction applied specifically to ethanol used as transportation fuel, this analysis considers only this portion in analyzing the tariff revenue from ethanol imports. Under two circumstances, the revenue generated by a tariff can be equal to zero or very near zero.

On the one extreme, a tariff can be so small that the revenue generated by the tariff is near zero. In this case, the tariff multiplied by the value or quantity of the good imported is near zero because the tariff is near zero. At the other extreme, a tariff can be so large that it is prohibitive for exporters of the good. In other words, when the tariff is applied to the value or price of the good, the cost of the imported good exceeds the cost of domestically produced goods. Thus, the quantity of the good imported is now zero (or near zero), so the tariff revenue will equal zero (the large tariff multiplied by a zero quantity imported will equal zero tariff revenue). This issue is particularly important when considering the import tariff on ethanol used as transportation fuel. On the one hand, the general tariff of 2.5 percent on all ethanol is near zero. Thus, the revenue generated by this general tariff would be nearly zero. On the other hand, the $0.54 per gallon duty on ethanol is nearly prohibitive at present on imports of ethanol for transportation fuel. Thus, the tariff revenue generated by this duty also is close to zero. The removal of the $0.54 import duty on ethanol, while leaving the 2.5 percent tariff in place, would move the market from one extreme to the other. That is, the tariff revenue generated by the $0.54 duty would be nearly zero because the imported quantity is nearly zero. If that duty is removed, the tariff revenue generated after that duty is removed would again be zero because the remaining 2.5 percent general tariff is near zero while the quantity imported would likely increase.

Analysis of the effect of removing the $0.54 duty on corn-grain ethanol after the Renewable Fuel Standard under the Energy Policy Act of 2005 went into effect but before RFS2 under EISA was enacted suggests that more ethanol would be imported (USITC, 2009). This analysis found that elimination of the duty would reduce the price of imported ethanol by 25 percent and increase the value of imports by 205 percent annually. This increase in imports would reduce domestic production of ethanol by 2 percent below the level produced under the $0.54 duty (USITC, 2009).[26]

---

[25]In particular, the change in imports resulting from a removal of trade restraints (tariffs) is likely to be determined by (a) the elasticity of substitution between imported and domestic goods, measuring the ability of users to substitute imported goods for domestic goods, (b) elasticity of import supply, measuring the responsiveness of domestic producers and consumers to changes in the price of that good, (c) elasticity of export demand, measuring the responsiveness of foreign producers and consumers to changes in the price of that good, (d) elasticity of substitution between inputs in production, measuring the ability of domestic and foreign producers of that good to substitute alternative inputs in the production of that good, and (e) income elasticity of domestic and foreign consumers, measuring the responsiveness of consumer demand to changes in consumer income (USITC, 2010a).

[26]The value of corn production would decrease by 0.6 percent and the value of corn exports would increase by 0.6 percent annually with the removal of the $0.54 duty (USITC, 2009).

Other Federal Programs Related to Biofuel Production

An additional category of budget costs related to RFS2 is that set of programs that provide subsidies for the production of cellulosic biofuel. Such subsidies take a variety of forms, and similar programs have been provided for the production of corn-grain ethanol, often on an intermittent basis, since 1980 (Duffield et al., 2008). In recent years, such subsidies have increased as part of the policy objective of increasing biofuel production.

The two largest costs associated with a biorefinery are the capital cost of the refinery facility and the cost of the feedstock processed in the facility (see Figure 4-6). Thus, federal biofuel production subsidies have taken a variety of forms, but most programs are designed to subsidize the production cost of the biofuel refining industry by lowering the capital cost of the construction of biorefinery facilities, reducing the variable cost of biofuel feedstock paid by biofuel refiners, providing implicit subsidies to the purchase price of cellulosic biofuel, or decreasing the total cost of biofuel production through the improvement of biofuel processing technology.

Subsidies to reduce the capital investment cost of constructing cellulosic biofuel refineries are typically provided in the form of tax credits, grants, loans, or loan guarantees that provide a rate of interest below that which investors could obtain from alternative financing sources (Table 4-6). For example, the Food, Conservation, and Energy Act of 2008 and the Energy Policy Act of 2005 included a variety of provisions to subsidize the capital cost of constructing cellulosic biofuel refineries. Some of these programs provided subsidies aimed at refineries using a particular form of biofuel feedstock (for example, municipal solid waste) while others provide capital subsidies for refineries without specifying the form of biofuel feedstock.

The second form of biofuel production subsidies, those that reduce the cost of feedstock purchased by cellulosic biofuel refineries, are typically provided in the form of payments per unit of feedstock purchased (Table 4-6). Such payments reduce the purchase price of biomass feedstock for biofuel refineries and, therefore, the production cost of biofuel products. The payment through BCAP for collecting, harvesting, storing, and transporting biomass is an example of this type of subsidy (see Box 4-4). By offsetting a portion of the variable cost of producing cellulosic biofuel, such subsidies transfer a portion of the production cost of biofuel from consumers to taxpayers.

A third form of subsidy for cellulosic biofuel production can be provided through the purchase price of biofuel produced by refiners. For example, Section 942 of the Energy Policy Act of 2005 establishes a "reverse auction" mechanism for the purchase of cellulosic biofuel (Table 4-6). This program provides an implicit subsidy to biofuel refineries by permitting refiners to submit bids to sell cellulosic biofuel to the federal government. Bids from refiners would specify the production incentive payment by the federal government that would be required for the refinery to supply a given quantity of cellulosic biofuel to the federal government. Bids would be accepted in reverse order (that is, from lowest incentive payment to highest) until a given quantity of biofuel is supplied. Because such payments would reflect the difference between the price of gasoline and the price of cellulosic biofuel and the variations in production costs across refineries with higher cost refineries receiving larger incentive payments until the desired quantity is reached, the incentive payments would constitute a subsidy of the total production cost of each refinery.

The final form of subsidy for cellulosic biofuel production is provided through research, development, and outreach programs designed to reduce the total production cost of cellulosic biofuel. Such programs can have a wide variety of effects on technical relationships in biofuel production. For example, research that increases the yield of biofuel

TABLE 4-6 Selected Federal Programs to Reduce Production Costs of Cellulosic
Biofuel Refineries

Programs	Millions of U.S. Dollars Authorized Annually, Unless Noted
Programs to Offset Total Production Cost of Cellulosic Ethanol	
Production Incentives for Cellulosic Biofuels Program (EPAct Section 942): Provides for federal purchase of biofuel via reverse auction format (producer supplies bid for production incentive payment needed to supply biofuel).	$100 million annually for 10 years.
Programs to Subsidize Capital Costs of Biorefineries	
Biorefinery Assistance Program (FCEA, Section 9003): Grants to assist in paying the costs of the development and construction of demonstration-scale biorefineries.	$150 million annually for 2009-2012.
Biorefinery Assistance Program (FCEA, Section 9003): Guarantees for loans made to fund the development, construction, and retrofitting of commercial-scale biorefineries.	$75 million in 2009 and $245 million in 2010.
Repowering Assistance Program (FCEA, Section 9004): Grants to existing biorefineries to replace fossil fuels used to produce heat or power for operation of biorefinery.	Available to any existing biorefinery, $35 million in 2009 and $15 million annually for 2009-2012.
Integrated Biorefinery Demonstration Projects (EPAct, Section 932(d)): Grants to demonstrate the commercial application of integrated biorefineries for producing biofuels or biobased chemicals.	$100 million to $150 million annually for 2007-2009.
Biomass Research and Development Initiative (EPAct, Sections 941(e) and (g)): Grants for demonstration of technologies and processes necessary for producing biofuels and other biobased products.	$100 million annually for 2006-2015.
Commercial Byproducts from Municipal Solid Waste and Cellulosic Biomass Loan Guarantee Program (EPAct, Section 1510): Provides loan guarantees for construction of facilities for converting municipal solid waste and cellulosic biomass to ethanol.	Such sums as needed by Department of Energy.
Cellulosic Biomass Ethanol and Municipal Solid Waste Loan Guarantee Program (EPAct, Section 1511): Provides loan guarantees for cellulosic biomass and sucrose-derived ethanol demonstration projects.	Loan guarantee of $250 million for no more than 4 plants.
Conversion Assistance for Cellulosic Biomass, Waste-Derived Ethanol, Approved Renewable Fuels (EPAct, Section 1512): Grants to producers of cellulosic ethanol derived from agricultural residues, wood residues, municipal solid waste, or agricultural byproducts.	$100 to $400 million during 2006-2008.
Sugar Ethanol Loan Guarantee Program (EPAct, Section 1516): Guarantees loans for construction of facilities to produce biofuel using sugarcane or byproducts of sugarcane.	$50 million per project.
Incentives for Innovative Technologies (EPAct, Section 1703): Provides loan guarantees for advanced energy projects, including advanced biofuels.	Such sums as needed.
Programs to Subsidize Feedstock Costs of Biorefineries	
Bioenergy Program for Advanced Biofuels (FCEA, Section 9005): Payment to producers of biofuel for proportion of feedstock purchased for biofuel production.	$300 million during 2009-2012.
Feedstock Flexibility Program for Bioenergy Producers (FCEA, Section 9010): Purchases sugar for use as biofuel feedstock and to prevent accumulation of government-owned sugar stocks.	Such sums as needed by USDA Commodity Credit Corporation.

TABLE 4-6 Continued

Programs	Millions of U.S. Dollars Authorized Annually, Unless Noted
Biomass Crop Assistance Program (FCEA, Section 9011): Payments to support establishment, production, and/or transportation of biomass feedstock crop and forest products.	Such sums as needed from USDA Commodity Credit Corporation.
Research Programs to Reduce Total Biofuel Production Costs	
Biomass Research and Development Program (FCEA, Section 9008): Research, development, and demonstration projects for biofuels and biobased chemicals and products.	$20 to $40 million mandatory annually for 2009-2012 and $35 million appropriated annually for 2009-2012.
Forestry Biomass for Energy Program (FCEA, Section 9012): Research and demonstration for use of forest biomass feedstock.	$15 million annually for 2009-2012.
Agricultural Bioenergy Feedstock and Energy Efficiency Research and Extension Initiative (FCEA, Section 7207): Grants to enhance biomass feedstock crops and on-farm energy efficiency.	$50 million annually for 2009-2012.
Sugar Cane Ethanol Program (EPAct, Section 208): Study production of ethanol from cane sugar, sugarcane, and sugarcane byproducts.	$36 million until expended.
Biomass Research and Development Initiative (EPAct, Sections 941(e) and (g)): Grants for applied fundamental research and innovation of technologies and processes necessary for production of biofuels and other biobased products.	$100 million annually for 2006-2015.
Regional Bioeconomy Development Grants (EPAct, Section 945): Grants to support growth and development of the bioeconomy through coordination, education, and outreach.	$1 million in 2006, such sums as needed thereafter.
Pre-Processing and Harvesting Demonstration Grants (EPAct, Section 946): Grants for demonstration of cellulosic biomass harvesting and preprocessing innovations for fuel or other energy.	$5 million annually for 2006-2010.
Education and Outreach on Biobased Fuels and Products (EPAct, Section 947): Education and outreach program for biomass feedstock producers or consumer education about biofuels and biobased products.	$1 million annually for 2006-2010.
Integrated Bioenergy Research and Development (EPAct, Section 971(d)): Funding for integrated bioenergy research and development programs, projects, and activities with federal agencies other than Department of Energy.	$49 million annually for 2005-2009.
Advanced Biofuels Technology Program (EPAct, Section 1514): Grants to demonstrate advanced technologies for alternative biomass feedstocks.	$110 million annually for 2005-2009.
Resource centers to further develop bioconversion technology using low-cost biomass for production of ethanol (EPAct, Section1511(c)).	$4 million annually for 2005-2007.
Renewable Fuel Production Research and Development Grants (EPAct, Section 1511(d)): Grants for research on renewable fuel production.	$25 million annually for 2006-2010.
Advanced Biofuel Technologies Program (EPAct, Section 1514): Grants to demonstrate advanced technologies for the production of alternative transportation fuels, including cellulosic ethanol.	$110 million annually for 2005-2009.
Bioenergy Research Centers (EISA, Section 233): Centers established to accelerate basic transformational research and development of biofuels, including biological processes.	Funding determined by the Department of Energy. Over $300 million a year in 2007.

NOTE: All years refer to Fiscal Years (October to September). "EPAct" refers to the Energy Policy Act of 2005. "FCEA" refers to the Food, Conservation, and Energy Act of 2008. "EISA" refers to the Energy Independence and Security Act of 2007.
SOURCES: 109 P.L. 58, 110 P.L. 140, 110 P.L. 234. Adapted from Koplow (2009).

obtained from a given unit of biomass can reduce either the capital cost of biofuel production (that is, the capital cost of producing a unit of biofuel) or the variable cost of biomass purchases (that is, the number of units of biomass feedstock purchased). Similarly, research that increases the on-farm yield of biomass feedstock crops (that is, units of feedstock produced per acre) would reduce the cost of feedstock purchased by biofuel refineries. To the extent that such programs reduce the cost of biofuel feedstock inputs, the benefits of such subsidies are likely to accrue to consumers of biofuel in the long run, not refiners of biofuel.[27] Three observations about the data reported in Table 4-6 were made in assessing the budget effects of biofuel policy. First, most of the programs listed are scheduled to end during the 2011-2012 time period and would not be in effect during the 2012-2022 period included in the RFS2 mandate. Because it is impossible to project the types of policies that could be in effect during the 2012-2022 period, the programs reported in Table 4-6 can only be considered as the type of programs that might be continued if RFS2 is to be met by 2022.

Second, federal budget expenditures arise in a three-step process. First, "enabling" or "organic" legislation is "legislation that creates an agency, establishes a program, or prescribes a function," such as an aspect of biofuel policy. Second, "appropriation authorization legislation" is legislation that "authorizes the appropriation of funds to implement the organic legislation" (GAO, 2004, p. 40 of Chapter 2). Organic and authorizing legislation may be combined in a single legislative action or may be separate legislation. Finally, appropriations legislation provides "legal authority for federal agencies to incur obligations and to make payments out of [the U.S.] Treasury for specified purposes" established in the organic and authorization legislation (GAO, 1993, p. 16). Thus, the dollar values reported in Table 4-6 are the *funding authorized* by the organic legislation, not the *funding appropriated*.

Because the funding appropriated is often much less than the funding authorized, the amounts reported in Table 4-6 can be assumed to be higher than the actual amount appropriated (that is, actual expenditures) for each program. For example, Koplow (2009) discounted these authorized funding levels by 50 percent to arrive at an estimate of the appropriated funding. Schick (2000) noted that the appropriated funding level for many programs "typically exceeds 90 percent of the authorized level" if authorization and appropriation was completed in the same fiscal year (that is, year one of a program). Schick also found that "there is often a widening gap between the authorized and the appropriated resources" if the program has a multi-year authorization as the overall budget environment continues to change with the passage of time (Schick, 2000, p. 171). Thus, no particular discount factor is applied to the authorized funding levels provided in Table 4-6, and the authorized amount can be considered as the upper-bound estimate of the total budget resources devoted to cellulosic biofuel programs.

Third, some of the programs listed in Table 4-6 could create a significant budget exposure if the mandate of RFS2 is accomplished by 2022. For example, BCAP provides an incentive payment of $45 per dry ton for the collection, harvest, storage, and transportation of each ton of biomass used for the production of cellulosic biofuel. Given an assumed refinery yield of 70 gallons per dry ton of biomass, fulfilling the RFS2 mandate of 16 billion gallons of ethanol-equivalent cellulosic biofuel in the year 2022 would require 229 million dry tons of biomass. Under the current rules of the BCAP program, payments can only be

---

[27]Economists have found, for example, that the long-run benefits from improvements in agricultural productivity primarily accrue to consumers of food products in the form of lower prices, not to farm producers or food processors (Ruttan, 1982). Since many of the economic characteristics of the agricultural sector (for example, inelastic demand) are similar to the energy sector, it is probably reasonable to conclude that in the long run the benefits of most forms of productivity-improving research on biofuel will accrue to the consumers of biofuel.

made for 2 years from a given source. Thus, with an incentive payment rate of $45 per dry ton of biomass, about 48 million dry tons of material would be paid for the incentive in 2022, and this would create a total budget cost of $2.1 billion for BCAP in 2022. Similarly, the achievement of RFS2 will require the expansion of biorefinery capacity. The existing budget resources devoted to grants, loans, and loan guarantees will likely be inadequate to support the launching of this industry. Thus, unless the economic viability of cellulosic biofuel production improves dramatically, the private investment needed to achieve such an expansion could be difficult to obtain without the availability of substantially larger capital-cost subsidies.

Social Value Effects Related to RFS2

Though difficult to monetize, Americans place a value on the environment. The expansion of the U.S. biofuel sector to meet RFS2 could have negative or positive environmental effects on a variety of resources that are highly appreciated by the American public. This section describes the basis for U.S. public support for first-generation and second-generation biofuel development, which is often grounded in concerns about climate change (Dietz et al., 2007; Solomon and Johnson, 2009; Johnson et al., 2011). It also reviews how people value environmental effects, such as water resources, forests and landscapes, and biodiversity, that interact with biofuel policy.

Role of the RFS Renewable Biomass Definition

While the Energy Policy Act of 2005 and EISA provide a great deal of the impetus for the push toward rapid biofuels development, in particular such as cellulosic biofuel, in other ways EISA was intentionally written to minimize negative greenhouse-gas emissions (for the definition in RFS2, see Chapter 1). The inclusion of the highly detailed and restrictive renewable biomass definition effectively limits woody feedstocks that can be used toward the cellulosic biofuel portion of RFS2 to forest residues from state and private forest plantations (timber stands composed of trees, usually of one species, that were physically planted by humans) or woody energy crops harvested from land that was not forested in 2007 (110 P.L. 140). Given the high level of fire risk many federal forests face due to overstocked stands and forest health issues (Becker et al., 2009), excluding all federal forests prevents the usage of an additional market that could facilitate thinning or residue removal.

Forestry professionals, organizations, industries, and some environmental groups have pushed for a relaxation of the definition to allow more forest types and materials to qualify, including residues and fire-control thinnings from managed federal forests and nonplantation forests. It is therefore important to acknowledge potential for the definition to be relaxed and outline the implications for effects on public values. Although the cost of harvesting "natural" forests for feedstock material and higher-value timber and paper may preclude substantial increases in natural forest harvesting in the short run (see Box 4-3) (Solomon et al., 2007), future markets could support such harvesting. Relaxing the definition could increase the value of harvesting and mill residues as coproducts, and thus enhance the marketability of pulp and timber, as well as provide environmental benefits.

Environmental Values and the American Public

Although environmental protection values are widely held by Americans, those values may be discussed in different terms and may be prioritized more or less highly in comparison to other values, such as economic development, national defense, or crime prevention.[28] Similarly, simply because individuals highly value their natural environment does not mean that their behavior, including support for policies such as climate-change mitigation, will always be consistent with the protection of these values. Many variables, such as knowledge, perceived norms, and structural barriers (for example, the lack of convenient public transportation leading to people driving personal vehicles more frequently), intervene between values and behavior. Environmental problems are usually complex, as are their solutions, and understanding any one problem and solution in detail requires a level of attention to that problem that is unlikely to be afforded by most Americans who do not rank environmental protection as a top concern.

Increasingly, environmentally related values are being discussed using the rubric of "ecosystem services" (de Groot et al., 2002; Kløverpris et al., 2008). The concept of ecosystem services describes components of the environment in terms of their value to humans and the larger ecosystem. For instance, temperate and tropical forests provide ecosystem services that include biodiversity preservation, watershed protection, and carbon sequestration. Insofar as biofuel development may influence incentives to maintain, for example, wetlands or forests in their natural species mixes, biofuel development could reduce or enhance the ability of these lands to provide ecosystem services.

Given that biological carbon sequestration is currently a goal primarily because it can assist with climate-change mitigation, concern exists that woody biofuels development could decrease overall carbon sequestration and impede efforts at mitigating climate change, thus putting mitigation through advanced biofuel development on a collision with mitigation through biological sequestration (see section "Interaction of Biofuel Policy with Possible Carbon Policies").

Climate Change and Public Values

Being able to link climate-change mitigation to bioenergy development requires a fairly detailed and sophisticated understanding of these problems and solutions.[29] This

---

[28]Widely recognized and cited research on public environmental values has been conducted by Dunlap and Van Liere (1978). Dunlap and Van Liere introduced the concept of the New Environmental Paradigm (NEP) (as opposed to the more conservative, less environmentally oriented Dominant Social Paradigm) operationalized through the NEP scale (Dunlap and Van Liere, 1978). NEP has become the gold standard for measuring environmental values, and the questions therein are frequently used or adapted to be included as a portion of a survey where one of the variables that the authors aim to measure is environmental orientation.

However, some critics argue that instead of reflecting the breadth of values held by the "general" American public, NEP instead is based in the philosophy, concepts, and terms of the environmental movement as represented by well-established groups like the Sierra Club and Greenpeace (Kempton et al., 1996). The critics argue that environmental values are widely shared and deeply held among the American public, even among political conservatives, but Dunlap and Van Liere's work suggests those are not shared values. Kempton et al. (1996) argue that many people hold environmental values but may not self-identify as "environmentalists" and may use different terms to present their concerns than those used by the more established environmental movement. They did not find a countervailing, antienvironmental "Dominant Social Paradigm" when they assessed environmentally related cultural models held by Americans.

[29]There is a growing field of literature that links support for climate-change mitigation policies to understandings of climate-change causes and effects (Kempton et al., 1996; Dietz et al., 2007; Solomon and Johnson, 2009; Dunlap, 2010; Johnson et al., 2011). Kempton et al. (1996) demonstrated that the American public's understanding of climate-change causes was confused and problematic—for example, many confused the hole in the ozone layer with climate change. At the time those articles were written, the authors suggested that this lack of understanding would likely reduce support for effective climate-change mitigation strategies because people would not understand how the solutions work. More than ten years later, researchers found that

need for in-depth understanding is also true of potential effects related to biofuel development. Being able to understand how biofuel development might affect water quality or biodiversity requires a level of attention to the details of environmental problems and solutions that is unlikely to be paid by average Americans, whether or not they consider themselves concerned about environmental protection. Instead, concerns about some of the more obscure issues, such as water-quality effects, will tend to be held by individuals who devote significant time to understanding environmental problems, including staff and members of national and local environmental groups. Other effects, such as the concern of biofuel development indirectly leading to increased timber harvesting and thus to biodiversity reductions or increased incentives to expand forest investments and biodiversity benefits, are complex and depend on many assumptions. Because the level of understanding required to see all of these linkages may not be common, studies assessing overall concern regarding particular values, independent of linkages to climate change or biofuel effects, are needed.

Water Quality and Public Values

The water quality-related effects of conventional corn production are discussed in the next chapter. Less documented are the conditions under which other biofuels such as soybean biodiesel and cellulosic biofuel can result in substantial benefits. Nonetheless, negative effects are conceivable under some scenarios such as one in which poorly executed timber harvesting to produce ethanol feedstock takes place on steep slopes proximate to coldwater fishery streams. On the other hand, to the extent that cellulosic biofuel market can reduce the rate of loss of America's farms and forests to industry and suburban sprawl, they may generate positive effects on water supplies, habitats, and viewscapes. Therefore, insofar as biofuel development alters water quality, it has the potential to affect highly valued resources, including aquatic habitats, drinking water sources, and recreationally valuable water bodies (Wilson and Carpenter, 1999).

Wilson and Carpenter (1999) performed a meta-analysis of the literature on public values and freshwater ecosystem services in the United States. They focused on studies using economic tools to attempt to establish dollar values for mostly nonmarket goods. Although they argued that many methods and approaches failed to establish consensus regarding specific value levels, they agreed that these services are highly valued. In addition, their findings support the notion that biofuel development can affect hydrologic services that are highly valued by the American public.

Forests and Public Values

Like hydrologic systems, forested systems have the potential to provide a wide variety of goods and ecosystem services highly valued by the American public. Although the EISA renewable biomass definition currently precludes the most highly valued "natural" forest ecosystems from usage for biofuel production, the relaxation of this definition to allow feedstock from a wider set of forest types could affect highly valued forests (Bengston and Xu, 1995; Xu and Bengston, 1997; Bengston et al., 2004). Xu and Bengston (1997) tracked forest-related values over time and found decreases in expressions of concern for

---

misconceptions are still common (Solomon and Johnson, 2009; Johnson et al., 2011). However, the same work showed that individuals with accurate understanding of causes and effects supported accurate understandings of appropriate solutions.

product-oriented values (such as timber and paper pulp) and increases in "life support" values, such as ecosystem services. Bengston et al. (2004) reanalyzed changes in American forest-related values over time and found declines in overall concern for product-oriented values and continued increases in concern for life-support values.

Increased public support for the protection of forests, among other environmental concerns, has translated over the past 40 years into a series of federal laws that specifically target forests or have broader environmental protection goals, of which one is forest protection. These 1970s-era laws include the National Environmental Policy Act (NEPA), the Endangered Species Act, and the National Forest Management Act (NFMA). In addition to including a variety of mechanisms designed to help ensure the protection of landscapes like forests, these laws provide tools for outside groups to challenge federal forest management through allowances for administrative appeals of federal decisions, public notice requirements related to important federal agency decisions, and public information mandates (for example, NFMA's publicly available forest planning documents and NEPA's Environmental Impact Statements) that are open to public comment and challenge. In some ways, federal forests may be the U.S. landscape type where management decisions are most open to outside challenge because tort law has become a part of the history of management of these forests. One outcome has been increased public awareness of forest protection goals and the inclusion in RFS2 of a definition of renewable biomass intended to protect those goals in preventing the use of federal forest materials as feedstock for RFS-compliant biofuels.

EFFECTS OF ADJUSTMENTS TO AND INTERACTIONS WITH U.S. BIOFUEL POLICY

As discussed in Chapter 1, biofuel policy exists to address three challenges faced by the U.S. economy: energy security, GHG emissions reduction, and rural development. However, because biofuels are not cost-competitive with fossil fuels, supporting their development and commercialization has direct costs, such as the tax credits, and possible indirect costs, such as the repercussions of any upward pressure on food prices. Although the committee was asked to recommend means by which the federal government could prevent or minimize adverse effects of RFS2 on the price and availability of animal feedstuffs, food, and forest products, it refrained from making policy recommendations. Policies have tradeoffs among goals and objectives. As such, policy recommendations reflect the recommenders' values of which tradeoffs are acceptable. This committee is not in the position of passing judgment on which tradeoffs are acceptable to society, but it can provide an assessment to inform decision-makers of the potential effects of several policy options. Therefore, some policy options that have been proposed to reduce or mitigate direct and indirect costs are discussed in this section without endorsement or criticism. This section also examines how biofuel policy may interact with a federal policy on reducing carbon because both policies have or would have GHG emissions reduction as an objective. Depending on how a carbon policy is implemented, it could reinforce GHG emissions reductions from biofuels; alternatively, it could compete with biofuels for land.

Potential Policy Alternatives

Though tax incentives for biofuels have been in place for over 30 years, securing funding for them in federal legislation is increasingly precarious. At the end of 2009, for example, the $1 per gallon tax credit for biodiesel was allowed to expire. It was eventually retroactively reinstated in legislation passed in December 2010, but the expiration had created uncertainty and caused production to stagnate (Abbott, 2010; Neeley, 2010; Stebbins, 2011). Furthermore, the biodiesel credit was only extended until the end of 2011. The tax credit for corn-grain ethanol and the tariff on imported ethanol also are scheduled to expire at the end of 2011. The credit for cellulosic biofuel is in place until the end of 2012.

Tax credits have almost always been extended—the need to reinstate the biodiesel credit is the exception rather than the rule. However, the uncertainty related to whether and for how long the credits will be extended deters long-term investors. Moreover, under the stress of a weaker economy, GAO (2011) has identified the VEETC as a redundant policy that could be eliminated to reduce pressure on the federal budget. GAO, some policymakers, and biofuel proponents have proposed amended or different forms of support to biofuels.

One option would be to link a biofuel subsidy with the price of crude oil or gasoline (GAO, 2011). This type of support mechanism would provide a payment when fossil fuel prices are low but decrease or cease when fossil fuel prices are high and biofuels are, thus, more competitive. The payment could be in the form of a blender's tax credit, or it could be paid directly to the producer, provided that the product is not subsequently exported.[30] This approach has the potential to reduce government spending on biofuel subsidies and diminish any upward pressure on agricultural commodity prices that could be caused by competition with biofuels when oil prices are high. However, like the other options discussed here, it is not tied directly to policy objectives such as reduction of GHG emissions.

Another possibility would be to direct the subsidy to the producer and to divide the payments into two parts. The first part would be a constant, per-gallon (or energy-equivalent) support payment. The second part would be a function of the GHG emissions reduction achieved by the producer. The policy objective would be to provide an incentive for biofuel producers to reduce GHG emissions as much as possible. The industry proposal for this option calls for accounting for emissions based on direct GHG emissions, not emissions that may occur through land-use change.

A subsidy could also be structured that favored the energy content of the fuel rather than the volume of fuel produced. For example, because drop-in fuels contain 1.5 times the energy of ethanol, they would receive a subsidy 1.5 times that of ethanol. Subsidies based on energy content instead of volume effectively level the playing field among competing technologies. RFS2 has already been converted from a volumetric standard to an energy standard as EPA has interpreted the standard as gallons of ethanol equivalent. Thus, if the mandate for 16 billion gallons of cellulosic biofuel were filled with drop-in fuels, only 10.7 billion gallons would be required. The subsidy payment could be in the form of a blender's tax credit or a payment to the producer.

A proposal has also been made to change the blender's credit to a production tax credit that would be applied to the producing firm. However, if the subsidy is redirected in this way, there would have to be a restriction on exporting the product to avoid violating agreements under the World Trade Organization related to export subsidies.

Instead of paying producers, processors, or blenders to make and use biofuels, another option would be to eliminate the need for the types of subsidies discussed above. This could be done by investing in research and development to make biofuel production and commercialization more cost-competitive with fossil fuels.

Subsidies in any form have a negative impact on the federal budget. An alternative is to increase taxes on fuels made from petroleum. The federal tax on gasoline was last increased in

---

[30]If the product were exported, the subsidy would essentially act as an export subsidy, which is a violation of World Trade Organization regulations.

1993. Inflation has decreased the real tax rate by one third between 1993 and 2011. Fuel taxes are dedicated to maintaining transportation infrastructure, of which biofuels could be a part.

Interaction of Biofuel Policy with Possible Carbon Policies

Because RFS2 was motivated in part by GHG emissions concerns, other governmental policies enacted to reduce carbon emissions will interact with the mandate. To the extent that biofuels result in liquid fuels with lower carbon intensity than fossil fuels, they would be favored in the carbon marketplace. Therefore, returning to the construct of the BioBreak model, carbon-reduction policies could encourage biofuel production by acting as a subsidy to close the price gap between a processor's WTP and a supplier's WTA. The mechanism for such a policy could take different forms, which are discussed below. However, carbon prices could also lead to shifts in land use that may favor carbon sequestration over the harvest of biomass (Wise et al., 2009), potentially favoring certain types of feedstocks or reducing the amount of feedstock available for fuel, and possibly food, production.

Biomass Costs with a Carbon Market

In addition to the subsidy options outlined above, another possible government intervention to encourage biomass production is to eliminate the price gap between the processor's WTP and the supplier's WTA by placing a price on carbon. The price would come from a carbon tax or carbon credit. The question is: What price would be required to establish a viable biomass fuel market? To derive the implicit price of carbon, a policy intervention in the cellulosic biofuel market that is motivated solely by the environmental benefits from GHG emissions reductions from biofuel relative to conventional fuel is assumed. Alternatively, the implicit price can be viewed as attributable to energy security and rural development benefits in addition to GHG reduction benefits.

BioBreak extends the breakeven analysis by using GREET[31] 1.8d GHG emissions savings from cellulosic ethanol relative to conventional gasoline along with the price gap to derive a minimum carbon credit or carbon price necessary to sustain a feedstock-specific cellulosic ethanol market. This carbon price can be thought of as either a carbon tax credit provided to the ethanol producer (or feedstock supplier) per dry ton of cellulosic feedstock refined or as the market price for carbon credits if processors are allocated marketable carbon credits for biofuel GHG reductions relative to conventional gasoline. Given the parameter assumptions of 2010 biorefining technology and 23.4 miles per gallon gasoline-equivalent (mpg$_{ge}$) fuel economy in the U.S. fleet of conventional and flex-fuel vehicles (E85), Figure 4-23 provides the carbon price needed to sustain each feedstock-specific biofuel market at an oil price of $111 per barrel. Only three feedstocks are considered: corn stover, wheat straw, and forest residue. The dedicated bioenergy crop feedstocks considered earlier in this chapter using BioBreak (that is, alfalfa, switchgrass, *Miscanthus*, and short-rotation woody crops) are not reported in this analysis because of the high degree of uncertainty surrounding their potential to reduce carbon emissions relative to petroleum-derived fuels. (See Chapter 5 for further discussion on life-cycle GHG emissions of biofuels produced from dedicated bioenergy crop feedstocks.) For the three feedstocks considered, the carbon price ranges between $118 and $138 per metric ton carbon dioxide equivalent (CO_2 eq). This carbon price can be interpreted as the carbon price needed to sustain feedstock-specific

---

[31] The Greenhouse Gases, Regulated Emissions, and Energy Use in Transportation Model by Argonne National Laboratory.

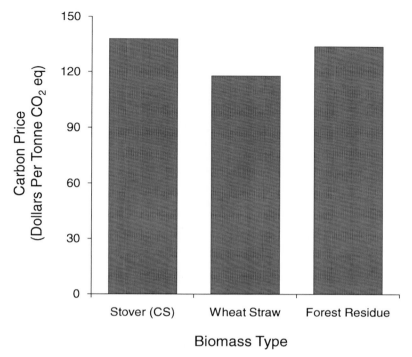

FIGURE 4-23 Projected carbon price needed for feedstock market ($ per metric ton) based on Bio-Break and using GREET 1.8d GHG emissions savings from ethanol relative to conventional gasoline along with the price gap to derive a minimum carbon credit or carbon price necessary to sustain a feedstock-specific cellulosic ethanol market.
NOTE: No policy incentives ($111 per barrel oil, 70 gallons per dry ton, 23.4 mpg_{ge}, 2010 biorefinery).

cellulosic ethanol production if carbon credits for GHG reductions were the only policy incentive. The reported carbon prices should not be interpreted as the carbon price needed to meet RFS2. Additional biomass feedstocks beyond corn stover, wheat straw, and forest residue will likely be needed to meet the RFS2 consumption mandate for cellulosic ethanol. Therefore, the carbon price needed to meet the mandate will depend on the price gap and the reductions in carbon emissions from additional feedstocks.[32]

Advancements in technology and efficiency will decrease the price gap. If biorefinery efficiency increases to GREET 2020 default assumptions, the ethanol conversion rate increases to 80 gallons per dry ton of feedstock, and the fuel economy increased to 25.4 mpg_{ge} for conventional and flex-fuel vehicles, then the carbon price drops by $18 to $24 per

---

[32]With the exception of short-rotation woody crops, dedicated bioenergy feedstocks considered earlier in this chapter have a significantly larger price gap than the three feedstocks considered in the carbon pricing analysis. If carbon emissions reductions from (more expensive) dedicated bioenergy crops are compared to the emissions reductions from corn stover, wheat straw, or forest residue, the carbon price needed to meet the RFS2 mandate will be significantly higher than the crop and woody residue values reported in Figure 4-23. Lower emissions reductions would only exacerbate this effect and result in a significantly higher carbon price while higher emissions reductions would reduce the carbon price needed to meet RFS2.

tonne for a resulting carbon price between $100 and $120 per tonne CO_2 eq.[33] The estimated carbon price is sensitive to technological progress, such as biorefinery and fuel efficiency, and to the parameter values influencing breakeven values for the biomass processor and supplier, including the oil price, regional biomass productivity, and parameter variability.

Interaction with Agricultural and Forestry Offsets

Under a national carbon policy, biofuels would not be the only way to reduce carbon. Another means that could be encouraged is using agricultural land or forestry to supply carbon credits to carbon markets. These credits have come be to known as "offsets," given that they are often assumed to be used to offset emissions of carbon dioxide from the energy sector. Offsets work by reducing GHG emissions from some activity in the agricultural or forestry sector or by increasing the carbon stored in soils or forest biomass. Examples of activities that potentially generate offsets include conversion of land to forests (afforestation), extending timber rotations, increasing forest management, shifting from conventional tillage to no tillage, reducing methane (CH_4) emissions from livestock operations, reducing fossil fuel use associated with agriculture, and reducing nitrogen oxide emissions from production agriculture (EPA, 2005). The amount of carbon offset by feedstocks differs with the type of crop, previous land uses, crop management, and agrichemical use.

Environmental policies that encourage carbon offsets in agriculture or forestry will have important interactions with biofuel policies because they could influence the total amount of land devoted to forest or agricultural production. A number of studies have examined the implications of offsets for land use in the United States. One of the most important effects relates to the potential for land-use change, and specifically for conversion of land into forests. A model by EPA (2005) suggested that up to 90 million acres of crop and pastureland could be converted to forestland if carbon prices are in the range of $15-$50 per tonne of CO_2 eq. Results from Sohngen (2010) suggest similarly large changes in land use, around 100 million acres of new forestland by 2050 with carbon prices of $30 per tonne CO_2 eq.[34] These studies assume one of two types of carbon payment regimes, either a carbon rental regime that makes rental payments as carbon is stored in forests and a payment for storage in wood products or a subsidy for storage and a tax for the net emission at harvest. Note that these two payment schemes are equivalent in present value terms for newly planted forests. In both these studies, the changes in land use described above are net of all underlying changes. New forests are derived from a combination of crop, pasture, and rangeland. Additional land-use changes outside the United States are captured by the global study by Sohngen (2010) but not by the EPA (2005) study. The scale of the changes in land use associated with carbon policies suggests that the overall value for land would increase dramatically if carbon policies were implemented.

Such large shifts in land use occur with carbon offset policies because offsets in forestry are particularly valuable. Consider a typical acre of cropland in the Eastern Corn Belt. The accumulation of carbon if land is converted to mixed hardwoods could be as much as 4 tonnes CO_2 eq per acre per year (Figure 4-24). Further, the carbon in the mixed hardwoods

---

[33]Further, if fuel cell vehicle technology operating on pure ethanol (E100) is available by 2020, the carbon price would decrease to range between $54 to $68 per tonne CO_2eq assuming a fuel economy of 44.3 mpg$_{ge}$ for fuel cell vehicles and a conversion yield of 80 gallons per dry ton of feedstock.

[34]Those model projections must be interpreted in the context of recent changes in land-use patterns in the United States. From 1982 to 2007, cultivated cropland declined by 70 million acres (from 375 million to 305 million acres), developed land increased by 40 million acres (from 71 million to 111 million acres), and there was little change in forestland, pastureland, rangeland, and noncultivated cropland (USDA, 2009).

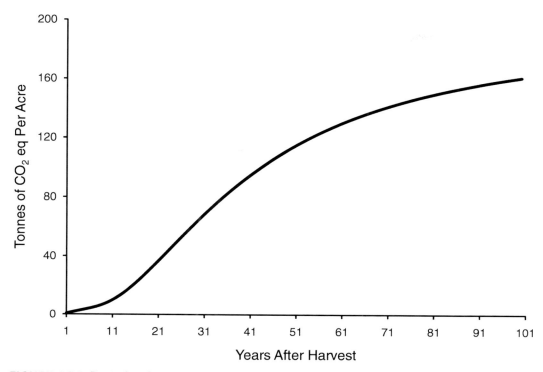

FIGURE 4-24 Typical carbon accumulation on mixed hardwood stands in the Eastern Corn Belt of the United States.

can be observed, as it is stored in trees growing on the landscape. The value of the carbon asset in an acre of trees with a carbon price of $15 per tonne CO_2 eq would be $1,350 per acre, assuming a discount rate of 5 percent. Compared to a timber value of about $250 per acre with current timber prices, carbon markets could provide strong incentives for land conversion. Note that these calculations of land value assume that timber is harvested with and without carbon values, although when carbon is valued, the economically optimal rotation age increases from 52 to 63 years of age.

Other types of offsets, such as those that reduce CH_4 emissions and nitrogen oxide emissions, are also valuable, but they do not have a strong influence on the competitive balance between cropland, pasture, and forestland. CH_4 recovery in livestock operations is done mainly in confined livestock operations where it makes economic sense to recover the CH_4. If the value of carbon offsets is to increase, such systems could influence the returns to animal operations so that the total number of animals could increase, thereby increasing the demand for feed. However, with current values for natural gas, CH_4 recovery is not profitable enough to have a large effect on projections of animal units.

In addition to afforestation, forest carbon offsets credits may be developed for other activities in forests, including increasing the rotation age in forests or for increasing the intensity of management in order to increase the total carbon stored on site. For example, in the Midwestern hardwood case examined above, a carbon price of $15 per tonne CO_2 eq increased the optimal rotation age from 52 years to 63 years. A number of studies have now shown that these actions are relatively low-cost options for storing carbon in forests

BOX 4-6
Comparison of Southern Pine and Hybrid Poplar in
Producing Timber and Biomass on Similar Sites

Unlike annual agricultural crops, forests produce output periodically. In North America, rotations can be fairly long, ranging from 8-10 years for short-rotation woody crops to 20-30 years for Southern pines to 40-50 years for Douglas Fir in the Pacific Northwest. An interesting question from the perspective of providing a stable supply of biomass relates to determining the optimal timber rotation. Optimal timber rotations are most often determined in terms of maximizing the value of the land and not in terms of maximizing timber supply. Due to discounting, maximizing value implies less output than maximizing output.

Southern pine and poplar, two alternatives for a productive site in the Southern United States, were compared to show the effect of maximizing value. For this example, it was assumed that productivity of the site is the same for both tree types, so they will have approximately the same maximum annual growth. Growth functions of the following functional form were developed for these two tree types:

Southern pine: Volume (dry tons/acre) = exp(5.19 − 25.39/AGE)

Poplar: Volume (dry tons/acre) = exp(4.76 − 16.44/AGE)

These growth functions are shown in Figure 4-25. To compare the two types of trees, several measures of growth are of interest. One measure of growth is the maximum periodic, or annual, growth. This is the maximum amount of biomass accumulated in a single year. Because the two types of trees have different growth characteristics, they accumulate biomass at different rates over time, and the maximum accumulation occurs at different time periods (Table 4-7). As noted above, the two tree types are assumed to be grown on the same site and thus subjected to the same soil and climatic conditions. As a result, the maximum rates of growth are constrained to be roughly the same.

Another measure of growth is the average annual growth, which is measured as the total volume divided by the age. This calculation provides a measure of the average annual material harvested per year if the stand were cut at the given age. The maximum average amount of material that could be harvested per year is 2.6 dry tons per acre per year. Because the growth functions have a different shape, the year at which the average annual harvest is maximized is different for each type. For Southern pine, the average annual growth is maximized at 25 years and for hybrid poplar it is maximized at 16 years.

As a landowner, maximizing annual flow is less important than maximizing the value of the land. The value of the land is maximized when the net present value of the stand is maximized. For this analysis, planting costs were assumed to be $243 per acre, and additional management costs were ignored to simplify the analysis. At rotation age, 90 percent of the growing stock biomass was assumed to be removed. For the Southern pine stand, 60 percent is used for sawtimber and is valued at $19.31 per dry ton ($36.50 per m³). The remainder is used for pulp and is valued at $5.15 per dry ton ($9.73 per m³). For hybrid poplar, all the material can be used for sawtimber, although the value of the sawtimber is assumed to be $14.81 per dry ton ($28 per m³). Land value is maximized in 24-year rotations with Southern pine and 14-year rotations with hybrid poplar. Despite lower value for the harvested material, the hybrid poplar has a larger stand value due to the shorter rotations and the ability to produce more solidwood in that shorter time period.

Interestingly, if all this material were converted to biomass feedstock supply at $4.76 per dry ton ($9 per m³), or roughly the pulpwood price, land value would be $277 per acre, or only about 25 percent of the potential value. Lower planting costs could increase the land value, but it is not clear if these lower planting costs could achieve the same stocking densities and biomass production.ea

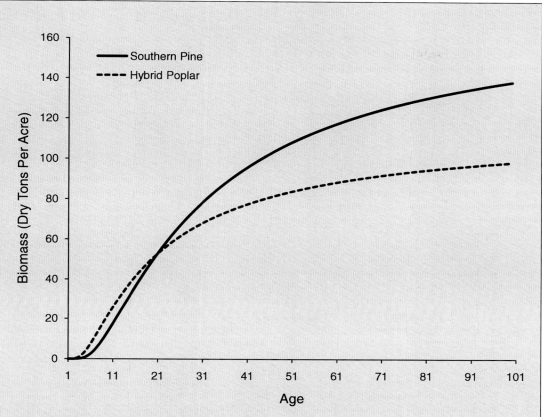

FIGURE 4-25 Growth of growing stock biomass.

TABLE 4-7 Growth Characteristics of Alternative Timber Types

	Southern Pine	Hybrid Poplar
Maximum annual growth in dry tons per acre (age maximized)	3.82 (13)	3.84 (9)
Average annual growth at year 10 in dry tons per acre	3.48	3.76
Maximum average annual growth in dry tons per acre per year (age maximized)	2.6 (25)	2.61 (16)
Maximum Net Present Value (NPV)	$854	$1,123
Rotation age that maximizes NPV	24	14
Average annual flow of biomass at the rotation age that maximizes NPV in dry tons per acre per year	2.6	2.57

in the United States, and they could constitute 30 to 50 percent of the total carbon seques-
tered in the next 30 years (EPA, 2005; Sohngen, 2010). Payments for these activities can be
provided to owners with standing timber stocks in order to generate carbon offset credits.
Since these incentives value the standing stock of timber, they would serve to increase the
value of forests and make standing forests more competitive with other types of land uses,
including feedstock production for biofuels. In other words, comprehensive carbon offset
policies that pay for offsets through management and increasing timber rotation ages will
increase the value of land and make the provision of tons of biomass for bioenergy markets
more expensive.

It is difficult at this time to determine what the net effect of both carbon offsets and RFS2
would be on land use. Market studies of carbon offsets imply that additional land converts
from livestock and crops to forests under most carbon price scenarios and that the returns
to all types of land uses increase, though these studies do not account for the expansion
of developed land. The recent EPA study on RFS2 suggests that the area of land used for
dedicated bioenergy crops, such as switchgrass, increase, while forestland and rangeland
decline. The combination of these results suggests that carbon offsets would compete with
cellulosic biofuel production for the same land; thus, environmental policies that encourage
carbon offsets could raise the costs of producing cellulosic biofuel feedstocks. Considered a
different way, however, this also implies that carbon offsets could limit the negative externalities associated with converting natural forests to dedicated bioenergy crops.

An important caveat to this conclusion occurs if cellulosic biofuel feedstocks are increasingly derived from forest sawtimber and pulpwood supplies rather than residues. In
this case, the demand for wood products would increase, timber prices would rise, and land
returns in forestry would increase. Because long-term, sustained increases in timber prices
raise timber rotation ages over time, and an increase in rotation age expands the supply of
timber, in some cases, biofuel outputs and carbon offsets may be complementary.

An additional possible response of markets to increasing biofuel demands is to shift
land towards shorter rotation species. Shorter rotations can increase land values, but they
do not necessarily increase timber supplies, as discussed in Box 4-6. The key way in which
shorter rotations can increase timber supplies occurs if managers are able to manage them
better to produce the desired outputs. For instance, there has been a long history of conversion of hardwoods to softwoods in the Southern United States. The key gain here has been
an increase in value on the landscape as managers have been better able to control conditions on softwood plantations, and they have been able to obtain higher value output per
acre with softwoods than hardwoods. However, net production of biomass on hardwoods
is typically greater, but less of it is suitable for high-value market products on a per acre
basis (Sohngen and Brown, 2006).

CONCLUSION

Because cellulosic biofuel is not yet commercially viable, the economics of this type of
fuel and its economic effects on other commodities and government programs are speculative. However, with the data that are available and the present state of technology, cellulosic
biofuel is not cost-competitive with fossil fuels without government support. Unless more
subsidies are used, the RFS2 mandate is enforced rigidly, taxes on petroleum products
are increased, or rapid technological advancements are made, cellulosic biofuel will not
substantially affect other commodity markets, though it could have repercussions for the
federal budget. If cellulosic biofuel becomes commercially viable, land prices will increase
due to competition with other agricultural or forestry uses, though the extent of the increase

due to biofuels will depend on the productivity of the land used for biomass production as well as demand for other uses of the land. Fossil fuel prices may decline slightly and imports will decrease, but this will also be influenced by improved fuel efficiency in the U.S. vehicle fleet and the capacity of the U.S. fleet to use biofuels. Because of its scarcity and its density, more woody biomass may be imported to meet the demand for biofuels and traditional uses.

Corn-grain ethanol and, to a lesser extent, soybean biodiesel are closer to being competitive with fossil fuels, particularly when combined with the tax credit and encouraged by RFS2. They have contributed to upward price pressure on agricultural commodities, food, and livestock feed; however, they are just one factor among many, including the growing global population, crop failures in other countries, decline in the value of the U.S. dollar, and speculative activity in the marketplace. The greater use of DDGS in animal feed to some extent has muted the unfavorable effects on the livestock industry.

If policies that were in place at the time this report was written are continued, it is extremely likely that meeting RFS2 will increase the federal budget, particularly in terms of subsidies spent on grants, loans, and loan guarantees to encourage cellulosic biofuel production and in terms of tax revenue forgone by the tax credits for blending biofuel with fossil fuels. To the extent that biofuel policy has raised food-related prices, it has affected federal spending in programs related to agriculture and food. Deciphering biofuels' contribution to increases or decreases in these programs is difficult, though, because of the number of variables. The effect of RFS2 on federal spending on conservation programs is uncertain. It remains to be seen whether biofuel feedstock production competes with acres or payments for CRP.

REFERENCES

Abbott, C. 2010. U.S. Biofuels Hurt if 2010 Tax Break Expires—Report. Available online at http://www.reuters.com/article/2010/03/09/biofuels-usa-taxbreak-idUSN0924548320100309. Accessed February 10, 2011.

Abbott, P.C., C. Hurt, and W.E. Tyner. 2009. What's Driving Food Prices? March 2009 Update. Oak Brook, IL: Farm Foundation.

Abbott, P.C., C. Hurt, and W.E. Tyner. 2011. What's Driving Food Prices in 2011? Oak Brook, IL: Farm Foundation.

Aden, A., M. Ruth, K. Ibsen, J. Jechura, K. Neeves, J. Sheehan, B. Wallace, L. Montague, A. Slayton, and J. Lukas. 2002. Lignocellulosic Biomass to Ethanol Process Design and Economics Utilizing Co-Current Dilute Acid Prehydrolysis and Enzymatic Hydrolysis for Corn Stover. Golden, CO: National Renewable Energy Laboratory.

Anex, R.P., A. Aden, F.K. Kazi, J. Fortman, R.M. Swanson, M.M. Wright, J.A. Satrio, R.C. Brown, D.E. Daugaard, A. Platon, G. Kothandaraman, D.D. Hsu, and A. Dutta. 2010. Techno-economic comparison of biomass-to-transportation fuels via pyrolysis, gasification, and biochemical pathways. Fuel 89(Supplement 1):S29-S35.

Babcock, B., S. Marette, and D. Treguer. 2011. Opportunity for profitable investments in cellulosic biofuels. Energy Policy 39:714-719.

Babcock, B.A., and J.F. Fabiosa. 2011. The Impact of Ethanol and Ethanol Subsidies on Corn Prices: Revisiting History. Ames: Iowa State University - Center for Agricultural and Rural Development.

Babcock, B.A., K. Barr, and M. Carriquiry. 2010. Costs and Benefits to Taxpayers, Consumers, and Producers from U.S. Ethanol Policies. Ames: Iowa State University - Center for Agricultural and Rural Development.

Baffes, J., and T. Haniotis. 2010. Placing the 2006/08 Commodity Price Boom into Perspective. Washington, DC: The World Bank.

Baier, S., M. Clements, C. Griffiths, and J. Ihrig. 2009. Biofuels Impact on Crop and Food Prices: Using an Interactive Spreadsheet. Washington, DC: Board of Governors of the Federal Reserve System.

Banse, M., H. van Mejil, A. Tabeau, and G. Woltjer. 2008. Will EU biofuel policies affect global agricultural markets? European Review of Agricultural Economics 35(2):117-141.

Becker, D., P. Jakes, D. Abbas, K.E. Halvorsen, P.J. Jakes, S.M. McCaffrey, and C. Moseley. 2009. Characterizing Lessons Learned from Federal Biomass Removal Projects. Boise, ID: Joint Fire Science Program.

Bengston, D.N., and Z. Xu. 1995. Changing National Forest Values: A Content Analysis. St. Paul, MN: U.S. Department of Agriculture - Forest Service.

Bengston, D.N., T.J. Webb, and D.P. Fan. 2004. Shifting forest value orientations in the United States, 1980–2001: A computer content analysis. Environmental Values 13(3):373-392.

Cai, X., X. Zhang, and D. Wang. 2011. Land availability for biofuel production. Environmental Science & Technology 45(1):334-339.

Capehart, T., and J. Richardson. 2008. Food Price Inflation: Causes and Impacts. Washington, DC: Congressional Research Service.

CBO (Congressional Budget Office). 2009. The Impact of Ethanol Use on Food Prices and Greenhouse-Gas Emissions. Washington, DC: Congressional Budget Office.

CBO (Congressional Budget Office). 2010. Using Biofuel Tax Credits to Achieve Energy and Environmental Policy Goals. Washington, DC: Congressional Budget Office.

Collett, S., and P. Villega, eds. 2005. The Poultry Informed Professional. Issue 85:July/August. Athens: The University of Georgia.

Collett, S., and P. Villega, eds. 2008. The Poultry Informed Professional. Issue 99:May/June. Athens: The University of Georgia.

Collett, S., and P. Villega, eds. 2011. The Poultry Informed Professional. Issue 118:May/June. Athens: The University of Georgia.

Collins, K. 2008. The Role of Biofuels and Other Factors in Increasing Farm and Food Prices: A Review of Recent Developments with a Focus on Feed Grain Markets and Market Prospects. Northfield, IL: Kraft Food Global, Inc.

Cotti, C., and M. Skidmore. 2010. The impact of state government subsidies and tax credits in an emerging industry: Ethanol production 1980-2007. Southern Economic Journal 76(4):1076-1093.

Cowan, T. 2010. Conservation Reserve Program: Status and Current Issues. Washington, DC: Congressional Research Service.

Crompton, J. 2008. Food Bank Crisis: Funding Cutbacks Cripple Efforts to Feed the Hungry. Available online at http://www.post-gazette.com/pg/08104/872253-58.stm. Accessed November 2, 2010.

Cubbage, F., S. Koesbandana, P. Mac Donagh, R. Rubilar, G. Balmelli, V.M. Olmos, R. De La Torre, M. Murara, V.A. Hoeflich, H. Kotze, R. Gonzalez, O. Carrero, G. Frey, T. Adams, J. Turner, R. Lord, J. Huang, C. MacIntyre, K. McGinley, R. Abt, and R. Phillips. 2010. Global timber investments, wood costs, regulation, and risk. Biomass and Bioenergy 34(12):1667-1678.

de Gorter, H., and D.R. Just. 2010. The social costs and benefits of biofuels: The intersection of environmental, energy and agricultural policy. Applied Economic Perspectives and Policy 32(1):4-32.

de Groot, R.S., M.A. Wilson, and R.M.J. Boumans. 2002. A typology for the classification, description and valuation of ecosystem functions, goods and services. Ecological Economics 41(3):393-408.

Dietz, T., A. Dan, and R. Shwom. 2007. Support for climate change policy: Social psychological and social structural influences. Rural Sociology 72(2):185-214.

Donohue, M., and D.L. Cunningham. 2009. Effects of grain and oilseed prices on the costs of U.S. poultry production. Journal of Applied Poultry Research 18(2):325-337.

Duffield, J.A., I. Xiarchos, and S.A. Halbrook. 2008. Ethanol policy: Past, present, and future. South Dakota Law Review 53(3):425-456.

Dunlap, R.E. 2010. Climate change and rural sociology: Broadening the research agenda. Rural Sociology 75(1):17-27.

Dunlap, R.E., and K.D. Van Liere. 1978. The "new environmental paradigm": A proposed measuring instrument and preliminary results. Journal of Environmental Education 9(4):10-19.

EIA (U.S. Energy Information Administration). 2008a. Annual Energy Outlook 2008. Washington, DC: U.S. Department of Energy - Energy Information Administration.

EIA (U.S. Energy Information Administration). 2008b. Federal Financial Interventions and Subsidies in Energy Markets 2007. Washington, DC: U.S. Department of Energy - Energy Information Administration.

EIA (U.S. Energy Information Administration). 2010a. Annual Energy Outlook 2010—With Projections to 2035. Washington, DC: U.S. Department of Energy - Energy Information Administration.

EIA (U.S. Energy Information Administration). 2010b. Annual Energy Review 2009. Washington, DC: U.S. Department of Energy - Energy Information Administration.

EIA (U.S. Energy Information Administration). 2010c. Petroleum and other liquids: Cushing, OK WTI spot price FOB (Dollars per Barrel). Available online at http://www.eia.doe.gov/dnav/pet/hist/LeafHandler.ashx?n=PET&s=RWTC&f=D. Accessed September 20, 2011.

EPA (U.S. Environmental Protection Agency). 2005. Greenhouse Gas Mitigation Potential in U.S. Forestry and Agriculture. Washington, DC: U.S. Environmental Protection Agency.

EPA (U.S. Environmental Protection Agency). 2010. Renewable Fuel Standard Program (RFS2) Regulatory Impact Analysis. Washington, DC: U.S. Environmental Protection Agency.

Espey, M. 1996. Explaining the variation in elasticity estimates of gasoline demand in the United States: A meta-analysis. Energy Journal 17(3):49-60.

Espey, M. 1998. Gasoline demand revisited: An international meta-analysis of elasticities. Energy Economics 20(3):273-295.

FAPRI (Food and Agricultural Policy Research Institute). 2010. U.S. Baseline Briefing Book: Projections for Agricultural and Biofuel Markets. Columbia: University of Missouri-Columbia.

Fight, R.D., B.R. Hartsough, and P. Noordijk. 2006. Users Guide for FRCS: Fuel Reduction Cost Simulator Software. Portland, OR: U.S. Department of Agriculture - Forest Service.

Fischer, G., E. Hizsnyik, S. Prieler, M. Shah, and H. van Velthuizen. 2009. Biofuels and Food Security. Vienna, Austria: OPEC Fund for International Development / International Institute for Applied Systems Analysis.

Forest2Market. 2010. Delivered Price Benchmark Service. Available online at http://www.forest2market.com/. Accessed February 14, 2011.

French, B.C. 1960. Some considerations in estimating assembly cost functions for agricultural processing operations. Journal of Farm Economics 42(4):767-778.

Galik, C.S., R. Abt, and Y. Wu. 2009. Forest biomass supply in the southeastern United States—Implications for industrial roundwood and bioenergy production. Journal of Forestry 107(2):69-77.

GAO (U.S. General Accounting Office). 1993. A Glossary of Terms Used in the Federal Budget Process. Washington: U.S. General Accounting Office.

GAO (U.S. General Accounting Office). 2004. Principles of Federal Appropriations Law. Washington: U.S. General Accounting Office.

GAO (U.S. Government Accountability Office). 2011. Opportunities to Reduce Potential Duplication in Government Programs, Save Tax Dollars, and Enhance Revenue. Washington: U.S. Government Accountability Office.

Gardner, B.L. 1975. Farm retail price spread in a competitive food-industry. American Journal of Agricultural Economics 57(3):399-409.

Gardner, B.L. 1987. The Economics of Agricultural Policies. New York: Macmillan.

Gehlhar, M., A. Winston, and A. Somwaru. 2010. Effects of Increased Biofuels on the U.S. Economy in 2022. Washington, DC: U.S. Department of Agriculture - Economic Research Service.

Glauber, J. 2008. Full Committee Hearing to Receive Testimony on the Relationship between U.S. Renewable Fuels Policy and Food Prices before the Senate Committee on Energy and Natural Resources, June 12.

Goodwin, P., J. Dargay, and M. Hanly. 2004. Elasticities of road traffic and fuel consumption with respect to price and income: A review. Transport Reviews 24(3):275-292.

Graham, D.J., and S. Glaister. 2002. The demand for automobile fuel: A survey of elasticities. Journal of Transport Economics and Policy 36(1):1-25.

Haynes, R.W. 2003. An Analysis of the Timber Situation in the United States: 1952 to 2050. Portland, OR: U.S. Department of Agriculture - Forest Service.

Haynes, R.W. 2008. Emergent Lessons from a Century of Experience with Pacific Northwest Timber Markets. Portland, OR: U.S. Department of Agriculture, Forest Service.

Heaton, E.A., S.P. Long, T.B. Voigt, M.B. Jones, and J. Clifton-Brown. 2004. Miscanthus for renewable energy generation: European Union experience and projections for Illinois. Mitigation and Adaptation Strategies for Global Change 9(4):433-451.

Hertel, T.W., and J.F. Beckman. 2010. Commodity price volatility in the biofuel era: An examination of the linkage between energy and agricultural markets. Paper read at the 2010 Annual Agricultural and Applied Economics Association Meeting, July 25-27, Denver, CO.

Hertel, T.W., W.E. Tyner, and D.K. Birur. 2010. The global impacts of biofuel mandates. Energy Journal 31(1):75-100.

Howard, J.L. 2007. U.S. Timber Production, Trade, Consumption, and Price Statistics 1965 to 2005. Madison, WI: U.S. Department of Agriculture - Forest Service.

Huang, H.-J., S. Ramaswamy, W. Al-Dajani, U. Tschirner, and R.A. Cairncross. 2009. Effect of biomass species and plant size on cellulosic ethanol: A comparative process and economic analysis. Biomass and Bioenergy 33(2):234-246.

IMF (International Monetary Fund). 2008. Is inflation back? Commodity prices and inflation in World Economic Outlook, October 2008: Financial Stress, Downturns, and Recoveries. Washington, DC: International Monetary Fund.

IMF (International Monetary Fund). 2010. IMF primary commodity prices. Available online at http://www.imf.org/external/np/res/commod/index.asp. Accessed September 22, 2010.

ISUE (Iowa State University Extension). 2010. Decision Tools—Crop Decisions. Available online at http://www.extension.iastate.edu/agdm/decisionaidscd.html. Accessed February, 1, 2011.

Jenkins, B.M., R.B. Williams, N. Parker, P. Tittmann, Q. Hart, M.C. Gildart, S. Kaffka, B.R. Hartsough, and P. Dempster. 2009. Sustainable use of California biomass resources can help meet state and national bioenergy targets. California Agriculture 63(4):168-177.

Jensen, H.H., and B.A. Babcock. 2007. Do biofuels mean inexpensive food is a thing of the past? Iowa Ag Review 13(3):4-5, 11.

Jiang, Y., and S.M. Swinton. 2009. Market interactions, farmer choices, and the sustainability of growing advanced biofuels: A missing perspective? International Journal of Sustainable Development and World Ecology 16(6):438-450.

Johnson, D.M., K.E. Halvorsen, and B.D. Solomen. 2011. Upper midwestern U.S. consumers and ethanol: Knowledge, beliefs and consumption. Biomass and Bioenergy 35(4):1454-1464.

Johnson, S. 2008. Commodity Prices: Outlook and Risks. Washington, DC: International Monetary Fund.

Kempton, W., J.S. Boster, and J.A. Hartley. 1996. Environmental Values in American Culture. 1st MIT Press, pbk. ed. Cambridge, MA: MIT Press.

Khanna, M. 2008. Cellulosic biofuels: Are they economically viable and environmentally sustainable? Choices 23(3):16-23.

Khanna, M., and B. Dhungana. 2007. Economics of alternative feedstocks. Pp. 129–146 in Corn-Based Ethanol in Illinois and the U.S.: A Report from the Department of Agricultural and Consumer Economics, University of Illinois. Urbana-Champaign: University of Illinois.

Khanna, M., B. Dhungana, and J. Clifton-Brown. 2008. Costs of producing *Miscanthus* and switchgrass for bioenergy in Illinois. Biomass and Bioenergy 32(6):482-493.

Khanna, M., X. Chen, H. Huang, and H. Onal. 2010. Supply of cellulosic biofuel feedstocks and regional production patterns. American Journal of Agricultural Economics 93(2):472-280.

Kior. 2011. SEC Filings. Available online at http://investor.kior.com/sec.cfm. Accessed August 1, 2011.

Kløverpris, J., H. Wenzel, and P. Nielsen. 2008. Life cycle inventory modelling of land use induced by crop consumption. The International Journal of Life Cycle Assessment 13(1):13-21.

Koplow, D. 2009. State and federal subsidies to biofuels: Magnitude and options for redirection. International Journal of Biotechnology 11(1/2):92-126.

Lazear, E. 2008. Hearing on Responding to the Global Food Crisis before the Senate Foreign Relations Committee, May 14.

Leibtag, E. 2008. Corn prices near record high, but what about food costs? Amber Waves 6(February):11-15.

Lipsky, J. 2008. Commodity Prices and Global Inflation: Remarks to the Council on Foreign Relations, May 8, New York City.

Lustig, N. 2008. Thought for Food: The Challenges of Coping with Soaring Food Prices. Washington, DC: Center for Global Development.

Manomet (Manomet Center for Conservation Sciences). 2010. Massachusetts Biomass Sustainability and Carbon Policy Study: Report to the Commonwealth of Massachusetts Department of Energy Resources. Brunswick, ME: Manomet Center for Conservation Sciences.

Miranowski, J., and A. Rosburg. 2010. Using cellulosic ethanol to "go green": What price for carbon? Paper read at the 2010 Annual Agricultural and Applied Economics Association, July 25–27, Denver, CO.

Miranowski, J., M. Khanna, and J.R. Hess. 2011. Economic of feedstock production, harvest, storage, and transport. Proceedings of the Sustainable Feedstocks for Advanced Biofuels Workshop.

Mitchell, D. 2008. A Note on Rising Food Prices. Washington, DC: The World Bank.

NAS-NAE-NRC (National Academy of Sciences, National Academy of Engineering, National Research Council). 2009a. America's Energy Future: Technology and Transformation. Washington, DC: National Academies Press.

NAS-NAE-NRC (National Academy of Sciences, National Academy of Engineering, National Research Council). 2009b. Liquid Transportation Fuels from Coal and Biomass: Technological Status, Costs, and Environmental Impacts. Washington, DC: National Academies Press.

Neeley, T. 2010. Biodiesel industry asks for reinstatement of tax credit before election. Available online at http://www.dtnprogressivefarmer.com/dtnag/view/blog/getBlog.do;jsessionid=6798D52B11E8AD50A8797C9AD6AB4F3A.agfreejvm2?blogHandle=ethanol&blogEntryId=8a82c0bc2a8c8730012b445bc23e08ec. Accessed February 10, 2011.

NHDRA (New Hampshire Department of Revenue Administration). 2010. Average Stumpage Value List. Concord: New Hampshire Department of Revenue Administration.

OMB (Office of Management and Budget). 2010. Budget of the U.S. Government: Fiscal Year 2011—The Budget. Available online at http://www.whitehouse.gov/omb/budget/Overview. Accessed October 26, 2010.

Perlack, R.D., and B.J. Stokes (Leads). 2011. U.S. Billion-Ton Update: Biomass Supply for a Bioenergy and Bioproducts Industry. Oak Ridge, TN: Oak Ridge National Laboratory.

Perlack, R.D., L.L. Wright, A.F. Turhollow, R.L. Graham, B.J. Stokes, and D.C. Erbach. 2005. Biomass as Feedstock for a Bioenergy and Bioproducts Industry: The Technical Feasibility of a Billion-Ton Annual Supply. Washington, DC: U.S. Department of Energy and U.S. Department of Agriculture.

Perrin, R.K. 2008. Ethanol and food prices: Preliminary assessment. Lincoln: University of Nebraska.

Reinbold, D. 2008. Price Increases Keep Local Food Banks Busy. Available online at http://voicesweb.org/node/1674. Accessed November 2, 2010.

Roland-Holst, D. 2010. Will emerging markets decide the food-fuel debate? Presentation to the Committee on Economic and Environmental Effects of Increasing Biofuels Production on October 7.

Rosegrant, M.W. 2008. Hearing on Biofuels and Grain Prices: Impacts and Policy Responses before the Senate Committee on Homeland Security and Governmental Affairs, May 7.

Ruttan, V.W. 1982. Agricultural Research Policy. Minneapolis: University of Minnesota Press.

Schick, A. 2000. The Federal Budget: Politics, Policy, Process. Revised ed. Washington, DC: Brookings Institution Press.

Schmit, T.M., R.N. Boisvert, D. Enahoro, and L.E. Chase. 2009. Optimal dairy farm adjustments to increased utilization of corn distillers dried grains with solubles. Journal of Dairy Science 92(12):6105-6115.

Sedjo, R.A. 2010. The Biomass Crop Assistance Program (BCAP): Some Implications for the Forest Industry. Washington, DC: Resources for the Future.

Sedjo, R.A., and B. Sohngen. 2009. The Implications of Increased Use of Wood for Biofuel Production. Washington, DC: Resources for the Future.

Smith, C.E. 2010. Special Report; Pipeline construction plans slow for 2010. Available online at http://www.ogj.com/articles/print/volume-108/issue-6/technology/special-report-pipeline.html. Accessed March 8, 2010.

Sohngen, B. 2010. Forestry carbon sequestration. Pp. 114-141 in Smart Solutions to Climate Change: Comparing Costs and Benefits, B. Lomborg, ed. Cambridge: Cambridge University Press.

Sohngen, B., and S. Brown. 2006. The influence of conversion of forest types on carbon sequestration and other ecosystem services in the south central United States. Ecological Economics 57(4):698-708.

Sohngen, B.L., and R.W. Haynes. 1994. The "Great" Price Spike Of '93: An Analysis of Lumber and Stumpage Prices in the Pacific Northwest. Portland, OR: U.S. Department of Agriculture - Forest Service.

Sohngen, B., J. Anderson, S. Petrova, and K. Goslee. 2010. Alder Springs Biomass Removal Economic Analysis. Arlington, VA: Winrock International and U.S. Department of Agriculture - Forest Service.

Solomon, B.D., and N.H. Johnson. 2009. Valuing climate protection through willingness to pay for biomass ethanol. Ecological Economics 68(7):2137-2144.

Solomon, B.D., J.R. Barnes, and K.E. Halvorsen. 2007. Grain and cellulosic ethanol: History, economics, and energy policy. Biomass and Bioenergy 31(6):416-425.

Spelter, H., and D. Toth. 2009. North America's Wood Pellet Sector. Madison, WI: U.S. Department of Agriculture - Forest Service.

Stebbins, C. 2011. Analysis: U.S. biodiesel on life support, but smiling. Available online at http://www.reuters.com/article/2011/02/04/us-usa-biodiesel-idUSTRE7130H920110204?feedType=RSS&feedName=GCA-GreenBusiness. Accessed February 10, 2011.

Steenblik, R. 2007. Biofuels—At What Cost? Government Support for Ethanol and Biodiesel in Selected OECD Countries. Winnipeg, Canada: International Institute for Sustainable Development.

Stubbs, M. 2010. Biomass Crop Assistance Program (BCAP): Status and Issues. Washington, DC: Congressional Research Service.

Swanson, R.M., J.A. Satrio, R.C. Brown, and D.D. Hsu. 2010. Techno-Economic Analysis of Biofuels Production Based on Gasification. Golden, CO: National Renewable Energy Laboratory.

Swinton, S.M., B.A. Babcock, L.K. James, and V. Bandaru. 2011. Higher US crop prices trigger little area expansion so marginal land for biofuel crops is limited. Energy Policy 39:5254-5258.

Taheripour, F., T.W. Hertel, and W.E. Tyner. 2010. Implications of biofuels mandates for the global livestock industry: A computable general equilibrium analysis. Agricultural Economics 10.1111/j.1574-0862.2010.00517.x:1-18.

TCPA (Texas Comptroller of Public Accounts). 2008. Government financial subsidies. Pp. 367-402 in The Energy Report 2008. Austin: Texas Comptroller of Public Accounts.

Timber-Mart South. 2010. South-wide Average Prices Available online at http://www.tmart-south.com/prices.html. Accessed February 14, 2011.

Timilsina, G.R., and A. Shrestha. 2010. Biofuels: Markets, Targets and Impacts. Washington, DC: The World Bank.

Tomek, W.G., and K.L. Robinson. 1983. Agricultural Product Prices. Ithaca, NY: Cornell University Press.

Trostle, R. 2008. Global Agricultural Supply and Demand: Factors Contributing to the Recent Increase in Food Commodity Prices. Washington, DC: U.S. Department of Agriculture - Economic Research Service.

Tyner, W.E. 2008. The U.S. ethanol and biofuels boom: Its origins, current status, and future prospects. BioScience 58(7):646-653.

U.S. Census Bureau. 2010. New Residential Construction: Building Permits, Housing Starts, and Housing Completions. Available online at http://www.census.gov/const/www/newresconstindex.html#. Accessed November 9, 2010.

USDA (U.S. Department of Agriculture). 2009. Summary Report: 2007 National Resources Inventory. Washington, DC, and Ames, Iowa: U.S. Department of Agriculture - Natural Resources Conservation Service and Center for Survey Statistics and Methodology.

USDA (U.S. Department of Agriculture). 2011. USDA Agricultural Projections to 2020. Washington, DC: U.S. Department of Agriculture.

USDA-ERS (U.S. Department of Agriculture - Economic Research Service). 2007. Major Land Uses: U.S. Cropland Used for Crops. Available online at http://www.ers.usda.gov/Data/MajorLandUses/. Accessed February 14, 2011.

USDA-ERS (U.S. Department of Agriculture - Economic Research Service). 2011a. Feed Grains Database: Custom Queries. Available online at http://www.ers.usda.gov/Data/Feedgrains/CustomQuery/. Accessed August 29, 2011.

USDA-ERS (U.S. Department of Agriculture - Economic Research Service). 2011b. Food CPI and Expenditures: Analysis and Forecasts of the CPI for Food. Available online at http://www.ers.usda.gov/Briefing/CPIFoodAndExpenditures/consumerpriceindex.htm. Accessed February 10, 2011.

USDA-ERS (U.S. Department of Agriculture - Economic Research Service). 2011c. Price Spreads from Farm to Consumer: Marketing Bill and Farm Value Components of Consumer Expenditures for Domestically Produced Farm Foods. Available online at http://www.ers.usda.gov/Data/FarmToConsumer/Data/marketingbilltable1.htm. Accessed February 20, 2011.

USDA-FAS (U.S. Department of Agriculture - Foreign Agricultural Service). 2010. Data: Current and Archived Reports. Available online at http://www.fas.usda.gov/data.asp. Accessed February 8, 2011.

USDA-NASS (U.S. Department of Agriculture - National Agricultural Statistics Service). 2007a. Agricultural Prices. Washington, DC: U.S. Department of Agriculture - National Agricultural Statistics Service.

USDA-NASS (U.S. Department of Agriculture - National Agricultural Statistics Service). 2007b. Agricultural Prices: 2006 Summary. Washington, DC: U.S. Department of Agriculture - National Agricultural Statistics Service.

USDA-NASS (U.S. Department of Agriculture - National Agricultural Statistics Service). 2010. Data and Statistics: Quick Stats. Available online at http://www.nass.usda.gov/Data_and_Statistics/Quick_Stats/index.asp. Accessed February 14, 2011.

USITC (U.S. International Trade Commission). 2009. The Economic Effects of Significant U.S. Import Restraints. Sixth Update, 2009. Washington, DC: U.S. International Trade Commission.

USITC (U.S. International Trade Commission). 2010a. Harmonized Tariff Schedule of the United States (2010) (Rev. 2). Washington, DC: U.S. International Trade Commission.

USITC (U.S. International Trade Commission). 2010b. USITC Interactive Tariff and Trade DataWeb. Available online at http://www.dataweb.usitc.gov/. Accessed February 14, 2011.

von Braun, J. 2008. High and rising food prices: Why are they rising, who is affected, how are they affected, and what should be done? Paper read at the U.S. Agency for International Development Conference on Addressing the Challenges of a Changing World Food Situation: Preventing Crisis and Leveraging Opportunity, April 11, Washington, DC.

Warren, D.D. 2010. Production, Prices, Employment, and Trade in Northwest Forest Industries, All Quarters 2009. Portland, OR: U.S. Department of Agriculture - Forest Service.

Wiggins, S., and S. Levy. 2008. Rising Food Prices: A Global Crisis. London: Overseas Development Institute.

Wilson, M.A., and S.R. Carpenter. 1999. Economic valuation of freshwater ecosystem services in the United States: 1971-1997. Ecological Applications 9(3):772-783.

Wise, M., K. Calvin, A. Thomson, L. Clarke, B. Bond-Lamberty, R. Sands, S.J. Smith, A. Janetos, and J. Edmonds. 2009. Implications of limiting CO_2 concentrations for land use and energy. Science 324(5931):1183-1186.

Wright, M.M., and R.C. Brown. 2007. Comparative economics of biorefineries based on the biochemical and thermochemical platforms. Biofuels, Bioproducts and Biorefining 1(1):49-56.

Wright, M.M., J.A. Satrio, R.C. Brown, D.E. Saugaard, and D.D. Hsu. 2010. Techno-Economic Analysis of Biomass Fast Pyrolysis to Transportation Fuels. Golden, CO: National Renewable Energy Laboratory.

Xu, Z., and D.N. Bengston. 1997. Trends in national forest values among forestry professionals, environmentalists, and the news media, 1982-1993. Society and Natural Resources: An International Journal 10(1):43-59.

5

Environmental Effects and Tradeoffs of Biofuels

Petroleum extraction, transport, refining, and combustion have many known negative environmental effects, including disruption of sensitive ecological habitats and high greenhouse-gas (GHG) emissions. Biofuels, too, have their environmental costs (NRC, 2003, 2010a), but displacing petroleum-based fuels with biofuels can reduce the nation's dependence on imported oil and potentially reduce overall environmental harm (Robertson et al., 2008). Each stage in a biofuel's life cycle uses nonrenewable resources and generates emissions that affect land, air, and water. Hence, the environmental benefits and negative effects over the life cycle of petroleum-based fuels and biofuels would have to be compared against each other so that policymakers can decide which tradeoffs are acceptable. There is neither a simple nor single means of comparing biofuels and petroleum-derived fuels over their full life cycles and over their entire suites of environmental effects, yet decades of research on this topic have revealed that some ways of producing biofuels from certain feedstocks offer distinct advantages over others and thus have greater potential for providing environmental benefits over petroleum-derived fuels. Furthermore, certain stages in the life cycle of biofuels have greater environmental effects than others, and thus deserve particular attention in targeting strategies for optimizing environmental outcomes.

This chapter covers the following topics on the potential environmental effects of increasing biofuel production:

- It provides an overview of the life-cycle assessment methodology typically used to assess environmental effects of biofuel production and use.
- It examines the current state of knowledge about key environmental effects. Each environmental effect is discussed, when applicable, in the context of feedstock production, conversion to fuels, and combustion and over the life cycle of biofuel production and use. Methods for assessing effects and the anticipated results or observed effects reported in the published literature are presented. Gaps in data availability and deficiencies in existing modeling platforms, each of which

181

contributes to uncertainty in assessing environmental effects, are also pointed out in the following areas:

- GHG emissions
- Air quality
- Water quality
- Water quantity and consumptive use
- Soil
- Biodiversity
- Ecosystem services

- It uses regional environmental assessments of biofuel production as an illustration because the effects of biofuel production are location-specific, and conclusions drawn from regional environmental assessments could differ from an assessment of cumulative effects across the nation.
- It discusses opportunities to minimize negative environmental effects at the end of the chapter.

Although the committee stresses the importance of comparing environmental effects of biofuels to petroleum-based fuels, environmental effects of petroleum-based fuels have been covered in other publications (NRC, 2003, 2010a) and are beyond the scope of this study.

LIFE-CYCLE APPROACH FOR ASSESSING ENVIRONMENTAL EFECTS: AN OVERVIEW

Biofuels affect the environment at all stages of their production and use. Some effects are easily noticed (for example, odors emanating from an ethanol plant). Others are less apparent, including those that result from activities along the biofuel supply chain (for example, nitrate leaching into surface waters as a result of nitrogen fertilizer application on corn fields) and those that could occur beyond the supply chain via market-mediated effects (for example, loss of biodiversity upon land-use change induced by higher corn prices). Different effects can occur at local, regional, national, or global scales. Some of these effects are easily quantified while others are difficult to measure.

To better understand the suite of environmental effects associated with biofuels, researchers commonly turn to the method of life-cycle assessment (LCA). At the outset, researchers need to define the goal and scope of LCA. For example, researchers need to consider whether the goal is to assess the effects of biofuel produced at an individual biofuel production facility, the average effect of biofuel produced for the entire nation, or the effect of biofuel produced as a result of a policy mandating additional production. Then, an inventory of the resources used and net quantities of substances emitted as a result of biofuel production and use is compiled. This inventory is used to prepare an impact assessment that quantifies the ultimate effects on human health, ecosystem function, and natural resource depletion. Numerous methods for compiling inventories and conducting impact assessments exist, all of which have particular strengths and limitations in their modeling of specific processes and the availability and quality of data used to populate these models.

LCA is a valuable tool for quantifying the environmental effects of biofuels, yet widespread misinterpretation of the results from studies using different assessment methods has led to great confusion. More often than not, this confusion arises when conclusions from these studies are reported without mention of the particular framework and assumptions under which the analyses were conducted. For example, statements such as "this biofuel releases less of this pollutant than gasoline" are by themselves meaningless and often misleading unless the goal and scope of the study cited in support of this statement are presented. (See Box 5-1 for a description of the importance of care when reporting results from LCA studies.)

A common problem is confusion over two different approaches of LCA—attributional and consequential—and their appropriate use when evaluating biofuels. Attributional LCA, the more traditional form, traces the material and energy flows of a biofuel supply chain and seeks to attribute environmental impact to a biofuel based upon these flows. Consequential LCA, on the other hand, considers the environmental effects of the cascade of events that occur as a result of a decision to produce or not to produce a given biofuel. Many differences between these two approaches of LCA arise because of their distinct applications (Ekvall and Weidema, 2004; Ekvall and Andræ, 2006). Attributional LCA makes use of process-specific or average data, while consequential LCA uses marginal data. Attributional LCA does not consider the market-mediated effects of a given biofuel, such as environmental effects caused by changes in crop or petroleum prices as a result of biofuel production. Consequential LCA, similar to a cost-benefit analysis, includes market-mediated effects. In essence, attributional LCA takes as a given the total environmental effect of all human activities and seeks to assign responsibility for a portion of the effect to a particular biofuel. Consequential LCA also takes as a given the total environmental effect of all human activities, but it assigns to a particular biofuel the change in total effect caused by a decision and the resulting action of whether to implement, expand, or contract biofuel production. As such, attributional LCA is useful in improving efficiency along a biofuel supply chain, and consequential LCA is appropriate in the evaluation of policy and regulation.

Both attributional and consequential LCA make use of knowledge of biofuel supply chains, but conducting the latter is far more complicated as it requires marginal data and modeling of market-mediated effects (Kløverpris et al., 2008; Finnveden et al., 2009). In addition, consequential LCA requires preparation of two alternate scenarios (that is, scenarios that represent "yes" and "no" to a decision) whereas attributional LCA requires only one scenario be described (that is, an actual or a projected scenario). Similarly, when measuring the direct environmental effects of supply chains themselves, attributional LCA can rely on actual, measured data, whereas consequential LCA requires that at least one set of data be estimated: When evaluating policies already fully implemented, one set would have to be estimated (that is, the scenario that did not occur) and when evaluating policies with future effects, two sets would have to be estimated (that is, the scenarios for both the "yes" and "no" to a decision). In total, the uncertainty surrounding the results from consequential LCA is compounded compared to attributional LCA, complicating its use in policy decisions, even where LCA is mandated such as in the Renewable Fuel Standard as amended in the Energy Independence and Security Act of 2007 (RFS2).

This discussion of LCA methodology is important to understanding the environmental effects of biofuels. To date, a large number of studies have used attributional LCA to evaluate individual biofuel production streams and the biofuels industry as a whole. Such studies are helpful for assessing the environmental performance of biofuel supply chains, but they do not consider the broader range of effects from increased biofuel production, such as the effects mediated by markets. Only studies that specifically estimate the environmental effects resulting from the marginal increase in fuel production caused by RFS2 are appropriate for assessing the environmental effects of increasing biofuel production due to its implementation. Studies that have used consequential LCA as a means of quantifying the marginal impact of increased biofuel production are sparse and much needed. In this chapter, results using both methods are presented, with the caveat that what might have been found under one set of circumstances may not hold under other conditions.

BOX 5-1
Illustration of How Different Approaches of Life-Cycle Assessment Are Used for Different Purposes

Many studies have been published comparing the environmental effects of biofuels and petroleum-based fuels, often with seemingly conflicting results. Nowhere has this been more evident than in the debate over whether corn-grain ethanol is a greater emitter of GHGs than gasoline. As such, it serves here as a basis for discussing LCA methodology. Careful examination of this debate shows that more often than not, seemingly conflicting results are not contradictory, but rather are simply the consequence of fundamental differences in goal and scope, assumptions, methodology, and underlying data.

Consider, for example, three different stakeholders who wish to know the quantity of GHGs released in corn-grain ethanol production. A manager of a corn-grain ethanol plant might be interested in estimating GHG emissions of his or her product for sale into California, which is regulated by its own Low Carbon Fuel Standard. An ethanol industry analyst might wish to know the average GHG emissions for corn-grain ethanol produced domestically so as to track industry improvement in efficiency on an annual basis. A federal regulator might wish to know the change in quantities of GHG emissions as a result of legislation mandating the production of additional ethanol.

Now consider how each might go about quantifying GHG emissions. For the plant manager, a static attributional LCA method for quantifying GHG emissions from his or her own facility's supply chain is most useful. This method would also suit the needs of the industry observer, albeit with a different focus on what might be considered a typical facility, on a subset of facilities representative of the industry, or on all facilities. For the federal regulator, however, a dynamic consequential LCA method that quantifies the net change in GHG emissions resulting from increased ethanol production is most appropriate. This includes market mediated effects extending well beyond the bounds of the ethanol and agribusiness industries themselves.

To populate their LCA models, the three stakeholders would choose data specifically well suited to their analysis. Consider, for example, the critical parameter of corn yield, or the weight of grain harvested from a given area of cropland. The ethanol plant manager may choose the average yield of grain delivered to the facility. The industry analyst may use the national yield average. The federal regulator may use a projected yield that accounts for both potential yield increases due to greater investment in crop production technology and potential yield decreases due to the disruption of existing crop rotations (for example, shifting from corn-soybean rotations to continuous corn) and the increased use of less productive lands (for example, use of idle cropland).

From these examples, it is clear that each of these three stakeholders could arrive at a different estimate of life-cycle GHG emissions from corn-grain ethanol, and each would be reasonable given the assumptions. An individual facility may produce ethanol with different life-cycle GHG emissions than the national average of all facilities. Producing additional ethanol as a result of a federal mandate would lead to a different amount of GHG emissions than what would have been generated in the mandate's absence. An important caveat to this discussion is that while all three ways of viewing the system are correct and each is useful to its own audience, each is an interpretation of a single reality (that is, the actual net quantity of GHG emissions released to the atmosphere), and as such the ultimate question to be answered is whether the decision to build and operate an ethanol production infrastructure leads to a reduction in global GHG emissions.

The scenario explored above is not provided merely as an academic exercise, but rather because it reflects the actual variation found in recent studies on the life-cycle environmental effects of biofuels. With respect to corn-grain ethanol GHG emissions specifically, some studies (for example, Liska et al., 2009) are site specific, focusing primarily on facilities situated in areas of exceptionally high corn-grain yields such as Iowa and Nebraska. Other studies, such as those of Farrell et al. (2006) and Wang et al. (2007), are concerned largely with average ethanol production at a national level. The U.S. Environmental Protection Agency's rulemaking for RFS2 is essentially focused on the additional ethanol that will be produced as a result of the Energy Independence and Security Act of 2007.

GREENHOUSE-GAS EMISSIONS

Feedstock Production

One of the most debated topics surrounding the environmental effect of biofuels is the net GHG emissions from producing various feedstocks. Potential GHG emissions from bioenergy feedstock production include carbon dioxide (CO_2), nitrous oxide (N_2O), and methane (CH_4).[1] As elaborated below, the key factors that affect GHG emissions from bioenergy feedstock production are site-specific and depend on the type of feedstocks produced, the management practices used to produce them, and any land-use changes that their production might incur.

Type of Feedstock and Management Practices

Potential bioenergy feedstocks mentioned in Chapter 2 can be categorized as annual, herbaceous perennial, short-rotation woody crops (SRWCs), and residue from other systems such as corn stover or forest residue. Choice of feedstock is an important factor in determining the GHG effect of biofuels. For example, perennial herbaceous biomass could increase soil carbon sequestration compared to annual crops (Anderson-Teixeira et al., 2009; Blanco-Canqui, 2010; NRC, 2010b). The GHG implications of a particular feedstock depend on the relationship between that feedstock and site properties such as soil type and climate. As with any agricultural crop, management practices affect the net GHG balance of bioenergy feedstock production in several ways: cropping patterns, amount of agrichemical use, tillage practices, and farm equipment use.

Farmers and foresters select management practices on the basis of crops grown, soil conditions, precipitation patterns, slope, exposure, available equipment, and their knowledge and preferences. In general, choices are made to maximize yield per dollar of input and are not made on the basis of GHG emissions. Yet, choices of management practices have a major influence on GHG emissions (NRC, 2010b). CO_2 released from fossil fuel combustion in the manufacturing, transport, and application of agricultural inputs (for example, fertilizers, pesticides, seed, and agricultural lime), N_2O released during nitrogen fertilizer production (Snyder et al., 2009), and N_2O released because of nitrification and denitrification stimulated by nitrogen fertilizer application (Bouwman et al., 2010) contribute to GHG emissions. Therefore, producers who choose to cultivate bioenergy feedstocks that require higher agrichemical input in place of crops that require less agrichemical input would incur increases in GHG fluxes. Some bioenergy energy feedstock such as forest residue would have no GHG contribution from agrichemical input.

Agricultural soil management accounted for about 68 percent of the total N_2O emissions in the United States in 2008 (EPA, 2010c). Emission of N_2O is predominantly a result of microbial processes of nitrification and denitrification; therefore, emission generally increases with nitrogen availability, or the extent to which nitrogen input exceeds crops' needs (Bouwman et al., 1993, 2002; McSwiney and Robertson, 2005). The type and timing of nitrogen fertilizer used also affects N_2O fluxes (Bavin et al., 2009). Technologies for precise application of fertilizers can potentially reduce fertilizer use without compromising yield (Snyder et al., 2009; Gebber and Adamchuk, 2010; Millar et al., 2010), but those technologies are not widely adopted because of socioeconomic, agronomic, and technological reasons

---

[1]Global warming potential of a GHG is the warming caused by emission of 1 ton of that GHG compared to 1 ton of CO_2 over a specific time interval. The global warming potentials over a 100-year period are 1 for CO_2, 25 for CH_4, and 298 for N_2O.

(Robert, 2002; Lamb et al., 2008; USDA-NIFA, 2009). Precision management of nitrogen fertilization can also improve biomass quality for cellulosic biofuels (Gallagher et al., 2011).

The environmental benefits of crop rotations include enhanced control of weeds, pests, and diseases; increased availability of nutrients; accumulation of soil carbon; and higher yields (NRC, 2010b). Those benefits, if combined with higher yields, contribute to reducing agrichemical input and GHG emissions. Increased diversity of crops planted in a field (either at once or over the course of a year) could also reduce the amount of pesticide application needed (GAO, 2009). For example, mixtures that include grasses and nitrogen-fixing legumes can also reduce nitrogen fertilizer needs (Tilman et al., 2006; Fornara and Tilman, 2008; NRC, 2010b). Gardiner et al. (2010) compared preexisting corn, switchgrass, and mixed prairie crops in Michigan and found that switchgrass and mixed prairie crops supported greater abundance of arthropod generalist natural enemies of crop pests. Even crop rotation between corn and soybean can help control pests and reduce the use of pesticides by breaking the pattern of pests and disease that can be present in monocultures. Integrated pest management can potentially contribute to reducing pesticide input (Trumble et al., 1997; Reitz et al., 1999; NRC, 2010b).

The effect of no-till and reduced tillage on soil organic carbon (SOC) storage is inconsistent and depends on depth of soil sampling and crop management (Dolan et al., 2006; Baker et al., 2007; Johnson et al., 2007; Luo et al., 2010; Kravchenko and Robertson, 2011). Studies that assess carbon content in the entire soil profile (0-60 cm) did not find higher soil carbon in no-till fields than in conventionally tilled fields (Blanco-Canqui and Lal, 2008; Christopher et al., 2009). Nonetheless, no-till and reduced tillage may contribute to reducing GHG emissions because those practices require less fossil-fuel inputs for machinery that perform the tilling (Adler et al., 2007) and emissions of N_2O might be lower (Omonode et al., 2011). No-till and reduced tillage also have other environmental benefits because they enhance soil water retention and microbial activity and diversity, reduce soil erosion and sediment runoff, and improve air quality compared to conventional tillage (NRC, 2010b).

Methods of Assessment

Over the past several decades, ecosystem ecologists have estimated carbon storage and GHG consequences of land-use management practices on regional and continental scales, using spatial databases to represent key driving variables, including soils (for example, STATSGO), average climatic data, satellite imagery (for example, MODIS), and current or projected land-use management, combined with simulation models. This strategy has been used to assess consequences of cropping (Campbell et al., 2005; Del Grosso et al., 2005; Izaurralde et al., 2006), forest management (Adams et al., 1999; Sohngen and Sedjo, 2000; Murray et al., 2005; Johnson et al., 2010), and climate change (Paustian et al., 1997; Lu and Zhuang, 2010). Notably, simulation results (and indeed the biological processes responsible for GHG fluxes) are very sensitive to site-specific factors that are variable. Those site-specific factors, including fertilization practices, cultivation and residue management, and forest age classes, are rarely available as input data. Thus, potential error increases for scaled-up estimates, based on the presence, accuracy, and spatial resolution of input data, and the ability of simulation models to accurately estimate fluxes.

Zhang et al. (2010) used this strategy to assess environmental effects including GHG emissions that might occur based on spatially explicit scenarios of bioenergy feedstock expansion, including annual crops, herbaceous perennial crops, SRWC, and residue harvest. They predicted locations for different bioenergy crops and management options in a nine-county region in southwestern Michigan that would minimize GHG emissions while maintaining certain minimum yields and maximum nitrate runoff levels. They presented sample results involving the minimization of GHG flux per unit area, although the flexibility of their framework allows for the calculation of other variables of interest, such as GHG flux per unit of energy produced, which may be more useful for integration with full LCAs. In addition, Zhang et al. (2010) noted that their framework could be extended into a spatially explicit LCA in which, for example, optimal locations for biorefineries could be modeled simultaneously with feedstock production locations.

Anticipated or Observed Results

As mentioned above, the effects of bioenergy feedstock production on GHG emissions depend on feedstock choice, management practices, and changes in land use and land cover so that any quantitative estimates of GHG emissions are site specific. This section discusses the anticipated or observed effects of feedstock production on GHG emission as organized by major feedstock categories.

For corn and soybean production, fertilizer use generates GHGs as a result of fossil-fuel input in manufacturing and transporting fertilizers and of nitrogen from fertilizers not taken up by plants and emitted as N_2O. In 2005, about 95 percent of the corn acreage in the United States received nitrogen fertilizer, and the average application rate was about 138 lb/acre (Table 5-1). Soybean requires less inputs (particularly nitrogen fertilizers) to produce than corn on a per-acre basis (Schnepf, 2004). However, a comparison of GHG contribution from fertilizer manufacture and use in feedstock production between biofuels have to account for crop yield per acre,[2] conversion yield from feedstock to biofuel,[3] and the energy content of biofuel.[4]

The opportunity offered by the future use of cellulosic feedstocks is that GHG emissions could be reduced, but that benefits can only be achieved in some situations. Corn stover, cereal straw, and other crop residues draw on existing crops so that their use as bioenergy feedstock under best management practices might not contribute much additional GHG emissions. However, overharvesting of crop residues could result in additional need for agrichemical inputs and the loss of soil organic matter, which is critical for maintaining soil structure and water retention capacity and for improving nutrient cycling and other soil processes (Wilhelm et al., 2007; NAS-NAE-NRC, 2009; NRC, 2010b). Any additional fuel use for collecting the residues that contributes to GHG emissions would also have to be accounted for.

TABLE 5-1 Fertilizer Use for Corn and Soybean Production in the United States

	Corn[a]	Soybean[b]
Acreage fertilized receiving nitrogen fertilizer (percent)	96	18
Average rate of nitrogen fertilizer application (lbs/acre)	138	16
Acreage fertilized receiving phosphate fertilizer (percent)	81	23
Average rate of phosphate fertilizer application (lbs/acre)	58	46
Acreage fertilized receiving potash fertilizer (percent)	65	25
Average rate of potash fertilizer application (lbs/acre)	84	80

[a]Latest data from source are for the year 2005.
[b]Latest data from source are for the year 2006.
SOURCE: USDA-ERS (2010c).

---

[2]Corn yield per acre is about 4 times higher than soybean yield (USDA-NASS, 2010).
[3]About 1 bushel of soybean produces 1.5 gallons of biodiesel, while 1 bushel of corn produces about 2.7 gallons of ethanol.
[4]The energy content of corn-grain ethanol is about two-thirds of that of soybean biodiesel.

Growing perennial dedicated bioenergy crops could have less direct GHG emissions than growing row crops because their root systems contribute to sequestration of carbon. Surveys of common agronomic practices for growing *Miscanthus* show a broad range in nitrogen fertilizer use, typically around 50-100 lbs per acre per year (Heaton et al., 2004; Khanna et al., 2008). In their review of published literature, Parrish and Fike (2005) reported that data on nitrogen requirements in switchgrass span a range of 0-200 lbs per acre, and that the variations can be partly attributed to different harvest practices, within-plant nitrogen recycling, and site-specific soil nitrogen mineralization rates and atmospheric deposition and microbial fixation of nitrogen. Liebig et al. (2008) measured changes in soil organic carbon (SOC) in the top 0-30 cm and 0-120 cm of soil in switchgrass fields on 10 farms that were previously used for annual crop production in the central and northern Great Plains. They reported accumulation of SOC over time, but the change in SOC varied considerably across sites from –2.2 to 16 Mg CO_2 eq per hectare per year in the top 0-30 cm. Garten et al. (2010) found that a single harvest of switchgrass at the end of the growing season increased SOC sequestration and system nitrogen balance on well-drained Alfisols in west Tennessee. SOC sequestration rates in the top 15 cm of reconstructed tall grass prairies on previously cultivated land in southern Iowa varied significantly with topography and age of the prairie stand (Guzman and Al-Kaisi, 2010).

Using woody residues as a bioenergy feedstock can result in relatively low GHG emissions compared to crops that are planted and harvested exclusively for bioenergy purposes if they are a byproduct of existing harvesting operations and do not require fertilizer input. In some regions of the United States, harvesting dead material from the forest floor and forest thinning could reduce the potential for wildfires (Fight and Barbour, 2005; Busse et al., 2009; Kalies et al., 2010) that also contribute much CO_2 to the atmosphere.

SRWC can sequester SOC depending on trees grown, soil types, and prior land use, according to a review of literature by Blanco-Canqui (2010). The author noted that nitrogen-fixing trees sequester more SOC than other trees. Fertilization and irrigation can increase SOC sequestration and yield increase, but CO_2 emissions associated with these activities may offset some SOC benefits (Blanco-Canqui, 2010).

Biofuel-Induced Land-Use Changes

Carbon is stored in soil and in above-ground and below-ground vegetation. Soil carbon storage depends on soil characteristics and past disturbances. The amount of carbon stored in vegetation depends on the vegetation type. Therefore, land-use changes that involve removing or planting of vegetation could either release a large amount of carbon from soil or store carbon depending on the conditions of the land prior to use, crop characteristics (Fearnside, 1996; Guo and Gifford, 2002b; Woodbury et al., 2006), and management practices (as discussed above). Similarly, land-use change could disrupt or enhance the future potential of land to store carbon.

Land use is defined by anthropogenic activities, such as agriculture, forestry, and urban development, that alter land-surface processes, including biogeochemistry, hydrology, and biodiversity. Land cover is the extent and type of physical and biological cover over the surface of land. Some authors have divided land-use changes into two types when considering biofuel policy: *direct* land-use change and *indirect* land-use change. Biofuel-induced land-use changes occur directly when land is dedicated from one use to the purpose of growing biofuel feedstock. Biofuel-induced land-use changes can occur indirectly if land use for production of biofuel feedstocks causes new land-use changes elsewhere through market-mediated effects. The production of biofuel feedstocks can constrain the supply of commodity crops and raise prices, thus triggering other agricultural growers to respond to market signals (higher commodity prices) and to expand production of the displaced commodity crop. This process might ultimately lead to conversion of nonagricultural land (such as forests or grassland) to cropland. Because agricultural markets are intertwined globally, production of bioenergy feedstock in the United States could result in land-use and land-cover changes elsewhere in the world. If those changes reduce the carbon stock in vegetation, carbon would be released in the atmosphere when land-use change occurs. In particular, transition from forest to cropland or pasture emits a large amount of CO_2 because of CO_2 releases from decomposition of woody debris and short-lived wood products (NRC, 2010c). Similarly, land-use change could disrupt or enhance the future potential of land to store carbon.

Many economic studies have shown the "unintended" consequences of policy (Stavins and Jaffe, 1990; Wu, 2000; Wear and Murray, 2004), and the principle from Wu's study is relevant to increasing biofuel production in United States. Wu (2000) showed cropland enrolled in the U.S. Department of Agriculture's (USDA) Conservation Reserve Program (CRP) had a 20-percent slippage. That is, for every 5 acres of cropland enrolled in CRP, 1 acre of noncropland is added to cropland elsewhere. That study did not account for carbon emissions, but it pointed out the rippling effects of shifting land uses. Other studies have linked land-use changes to carbon changes and showed that projects and policies intended to mitigate GHG emissions in the forestry or agricultural sector could lead to "leakage,"[5] or responses to those projects and policies by other parties that also cause GHG emissions (Sohngen and Brown, 2004; Murray et al., 2007).

Methods of Assessment
Land-Use and Land-Cover Changes. Remote sensing using satellite and aircraft sensors can be used to map land cover and land use and provide information on above-ground vegetation and residue cover (NRC, 2010c). Data from remote sensing can be coupled with land monitoring to estimate GHG fluxes from land-use changes (Houghton, 2010; NRC, 2010c; West et al., 2010). Uncertainties of annual carbon fluxes from deforestation, reforestation, and forest degradation based on remote sensing vary from 25 to 100 percent (NRC, 2010c). Variations in plant residue, along with soil moisture and mineralogy and vegetation cover, are a problem in estimating soil surface carbon. Even so, progress has been made in assessing crop residue coverage using space-borne hyperspectral instruments (Daughtry et al., 2006; NRC, 2010c). Estimates of N_2O emissions from managed lands have about 50-percent uncertainty even with the best inventory methods, and those estimates are even more uncertain in developing countries than in developed countries (NRC, 2010c).

Market-Mediated Effects. A number of different types of economic models have been used to calculate the global indirect effects of increasing biofuel production. An important aspect emphasized by these models is global interaction. For example, shocks to supply and demand in one region have well-defined price effects on global markets, as illustrated by the market price fluctuations as a result of drought in Russia in 2010. Economic models have been developed to capture this phenomenon. The short-term and long-term effects of biofuel policy on global commodity markets are discussed in Chapter 4.

A second aspect emphasized by these models is the competition among different land uses. Economic models are often best suited to account for the behavior of different

---

[5]GHG leakage is the term that was introduced to refer to the conditions when an activity displaces GHG emissions outside the boundaries of the activity area (Murray et al., 2007). For example, afforestation efforts in one country could lead to market forces that encourage deforestation in another country (Meyfroidt et al., 2010).

competing demands for land, as well as the supply of land. A number of different economic models, including general-equilibrium and partial-equilibrium models, have been used to study indirect land changes, and the advantages and disadvantages of several approaches have been discussed elsewhere (Kretschmer and Peterson, 2010). The estimates of indirect land changes are then added to direct GHG models, such as GREET,[6] to estimate total direct and indirect GHG emissions. Although such analyses consider emissions as a result of market-mediated effects on land use, they are not, strictly speaking, consequential LCAs. Rather, they represent a hybrid approach in which marginal data for a specific parameter (land use) are incorporated into an attributional LCA model. Among many differences, comprehensive consequential LCA would, for example, also consider elasticity of petroleum markets.

GHG Emissions Estimated from Market-Mediated Land-Use Changes. GHG emissions from indirect land-use or land-cover changes can be estimated by coupling estimates of market-mediated land-use or land-cover changes with estimates of GHG emissions from those projected land-use or land-cover changes. The resulting projection of GHG emissions from indirect land-use changes has large uncertainty because of difficulty to establish a causal link between direct-use changes and indirect-use changes that are separated spatially and temporally. For example, many factors influence land-use changes, and showing precisely that a price change induced by biofuel policy as the precipitating cause is difficult. Even if an economic linkage can be shown, calculating the carbon change is difficult because there is substantial heterogeneity in carbon on the landscape. If the indirect land-use change involves removing tropical forests, the carbon emissions could be high, but if the indirect land-use change involves converting pasture or fallow land to cropland, then the carbon effects could be smaller.

Several concerns have been raised about the existing estimates of the indirect effects of land use. One concern relates to the many steps that need to be undertaken to show indirect land-use change and uncertainty associated with all those steps. For example, the first step in any analysis of the effects of U.S. policy is to determine what crops besides corn are displaced as a result of increased biofuel production. The second step is to determine how much these changes in U.S. markets influence prices in other countries (Babcock, 2009; Zilberman et al., 2010). The key concern with these calculations is that U.S. economists have an idea of U.S. farmers' responses to price change on the basis of historic trends, but Babcock (2009) argued that the response of farmers in other parts of the world to price changes is much less certain. Similar concerns have been raised by Kim and Dale (2011), who were unable to find correlative evidence between increased demand for corn and land-use change from 2001-2007. O'Hare et al. (2011) argued that Kim and Dale's analysis was flawed. The committee advocates that additional data and analyses are needed to assess net changes in land use as a result of market-mediated effects of feedstock production for biofuels. A second concern is that simulations from economic models use point estimates of various parameters, each of which varies temporally and spatially (Zilberman et al., 2010). A third concern is that other factors that contribute to land-use change decisions, including cultural, political, and ecological factors (Geist and Lambin, 2002; Turner et al., 2007), are not accounted for in economic models. Finally, one response to rising prices is intensification of existing croplands. The different models discussed later account for cropland intensification to different extents. For example, the study by Searchinger et al. (2008) assumes that increased yields from intensification will be offset by lower yields on lower-quality lands

---

[6]The Greenhouse Gases, Regulated Emissions, and Energy Use in Transportation Model by Argonne National Laboratory.

brought into production. The results from Hertel et al. (2010) directly incorporate intensification of crop management as a result of rising prices. Cropland intensification helps reduce the overall indirect effects.

Anticipated Effects

Direct conversion of native ecosystems to producing corn for ethanol releases large amounts of GHG into the atmosphere (Fargione et al., 2008; Gibbs et al., 2008; Ravindranath et al., 2009). Based on the definition in RFS2, only planted crops and crop residue from agricultural land cleared prior to December 19, 2007, and actively managed or fallow on that date are considered compliant feedstocks. This definition discourages land clearing of native ecosystems for bioenergy feedstock production so that GHG emissions from direct land-use change could be minimized. However, some farmers could use existing cropland to produce bioenergy feedstocks.

Conversely, converting from annual to perennial bioenergy crops can enhance carbon sequestration on that piece of land (Fargione et al., 2008). The perennial bioenergy crops are considered RFS-compliant feedstock. However, the carbon storage could be offset by market-mediated effects on land-use and land-cover changes elsewhere as a result of biofuel production in the United States.

A few authors estimated GHG emissions from indirect land-use change as a result of increasing corn-grain ethanol production in the United States. Their simulations represent changes in GHG emissions from land-use changes with or without U.S. biofuel production. Other drivers of land-use changes were not considered. Searchinger et al. (2008) estimated that GHG emissions from indirect land-use change in Brazil, China, India, and the United States from U.S. corn-grain ethanol production to be 104 g CO_2 eq per MJ. Searchinger et al. (2008) projected land-use changes on the basis of historical data from 1990 to 1999. They estimated GHG emissions from the land-use change would be offset by GHG benefits accrued from substituting gasoline with corn-grain ethanol only after 167 years.

Dumortier et al. (2010) demonstrated that differences in the economic model and data source did not alter the estimate of GHG emission from indirect land-use change much when they used the same assumptions of increase in ethanol production over time and types of land cover converted as Searchinger et al. (2008). In contrast, changes in assumptions on the type of land converted, net land displacement factor,[7] crop yield, and increase in ethanol production had large effects on estimated GHG emissions (Dumortier et al., 2010; Plevin et al., 2010).

The model of the Global Trade Analysis Project (GTAP) has been used to estimate biofuel-induced land-use change emission estimates for the California Air Resources Board (Tyner et al., 2010). To evaluate the land-use implications of U.S. ethanol production, they developed three groups of simulations. In the first group, they calculated the land-use implications of U.S. ethanol production off the 2001 database. This is version 6 of the GTAP global database, which is updated every 2-3 years. This approach isolates effects of U.S. ethanol production from other changes that shape the world economy. In the second group of simulations, Tyner et al. (2010) first constructed a baseline that represents changes in the world economy during the time period of 2001-2006. Then they calculated the land-use impact of U.S. ethanol production based on the updated 2006 database. Finally, in the third group of simulations, they used the updated 2006 database obtained from the second group

---

[7]Net land displacement factor is the ratio of land acreage brought into crop production anywhere in the world as a result of market-mediated effects of bioenergy feedstock production to land acreage dedicated to bioenergy feedstock production.

of simulations but assumed that during the time period of 2006-2015, population and crop yields would continue to grow. They estimated that the average land requirement for the incremental ethanol production was 0.32 acres of land to produce 1,000 gallons of ethanol. Twenty-four percent of the land-use change was estimated to occur in the United States and 76 percent in the rest of the world. Forest reduction was estimated to account for 33 percent of the global change and pasture 67 percent. In the GTAP database, grassland is included in pasture, and CRP lands were excluded from this analysis.

The range of estimates shown in Table 5-2 illustrates how the changes in assumptions that form a particular scenario affect GHG emissions from indirect land-use changes. Any of those scenarios in Table 5-2 are possible, and the GHG emissions from indirect land-use changes will depend on which ones of those or other alternative scenarios play out. Using a reduced-form model and a range of scenarios, Plevin et al. (2010) estimated that the range of GHG emissions from indirect land-use change as a result of increasing U.S. corn-grain ethanol production to be 10-340 g CO_2 per MJ, with a 95-percent central interval between 21 and 142 g CO_2 per MJ. If dedicated bioenergy crop production displaces commodity crops in the United States and if the displacement affects global markets, economic models project that indirect land-use change and associated changes in GHG emissions can be expected.

Expanding production of biofuels in the United States increases pressure on land supply and causes land-use changes elsewhere in the world through market-mediated effects (Melillo et al., 2009; Bowyer, 2010; Overmars et al., 2011). In the United States, the proportion of corn-grain used for ethanol has increased from less than 10 percent in 2000 to 40 percent in 2010 (see Figure 2-5 in Chapter 2), though net exports have held steady for corn, increased for soybean, and declined for wheat (Chapter 4). The extent of biofuel market-mediated land-use changes are uncertain because there are different ways farmers around the world could respond to changes in land-use pressure and market price signals. Other than expanding cultivated land, farmers also could respond to price signals by intensifying the use of existing agricultural lands—for example, increasing fertilization, double cropping, decreasing fallow periods, or using new technologies to increase agricultural outputs per unit cultivated land (Fischer et al., 2009; Melillo et al., 2009; Searchinger, 2010). Improving crop productivity per unit land cultivated can have a profound influence on land-use change emissions in that it changes the land base required for agricultural production for food, feed, and biofuels (Wise et al., 2009).

The Hertel et al. (2010) study attempted to do a systematic analysis of land-use change that was induced by emissions from U.S. biofuel production. They concluded that the corn-grain ethanol-induced emissions from land-use change range between 2 and 51 g CO_2 per MJ.

The range of estimates for GHG emissions from indirect land-use changes is wide (that is, precise value is highly uncertain) largely because it is difficult to separate market-mediated effects of land-use change as a result of increasing biofuel production from other drivers of land-use changes. However, a key point is that land-use and land-cover changes can have profound effects on GHG emissions. The extent of biofuel-induced land-use change emissions are highly uncertain, but with 40 percent of the corn crop in the United States in 2010 (about 27 percent after accounting for dried distillers grains with solubles [DDGS]) going to biofuels, GHG emissions from land-use changes cannot be ignored.

Next Steps

In coming years, scientists will undoubtedly continue to refine their models to improve estimates of GHG emissions as a result of land-use changes. However, uncertainty of GHG emissions from land-use and land-cover changes can be expected to remain large because

TABLE 5-2 GHG Emissions from Market-Mediated Indirect Land-Use Changes as a Result of Expanding Corn-Grain Ethanol Production in the United States Estimated by Various Authors

Economic models used to estimate market-mediated effects	Land-cover change data used	Emission factors used	GHG emissions from indirect land-use change (g CO$_2$eq per MJ)	Target year	Increase in ethanol production (million liters)	Key assumptions	Reference
FAPRI	Woods Hole (1990s)	Woods Hole	104	2016	56	• Net land displacement factor = 72 percent. • Different types of forests, savannah, or grassland are converted to cropland in Brazil, China, India, and the United States. • Percent forest land converted = 52; percent grassland converted = 48.	Searchinger et al., 2008
GTAP	GreenAgSiM	IPCC	118	2018 or 2019	56	• Same assumptions as Searchinger et al., 2008.	Dumortier et al., 2010
GTAP	GreenAgSiM	IPCC	91	2018 or 2019	56	• Same assumptions as Searchinger et al., 2008, except no conversion of U.S. forest land to cropland.	Dumortier et al., 2010
CARD Agricultural Outlook Model	GreenAgSiM	IPCC	75	2018 or 2019	30	• No conversion of U.S. forest land to cropland. • Crop yield 1 percent higher than the slope of trend yield compared to Searchinger et al., 2008.	Dumortier et al., 2010
CARD Agricultural Outlook Model	GreenAgSiM	IPCC	21	2018 or 2019	30	• No conversion of U.S. forest land to cropland. • Crop yield 1 percent higher than the slope of trend yield compared to Searchinger et al., 2008.	Dumortier et al., 2010
GTAP	Woods Hole (1990s)	Woods Hole	27	2010	50	• Net land displacement factor = 28 percent. • Percent forest land converted = 19; percent grassland converted = 81.	Hertel et al., 2010
FASOM, FAPRI	MODIS-5	Winrock International	82	2012	7.5	• Net land displacement factor = 89 percent.	EPA, 2010d
FASOM, FAPRI	MODIS-5	Winrock International	58	2017	14	• Net land displacement factor = 55 percent.	EPA, 2010d

continued

TABLE 5-2 Continued

Economic models used to estimate market-mediated effects	Land-cover change data used	Emission factors used	GHG emissions from indirect land-use change (g CO$_2$eq per MJ)	Target year	Increase in ethanol production (million liters)	Key assumptions	Reference
FASOM, FAPRI	MODIS-5	Winrock International	34	2022	10	• Net land displacement factor = 29 percent.	EPA, 2010d
GTAP	GTAP database	Woods Hole	14.5	2022	13	• Percent forest land converted = 33; percent grassland converted = 67.	Tyner et al., 2010
Reduced-form modeling			10-340	2025-2055		• Net land displacement factor = 25-80 percent. • Percent forest land converted = 15-50; percent grassland converted = 45-85; percent wetland converted = 0-2.	Plevin et al., 2010

NOTE: Land-use change amortized over 30 years.
SOURCE: Adapted from Plevin et al. (2010).

actual land changes and their relation to increasing biofuel production in the United States will only be observed as markets adjust to increased biofuel production. Even with long-term empirical data on land-use and land-cover changes, measurements of associated GHG emissions, and data on agricultural markets, estimating the global GHG benefits or emissions from U.S. biofuel production will require a comparison with a reference scenario, which inevitably is a simulation of what would have happened absent biofuels. Such a reference scenario may include GHG emissions resulting from any change in the use of oil sands and other nonconventional sources of petroleum (Jordaan et al., 2009; Yeh et al., 2010). To improve GHG estimates from indirect land-use changes as a result of U.S. biofuel policy, data would have to be collected continuously and models would have to be refined for as long as biofuels are produced. Additional data and information to be collected include:

- Global land-cover change to assess changes in carbon stocks;
- Global commodity market and land use to observe any market-mediated effects on land changes from RFS2;
- Drivers of land changes to parse out the market-mediated effects on land changes from other factors that affect land-use decisions.

Additional research is needed to better understand the socioeconomic processes of land-use change and to integrate that process understanding into models for estimating market-mediated effects and for GHG emissions to better inform the GHG effects of biofuel-induced land-use and land-cover changes.

Conversion to Fuels

The conversion of feedstocks into biofuels at biorefineries results in GHG emissions from on-site combustion of fossil fuel or biomass, from production of process chemicals and enzymes, from process emissions including those from fermentation, and more broadly from transport of inputs and products and from generation of purchased electricity. Continuous emission monitoring systems can provide measurements of CO_2 in biorefineries in operation. CO_2 emissions also can be estimated using a mass balance approach (Huo et al., 2009; NAS-NAE-NRC, 2009; DOE-NETL, 2010). Although total biorefinery emissions can be measured or estimated, it is important to distinguish between GHG emissions from fossil sources and those from biogenic sources for purposes of GHG accounting. Biomass, a biogenic source of carbon, is commonly assumed to be carbon neutral because the carbon emitted when burning had previously been removed from the atmosphere as CO_2 during plant growth. Although biomass itself can be treated as carbon neutral, the processes used to grow and collect biomass, including any associated land-use change, can incur GHG emissions.

In general, for corn-grain ethanol production, using natural gas at biorefineries has lower GHG emissions than using coal, and using biomass to provide heat, power, or both may have lower emissions still (Kaliyan et al., 2011; Wang et al., 2011a). In corn-grain ethanol refineries, the amount of DDGS coproduct that is dried and the extent to which it is dried further affect energy use, and hence biorefinery CO_2 emissions. In 2011, the Renewable Fuels Association estimated that about 60 percent of DDGS was dried. For biodiesel production, GHG emissions at locations where transesterification occurs are minimal compared to corn-grain ethanol. In cellulosic-ethanol refineries as they are typically proposed, burning lignin and other residues to generate steam and power results in the release of biogenic CO_2 rather than the fossil CO_2 that would be released from natural gas or coal, and any excess electricity generated can be sold to the grid (NAS-NAE-NRC, 2009). Variations in CO_2 emissions from a biorefinery that converts corn stover to ethanol biochemically compared to one that converts wood chips to ethanol thermochemically are estimated to be small (Foust et al., 2009; NAS-NAE-NRC, 2009), particularly when they are compared to variations in CO_2 emissions in other parts of the fuel production life cycle. However, actual quantities of emissions from different types of facilities can only be verified once they are in operation. GHG emissions from manufacturing of fertilizers could potentially be reduced if biochar, a coproduct from pyrolysis, is used as soil amendment for biomass feedstock production. However, the effects of biochar on plants (for example, phytotoxicity and nutrient availability) and soil (carbon mineralization) are uncertain and require further examination (Lee et al., 2010; Gell et al., 2011; Nelson et al., 2011; Zimmerman et al., 2011).

Life-Cycle GHG Emissions

The amount of GHG emitted over the life cycle of biofuels is a subject of intense research interest and public debate. This section discusses the potential GHG emissions over the life cycle of biofuels and the potential changes in global GHG emissions as a result of increasing biofuel production in the United States.

Methods for Assessing Effects

As discussed earlier in this chapter, two approaches can be used for life-cycle assessments—attributional and consequential—each of which suits a different purpose. Attributional LCA sums up the GHG emissions along a static biofuel supply chain. Consequential LCA describes the net overall GHG emissions as a result of increasing or decreasing biofuel production.

Models that have been developed for attributional LCA of GHG for biofuels commonly used in the United States include GREET (Wang et al., 2011a), BESS (Liska et al., 2009), and EBAMM (Farrell et al., 2006), among others. Plevin (2009b) found that using different models for attributional LCA does not result in drastically different outcomes if system boundaries and input data are consistent.[8] In contrast, differences in methodological choices, such as treatment of coproducts, treatment of time, and assumptions of displaced energy, further complicate the comparison among studies (Box 5-2). Differences in estimates of key parameters, such as CO_2 emissions from land-use change and N_2O emissions from fertilization (Ogle et al., 2007; Erisman et al., 2010), have further led to discrepancies (Börjesson, 2009; Hoefnagels et al., 2010; Hsu et al., 2010).

Comprehensive consequential LCA studies that consider all GHG effects as a result of increased biofuel production, let alone RFS2 specifically, have been elusive to date. Given the importance of indirect land-use change in GHG accounting of biofuels, however, many attributional models such as GREET now add on some estimate of this parameter (Table 5-2). Another market-mediated effect that has yet to be incorporated into most modeling exercises is the "rebound effect" where the addition of biofuels, into the market leads to a less than complete displacement of petroleum-derived fuels (Fargione et al., 2010; Hochman et al., 2010).

---

[8]See Plevin (2009a,b), Liska and Cassman (2009a,b), and Anex and Lifset (2009) for an informative exchange on system boundaries and data choice in site-specific attributional LCA of corn-grain ethanol.

BOX 5-2
Methodological Assumptions Affecting GHG LCA Analyses

The practice of LCA seeks to model processes or decisions using empirical data, but in addition to dealing with the uncertainty surrounding these data, the modelers have to make a series of methodological choices and assumptions. Three examples of such choices are the treatment of coproducts, the consideration of time, and the consideration of displaced products. The choices researchers make can have dramatic effects on their results.

When a production stream leads to multiple products, modelers have to decide how to allocate the resource use and generation of pollution. Options include allocating according to the value of these products, their mass, or even by what other products they displace in the market. In the modeling of biofuel production, the treatment of coproducts such as animal feed (for example, DDGS from corn-grain ethanol and soybean meal from soybean diesel) or energy (for example, electricity cogenerated from lignin combustion when producing cellulosic ethanol) requires careful consideration as different methods may lead to very different results (Pradhan et al., 2008; Morais et al., 2010; Singh et al., 2010; van der Voet et al., 2010; Wang et al., 2011b; Börjesson and Tufvesson, 2011).

The treatment of time in an LCA also is subject to the modeler's judgment (Delucchi, 2011; McKone et al., 2011). A carbon debt from land-use change could be incurred largely immediately following land conversion, but the offset of fossil GHG emissions might continue to occur for many years after land clearing (Marshall, 2009; McKechnie et al., 2010; Anderson-Teixeira and Delucia, 2011). Carbon debt is commonly amortized over 30 years, as in Table 5-2, but 30 years is often chosen more-or-less arbitrarily to reflect an expected life of a biorefinery. For a 30-year amortized value to be valid, converted land would have to be used continuously for biofuel production for 30 years after conversion. What is more, carbon released upon conversion is in the atmosphere for 30 years longer than carbon displaced in the 30th year of production, but a correction factor for this phenomenon is not applied consistently in LCA studies (Kendall et al., 2009; O'Hare et al., 2009; Levasseur et al., 2010).

When calculating GHG savings from biofuel production, LCA modelers also have to decide which energy sources the biofuels are displacing. Furthermore, modelers have to take into account any opportunity cost of using biomass for liquid fuels rather than electricity production. Indeed, net reductions in GHG emissions from other uses of biomass such as electricity may be higher (Campbell et al., 2009; Ohlrogge et al., 2009; Campbell and Block, 2010; Khanna et al., 2010; Lemoine et al., 2010; Melamu and von Blottnitz, 2011), but this needs to be weighted against what society desires such as liquid fuels to improve national energy security, for example. From these examples, the need for transparency when performing GHG accounting on biofuels becomes exceedingly important. More generally, the International Organization for Standards (ISO) stresses the necessity for clarity over a single methodology in its ISO 14040:2006 standard for life-cycle assessment.[1]

---

[1]"ISO 14040:2006 describes the principles and framework for life-cycle assessment (LCA) including: definition of the goal and scope of the LCA, the life-cycle inventory analysis (LCI) phase, the life-cycle impact assessment (LCIA) phase, the life-cycle interpretation phase, reporting and critical review of the LCA, limitations of the LCA, the relationship between the LCA phases, and conditions for use of value choices and optional elements" (ISO, 2006).

Anticipated Results

Biofuels from Food-Based Feedstocks

Ethanol production efficiency has shown great improvement over the decades (Figure 5-1) (Hettinga et al., 2009; Wang et al., 2011a). Most GHG accountings of corn-grain ethanol conducted before 2008 found a reduction in GHG emissions relative to gasoline of about 20 percent[9] (Farrell et al., 2006; Hill et al., 2006; Wang et al., 2007). Such analyses

---

[9]Comparisons between biofuels and petroleum-derived fuels are commonly expressed using phraseology such as "X reduces emissions relative to Y by Z%," but there is an important caveat to such usage. Comparisons of this sort are typically made on a per unit of energy (for example, MJ) or per vehicle distance traveled (for example, mile or km) basis and assume a 1:1 displacement. The "rebound effect," a decision to produce more biofuels, increases

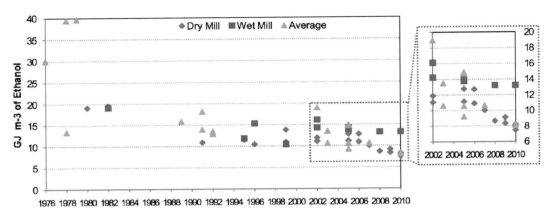

FIGURE 5-1 Historical trend in corn-grain ethanol biorefinery energy use.
SOURCE: Wang et al. (2011a). Reprinted with permission from Elsevier.

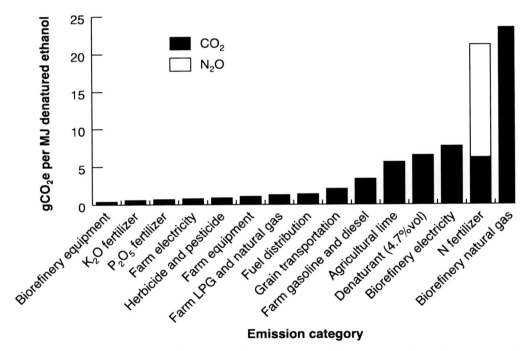

FIGURE 5-2 Contribution of different emission categories to life-cycle emissions of corn-grain ethanol production as measured in an attributional GHG accounting of a Midwestern U.S. facility.
NOTE: Emissions from land-use change are not included. Results are from GREET as calibrated using BESS model
inputs.
SOURCE: Adapted from Plevin (2009b) with permission from John Wiley and Sons.

TABLE 5-3 Published Estimates of and Some Assumptions Used in Estimating Life-Cycle Greenhouse-Gas (GHG) Emissions of Corn-Grain Ethanol

Life-cycle GHG (g CO_2 eq per MJ)	Region for which the estimate was made	Indirect land-use change included	Reference
77	U.S. average	No	Farrell et al. (2006)
85	U.S. average	No	Hill et al. (2006)
177	U.S. average	Yes	Searchinger et al. (2008)
52	Individual facility in the Midwest	No	Bremer et al. (2010)
104	U.S. average	Yes	Hertel et al. (2010)
101	U.S. average	Yes	Mullins et al. (2010)
69	U.S. average	Yes	Wang et al. (2011a)

NOTE: None of the studies listed in this table estimated specifically the extent to which volumes of corn-grain ethanol produced to meet the RFS2 consumption mandate would change GHG emissions.

typically considered only emissions resulting from the supply chain (Figure 5-2). During 2008, the rapid increase in the amount of corn being used for ethanol resulted in a number of new studies being published that account for market-mediated effects of increased ethanol production on land-use change. A sample of modeled estimates of life-cycle greenhouse-gas emissions published from 2006 to 2011 spans 52-177 g CO_2 eq per MJ (Table 5-3). The estimates vary, and some of the key drivers in differences include

- The geographic range considered;
- Whether direct or indirect land-use changes were included in the estimates;
- Assumptions used in estimating indirect land-use changes as shown in Table 5-2;
- Flux values used for N_2O emissions;
- How GHG credits from coproduct production were estimated;
- Technologies and fossil fuel used in the biorefineries;
- The fraction of DDGS that is dried versus fed wet to livestock; and
- Baseline volume of ethanol production.

When the life-cycle GHG emissions in Table 5-3 are compared against the 2005 baseline GHG emissions (as in the case for RFS2), corn-grain ethanol might not have lower values than petroleum-based gasoline. Indeed, studies such as those of Mullins et al. (2010) and Plevin et al. (2010) that address uncertainty in modeled results directly have revealed plausible scenarios in which GHG emissions from corn-grain ethanol are much higher than those of petroleum-based fuels (Figure 5-3). Similar analyses that considered alternate scenarios in which corn-grain ethanol is not produced also found that corn-grain ethanol may have higher GHG emissions than petroleum-based fuels when global system boundaries are used (Feng et al., 2010).

In its Final Regulatory Impact Analysis for RFS2, the U.S. Environmental Protection Agency (EPA) (2010d) conducted what is best described as a hybrid attributional-consequential LCA approach toward assessing life-cycle GHG emissions of corn-grain ethanol

---

availability, which depresses fuel prices and leads to greater overall consumption. As such, a 1:1 displacement is the maximum, and the actual amount of GHG emissions released as a result of increased biofuel production from a policy such as RFS2 is likely to be higher than would be calculated using a 1:1 energy-adjusted volumetric displacement (that is, GHG reductions from biofuels are likely to be exaggerated when market elasticity is ignored). This issue is closely tied to differences in attributional and consequential LCA.

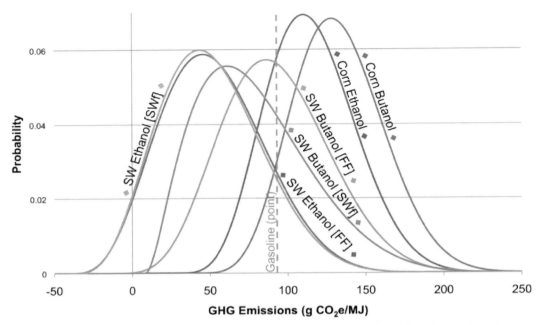

FIGURE 5-3 Probability distributions for U.S. industry greenhouse-gas (GHG) emissions for corn and switchgrass (SW) biofuels. [FF] refers to the burning of fossil fuels at the biorefinery, while [SWf] refers to the combustion of switchgrass for heat and electricity.

NOTE: Uncertainties in GHG emissions from land-use changes contribute most to the extending range of estimates for life-cycle GHG emissions from biofuels.

SOURCE: Mullins et al. (2010). Reprinted with permission from *Environmental Science and Technology* 2010, 45(1):132-138. Copyright 2010 American Chemical Society.

and other biofuels. That is, EPA included GHG emissions from land-use change (consequential approach) but only assessed the industry at given points in time (attributional approach) rather than over the entire duration of EISA, as would be called for in a consequential LCA. A thorough review of EPA's assumptions and calculations behind its estimates of GHG emissions for various biofuels is beyond the scope of this report, but EPA's assessment is presented as a comparison with the studies mentioned above. For a conventional biofuel such as corn-grain ethanol to qualify for RFS2, it has to meet the compliance thresholds of a 20-percent reduction in life-cycle GHG emissions compared to a 2005 gasoline baseline. The term "life-cycle greenhouse gas emissions" is defined as follows:

> The term "life-cycle greenhouse gas emissions" means the aggregate quantity of greenhouse gas emissions (including direct emissions and significant indirect emissions such as significant emissions from land use changes), as determined by the Administrator, related to the full fuel life cycle, including all stages of fuel and feedstock production and distribution, from feedstock generation or extraction through the distribution and delivery and use of the finished fuel to the ultimate consumer, where the mass values for all greenhouse gases are adjusted to account for their relative global warming potential. (110 P.L. 140)

EPA estimated that corn-grain ethanol reduces GHG emissions by 21 percent relative to gasoline, allowing it to qualify for RFS2 over 2008-2022. EPA's determination was based on its evaluation of corn-grain ethanol and other biofuels at three points in time: 2012, 2017, and 2022. Industry average emissions were calculated at each of these 3 years, as shown for corn-grain ethanol in Table 5-4. EPA found corn-grain ethanol, regardless of whether the

TABLE 5-4 Greenhouse-gas (GHG) Emissions from Corn-Grain Ethanol Relative to Gasoline as Determined by EPA in Its Final Rule for RFS2

Biorefinery Heat Source	Dried distillers grain with solubles (DDGS)	2012	2017	2022
Natural Gas	Dry	33	10	−17
	Wet	21	−2	−27
Coal	Dry	66	41	12
	Wet	41	17	−10
Biomass	Dry	6	−15	−40
	Wet	−3	−16	−41

NOTE: Positive values indicate higher emissions than gasoline, and negative values are lower. As of 2011, nearly all corn-grain ethanol biorefineries use natural gas or coal as heat sources. Most facilities would have to be retrofitted in short order to achieve the GHG emissions from corn-grain ethanol produced by facilities using biomass as the heat source listed in the table.
SOURCE: EPA (2010b).

coproduct is sold wet or dry, to have life-cycle GHG emissions higher than gasoline in 2012 or 2017 unless it is produced in a biorefinery that uses biomass as a heat source (Table 5-4). EPA calculated its 21-percent GHG reduction as a weighted average of projected biorefinery and corn production efficiencies that could be realized in 2022 (Plevin et al., 2010). Thus, according to EPA's own estimates, corn-grain ethanol produced in 2011, which is almost exclusively made in biorefineries using natural gas as a heat source, is a higher emitter of GHG than gasoline. Nevertheless, corn-grain ethanol produced at the time this report was written still qualified for RFS2 based on EPA's industry-weighted average of projected 2022 industry. The discrepancy between how RFS2 is implemented (under the assumption of 21-percent reduction of GHG emissions by corn-grain ethanol compared to gasoline) and EPA's own analysis suggests that RFS2 might not achieve the intended GHG reductions. According to EPA's results (Table 5-4), atmospheric GHG concentrations will be higher in the presence of RFS2 due to the cumulative GHG effect of corn-grain ethanol produced over 2008-2022 than in the absence of RFS2, in which case gasoline would be used. EPA's evaluation of other biofuels follows a similar methodology. Therefore, the GHG reductions in other types of biofuels described in the RFS2 Final Rule also deserve similar scrutiny as the industry develops.

For food-based biofuels other than corn-grain ethanol, a consensus on whether biodiesel from oilseeds reduces GHG emissions has not been reached within the scientific community. Although GHG emissions in the direct supply chain tend to be small (Hill et al., 2006; Huo et al., 2009), those associated with land-use change far dominate the life-cycle emissions because feedstocks with low energy yields, such as soybean, tend to require large amounts of land (Miller, 2010).

Biofuels from Wastes and Residues

Biofuels produced from wastes such as agricultural and forestry residues, municipal solid waste (MSW), and waste grease have consistently been shown to have lower life-cycle GHG emissions than petroleum-based fuels. For agricultural and forest residues, low life-cycle GHG emissions will only be realized under conditions that do not interfere with land productivity or soil carbon storage (Cherubini and Ulgiati, 2010; Karlen et al., 2010). Based on the potential volume of wastes, biofuels from MSW were estimated to be able to replace about 2 percent of petroleum-based fuels in the United States (Kalogo et al., 2006) and about 5 percent globally (Shi et al., 2009).

Biofuels from Dedicated Energy Crops

The use of herbaceous and woody dedicated energy crops for biofuels could lower or raise GHG emissions depending on how and where these crops are grown. If land already in food crop production or in pasture is converted to dedicated energy crops, the resulting carbon debt from market-mediated effects might be sufficiently high to offset any carbon savings otherwise realized (Roberts et al., 2010). Similar uncertainty lies in the use of agricultural land not currently in agricultural production, such as abandoned land or reserve land, because the fossil carbon saved by displacing petroleum would need to exceed the carbon storage that would have occurred on that land in the absence of biofuel production (carbon opportunity cost). Lands that are currently uneconomic for crop production because of one or more limiting characteristics, whether in production or not (Wiegmann et al., 2008), could also be used if they meet EISA's land requirements. In those cases, the same considerations of direct and indirect carbon debts and carbon opportunity costs apply.

The relative uncertainty surrounding GHG emissions from biofuels from dedicated energy crops was highlighted by Spatari and MacLean (2010). They used a Monte Carlo simulation to show potentially high and uncertain GHG emissions for switchgrass ethanol largely as a result of CO_2 flux from land-use change and N_2O flux from nitrogenous fertilizer use. In comparison, the authors demonstrated much greater confidence in ethanol from corn-stover biofuels for reducing GHG emissions. In any case, GHG emissions from a given piece of land producing cellulosic biofuels are expected to be lower than those from lands producing corn-grain ethanol or soybean biodiesel (Hammerschlag, 2006; Williams et al., 2009).

Estimating Effects of Achieving RFS2 on GHG Emissions

From the assessment of the literature above, the committee concluded that

- Food-based biofuels such as corn-grain ethanol have not been conclusively shown to reduce GHG emissions and might actually increase them.
- Biofuels from agricultural and forestry residues and municipal solid waste are most likely to reduce GHG emissions.
- Biofuels from dedicated bioenergy crops such as switchgrass may either reduce or increase GHG emissions depending on how and where biomass is grown.

These conclusions do not provide a complete evaluation of the effect of achieving the RFS2 consumption mandate on GHG emissions. Indeed, the published studies mentioned in this report do not and cannot address that issue. Understanding the effect of RFS2 on global GHG emissions would require preparation of a consequential LCA that assesses cumulative effects over time (that is, all years up to 2022 would be considered rather than considering the GHG effects in the year 2022 only). As in all LCAs, GHG released and stored throughout the many steps in the supply chain—from biomass production, harvesting and transport, conversion to fuels in biorefineries, to distribution and use—are considered. In addition, any market-mediated effects on land-use change and petroleum markets as a result of U.S. biofuel policy would have to be accounted for. Such consequential LCA would require the following information to be collected over time or estimates to be made:

- Information on and estimates of what biofuels are produced, how they are produced, and how they affect and are affected by agricultural and energy markets. As mentioned in an earlier section, these factors have large effects on net GHG emissions of biofuels;

- Data and estimates of market-mediated effects of land use, commodity markets, and energy markets over time; and
- Information on the extent to which the introduction of new biofuels into fuel markets displaces petroleum-based fuel production, so as to verify the assumption of complete displacement of petroleum-based fuel by biofuels used in attributional LCAs.

In preparing a complete LCA for assessing the future effects of achieving RFS2 on global GHG emissions, two sets of scenarios have to be evaluated and compared with each other. In the first set of scenarios, the functional unit would be defined as the volume of biofuel produced as a result of RFS2 given all the other factors that influence global biofuel and conventional fuel production. Scenarios in this set could include, for example, various market conditions and levels of technology. In the second set of scenarios, RFS2 would not be enacted and some greater amount of petroleum-based fuel is used and less land is repurposed for biofuel production. Scenarios in this set would be matched to the various market conditions and levels of technology evaluated in the first set. Compared to each other, the two sets of scenarios would provide an indication of whether enacting RFS2 leads to a net decrease in global GHG emissions. For policy evaluation and design, a third set of scenarios may be used in which alternative means of reducing GHG emissions are considered, including the use of biomass for bioelectricity, bioproducts, or building materials.

AIR QUALITY

Production and use of biofuels release air pollutants other than GHG that affect people and their surroundings. Air pollutants from biofuels include criteria air pollutants (for example, carbon monoxide [CO], sulfur dioxide [SO_2], nitrogen oxides [NO_x], particulate matter [PM], and ozone [O_3]); precursors to the atmospheric formation of PM or O_3 (including ammonia [NH_3] and volatile organic compounds [VOCs]); and other hazardous air pollutants, many of which are themselves VOCs (for example, acetaldehyde, benzene, 1,3-butadiene, and formaldehyde). These pollutants have varied effects, including damage to human health (for example, cancer, cardiovascular disease, respiratory irritation, and birth defects) and the environment (for example, reduced visibility, acidification of water and soils, and damage to crops) (Aneja et al., 2009; Uherek et al., 2010).

Emissions from Biofuel Use

On-road vehicles are a major source of many pollutants affecting air quality (Abu-Allaban et al., 2007; Frey et al., 2009). The use of biofuels in vehicles is responsible for emissions of pollutants through evaporation and combustion. The quantity of these emissions depends on various factors, including combustion technologies, emission controls, temperature, and the level at which biofuels are blended into petroleum-based fuels. Reviews of the literature have revealed that relative to petroleum-based fuels, the use of biofuels tends to decrease emissions of some pollutants while increasing those of others. In general, low-level blends of ethanol into gasoline, such as E10 typically lead to lower CO emissions but higher emissions of other species such as nonmethane hydrocarbons (NMHCs), nonmethane organic gas, acetaldehyde, benzene, and 1,3-butadiene (Table 5-5) (Durbin et al., 2007; Jacobson, 2007; Graham et al., 2008; Ginnebaugh et al., 2010). The use of ethanol as an oxygenate in reformulated gasoline does little to reduce ozone levels and may even increase them in areas (NRC, 1999). Higher ethanol blends such as E85 tend to have lower emissions

TABLE 5-5 Average Percent Change in Tailpipe Emissions Compared to a Reference Fuel Containing No Ethanol

	E10	E85
Nonmethane hydrocarbons (NMHCs)	+9	–48
Nonmethane organic gas	+14	ND[a]
Acetaldehyde	+108	+2540
1,3-Butadiene	+16	–77
Benzene	+15	–76
Nitrous oxides (NO_x)	ND[a]	–45
Formaldehyde	ND[a]	+73
Carbon monoxide (CO)	–16	ND[a]

[a]No statistical difference at $p = 0.05$
SOURCE: Graham et al. (2008).

of NO_x, NMHCs, 1,3-butadiene, and benzene, but higher emissions of acetaldehyde and formaldehyde (Graham et al., 2008; Anderson, 2009; Yanowitz and McCormick, 2009). In general, use of biodiesel blended into diesel reduces PM, CO, and hydrocarbon emissions, but increases those of NO_x (McCormick, 2007; Pang et al., 2009; Traviss et al., 2010). Other biofuels such as biobutanol could reduce certain tailpipe emissions (Mehta et al., 2010).

Emissions from Biofuel Production and the Full Life Cycle

Much effort has gone into estimating tailpipe emissions from biofuels, but such a narrow focus misses emissions elsewhere in the life cycle. For example, for corn-grain ethanol produced using natural gas at a dry-mill biorefinery, the vehicle use phase, which includes tailpipe emissions and evaporative emissions from vehicles and filling stations, is responsible for over 90 percent of CO emissions, but only 68 percent of VOC, 22 percent of primary $PM_{2.5}$, 17 percent of NO_x, 13 percent of NH_3, and less than 1 percent of SO_x emissions (Hill et al., 2009). The importance of considering supply chain air pollutant emissions when evaluating transportation options is not unique to biofuels. In a survey of automobiles, buses, trains, and airplanes, Chester and Horvath (2009) found criteria air pollution emissions from the nonoperational stages of a vehicle's life cycle (for example, fuel production, vehicle manufacture, infrastructure construction, maintenance, and operation) to be between 1.1 and 800 times larger than vehicle operation.

For corn-grain ethanol, life-cycle emissions of major air pollutant species (for example, CO, NO_x, $PM_{2.5}$, VOC, SO_x, and NH_3) are higher than for gasoline (Figure 5-4) (Wu et al., 2006; Hess et al., 2009; Hill et al., 2009; Huo et al., 2009). Cellulosic ethanol from either corn stover or dedicated bioenergy crops (such as switchgrass or *Miscanthus*) shows a similar pattern, although SO_x life-cycle emissions could be lower than that of gasoline depending on the extent to which cogenerated electricity produced at the biorefinery offsets fossil electricity, mainly from coal (Wu et al., 2006; Hill et al., 2009). Further improvements in efficiency and pollution control throughout the life cycle, including at biorefineries (Jones, 2010; Spatari et al., 2010), would reduce biofuel life-cycle emissions. Although GHG emissions from land-use change as a result of bioenergy feedstock production have been widely discussed, land-use change also affects air quality directly. Such effects from changes on the U.S. and global landscape could potentially be appreciable, as has been estimated in the conversion of tropical rainforest to palm oil plantations leading to greater emissions of VOC and NO_x, and thus higher ground-level ozone (Hewitt et al., 2009).

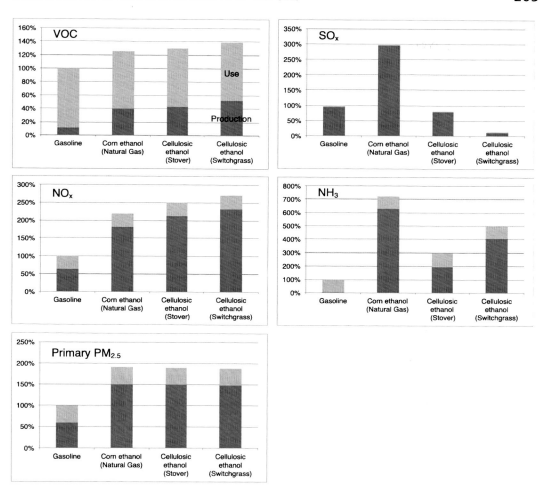

FIGURE 5-4 Life-cycle emissions of volatile organic compounds (VOCs), sulfur oxides (SO_x), nitrous oxides (NO_x), ammonia (NH_3), and primary particulate matter$_{2.5}$ ($PM_{2.5}$) from gasoline, dry-mill corn-grain ethanol produced using natural gas at the biorefinery, cellulosic ethanol from corn stover, and cellulosic ethanol from switchgrass.
NOTE: For each pollutant, values are scaled to life-cycle emissions of gasoline at 100 percent.
DATA SOURCE: Hill et al. (2009).

Effects on Human Health and Environmental Effects

Unlike GHGs, which are mixed in the atmosphere and affect climate change at a global level, air-quality pollutants affect the environment on local and regional scales. As such, life-cycle inventories of quantities of air-quality pollutants, such as those discussed in the previous section, do not themselves describe the ultimate effect of these pollutants. Such methods as impact pathway analysis could be used to assess the ultimate effect. Studies that have considered the ultimate impacts of biofuels have consistently found corn-grain ethanol to have human health damage costs equal to or higher than gasoline (Figure 5-5) (Hill et al., 2009; Kusiima and Powers, 2010; NRC, 2010a). Conversely, the same studies found that human health damage costs from cellulosic ethanol are likely to be lower than those of corn-grain ethanol and could be marginally better than those of gasoline.

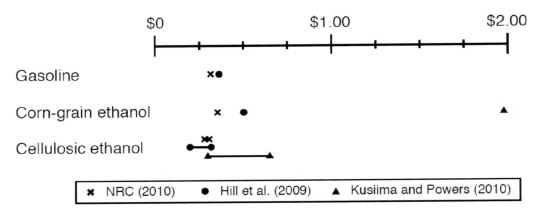

FIGURE 5-5 Human health damage costs (dollars per gallon of gasoline equivalent) of life-cycle air-quality impacts of gasoline, corn-grain ethanol, and cellulosic ethanol.
NOTE: The three studies did not consider exactly the same pollutants. For example, unlike Hill et al. (2009) and Kusiima and Powers (2010), NRC (2010a) did not consider NH_3 emissions. Horizontal lines represent the ranges of estimates.

Air-Quality Effects of RFS2 Estimated by EPA

EPA's assessment of RFS2 is summarized in its regulatory announcement:

> The increased use of renewable fuels will also impact emissions with some emissions such as hydrocarbons, nitrogen oxides (NO_x), acetaldehyde and ethanol expected to increase and others such as carbon monoxide (CO) and benzene expected to decrease. However, the impacts of these emissions on criteria air pollutants are highly variable from region to region. Overall the emission changes are projected to lead to increases in population-weighted annual average ambient PM and ozone concentrations, which in turn are anticipated to lead to up to 245 cases of adult premature mortality. (EPA, 2010a)

EPA began publishing its assessment of RFS2 in the peer-reviewed literature in 2010. Cook et al. (2011) considered changes in concentrations of various pollutants and found that increased ethanol use as a result of RFS2 would increase O_3 concentrations over much of the United States by as much as 1 part per billion (ppb) by 2022. Certain highly populated areas are projected to show decreases in O_3 concentrations due to increased NO_x emissions in VOC-limited areas. Changes in concentrations of other species are mixed (Table 5-6).

WATER QUALITY

Feedstock Production

Effects on water quality from increased biofuel production are caused by changed use of land to produce crops for feedstock, the use of water for irrigating crops, and conversion of crops to fuel in the production process itself. (Water use in conversion of biomass to fuel is discussed later.) Water quality is affected by original soil and land-cover conditions; amount, type, and timing of fertilizer applications; management practices such as tillage; and prevailing weather, particularly the amount and duration of heavy rainfall in relation

TABLE 5-6 Nationwide Emission Inventories for 2022 for the Renewable Fuel Standard (RFS) and RFS2

Pollutant	U.S. Total RFS Annual Tons	U.S. Total RFS2 Annual Tons	RFS2 versus RFS Percent Change
Nitrous oxides (NO$_x$)	11,415,147	11,781,115	3.21
Hydrocarbons (HC)	10,292,785	10,412,658	1.16
Particulate Matter$_{10}$ (PM$_{10}$)	11,999,983	12,068,629	0.57
Particulate Matter$_{2.5}$ (PM$_{2.5}$)	3,371,024	3,389,223	0.54
Carbon monixide (CO)	51,631,075	47,011,171	-8.95
Benzene	226,683	217,021	-4.26
1,3-Butadiene	14,458	14,264	-1.34
Acetaldehyde	58,405	65,722	12.53
Formaldehyde	140,156	140,330	0.12
Acrolein	6,399	6,477	1.23
Ethanol	457,071	906,719	98.37
Sulfur dioxide (SO$_2$)	8,878,706	8,936,086	0.65
Ammonia (NH$_3$)	4,213,048	4,213,189	0.00

NOTE: RFS enacted by the Energy Policy Act of 2005. RFS2 refers to amendments made under EISA.
SOURCE: Cook et al. (2011).

to fertilizer applications (Engel et al., 2010). The portion of nitrogen fertilizer that becomes nitrified to its most mobile form (nitrate) leaches from fields during precipitation events, creating runoff to streams and infiltration to groundwater. As with all other environmental effects, water-quality implications of biofuel production need to be compared to alternate uses of the land and to effects of fossil fuel exploration, extraction, production, and delivery.

The effects of producing bioenergy feedstock on water quality depend largely on the choice of feedstock and its management. Corn requires higher levels of inputs than most annual crops (NRC, 2008), including large amounts of nitrogen fertilizer. (See also Table 5-1 earlier.) Crops that could serve as feedstocks for cellulosic biofuels are expected to exert less harmful effects on water quality than corn and to reduce nutrient runoff because of less intensive land management practices. For example, perennials (switchgrass, *Miscanthus*, prairie polyculture, poplar, willow, pine, and sweet gum) affect water quality less than annual crops because of lower fertilization requirements and reduced need for tillage, which exposes the soil to wind and water erosion and to microbial oxidation. But even perennial crops such as switchgrass and hybrid poplar trees (grown in a short rotation of 4-6 years) can benefit from fertilization, in most cases, to maximize yields for feedstock production. Fertilizing these perennials can cause some nutrient runoff, although less than fertilizing row crops such as corn and soybean because of perennials' superior nutrient uptake efficiency. Planting perennial bioenergy crops in sites with high erosion, or using perennials as buffer strips between annuals and riparian zones, could offer net improvements in water quality as deep-rooted perennials absorb excess nutrients from annuals; reduce erosion, runoff, and other downstream effects; and reduce requirements for pesticides.

Water quality effects discussed in this section include

- Nutrient runoff to surface waters (nitrogen, phosphorus, silica)
- Pesticide runoff (herbicides and insecticides)
- Soil erosion and runoff (sedimentation of habitats and increased turbidity)
- Nutrient percolation, infiltration, and contamination (nitrate).

The qualitative effects of growing bioenergy feedstocks are not different than existing agriculture for the same crops. If growing bioenergy feedstocks increases the extent of agriculture of annual crops within a given basin, it could cause greater effects on water quality. To date, corn grain has been used to produce ethanol, and soybean has been used to produce biodiesel. Acres of corn and soybean planted in the United States have increased during the growth of the biofuel industry (2000-2009) from about 73 to 93 million acres for corn, and from 72 to 77 million acres of soybean (see Figure 4-8 in Chapter 4). Increased acreages of corn have been planted in Iowa and Nebraska, the leading states in ethanol production (Agnetwork, 2010). In addition to increased acres planted in corn, the average yield across the nation has increased from 137-156 bushels per acre from 2000-2010 (USDA-ERS, 2010b). (See also Figure 2-3 in Chapter 2.) The long-term trend for corn yields from 1990-2010 was an increase of about 2 bushels per acre per year. In 2010, 88 million acres of corn were planted, from which 13 billion bushels of corn were harvested.

Methods to Assess Effects

Methods to assess the effects mentioned above include monitoring and modeling of water quality. Monitoring is used by states to determine whether surface waters are meeting their designated uses under Section 303(d) of the Clean Water Act. These designated uses generally fall into three broad categories, which may be subdivided further:

- Aquatic life (aquatic ecosystems health)
- Primary and secondary contact recreation (swimming and boating)
- Drinking water (human health).

If designated uses of the water are not met, the water is considered "impaired." The state then needs to list the lake, reservoir, river, or stream on its list of impaired water bodies and make a calculation and a plan of how to restore it. This process includes establishing Total Maximum Daily Loads (TMDLs) for the water body.

Interestingly, states have been reluctant to promulgate nutrient water quality criteria for their surface waters. In some cases, the reluctance might be a result of stringent criteria recommended by EPA (Heltman and Martinson, 2011). Many lakes and streams would be considered impaired as a result of strict application of such criteria. In states such as Iowa, where nearly 90 percent of the land is already in agricultural use, more than half of all water bodies would be designated as impaired because of nutrient runoff. Solving the problems caused by nutrient runoff would require a detailed TMDL to be developed for all impaired waters and a management plan formulated. However, because runoff from agriculture is not considered a "point source" in the Clean Water Act, permits are not required for farmers to release runoff while producing agricultural crops. Thus, there is no easy way to mitigate the nutrient runoff problem, although integrating perennial biomass feedstock crops into these landscapes to protect water resources could help.

Currently, long-term data are collected and maintained by the U.S. Geological Survey (USGS) in the National Stream-Quality Accounting Network (NASQAN) and National Water-Quality Assessment (NAWQA) programs. These data provide baseline and continuing comparable data to evaluate changes that could then be correlated with regional dynamics in land use, land cover, weather, and climate (for example, Sprague et al., 2011). As with any regional-scale study that integrates across watersheds, water-quality effects are attributed to multiple causes. Further experiments and monitoring designed at spatial and temporal resolutions to assess the effects of biofuel production on water-quality would be useful.

Models to assess the effects of changes in land use and stream or lake quality are many, and they differ in their goals, assumptions, approaches, complexity, and amount of input data required to analyze the problem. Some of the leading watershed and stream models include the Soil and Water Assessment Tool (SWAT, USDA),[10] River and Stream Water Quality Model (QUAL2K, EPA),[11] Water Quality Analysis Simulation Program (WASP, EPA),[12] CE-QUAL (U.S. Army Corps of Engineers),[13] Spatially Referenced Regressions On Watershed Attributes (SPARROW, U.S. Geological Survey),[14] and Hydrological Simulation Program—FORTRAN (HSPF, EPA).[15] Soil erosion is a key part of several of these models. The Revised Universal Soil Loss Equation (RUSLE, USDA)[16] has been extensively used over large areas and long time frames (annual averages) to determine soil erosion. Delivery of soil to the stream is more complicated, and few models other than SWAT and HSPF perform such operations. Only SWAT and HSPF represent processes for an entire agricultural watershed including erosion and runoff from the field to the stream and also in-stream transport and reactions. These two models require considerably more input data than other models for their simulations. Groundwater models include the Groundwater Monitoring System (GMS) with submodels MODFLOW, MT3D, and others. USGS uses land physiographic, hydrologic, and applications factors in multiple linear regression models for both groundwater and surface water projections.

Anticipated and Observed Effects

Nitrogen loads are measured and modeled to be in excess of 5,650 lbs/mi² per year (or 1,000 kg/km² per year as shown in Figure 5-6) in the Corn Belt of the Midwest. This loading

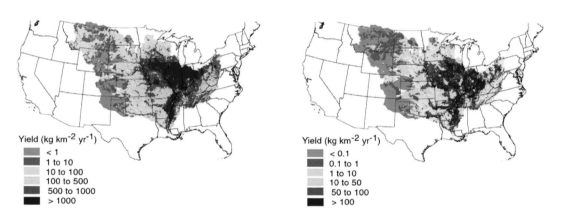

FIGURE 5-6 Model estimates of nitrogen and phosphorus yield from runoff in the Mississippi River Basin for 1992-2002.
NOTE: The largest yields emanate from areas where significant areas of land are planted in corn row crop.
SOURCE: Alexander et al. (2008) based on historical data and the SPARROW model. Reprinted with permission from *Environmental Science and Technology* 2008, 42(3):822-830. Copyright 2008 American Chemical Society.

---

[10] http://swatmodel.tamu.edu/.

[11] http://www.epa.gov/athens/wwqtsc/html/qual2k.html.

[12] http://www.epa.gov/athens/wwqtsc/html/wasp.html.

[13] http://www.pdc.pdx.edu/ce-qual/.

[14] http://water.usgs.gov/nawqa/sparrow/.

[15] http://water.usgs.gov/software/HSPF/.

[16] http://www.iwr.msu.edu/rusle/.

represents 5 to 10 percent of the nitrogen applied to corn and a significant economic loss for the farmer. But it also impairs downstream uses all the way from the farm to its ultimate discharge in the Gulf of Mexico. Discharges from the Mississippi River-Atchafalaya Basin exacerbate hypoxia[17] during July to October in the Gulf that threatens shrimp, crab, and oyster fisheries over an area of 7,800 mi^2 (the approximate area of hypoxia in 2007 and 2008). Hypoxia occurs naturally in many coastal waters, and the occurrence and extent of hypoxia are the collective result of a complex combination of basin morphology, climate, weather, circulation patterns, water retention times, freshwater inflows, stratification, mixing, and nutrient loadings (Dale et al., 2010b). Several hypoxic events occurred from 1870 to 1910 prior to widespread fertilizer use and were attributed to natural variation in river flow (Osterman et al., 2005). However, the increase in the area of hypoxia (Rabotyagov et al., 2010) and its sensitivity to nutrient loads (Liu et al., 2010) have been largely attributed to nitrogen loadings, phosphorus fluxes, and cultural eutrophication (Bricker et al., 1999; Rabalais and Turner, 2001; Scavia et al., 2003; Turner et al., 2006; Scavia and Liu, 2009). The observed export of nitrate into aquatic systems varies annually because of variations in nitrate fertilizer application rates and because of the effect of hydrology and weather on the storage of nitrate in soil versus leaching (Donner et al., 2002; Donner and Kucharik, 2003).

There is evidence that EISA and the push for biofuels has caused more land to come into corn production (USDA-ERS, 2010b). The area of corn planted in the United States peaked in 2007 (93.5 million acres; see also Figure 4-8 in Chapter 4) when corn prices were high and corn-grain ethanol production was rapidly increasing (NCGA, 2010), and acreage planted in 2011 is projected to be the second highest in the United States since 1944 (USDA-NASS, 2011). Increased cropping area of corn for ethanol production is assumed to exacerbate eutrophication and hypoxia due to the high inputs of nitrogen, phosphorus, and pesticides required for corn production (NRC, 2008).

A recent analysis of NAWQA data by Sprague et al. (2011) (Table 5-7) found that since 2000 most of the drainages associated with the Mississippi River increased in flow-normalized concentration and flux of nitrate. Nitrate fluxes are affected by several factors including input and discharge rate associated with weather dynamics. Moreover, additional land-cover change associated with corn-grain ethanol has occurred since 2008. Therefore, monitoring designed to assess the effects of biofuels on water quality is needed to ascertain the effects of increasing biofuel production on water quality.

Measured nutrient loadings coming from land with a higher percentage of land planted in corn tend to have greater nutrient loadings as modeled for the Mississippi River Basin by Alexander et al. (2008) (Figure 5-6). Models by Scavia and colleagues (Scavia et al., 2003; Scavia and Liu, 2009), Rabalais and Turner (2001), and Turner et al. (2006) relate the hypoxic area in July to August to the nitrogen loading emanating from the Mississippi River and Atchafalaya River from May to June. Thus, increases in nitrogen runoff serve to increase gulf hypoxia according to the models.

Donner and Kucharik (2008) projected annual mean dissolved inorganic nitrogen flux to the Gulf of Mexico to increase by 10 to 18 percent if an additional 15 billion gallons of corn-grain ethanol is to be produced. They used two land-use scenarios—one that combines land-use shifts and corn planting on CRP land in corn-growing counties and another that produces corn-grain for 15 billion gallons of ethanol without the diversion of any corn from other current uses. The two scenarios were compared to a control case based on mean land use and land cover from 2004 to 2006. Donner and Kucharik's estimates do not directly

---

[17]"Hypoxia is the condition in which dissolved oxygen is below the level necessary to sustain most animal life–generally defined by dissolved oxygen levels below 2mg/l (or ppm)" (CENR, 2000).

TABLE 5-7 Net Change in Flow-Normalized Nitrate Concentration and Flux Between 1980 and 2008

Site	Flow-Normalized Concentration of Nitrate as N			Flow-Normalized Flux of Nitrate as N			
	Annual Mean Flow-Normalized Concentration in 1980, mg/L (10^{-4} oz/gal)	Change, 1980-2008, mg/L (10^{-6} oz/gal)	Change, 1980-2008, Percent	Total Annual Flow-Normalized Flux in 1980, 10^8 kg/yr (10^8 lbs/yr)	Total Annual Flow-Normalized Yield (Flux Per Unit Area) in 1980 (kg/km^2/yr) (lbs/mi^2/yr)	Change, 1980-2008, 10^8 kg/yr (10^8 lbs/yr)	Change, 1980-2008, Percent
Mississippi River at Clinton, IA	1.13 (1.51)	0.86 (115)	76	0.66 (1.46)	297 (1700)	0.44 (0.97)	67
Iowa River at Wapello, IA	5.02 (6.70)	0.17 (22.7)	3	0.59 (1.30)	1813 (10400)	-0.02 (-0.04)	-3
Illinois River at Valley City, IL	3.81 (5.09)	-0.04 (-5.34)	-1	0.99 (2.18)	1433 (8180)	-0.01 (-0.02)	-1
Mississippi River below Grafton, IL	2.56 (3.42)	0.49 (65.4)	19	3.33 (7.34)	751 (4290)	0.47 (1.04)	14
Missouri River at Hermann, MO	0.96 (1.28)	0.72 (96.1)	75	0.90 (1.98)	67 (382)	0.51 (1.12)	57
Mississippi River at Thebes, IL	1.93 (2.58)	0.38 (50.7)	20	4.74 (10.4)	257 (1470)	0.44 (0.97)	9
Ohio River at Dam 53 near Grand Chain, IL	0.99 (1.32)	0.03 (4.01)	3	3.10 (6.83)	590 (3370)	-0.04 (-0.09)	-1
Mississippi River above Old River Outflow Channel, LA	1.25 (1.67)	0.13 (17.4)	10	8.11 (17.88)	278 (1590)	0.69 (1.52)	9

SOURCE: Sprague et al. (2011). Reprinted with permission from *Environmental Science & Technology* 2011, 45(17):7209-7216. Copyright 2011 American Chemical Society.

estimate the effect of increasing biofuel production as a result of RFS2, as their baseline scenario is no ethanol production. Before RFS went into effect in 2005, 3.9 billion gallons of ethanol were produced. Modeling two scenarios, one that uses corn grain and cellulosic biomass to meet the consumption mandate of RFS2 and another that uses only cellulosic biomass to meet the consumption mandate, Costello et al. (2009) found that using only cellulosic biomass for biofuel production could reduce nitrate output from the Mississippi and Atchafalaya River basins by an average of 20 percent.

Nutrient runoff increases nitrogen concentration in surface waters, which causes excessive algal and plant growth and loss of transparency in the water column. Those effects, in turn, change habitats for biota and cause taste and odor problems for drinking water supplies. To the extent that RFS2 increases corn and soybean production, it is expected to increase nutrient runoff (Donner and Kucharik, 2008). To the extent that cellulosic feedstock production under RFS2 accelerates change from traditional cultivation to well-managed perennials and reduces runoff, it can provide water-quality benefits by reducing nutrient and sediment runoff. The effects of second-generation biofuel policies on water quality could be positive or negative depending on the location of the feedstocks, choice of feedstock, management practices used, and overall land-use changes. Given U.S. biofuel production goals under RFS2, data need to be collected to document how these shifts in land use actually affect water quality.

As corn acreage and yields increase, greater nitrogen fertilizer is required to replace the nitrogen taken off the land in the crop. Thus, there is a tendency for greater runoff and loadings to streams and rivers from increased corn production (Donner et al., 2002; Donner and Kucharik, 2003). USDA (Malcolm and Aillery, 2009) used a national agricultural sector model to estimate the expected market and environmental outcomes of producing 15 billion gallons of corn-grain ethanol in 2016 (as reflected in EPA's RFS2) compared to a baseline of 12 billion gallons of corn-grain ethanol. They projected an increase of 3.7 million acres of corn including 1.7 million acres of continuous corn as a result of achieving the mandate for conventional biofuels in 2016. The projected increase in corn acreage was estimated to cause a 2.1-percent increase in sheet erosion of soil, 2.5-percent increase in nitrogen runoff (29,200 tons of nitrogen), and a 2.8-percent increase in runoff of pesticides. Simpson et al. (2008) estimated that a long-term increase of 16 million acres of corn could be added to account for future biofuel production.

Mubako and Lant (2008) used nitrogen, phosphorus, and pesticide application rates from Hill et al. (2006) to estimate total water-quality effects of corn-grain ethanol. Application rates were assumed to average 130 lbs N per acre, 47 lbs P per acre, and 2.0 lbs pesticides per acre (or 146 kg N/ha, 53.1 kg P/ha, and 2.3 kg pesticides/ha, as shown by Mubako and Lant, 2008), yielding estimates of applications on a volumetric basis: 65.5 g N/L, 23.8 g P/L, and 1.03 g pesticides/L of ethanol produced (Table 5-8). Assuming 10.6 t/ha of soil erosion, then 4.8 kg of soil are eroded per liter of ethanol produced. Further, using 21.1 MJ as the energy value of a liter of ethanol and a net energy return on energy invested of 1.25 (Hill et al., 2006), Mubako and Lant (2008) concluded that 15.5 g N, 5.65 g P, 0.24 g pesticides, and 1.13 kg of eroded soil are required per (net) MJ of energy gained from ethanol.

Using crop residues, such as corn stover, could cause greater or less soil erosion than other options. But in cases of high crop yield, excessive residues could reduce performance of no-till drill techniques and reduce crop production. In those cases, some residue removal could enhance no-till management (Siemens and Wilkins, 2006; Edgerton, 2010). Corn stover is likely to be supplied from the heart of the Corn Belt, centered in Iowa and Illinois at locations near to ethanol production facilities (Chapters 2 and 3). Regarding utilization of corn stover, the Water Erosion Prediction Project (WEPP) computer model has been used

TABLE 5-8 Application Rates onto Land for Nitrogen, Phosphorus, and Pesticides (and Soil Erosion) as a Result of Growing Corn as Feedstock for Ethanol Production

	Nitrogen (N)	Phosphorus (P)	Pesticides	Erosion
Application rate, kg/ha (lbs/acre)	146.1 (130.4)	53.1 (47.4)	2.3 (2.05)	10,620 (9,475)
Application rate, kg per tonne of corn produced (lbs/bushel)	15.46 (0.866)	5.62 (0.315)	0.243 (0.014)	1124 (62.9)
Application, kg per tonne of ethanol produced (lbs per ton of ethanol produced)	51.71 (114.0)	18.80 (41.45)	0.814 (1.79)	3759 (8,287)
Mass ratio of N, P, pesticides, and erosion to ethanol	0.052	0.019	0.0008	3.76
Application rate, g/L ethanol (oz/gallon)	65.54 (8.75)	23.83 (3.18)	1.03 (0.137)	4764 (636.12)
Application rate g per MJ net energy gain (oz/BTU)	15.53 ($5.78 \times 10{-04}$)	5.65 ($2.10 \times 10{-04}$)	0.24 ($8.93 \times 10{-6}$)	1129 (0.042)

NOTE: Values in U.S. standard units are shown in parentheses.
SOURCE: Mubako and Lant (2008).

to simulate soil loss in Iowa at 17,848 sites based on the 1997 National Resources Inventory (Newman, 2010). Prepared soil erosion hazard maps indicate that harvesting of any corn stover is not recommended on the most steeply sloped soils. However, most sites can withstand some removal, and many sites can sustain 40- to 50-percent removal of corn stover or more, based on soil erosion considerations only.

The effects of RFS2 on those environmental qualities can be estimated by a consequential LCA. However, neither attributional nor consequential LCAs that project the effects of biofuel production on quality of surface water, streams, and groundwater have been completed (Secchi et al., 2011). Precise information on the location of feedstock production, type of feedstock grown, management practices used, and any changes in land cover is necessary for such analysis. However, the SWAT model and SPARROW model have been used to determine differences in water quality attributable to various crop covers, including corn, soybean, sugarcane, switchgrass, woody crops, and other grasses. For example, the SWAT model was used to examine nitrogen loadings in various subwatersheds and land covers of the Raccoon River in central Iowa (Schilling and Wolter, 2008). As expected, corn yielded the greatest nitrogen loadings to the Raccoon River watershed (about 7,800 mi^2) of any land cover studied in Iowa. However, the model was also able to apportion the nitrogen loading to the exact practice or process from which it emanated. Mineralization from stored soil nitrogen was the greatest input causing long-term delivery of nitrogen into streams long after other nitrogen applications ceased. Fertilizer inputs were the second most important, followed by manure applications and atmospheric deposition. The relationship between fertilizer application rates and nitrogen loading into surface water is not linear: that is, decreasing fertilizer application rates do not decrease nitrate loading delivered to the receiving stream by the same magnitude. A decrease in fertilizer application rate from 152 to 45 lbs per acre only reduced the nitrate loading at the watershed outlet by 30 percent. Changing land cover, as in putting land enrolled in CRP to row crop in the Raccoon River watershed, increased nitrate loads at about a 1:1 ratio. For example, if 9.5 percent of the land in the watershed were changed to row crop from CRP, it would result in an 8.9-percent increase in nitrate load. There was a larger effect on nitrate loadings from converting floodplain alluvial soils to corn than from upland sloped soils, so the location of the land conversion within the watershed is important (Schilling and Wolter, 2008).

In general, using cover crops of legumes, cereals, or grasses in fields during noncrop periods to reduce nitrate leaching during vulnerable fall and spring periods was the most effective practice to decrease nitrogen loadings, especially from baseflow or tile drainage (Schilling and Wolter, 2008). Changing from conventional anhydrous ammonia application on corn to innovative subsurface injection methods was the most effective management practice to reduce nitrate loadings from surface runoff. Thus, to the extent that biofuel policies successfully promote use of cover crops and more efficient agricultural practices, they will improve water quality.

The SWAT model indicates that perennial crops with lower nitrogen inputs, no tillage, and perennial root systems can be used to decrease nitrogen loadings to streams as compared to other crops and management regimes. Sahu and Gu's modeling results (2009) showed that planting switchgrass as a contour or riparian buffer in the Walnut Creek watershed in Iowa can reduce nitrate outflow. The extent of nitrate reduction depends on the size and location of the buffer strips. Ng et al. (2010) simulated the effects of planting *Miscanthus* in place of conventional row crops on nitrate loading in Salt Water Creek, Illinois. They found that nitrate loading was projected to decrease as the amount of land converted to *Miscanthus* increases. The extent of nitrate loading also depended on the amount of nitrogen fertilizer applied to *Miscanthus*. Percolation of nutrients and fecal coliform to groundwater is a major problem from row crop agriculture in areas where soils are sandy and permeable (Nolan et al., 2002). Nitrate, atrazine, and coliform bacteria are known to be affecting surficial groundwater supplies from corn and soybean agriculture (Gilliom et al., 2007). To the extent that cellulosic crops are used for the remainder of RFS2, water-quality effects on groundwater could be reduced and would likely be less than those of equivalent row crops. Grasses and perennial crops have deep, dense root systems year-round that serve to hold nutrients in place. Some authors suggested planting short-rotation woody crops as buffer strips because of their high nutrient uptake ability (Adegbidi et al., 2001; Fortier et al., 2010). Half of the crops can be harvested as bioenergy feedstock at a given time while the other half continues to serve as a vegetative buffer (Berndes et al., 2008).

Next Steps Needed

It would be desirable to develop scenarios and apply them in watershed models to predict changes in water quality resulting from implementation of the RFS2 schedule. The models would have to include the physiological traits of perennial crops (such as above- and below-ground biomass existing for more than one year) (Baskaran et al., 2009). Furthermore, empirical data need to be collected at watershed scales to validate these models. At present, there are no watershed-scale data comparing effects of cellulosic bioenergy feedstock production to traditional row-crop production. Thus, model projections cannot be substantiated.

Literature studies indicate that substantial reductions in nutrient loadings would be realized if large areas of land are converted to perennial crops from row crops to produce cellulosic biofuels (Schilling and Spooner, 2006; Costello et al., 2009). Improvements can also be realized by integrating appropriate perennials into larger crops systems where they can be most effective in capturing excess nutrients and protecting water supplies (Anex et al., 2007; Johnson et al., 2008).

Conversion to Fuels

Waste streams from ethanol distilling plants include salts, which are formed by scaling and evaporation in the cooling towers and boilers. If these deposits are not removed (a process called "blow down"), the efficiency of the system decreases dramatically. Blow down results in high-concentration discharges of these salts. The biorefining process requires very pure water, and the use of osmotic purification systems that remove impurities from either surface or groundwater result in additional salt discharges. The Clean Water Act's National Pollutant Discharge System (NPDES) permitting process is required for any facility to discharge this effluent.

Water-quality effects associated with fatty acid methyl esters (FAME) biodiesel production include discharges from oil extraction, chemical reaction processes, separation, purification, and conditioning. The pretreatment of lignocellulosic biomass, which requires water and chemicals, and production of other waste streams that could alter biological oxygen demand (BOD) and concentrated organic content loadings, could affect water quality.

Methods for Assessing Effects

Unlike discharges from feedstock cultivation, biorefineries for converting biomass to fuels are point sources of pollutants to waters. NPDES requires biorefineries to obtain federal permits from EPA or from state agencies authorized by EPA to implement the NPDES program. Under NPDES, a point source can discharge specific pollutants into federally regulated waters under specific limits and conditions. Effects of discharges from biorefineries on water quality can be measured or estimated with less uncertainty than effects of increasing feedstock production on water quality.

Few simulations focus on water effects of refinery operations. Mitigation practices for disposal and treatment of refinery effluents are commercially available and employed. The quality of water discharged by existing corn-grain ethanol and biodiesel biorefineries are monitored by state officials. The wastewater is treated to a high quality prior to discharge (GAO, 2009). Because there are no commercial-scale cellulosic-biofuel refineries, water discharges can only be modeled or extrapolated from demonstration-scale refineries.

Anticipated and Observed Effects

Water-quality effects have been identified for corn-grain ethanol production and include discharges from feedstock processing, pretreatment, sachariffication, fermentation, and effects from boilers and cooling towers. A sample compositional analysis of these discharges is presented in Table 5-9.

Dry-mill corn-grain ethanol effluent and solids are also part of the refining process, and 4-6 gallons of stillage are produced for every gallon of ethanol (Khanal et al., 2008). The process of concentrating solids to produce distillers dry grain (DDG) is done by centrifugation and produces a product with 90-percent total solids content (Rausch and Belyea, 2006). The remaining fluids are composed of concentrated organic content, in the range of 11-13 oz per gallon, a pH of 3.3 to 4.0, and a total solids loading of 7 percent (Wilkie et al., 2000). The remaining fluids (termed "thin stillage") are recirculated as process water, and a portion is evaporated to produce a syrup containing 30-percent solids that is blended with DDG to produce DDGS, a coproduct that can be used as livestock feed. In a survey of dry-mill corn-grain ethanol biorefineries, 55 biorefineries reported water discharge of 0.46 gallons per gallon of anhydrous ethanol (Mueller, 2010). Some plants recycle their

TABLE 5-9 Compositional Analysis of Two Ethanol Plant Discharges Adapted from NRC 2008

Constituent, mg/L (oz/gallon × 10⁻³)	Siouxland Ethanol Facility (Sioux Center, Iowa) Raw Ground W	Little Sioux Ethanol Facility Simulated Blowdown Big Sioux RO reject water	Surface water	Tower efficiency
Total dissolved solids (TDS)	2,113 (257)	7,288 (887)	703 (85.6)	3,240[a] (394)
Calcium ion (Ca²⁺)	305 (37.1)	1,033 (126)	129 (15.7)	638 (77.6)
Magnesium ion (Mg²⁺)	138 (16.8)	458 (55.7)	58 (7.06)	185 (22.5)
Potassium ion (K⁺)	0 (0)	0 (0)	2 (0.24)	33 (4.02)
Sodium ion (Na⁺)	148 (18.0)	485 (59.0)	20 (2.43)	297 (36.1)
Chlorine ion (Cl⁻)	23 (2.80)	131 (15.9)	35 (4.26)	27 (3.29)
Sulfate ion(SO₄²⁻)	1,420 (173)	4,716 (574)	107 (13.0)	2,265 (276)

[a]Concentration in milligrams per liter as calcium carbonate ($CaCO_3$).
NOTE: Values in U.S. standard units are shown in parentheses.
SOURCE: Parkin et al. (2007).

water completely through a combination of centrifugation and evaporation and have no wastewater discharge (Aden, 2007; Mueller, 2010).

In biodiesel refineries, the production of FAME releases glycerin (the backbone of the original fatty acid) in the water stream as part of the transesterification process. Glycerin and unreacted methanol are often found in the effluents of biodiesel refineries that are not designed to recover those byproducts. Those compounds make their way into local municipal wastewater treatment facilities, increasing BOD, or required oxygen level needed to break down the material. The BOD contribution from those biorefineries' effluents may be about 10 ounces per gallon (GAO, 2009). Newer and larger biodiesel refineries are able to extract and purify glycerin to be further used in coproducts including cosmetics and animal feed, and as new technologies are deployed, the recovery of glycerin will be more efficient, ameliorating any negative effects.

Data on cellulose to ethanol effluent and solids from commercial operations are not yet available, and literature regarding their potential composition is limited to laboratory-scale reactions. In one such study, stillage was generated in the range of 11.1 ± 4.1 gallons per gallon of ethanol. Concentrated organic content of the stillage was estimated as 7.46 ± 4.87 oz per gallon, BOD as 3.36 ± 1.85 oz per gallon, total nitrogen as 0.34 ± 0.56 oz per gallon, total phosphorous as 3.41 ± 3.65 oz per gallon, sulfates as 0.079 ± 0.0148 oz per gallon, and pH as 5.35 ± 0.53 (Khanal et al., 2008). Stillage could contain phenolic compounds from the lignocellulosic feedstock and furfurals from acid hydrolysis (Wilkie et al., 2000). Estimates of wastewater discharges also were reported in environmental assessments of planned cellulosic biorefineries (DOE, 2005; ENSR AECOM, 2008; DOE-EERE, 2010b).

Use of Coproducts

Use of a large proportion of DDGS in diets for livestock also raises safety (see Appendix N) and environmental concerns. Environmental problems arise from a mismatch in the nutrient balance in DDGS relative to that needed for animals that consume them. DDGS have roughly three times the amount of nitrogen and phosphorus as corn. They are commonly used in ruminant diets in place of corn, but this can result in levels of nitrogen, phosphorus, and sulfur that are in excess of an animal's needs (Schmit et al., 2009). When DDGS are fed in place of corn and soymeal to broiler chickens, they result in greater excretion of nitrogen due to their poor amino acid balance and poor protein digestibility (Applegate et al., 2009). However, phosphorus excretion does not increase when the diets are appropriately formulated. When fed to laying hens or pigs, DDGS result in greater excretion of nitrogen and phosphorus. In each case, excess nutrients are excreted into manure. When this manure is used as fertilizer, the higher levels of nutrients may result in N and P loading on croplands, depending on agronomic conditions (Benke et al., 2010). Moreover, the solubility of excreted phosphorus in laying hens fed with DDGS is higher (Leytem et al., 2008), though the amount of ammonia released from their manure is lower (Wu-Haan et al., 2010). Proper formulation of diets to minimize nutrient excesses and the use of enzymes such as phytase and xylanase can mitigate nutrient excesses. However, these solutions are not always economically advantageous, depending on ingredient costs and environmental restrictions on manure application rates.

WATER QUANTITY AND CONSUMPTIVE WATER USE

Feedstock Production

Water withdrawals in the United States have not increased substantially in recent decades. In fact, some states (for example, California) have continued to gain population while using less water. Progress in water-use efficiency and conservation is encouraging. However, if production of feedstocks for increased biofuels requires more water from unsuitable sources either for feedstock production or for withdrawals required at production facilities, then increases in consumptive water use (NRC, 2008) could result in competition for freshwater with other uses. Future biofuels under RFS2 might adopt crops that are less water-demanding than corn and soybean, and therefore might not require irrigation. Widescale placement of perennial bioenergy crops across the central United States could also have large effects on evapotranspiration, affecting the availability of water stored in soils (VanLoocke et al., 2010; Georgescu et al., 2011).

Methods to Assess Effects

Streamflow is gauged by USGS for the nation's streams and rivers, but the spatial distribution of the gauges vary across the country (USGS, 2011). Lake levels are monitored by USGS and the U.S. Army Corps of Engineers (reservoirs), and groundwater is monitored sporadically by the states and in special studies by USGS. Tipping-bucket rain gauges are measured in state networks and at airports; Next Generation Radar (NEXRAD) laser-doppler network and models are used for weather forecasting and rainfall-runoff modeling by the National Weather Service. These data are used as input for various hydrometeoro-logical models and rainfall-runoff models involving agriculture. Agriculture crop models use those basic data as input for models such as CENTURY, DeNitrification DeComposi-tion (DNDC), and Photosynthetic/EvapoTranspiration (PnET); these models are in turn linked with water quality models mentioned earlier such as SWAT and HSPF. The results of water-use models are often coupled with basic agriculture yield data from USDA in life-cycle assessment models to assess the performance of various crops for feedstocks in biofuel production.

Empirical Evidence

Measures of water quantity effects due to increased production of biofuels have been concentrated in a few locations where corn is irrigated or production facilities are with-drawing water from depleting groundwater sources. As a case study, Nebraska is among the states with the largest water withdrawals for irrigation, and its usage has continued to increase in recent years, largely driven by the need to irrigate corn for ethanol. Corn acre-age in Nebraska averaged 8.3 million acres during 2000-2006, but it increased to 9.4 million acres in 2007, 8.8 million acres in 2008, and 9.2 million acres in 2009. About 70 percent of the corn in Nebraska is irrigated (Nebraska Corn Board, 2011). Thus, irrigation require-ments result in considerable withdrawals from the High Plains Aquifer. Figure 5-7 shows the drawdown in the High Plains (or Ogallala) Aquifer in Nebraska since predevelopment.

On average, about 70 percent of irrigated water is consumed in the process of irrigating corn (consumptive use) (Wu et al., 2009a). It is not returned to the stream or groundwater, but rather it returns to the atmosphere as evapotranspiration from crops. Figure 5-8 shows the areas of the country where corn is irrigated. Corn acreage that requires irrigation and the quantity of water use vary across the United States. In arid regions such as North

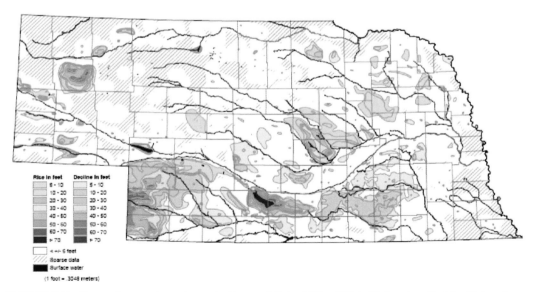

FIGURE 5-7 Groundwater drawdown in the surficial aquifer (High Plains Aquifer) in Nebraska as a result of years of agricultural and municipal withdrawals.
SOURCE: Conservation and Survey Division, School of Natural Resources, UNL (2009). Reprinted with permis-sion from Conservation and Survey Division, School of Natural Resources, University of Nebraska, Lincoln.

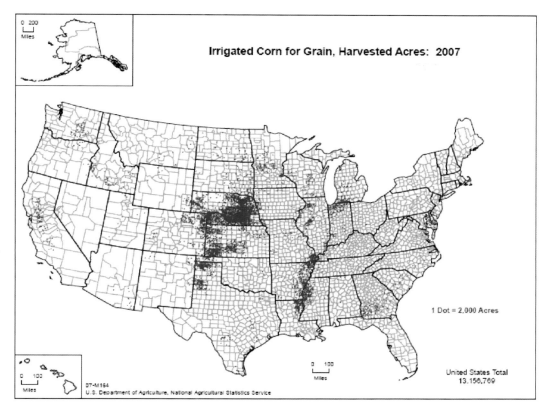

FIGURE 5-8 U.S. irrigation corn for grain.
SOURCE: USDA-NASS (2009).

Dakota, South Dakota, Nebraska, and Kansas, the estimated use of freshwater to irrigate corn is 865 gallons per bushel. There is hydraulic connection between the High Plains Aquifer and surface waters. The Republican River runs from Colorado through Nebraska and into Kansas, and the river loses water along its entire stretch. Consumptive water use for corn production could be high irrespective to which purpose the corn would be dedicated.

Total water withdrawals (agriculture + municipal + industrial) are summed and mapped in Figure 5-9. Areas colored in brown indicate water withdrawals of 9.84-98.4 inches of water averaged over the land area of each county. Precipitation of that amount would be needed to replenish aquifers and maintain groundwater levels. In the United States, 55 million acres of cropland are irrigated, mainly in the West, the Mississippi Delta region, and Florida. Agriculture uses about one-third of all water use and 80 percent of U.S. consumptive water use (USDA-ERS, 2004). Irrigation for agriculture has been increasing in states that are large producers of corn-grain ethanol.

Results

Stone et al. (2010) assessed the bioenergy production goals outlined in the report *Biomass as Feedstock for a Bioenergy and Bioproducts Industry: The Technical Feasibility of a Billion-Ton Annual Supply* (Perlack et al., 2005) relative to water resource effects and climate change (Table 5-10) and found that consumptive water use depends largely on the choice

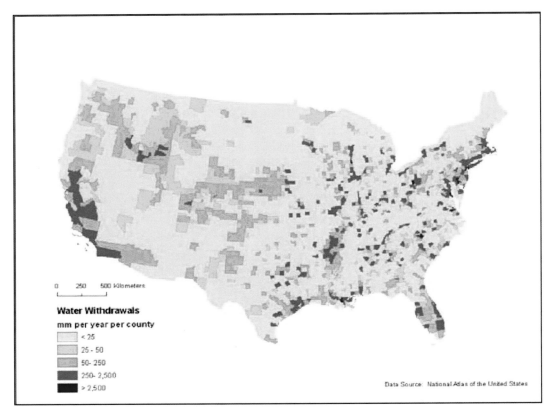

FIGURE 5-9 Total water withdrawals (from agriculture, municipalities, and industry) in the United States by county in 2000.
SOURCE: Hutchinson (2008). Reprinted with permission from K. Hutchinson, USGS Iowa Water Science Center.

of feedstock and where it is grown. Corn grain, corn stover, and grain sorghum used the most water among ethanol feedstocks, and water use by soybean and canola were also high. Sugarcane, switchgrass, and sweet sorghum were superior-performing crops as ethanol feedstocks with respect to water use.

Next Steps Needed

Biofuel is part of the nation's strategy for energy independence. However, water availability is a critical aspect of increasing feedstock production. Empirical data need to be evaluated to ensure increasing bioenergy feedstock production does not result in continuous depletion of groundwater. Improved analysis of empirical data that are now becoming available needs to be incorporated to improve the modeling of the nation's water resources and to inform regulators and the public of the environmental implications of increased biofuel production.

TABLE 5-10 Comparison of Water Requirements for Ethanol Production from Corn Grain, Sugarcane, and Other Potential Energy Crops

Crop	Water requirements, m^3 water/Mg crop (gallons water/ton)	Biofuel conversion, L fuel/Mg crop (gallons fuel/ton crop)	Crop water requirement for biofuel, m^3 water/Mg fuel (gallons water/ton fuel)	Crop water requirement per unit energy, m^3 water/GJ (gallons water/Btu)
Ethanol				
World corn (grain)	833 (200,000)	409 (98.0)	2,580 (618)	97 (0.027)
World sugarcane	154 (36,906)	334 (80.0)	580 (139)	22 (0.006)
Nebraska corn (grain)	634 (152,000)	409 (98.0)	1,968 (472)	74 (0.021)
Corn stover	634 (152,000)	326 (78.1)	2,465 (591)	92 (0.026)
Corn stover + grain	634 (152,000)	735 (176)	1,093 (262)	41 (0.011)
Switchgrass	525 (126,000)	336 (80.5)	1,980 (475)	74 (0.021)
Grain sorghum	2,672 (640,000)	358 (85.8)	9,460 (2,270)	354 (0.099)
Sweet sorghum	175 (41,900)	238 (57.1)	931 (223)	35 (0.010)
Biodiesel				
Soybean	1,818 (436,000)	211 (50.6)	9,791 (2,350)	259.0 (0.072)
Canola	1,798 (431,000)	415 (99.5)	4,923 (1,180)	130 (0.036)

NOTE: Values in U.S. standard units are shown in parentheses.
SOURCE: Stone et al. (2010).

Conversion to Fuels

Water use in biorefineries depends on the feedstock and conversion process used. Process water for biofuel production raises site specific and region-specific concerns about water availability. In general, however, the overall volume of water consumed in the processing of feedstock to fuel is small compared to the volume of water needed to grow the biomass feedstocks (NRC, 2008).

Water use in biorefineries for corn-grain ethanol production includes the hydration of biomass flour (ground corn, wheat, or any other grain used) for the mixture of enzymes and its subsequent high temperature breakdown to release glucose monomers. The slurry mix of mash and yeast is fermented in tanks, producing ethanol and CO_2. The fermented mash (termed "beer") is fractionally distilled to separate water from the ethanol, and the solids (termed "stillage") are processed and sold for added-value product lines. Ethanol has a high affinity to water, so that an additional dehydration step is needed to remove trace amounts of water from the produced ethanol. Water consumption outside of the processing itself includes evaporative losses from cooling tower circulation and during the drying process for stillage.

Water use for biodiesel refineries includes water used in processing the feedstock, separation of products and coproducts, and conditioning. There are several liquid streams involved in processing suitable biomass feedstocks into biodiesel using transesterification. For oil processing and extraction, the liquid removed from the solids, called miscella, consists of hexane, soybean oil, and water. The miscella is separated into its components using distillation. The hexane is reused, the water is disposed, and the oil is processed into biodiesel.

At a cellulosic ethanol biorefinery that uses biochemical pathways, water is used for hydrolysis of cellulosic material, boiler makeup and blowdown, cooling, and cleaning of filters and other equipment (Jones, 2010).

Anticipated and Observed Results

Water uses are varied as a result of the different conversion pathways that can take place. Water use for processing corn grain to ethanol and soybean to biodiesel are estimated to be lower than water use for processing cellulosic biomass to ethanol, in part because production of cellulosic ethanol has not been commercialized or undergone the process improvement that the production of corn-grain ethanol and biodiesel has.

The NRC report *Water Implications of Biofuels Production in the United States* (2008) summarized existing peer-reviewed publications. The authoring committee of that report concluded at that time that a corn-grain ethanol refinery consumed an average of 4 gallons of water for every gallon of ethanol produced. In other words, a 100-million-gallon per year refinery would consume a little over 400 million gallons of water every year it operates, a volume that would need to be removed from surface waters or aquifers. Optimization of water use in corn-grain ethanol biorefinery continued to be improved. Corn-grain ethanol refineries that participated in a survey conducted in 2008 reported water use of 2.7 gallons of water per gallon of ethanol (Mueller, 2010). From 1998 to 2007, water use in corn-grain ethanol biorefineries was estimated to have decreased by 48 percent in volume (Wu et al., 2009a). The decrease in water consumption is related to more process water being recycled in cooling and other refinery-related activities.

Figure 5-10 shows the locations of existing and planned ethanol biorefineries in the United States as of 2007. (An updated map is shown in Figure 2-5 in Chapter 2.) Most biorefineries were built or planned in corn-growing regions to be near the feedstock crop. In the eastern half of the country, rainfed agriculture is used to grow the corn. In the West, irrigation water, mostly from groundwater, is used. The ethanol biorefineries are shown as black dots in Figure 5-10, and the size of the dots reflect total water use each day. Major aquifers are also shown in Figure 5-10, which shows the unconfined High Plains Aquifer (Ogallala Aquifer) stretching from South Dakota to the panhandle of Texas. Throughout the Corn Belt, glacial (confined) aquifers are used frequently for the source water in ethanol biorefineries, and many of these aquifers have been overdrawn (UNL, 2007).

Reported averages of water consumption in biodiesel refineries vary between 1 to 3 gallons of water for every gallon of biodiesel produced (NRC, 2008; GAO, 2009). Much of the water use is a result of water loss in evaporation and feedstock drying processes.

Data on water consumption at cellulosic-ethanol refineries are only available for demonstration facilities. The water use rates in the permits of three demonstration facilities that convert cellulosic feedstock to ethanol using biochemical conversion range from 6-13

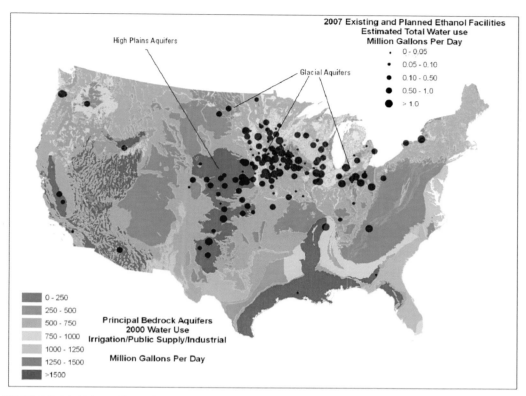

FIGURE 5-10 Ethanol biorefineries superimposed on a map of the major bedrock aquifers and their water usage rates.
SOURCE: NRC (2008), courtesy of Janice Ward, USGS.

gallons of water per gallon of ethanol produced. However, actual water use could be lower (Jones, 2010).

Thermochemical processes for cellulosic feedstocks could be optimized so that the water requirement would be 1.9 gallons of water for every gallon of ethanol produced (Phillips et al., 2007). Pate et al. (2007) estimated between 2-6 gallons of water per gallon of ethanol produced as a range representative of several potential conversion pathways. In its demonstration facility, Range Fuels reported water requirement of 1 gallon of water per gallon of ethanol produced (DOE, 2009).

Next Steps Needed

The volume of process water needed to operate a 100-million-gallon per year ethanol biorefinery using current technology is estimated at 300-400 million gallons of water per year. Therefore, careful assessment of local and regional water availability is critical in the siting of biorefineries to avoid depletion of water resources. The quality of water used in biochemical-conversion biorefining affects the performance of key plant components, including boiler efficiency and biochemical process inhibition. Water pretreatment, through the use of osmotic membranes, is often not accounted for in the published reports of water consumption (GAO, 2009), leading to values that may under-represent actual water volumes. Because much water is lost by evaporation in biorefineries, development and implementation of new technologies to reduce evaporative loss from processing biomass to fuel provide opportunities to reduce consumptive water use in biorefineries (Huffaker, 2010).

Life-Cycle Consumptive Water Use

Although consumptive water use has been estimated in various stages of biofuel production, few studies on water use over the life cycle of biofuel production exist.

Methods of Assessment

Most studies use attributional LCA to assess life-cycle consumptive water use (King and Webber, 2008; Wu et al., 2009a; Fingerman et al., 2010). Harto et al. (2010) used a life-cycle assessment that combines a materials-based process method and an economic input-output method. The materials-based process method describes elements in a supply chain and includes data collected or estimated at site and facility. The economic input-output method uses national sectoral data to describe material use.

The system boundaries of the different analyses varied. All studies included water inputs for crop production and ethanol production in their analyses. However, the level of detail included and the method for estimating water use for crop production differed (Table 5-11). Irrigation water consumed is included in all studies, but only Fingerman et al. (2010) included water from rainfall estimated to be consumed by crops. They and others (Powers et al., 2010) argue that including irrigation water only in LCA infers that rainfed agriculture does not consume any water. Precipitation that is not taken up by crops can contribute to groundwater recharge or can be used for other purposes. Some studies also included water embedded in the manufacture of farm inputs or in the production of gasoline,

TABLE 5-11 Inputs Used for Life-Cycle Analysis of Biofuel Consumptive Water Use in Different Studies

Inputs for Life-Cycle Analysis of Biofuel Consumptive Water Use	Reference
Crop production	Chiu et al., 2009

Crop production
   Crop evapotranspiration (including irrigation and precipitation)
   Conversion to fuel
	Fingerman et al., 2010

Crop production
   Consumption of irrigation water, if applicable
   Water use in manufacturing farm inputs (for example, fertilizers)
Biorefinery construction
Conversion to fuel
Distribution and marketing
Water credits from coproducts
	Harto et al., 2010

Crop production
   Consumption of irrigation water, if applicable
   Water consumed to make the gasoline, diesel, or electricity used during farming
Water credits from coproducts
	King and Webber, 2008

Crop production Consumption of irrigation water, if applicable Conversion to fuel	Wu et al., 2009

diesel, or electricity used for farming, but those water inputs are small proportions of the life-cycle water use (King and Webber, 2008; Harto et al., 2010). In addition to water use for crop production and at biorefineries, one or more studies included water use in distribution and marketing and water credits from coproducts (Table 5-11).

The effect of consumptive water use of biofuel on water resources would have to be considered in the context of water availability and demand. In one sense, using water for biofuel production and carbon sequestration in feedstocks is a purposeful trade of water for carbon (Jackson et al., 2005). Production of all crops and bioenergy feedstock results in evapotranspiration. Whether the water for plant growth originates from rainfall or from groundwater aquifers does not affect the amount of evapotranspiration from the land use. If consumptive water use results in precipitation within the same region, it returns to the regional water balance. When local evapotranspiration falls as precipitation within the same basin, it is often referred to as the local "recycle ratio." A recycle ratio of 1 indicates that all the local water that was evapotranspired returned to the same basin. Whether a recycle ratio of 1 is desirable depends on local conditions, such as whether the land is too moist already or in a drought condition, or whether flooding is a concern, or if the returned precipitation affords other precious water uses. Consumptive water use by irrigation of feedstock crops from an overdraft[18] aquifer is a serious condition. Under prolonged over-draft, the depth of an aquifer could become depleted to a level that is not economically feasible to pump. The time required to recharge a natural aquifer to a level feasible for use is much longer than the turnover of water by the hydrologic cycle of evapotranspiration, precipitation, and recycle.

Results

Estimates for life-cycle water use for corn-grain ethanol vary widely mostly because of regional variability in irrigation of the crop. Chiu et al. (2009) used regional time-series agricultural and ethanol production data in the United States to estimate the current state of water requirements for ethanol (embodied water on a liter of water per liter of ethanol produced basis as shown in Table 5-12). The total water embodied (EWe) was greatest in Western states (California, New Mexico, Wyoming, Colorado, Kansas, and Nebraska), and most of the inputs were from groundwater, except for the states of Kentucky, Colorado, California, New Mexico, South Dakota, and Wyoming.

Other estimates of consumptive water use over the life cycle of corn-grain ethanol production range from 10-1,600 gallons of water per gallon of ethanol or 15-2,400 gallons of water per gallon of gasoline equivalent (as shown in Table 5-13 for comparison with life-cycle water use for gasoline production). Most studies identify the water resource need of feedstocks as an important factor in determining consumptive water use over the life cycle of biofuels (Dominguez-Faus et al., 2009; Wu et al., 2009b). If the crop is not irrigated and water use from precipitation is not taken into account, life-cycle water use was estimated to be as low as 15 gallons of water per gallon of gasoline equivalent, which is over twice as high as any estimates of life-cycle water use for petroleum-based fuel.

Water use for biofuel production from switchgrass could be comparable to that of petroleum-based fuel if switchgrass is not irrigated and if it is converted to fuels by ther-mochemical conversion (Wu et al., 2009b; Harto et al., 2010). However, studies have shown switchgrass yields respond positively to precipitation and irrigation (Heaton et al., 2004;

---

[18]An aquifer is said to be in a state of overdraft when its rate of extraction exceeds its rate of recharge by natural processes.

TABLE 5-12 Embodied Water in Ethanol (EWe) and Total Consumptive Water Use (TCW) in the 19 Ethanol-Producing States in 2007, Ranked According to Each State's EWe

State	Ethanol Production, million liters (million gallons)	EWe, L L⁻¹ (gallons of water per gallon of ethanol)	EWe Ground Water, L L⁻¹ (gallons of water per gallon of ethanol)	EWe Surface Water, L L⁻¹ (gallons of water per gallon of ethanol)	Irrigated Water, million liters (million gallons)	Process Water, million liters (million gallons)	TCW, million liters (million gallons)	Corn Processed into Ethanol (percent)
Ohio	11 (2.91)	5	4	1	11 (2.91)	41 (10.8)	52 (13.7)	0.20
Iowa	6,857 (1,810)	6	6	0	17,288 (4,570)	24,745 (6,540)	42,032 (11,100)	28
Kentucky	134 (35.4)	7	4	4	472 (125)	484 (128)	956 (253)	7
Tennessee	254 (67.1)	10	6	5	1,681 (444)	915 (242)	2,597 (686)	29
Illinois	3,486 (921)	11	11	0	27,389 (7,240)	12,581 (3,320)	39,970 (10,600)	15
Indiana	954 (252)	17	11	6	12,539 (3,310)	3,442 (909)	15,981 (4,220)	9
Minnesota	2,296 (607)	19	16	3	34,589 (9,140)	8,286 (2,190)	42,875 (11,300)	19
Wisconsin	1,067 (282)	26	26	0	24,208 (6,400)	3,852 (1,020)	28,060 (7,410)	23
Michigan	587 (155)	47	31	16	25,177 (6,650)	2,117 (559)	27,295 (7,210)	19
Missouri	587 (155)	57	55	2	31,156 (8,230)	2,117 (559)	33,273 (8,790)	12
North Dakota	505 (133)	59	31	28	28,146 (7,440)	1,824 (482)	29,970 (7,920)	18
South Dakota	2,203 (582)	96	38	58	203,762 (53,800)	7,950 (2,100)	21,712 (5,740)	39
Georgia	2 (0.5)	128	85	42	188 (49.7)	5 (1.32)	194 (51.2)	0.25
Nebraska	2,481 (655)	501	422	80	1,235,128 (326,000)	8,954 (2.370)	1,244,082 (329,000)	16
Kansas	804 (212)	528	486	42	421,840 (111,000)	2,903 (767)	424,743 (112,000)	15
Colorado	322 (85.1)	1,176	226	950	377,082 (99,600)	1,161 (307)	378,243 (99,900)	20
Wyoming	19 (5.02)	1,354	125	1,229	25,547 (6,750)	68 (18.0)	25,615 (6,770)	23
New Mexico	114 (30.1)	1,427	615	812	161,587 (42,700)	410 (108)	161,997 (42,800)	113
California	257 (67.9)	2,138	814	1,323	549,240 (145,000)	929 (245)	550,169 (145,000)	68
Average[a]		142	91	51				23

[a]Average is weighted by ethanol production in 2007 and calculated for the purpose of comparison only. Because of the large variation between regions, significance of the average for representing the nation's EWe is limited.
NOTE: Values in U.S. standard units are shown in parentheses.
SOURCE: Chiu et al. (2009).

TABLE 5-13 Consumptive Water Use over the Life Cycle of Biofuel and Petroleum-Based Fuel Production Estimated by Different Studies

Life-cycle consumptive water use, gallons of water per gallon of gasoline equivalent			Reference
Corn-grain ethanol	Switchgrass ethanol	Petroleum-based fuel	
1500	923-1307	Not estimated	Fingerman et al., 2010
42-640	2.9-640	1.9-5.9[a]	Harto et al., 2010
62-2400	Not estimated	1.4-2.9[a]	King and Webber, 2008
15-490	2.9-15	3.4-6.6[b]	Wu et al. 2009

[a]Petroleum-based fuel considered is conventional gasoline.
[b]Petroleum-based fuels considered include gasoline and oil sands.

Robins, 2010). Harto et al. (2010) showed that irrigation of switchgrass under drought conditions could increase the quantity of water consumed substantially (Table 5-13).

Next Steps

Precise estimates of the effect of producing biofuels to meet RFS2 on consumptive water use nationwide compared to that of producing petroleum-based fuel are not possible because the result depends on multiple factors. The key determining factors are

- Evapotranspiration during feedstock production and water availability in production locations;
- Whether the bioenergy feedstock is irrigated;
- Method used for crude oil exploration and recovery;
- Whether biochemical or thermochemical conversion is used to produce biofuels.

The quality of information for the life-cycle estimates could be improved as cellulosic biofuel facilities become operational on a commercial scale. Information on water use by onshore or offshore oil recovery in the United States was collected in the 1980s and 1990s and could be outdated (Wu et al., 2009b). However, a nationwide estimate of water use in biofuel production might not be as important as regional estimates of consumptive water use by biofuels and an assessment of other water needs and water availability in each region. Because biorefineries are typically located in close proximity to bioenergy feedstock production, water use will be concentrated in one locale and could be of particular concern in water-scarce areas. Likewise, using freshwater in petroleum refining in arid regions that experience seasonal scarcity could pose a strain on that resource base even though the water use per unit product is only 1.5 gallons per gallon of petroleum-based fuel (Fingerman et al., 2010). Regional assessments of consumptive water use over the life cycle of biofuel production would be helpful in ensuring that biofuel production does not incur undue stress on water availability or result in groundwater overdraft.

EFFECTS ON SOIL

Metrics of soil quality include bulk density, erodability, soil carbon storage, and the rate of nitrogen and phosphorus turnover as they influence denitrification and nitrogen and phosphorus leaching. Processes in biofuel production that could affect soil quality include management practices in feedstock production, feedstock and residue removal, and discharges from conversion to fuels.

Feedstock Production

Among the key debates associated with biofuel production are the net effects of feedstock production on ecosystem carbon storage, particularly with respect to soil carbon storage. The extent to which biofuels represent a biologically renewable resource with respect to ecosystem carbon or soil carbon, one of the key storage pools, depends on two factors: the rate of net carbon uptake by the ecosystem (net ecosystem production) and the rate of physical removal of carbon for bioenergy feedstock production. The rates of carbon uptake and removal are contingent upon a number of natural and human-influenced factors, such that the net effects depend largely on local conditions. The contingencies and ranges of carbon loss or gain to be expected from bioenergy feedstock production are discussed in this section.

The net rate of soil carbon gain or loss for a particular system reflects the rate of net primary production and litterfall from plants (above and belowground) minus heterotrophic respiration. Net primary production is closely coupled to water availability (precipitation and irrigation), temperature (length of growing season as well as influence on evaporation), and, in some areas, nutrient availability (primarily nitrogen on land). Heterotrophic respiration depends on the same factors, in addition to those that increase biological decomposition rates of detritus, such as tillage, or other soil disturbances (for example, irrigated and tilled corn), or changing environmental conditions that shift a system from anaerobic, or slow-decomposition systems, to aerobic, rapidly decomposing systems (for example, drained wetlands).

Undisturbed, or natural systems, are generally in equilibrium with respect to ecosystem carbon, with net primary production equaling heterotrophic respiration, while those undergoing succession accumulate carbon (Odum, 1969; Wilkie et al., 2000) in live biomass and detritus. Disturbances of many types, and in particular land-use shifts to agricultural practices representing transition from perennial ecosystems to annual crops, result in losses of major stores of soil carbon (Burke et al., 1989; Davidson and Ackerman, 1993; Lal, 2004) due to cultivation-induced increases in heterotrophic respiration and erosion. While rates of loss are highly variable, Lal et al. (2007) estimate that prime agricultural soils worldwide have lost 20-80 tonnes of carbon per hectare.

The extent to which bioenergy feedstock production represents changes in ecosystem carbon or in soil carbon stores depends on the circumstances. In general, biofuel production systems that include shifts to perennial plants that are located in high precipitation areas have the greatest possibility of net carbon storage (Guo and Gifford, 2002; Lal, 2004; Lal et al., 2007). In addition, practices that incorporate reduced tillage and conserve soil water may be important for decreasing losses over the long term (Morgan et al., 2010). Degraded soils may provide the highest possibility for carbon storage, as they represent succession systems recovering to maximum potential carbon. Lal et al. (2007) suggested that levels exceeding natural soil carbon storage could be achieved only where total plant production is enhanced over natural conditions, for instance, in systems that are fertilized, irrigated, or both.

The influence of feedstock production on soil nitrogen and phosphorus cycling depends on fertilization and irrigation strategies. Where nitrogen and phosphorus are added in excess of plant uptake, the probability of nitrogen losses via denitrification and of nitrate and phosphate leaching increases if the amount of surplus nutrient exceeds a certain threshold (Van Groenigan et al., 2010) (see also "Water Quality" section above). Cellulosic feedstocks can be derived from perennial plants that may require less fertilizer than annual row crops and conserve soil nitrogen and phosphorus in soil organic matter and roots (Robertson et al., 2011b).

Residue Removal

Removal of residues from either forested or agricultural lands affects soils in two ways. First, it represents removal of detrital biomass, containing carbon and associated nutrients that otherwise represent a storage pool that would be slowly decayed. Second, the process of residue removal could have physical effects on soils through mechanical disturbance that increase erosive potential and through removal of residues that stabilize soil surfaces from wind and water-driven erosion. Long-term experiments and simulation analyses (Gollany et al., 2011; Huggins et al., 2011; Machado, 2011) indicate that the effects of residue removal on agricultural lands on soil carbon depend highly on site factors (for example, productive potential and current soil carbon) and on the management strategy used (for example, tillage intensity and crop rotations). A review of the field experiments testing the effects of residue removal on forest soils (Eisenbies et al., 2009) suggests that increases in fertilization may be necessary to replace soil nutrients from intensive combined harvest and residue removal, but long-term effects on site productivity or soil quality of forest residue removal are unclear.

Conversion to Fuels

Most biorefineries in operation or proposed have a near zero discharge design so that the effect of conversion of biomass to fuel on soil quality is small.

Soil quality effects from biochemical refineries include the solid waste streams from enzymatic production, brine disposal from accumulated solids in cooling towers and boilers, as well as those originating from water conditioning. Soil quality effects from FAME biodiesel refinery include the solid waste streams from water conditioning and oil extraction. Oils, grease, and saponified materials are screened and skimmed from the reactor vessels and sent to landfills. The amounts vary based on the feedstock source, process used, type of catalyst used, and efficiency of the reaction system. The largest effect on soil at the biorefinery is the construction of the facility.

BIODIVERSITY

Biodiversity refers to "the variety and variability among living organisms and the ecological complexes in which they occur" (U.S. Congress Office of Technology Assessment, 1987). Biodiversity encompasses the variety and variability of animals, plants, and microorganisms that are necessary to sustain key functions of an ecosystem and has been referred to as the "foundation of ecosystem services" (Cassman et al., 2005). The complex role of biodiversity for agriculture has been discussed in other reports (Cassman et al., 2005; Foley et al., 2005; NRC, 2010b; and references cited therein).

Bioenergy feedstock production could threaten or enhance biodiversity depending on feedstock type, agricultural management practices, and land-cover change (Bies, 2006; Fargione et al., 2009). Monocultures, as in the case of growing corn continuously, threaten biodiversity, as a homogeneity of crop species often leads to intensive farming practices through increased fertilizer and pesticide application and tillage and has been shown to lead to a decline in biodiversity (GAO, 2009). Changing from a practice of growing diverse crops in one area to a single crop not only reduces the biodiversity of farm acreage, but also reduces biodiversity benefits (such as pest control) for the surrounding landscape (Dale et al., 2010a). This section focuses on the potential effects of bioenergy feedstock production and harvest on biodiversity.

Methods of Assessment

Direct observations of species richness and abundance can be made to compare biodiversity under different vegetation cover (for example, row crops versus perennial grasses), as discussed in the next section. Models can be used to project the effects of land-cover change on biodiversity. Invasive species could threaten biodiversity of nearby ecosystems. Whether a species of herbaceous perennial is likely to be an invasive species can be assessed using an invasive species assessment protocol (Randall et al., 2008) or a weed risk assessment protocol (Buddenhagen et al., 2009).

Anticipated and Observed Results

Vegetation types on agricultural land have been shown to affect species richness of beneficial insects and birds. Gardiner et al. (2010) measured the number of species of beneficial insects (including bees, lady beetles, and flies) in 10 replicates of three vegetation types— corn, switchgrass, and mixed prairie. The corn fields studied were actively managed for grain production. The switchgrass and mixed prairie fields were not actively managed and the switchgrass stands included 13-38 plant species, whereas the mixed prairie stands included 25-49 plant species. The abundance and diversity of bees were reported to be 3-4 times higher in the switchgrass and mixed prairie fields than in the corn fields. However, the diversity of bees could be reduced if switchgrass is managed as a monoculture.

Land-cover change from grassland to corn was shown to be correlated with reduced grassland bird diversity and population (Brooke et al., 2009). Using geographic information systems mapping, Brooke et al. constructed a series of maps that show areas where increased corn plantings coincide with loss of grassland habitats. They found a statistically significant decrease in the number of grassland bird species and number of grassland birds sighted from 2005 to 2008 in areas where corn plantings increased by over 3 percent from 2004 to 2007. In the southern peninsula of Michigan, Robertson et al. (2011a) also observed higher bird species richness in mixed prairie grass or unmanaged switchgrass fields than in corn fields (n = 20 each).

Another study used data from the Northern American Breeding Bird Survey in a model to project changes in species richness of birds and the number of bird species of conservation concern under two scenarios of land-cover change (Meehan et al., 2010). In one scenario, 23 million acres of land in the Upper Midwest that contain low-input high-diversity (LIHD) crops (for example, mixed perennial grasses) were converted to high-input low-diversity (HILD) crops (for example, corn and soybean), and bird species richness was projected to decrease by 7 to 65 percent in 20 percent of the region as a result of the land-cover change. In contrast, 21 million acres of land that contains HILD crops were converted to LIHD crops in another scenario, under which bird species richness was projected to increase by 12 to 207 percent in 20 percent of the region. The magnitude of change in species richness was even more pronounced if only bird species of conservation concern are considered (Meehan et al., 2010). The study represents two extreme scenarios because the 23 million acres of land in the first and second scenarios included 5 million and 1.5 million acres of land that are not suitable for crop production. In addition, some acreage of herbaceous perennial grasses for biofuels is likely to be managed monoculture, which is likely to support lower diversity than a mixed grass stand. Nonetheless, it illustrates how land-cover change can affect species richness of grassland birds and that bird species of conservation concern in that region tend to be more sensitive to land-cover changes than other bird species.

Because of the influence of vegetation type on animal biodiversity, the potential for taking CRP land out of retirement to grow corn for corn-grain ethanol raises biodiversity concerns. CRP increases wildlife habitat by removing land from crop production and by requiring a land cover of perennial vegetation for a portion of the land. Several grassland bird species and ducks that have declined elsewhere in recent decades have increased in abundance on lands enrolled in CRP (Dale et al., 2010a). On the basis of the studies mentioned above, the abundance and diversity of beneficial insects and grassland birds are likely to decline if CRP land, on which native grasses have developed over time, are taken out of enrollment to plant row crops for biofuels. Although data on CRP enrollment are available, information on the use of the lands that came out of retirement is not collected. Thus, whether or how often expired CRP lands are put back into row crop production for biofuels is unknown.

Dedicated bioenergy crops such as switchgrass and mixed prairie grasses have the potential to increase animal biodiversity relative to corn-grain ethanol, and the type of management and harvesting and the placement of these crops on the landscape are important to promoting biodiversity (Robertson et al., 2011a). Grass stands that are less diverse are likely to have fewer animal species as discussed earlier. Harvesting prairie fields for biomass could disrupt habitats for some animal species (Flaspohler et al., 2009; USDA-NRCS, 2009). Harvesting after frost can mitigate the negative biodiversity impacts of harvesting these fields. Partial harvesting can provide winter cover for wildlife (USDA-NRCS, 2009). Harvesting practices can consider timing of bird nesting and other temporal values (for example, seasonal water regulation, scenic values, migrations, and other wildlife requirements) (Tolbert, 1998; Tolbert and Wright, 1998). A study of switchgrass fields in Iowa found bird diversity to be generally low, but bioenergy harvest influenced bird distribution (Murray and Best, 2010). Generalist species were nearly equally abundant in harvested and nonharvested fields, whereas grasshopper sparrows (*Ammodramus savannarum*) were more abundant in the shorter, sparser vegetation of harvested fields than in nonharvested fields. Roth et al. (2005) suggested that partial harvest of switchgrass fields in Wisconsin could enhance grassland bird diversity because different vegetation structure (that is, grass height) attracts different species.

Feedstocks for cellulosic biofuels also include timber residues. Some effects on biodiversity are likely to be associated with removal of forest residues that were previously left in place. For example, removing tree tops, branches, and other woody material that were previously left on site for bioenergy feedstock could result in loss of ground-level habitat for arthropods and amphibians. Those organisms require the dark, moist, and cool habitats underneath woody residues. Moreover, if timber harvesting removes large amounts of canopy coverage, the ground will be exposed to the warm, drying effects of increased sunlight (Janowiak and Webster, 2010). Because arthropod communities can be richly diverse and form a major lower trophic level that feeds higher trophic levels, this habitat loss can have associated negative effects on birds and mammals (Castro and Wise, 2009). Forest harvest operations associated with bioenergy feedstock removal can use the coarse woody debris (CWD) (snags and downed logs) that provides for a variety of organisms during critical stages of their life history (such as breeding, foraging, and basking). Riffel et al. (2010)

reviewed 25 studies involving manipulations of CWD (that is, removed or added downed woody debris and snags) and found that diversity and abundance of cavity-nesting and open-nesting birds and of invertebrates were substantially and consistently lower in treatments with less CWD. However, they also found that biomass harvests on a pilot scale reduce CWD levels to a lesser extent than the experimental studies they analyzed. The effect of CWD removal on biodiversity could be less severe if less CWD is removed.

Although loss in aquatic biodiversity in some coastal waters is a result of years of eutrophication contributed by agriculture and not a specific result of bioenergy feedstock production, bioenergy feedstock production can contribute to worsening or mitigating eutrophication. As discussed earlier (in the section "Water Quality"), increasing corn production is likely to increase sheet erosion of soil and runoff of nitrogen and pesticides (Secchi et al., 2011). Decreasing water quality and increasing areas, severity, and duration of hypoxia in coastal waters lead to loss in aquatic species (Rabalais et al., 2002; Vaquer-Sunyer and Duarte, 2008). In contrast, reducing nutrient outflow to surface water from crops by planting herbaceous perennials can improve water quality (Sahu and Gu, 2009; Ng et al., 2010). A preliminary study by Schweizer et al. (2010) suggested that converting some cropland and pasture in the White River Basin and the Red River Basin watersheds to switchgrass could improve water quality and species richness of fish in those river basins.

In addition to animal biodiversity, growing bioenergy crops can also affect plant biodiversity. If a bioenergy feedstock has invasive potential, it could expand into noncrop areas and drive out native vegetation (NAS-NAE-NRC, 2009). The herbaceous perennials chosen as dedicated bioenergy crops are selected, bred, or genetically modified to be fast-growing, productive on marginal lands, and resilient, which are traits of successful invasive species (Raghu et al., 2006; Barney and DiTomaso, 2008; Fargione et al., 2009). Cultivars improved by either conventional or genetically engineered breeding also have the potential to become invasive within their own native species range. Using a weed risk-assessment protocol, Barney and DiTomaso (2008) estimated the invasive potential of switchgrass, giant reed, and a sterile hybrid of *Miscanthus*. Their assessment suggested that switchgrass has high invasive potential in California and giant reed has high invasive potential in Florida. Lack of seed production from the *Miscanthus* sterile hybrid substantially reduces its invasive potential (Lewandowski et al., 2003; Barney and DiTomaso, 2008), but reversion to seed production in hybrids can occur. Careful screening and testing of bioenergy feedstock to demonstrate low invasiveness in target regions of feedstock production could reduce the likelihood of bioenergy feedstock invading nearby ecosystems (Davis et al., 2010; DiTomaso et al., 2010; Quinn et al., 2010; Barney and DiTomaso, 2011).

Next Steps Needed

Bioenergy feedstock production could reduce plant and animal biodiversity or provide opportunities to improve it (Webster et al., 2010). The precise effect of increasing production of bioenergy feedstock requires regional assessment of compatibility of feedstock type, management practices, timing of harvest, and input use with plants and animals in the area of production and its surroundings (Fargione et al., 2009; Landis and Werling, 2010). To reduce the potential of next-generation bioenergy feedstocks becoming invasive species, Barney and DiTomaso (2008) suggested a system of preintroduction screening for each proposed bioenergy feedstock in specific target regions, and an analysis of risk assessment, climate-matching modeling, and cross-hybridization potential. Landis et al. (2010) mentioned the need for research on arthropod dynamics within biofuel crops, their spillover into adjacent habitats, and their implications on the entire landscape. As in the case of GHG emissions, biodiversity could be affected by land-cover change, associated with bioenergy feedstock production. Monitoring of land-cover change, including identifying land taken out of CRP that was subsequently used for producing bioenergy feedstocks, would help identify and assess any effects of increasing biofuel production on biodiversity. The effect of indirect land-use changes as a result of bioenergy feedstock production, corn grain in particular, has not been extensively studied and needs to be considered. The implications of biofuel production and feedstock choices for biodiversity are complex so that a systematic approach is needed to discern interactive effects of bioenergy crop production and other forces on biodiversity.

ECOSYSTEM SERVICES

Ecosystem services are the beneficial processes that ecosystems provide to humankind (Costanza et al., 1997; Millennium Ecosystem Assessment, 2005). While ecosystem services often are not valued in the marketplace, they provide crucial regulating (for example, flood regulation), supporting (for example, soil formation), provisioning (for example, fish for food), or cultural (for example, recreational) services (Brauman and Daily, 2008). Much of this chapter focuses on environmental impacts of biofuel production related to ecosystem services that are associated with GHG, air quality, and water quality and quantity. As mentioned before, bioenergy feedstock production can both enhance and decrease different ecosystem services depending on the scale of biofuel production, changes in land-use management relative to prior conditions, and land-use practices. Some reports suggested that landscapes for bioenergy feedstock production could be designed to maximize ecosystem benefits (Figure 5-11) (Foley et al., 2005; Johnson et al., 2008; NAS-NAE-NRC, 2009; NRC, 2010b).

Methods to Assess Effects

Valuing ecosystems and their services generally occurs through biophysical or economic valuations (Boyd and Wainger, 2003). For biophysical valuations, there are no standard methods of assessment or agreed-upon indicators that measure ecosystem quality. Often, areas that contain the most native species diversity are considered the most valuable (Boyd and Wainger, 2003).

For economic valuation, ecosystem services are often measured through nonmarket valuation methods, as ecosystem services are often not bought or sold in the marketplace. These methods include the avoided cost method, contingent valuation method, travel cost method, and others. The avoided cost method measures the value of the replacement service, such as insecticide needed to be applied if the natural biocontrol of pests is reduced in an area. Contingent valuation studies reveal the "willingness to pay" of society for an ecosystem service, such as the aesthetic value of a national forest. Similarly, the travel cost method evaluates how far people have travelled and how much money they spend to access or enjoy a resource. While these nonmarket valuation methods have been used for many years, they are at times controversial and are technically challenging to undertake (Boyd and Wainger, 2003).

Empirical Evidence

Though there have been few specific studies on the gain or loss of ecosystem services due to biofuel development, one recent study measured the effects of increased monoculture corn crops and the consequent loss of natural pest control of the soybean aphid.

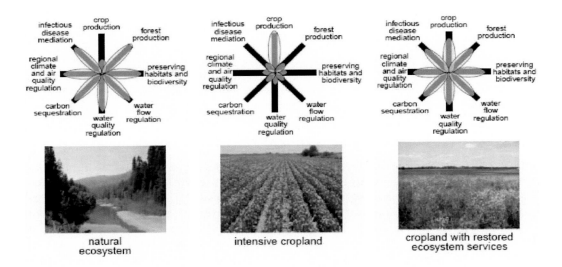

FIGURE 5-11 Conceptual framework for comparing tradeoffs of ecosystem services under different land uses.
NOTE: "The provisioning of multiple ecosystem services under different land-use regimes can be illustrated with these simple 'flower' diagrams, in which the condition of each ecosystem service is indicated along each axis. (In this qualitative illustration, the axes are not labeled or normalized with common units.) For purposes of illustration, we compare three hypothetical landscapes: a natural ecosystem (left), an intensively managed cropland (middle), and a cropland with restored ecosystem services (right). The natural ecosystems are able to support many ecosystem services at high levels, but not food production. The intensively managed cropland, however, is able to produce food in abundance (at least in the short run), at the cost of diminishing other ecosystem services. However, a middle ground—a cropland that is explicitly managed to maintain other ecosystem services—may be able to support a broader portfolio of ecosystem services" (Foley et al., 2005, p. 573).
SOURCE: Foley et al. (2005). Reprinted with permission from the American Association for the Advancement of Science.

Findings included that the natural pest control of the aphid was worth $239 million in four states (Iowa, Michigan, Minnesota, and Wisconsin). Increased monoculture of corn, however, reduced natural pest control services and cost soybean producers in these four states about $58 million per year through reduced yield and increased pesticide use (Landis et al., 2008).

REGIONAL AND LOCAL ENVIRONMENTAL ASSESSMENTS

Earlier sections outlined some of the measured or anticipated environmental effects that could result from increasing biofuel production in the United States. Most studies cited focus on measured or estimated effects from corn-grain ethanol and soybean biodiesel. Because cellulosic biofuels have not been deployed on a commercial scale, environmental effects from producing 16-20 billion gallons of cellulosic biofuels cannot be directly measured and can only be estimated. Another limitation is that environmental effects are specific to location, feedstock, and technology. Therefore, environmental effects can be estimated with greater confidence at a local or regional scale than on a national scale, even though comprehensive estimates at national and international scales are necessary to inform decision-making. In addition, an individual biorefinery might have localized environmental effects that are not of concern beyond the local scale.

The locations where cellulosic bioenergy feedstock will be grown depend on agronomic and economic conditions. Therefore, the committee used the National Biorefinery Siting Model (NBSM) (Parker et al., 2010) to identify specific biorefinery locations in the United States and associated biomass supplies and counties of origins (see Tables 3-1 and 3-2 in Chapter 3). The locations identified coincide with other model or biorefinery-siting projections of the most likely places for cellulosic and crop residue-based biofuels to be produced in the United States in the future. This section describes the U.S. Department of Energy and other government assessments of some local or regional environmental effects of cellulosic biofuel production in regions of the United States where cellulosic biofuel production is planned or projected to occur. Results of local environmental assessments for some planned or proposed cellulosic biorefineries in these regions are used as illustrations. However, the environmental effects of operating these facilities extend well beyond their immediate footprint. The large-scale effects include life-cycle environmental effects resulting from changes in resource use and land use and from pollutants emitted elsewhere in the supply chain or as a result of market-mediated effects.

Corn Belt Case Study

The Corn Belt has great potential to contribute crop residues for cellulosic ethanol. POET Project Liberty, LLC, proposed to expand an existing corn-grain ethanol facility near Emmetsburg, Iowa, into "a biorefinery that integrates advanced corn dry milling and lignocellulosic conversion technologies to produce ethanol and byproducts" (ENSR AECOM, 2008, p. i). An environmental assessment of a proposed cellulosic ethanol refinery was conducted as required by part of a grant program supporting facility development and consistent with requirements of the National Environmental Quality Act (NEQA) (ENSR AECOM, 2008). The NEQA requires assessment of air, water, soil, endangered wildlife and plant species, traffic, and social consequences of DOE-funded projects. Such assessments provide information about potential effects of new cellulosic production facilities on the towns and landscapes where they are planned be sited. Many existing corn-grain ethanol refineries are located in the Corn Belt (see Figure 2-5 in Chapter 2). Assessment of the local effects of the proposed Emmetsburg facility reflects potential effects in other locations where corn-grain ethanol refineries might expand to include processing of cellulosic feedstocks.

Feedstock

Existing land use is almost entirely devoted to annual crop production, with corn and soybean dominating (ENSR AECOM, 2008). The corn-grain portion of the facility was estimated to require about 55 percent of existing grain production in the immediate area of the biorefinery. The intended cellulosic feedstock is corn cobs, but the facility might use some corn fiber separated from the corn kernel. Removal of cobs was estimated to amount to about 6 percent of the carbon in corn residues on average and a small amount of nutrients per acre. Removal of cobs was determined to have no obvious short-term or long-term effects on the productivity of farmland in the region (ENSR AECOM, 2008).

Direct Facility Effects

DOE judged that impacts from construction and operation of the cellulosic biofuel facility in Emmetsburg, Iowa, would not exceed national or local environmental standards, including those requiring accounting for social effects. There would be positive effects on employment (ENSR AECOM, 2008).

Water Resources

Farming in the region surrounding the Emmetsburg facility is entirely rainfed. Much of the original area was wet prairie, and there are areas of somewhat-poorly drained to poorly drained soils throughout the region. Many fields are tile drained, and these tiles convey water and nutrients to surface water channels, to the West Des Moines River, and then to the Mississippi River. (See earlier section "Water Quality.") Nutrient loss in the Mississippi River drainage area was not discussed in the environmental assessment because no expansion of the farmed area or significant alternation of local crop rotations or farming practices as a result of the cellulosic biorefinery was anticipated. However, as discussed earlier in this chapter, increased corn-on-corn production would likely contribute to higher nutrient loss. Under this assumption, DOE found that there are no significant environmental effects from the operation of a potential corn grain-cellulosic residue (cobs) biorefinery (ENSR AECOM, 2008).

Wildlife

The U.S. Fish and Wildlife Service identified two federally protected plant species that might be present in Palo Alto County where the facility will be built. However, occurrence of either species was not observed. Therefore, no adverse effects from the cellulosic facility on any endangered or threatened species of plants or wildlife in the surrounding landscape were identified (ENSR AECOM, 2008).

Southern High Plains Region Case Study

The NBSM (Parker et al., 2010) identified Garden City, Kansas; Guymon and Keyes, Oklahoma; and Dumas, Texas, as some of the likely sites for cellulosic biorefineries in the Southern High Plains region (see Table 3-1 in Chapter 3). This cluster of biorefineries represents the agroecological and environmental conditions common in the region. Several grain-based ethanol operations are in the area, including in Liberal and Garden City, Kansas, and in Plainview and Levelland, Texas (RFA, 2011). Cropping systems typical of the area emphasize a combination of dry land and irrigated crops, principally wheat-fallow or wheat-sorghum-fallow in dryland areas, and corn, sorghum, and wheat with smaller amounts of soybeans and wheat where irrigation is possible. In Texas, cotton is produced under dryland and irrigated conditions, so cotton residues would be available in those regions. Although switchgrass was used as the dedicated bioenergy crops in NBSM, other adapted species will likely be included in these regions (DOE-EERE, 2010a; Nelson, 2010).

Abengoa Corporation has proposed locating a new facility near Hugoton, Kansas, to use cellulosic and grain feedstocks. Because Hugoton is somewhat centrally located in the area discussed here and an environmental assessment was published by the Department of Energy (DOE-EERE, 2010a) for this facility, it is used as an illustration in the region. As with all environmental assessments, there could be unique circumstances including special issues associated with soil erosion, water supplies, or wildlife that vary from one location to another within the region.

Feedstocks

The proposed Abengoa facility would initially use cellulosic biomass consisting primarily of corn and grain-sorghum stover and wheat straw produced principally or primarily on farmland classified as highly productive, with some residues from lower classified soils (DOE-EERE, 2010b). DOE concluded that sufficient crop residues could be derived from the most productive soils without depleting soil organic matter or causing erosion. Up to 50 percent of the available residues from highly productive farmland was estimated to be removable. Total amount of residues, including those from less productive fields, equaled about 33 percent of all crop residues available in the region. Overall, DOE estimated that the region surrounding the Abengoa facility produced five times the residue requirements needed by the proposed facility. At least some corn-residue removal is beneficial for farming practices in some circumstances (Edgerton, 2010).

In addition, some dedicated bioenergy crop harvests from nonirrigated or marginal croplands or from expired CRP land would be used. Abengoa proposed that in time, dedicated bioenergy crops would constitute three-quarters of all cellulosic biomass used by the facility. DOE assumed that crop residues would be more available than dedicated bioenergy crops at the outset, but that increasing production of dedicated bioenergy crops in the future could have largely beneficial effects on the landscape, particularly if highly erodible croplands were converted to perennial grasses.

The Abengoa proposal also includes a grain-based ethanol facility. DOE estimates that the proposed facility would require 2 to 3 percent of the grains produced in the region. Grain sorghum would meet a large proportion of the feedstock need of the facility (DOE-EERE, 2010b). The Renewable Fuels Association identifies two other grain-ethanol refineries operating in the region near Garden City and Liberal, Kansas, that produced nearly 100 million gallons of ethanol per year in 2010, using about 270 million bushels per year of corn-grain equivalent, though some of the grain used is sorghum.

DOE concluded the Abengoa facility could flexibly secure its feedstock supplies because crop residues and grain supplies exceed facility requirements. Because supplies are abundant, there would be little to no pressure on existing land use or need for land alteration or expansion of cultivated area (DOE-EERE, 2010b). Some less productive farmland could be converted to perennial grasses with positive consequences for conservation and perhaps for wildlife. Independent assessments of the region support the notion that harvesting limited amounts of residues and producing perennial bioenergy crops would not increase soil erosion or undermine future productivity of farmland (Nelson et al., 2006; Nelson, 2010).

Water Use

In the region as a whole, precipitation amounts tend to limit crop yields in most years. Water used for irrigation is derived mostly from the Southern High Plains (or Ogallala) Aquifer. The Ogallala Aquifer has been overdrafted in large parts of this region (Lamm et al., 1995), and irrigation is significantly curtailed. However, saturated thickness of the Southern High Plains Aquifer is locally and regionally variable, and some areas are stable or increasing in depth (Figure 5-12). The availability of grains and residues from irrigated

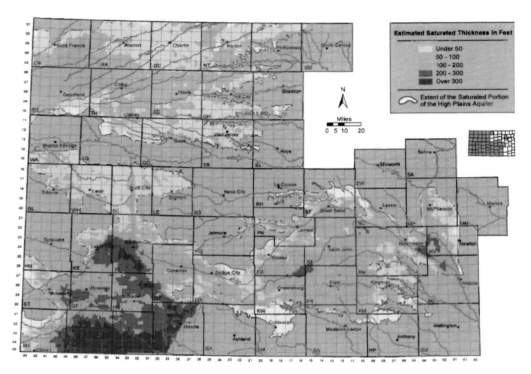

FIGURE 5-12 Average 2004-2006 saturated thickness for the High Plains Aquifer in Kansas.
SOURCE: DOE-EERE (2010b).

agriculture depends on the continued viability of irrigation in some parts of the larger feed-stock supply region. All regions of Kansas are governed with a groundwater appropriation and permit system that protects the rights of current users and seeks to sustain water use into the future. The areas first developed for irrigation have been largely overappropriated, while newer areas still have abundant water supplies. The area around the proposed facility has supplies in the aquifer estimated to sustain current irrigation demands for 100 to 200 years (Figure 5-13). However, the saturated thickness of the aquifer will decrease over time if current rates of use are maintained, and the water drawdown will be permanent because the rate of water extraction exceeds the rate of replenishment. Future improvements in water-use efficiency and a shift to dedicated bioenergy crops that require less irrigation will likely reduce further water demands and could extend the lifetime of the aquifer.

Wildlife

The wildlife species of concern are mostly nesting bird species, particularly the lesser prairie chicken, as well as two species of prairie dogs and the black-footed ferret. Migratory waterfowl use wet potholes and permanent wetlands seasonally. No important land-use changes are anticipated because most of the landscape is used for agriculture and grazing; therefore, current wildlife populations are thought to be unaffected by the proposed facility (DOE-EERE, 2010b).

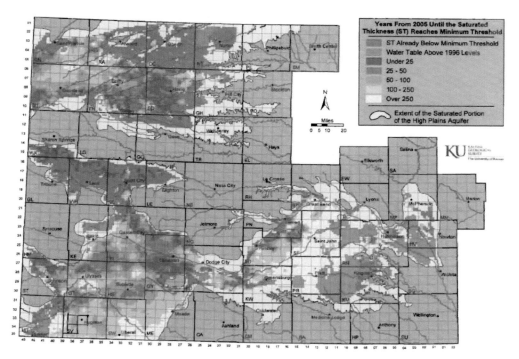

FIGURE 5-13 Usable lifetime of the High Plains Aquifer in Kansas estimated on the basis of groundwater trends from 1996 to 2006 and the minimum saturated requirements to support well yields of 400 gallons per minute under a scenario of 90 days of pumping with wells on ¼ sections.
SOURCE: DOE-EERE (2010b).

Direct Facility Impacts

All direct facility impacts, such as water consumption, wastewater and other waste generation, and air emissions, were judged to be potentially within existing national and local standards in the Hugoton region and would not impair the environment around the facility. The planned facility will produce not only ethanol, but also sufficient biopower (electricity) to meet the needs of the facility and some excess electricity for sale to the regional power grid (DOE-EERE, 2010b).

Northern Great Plains Region

NBSM identified several likely cellulosic biorefinery locations in the Northern High Plains (see Table 3-1 in Chapter 3). The sites identified are located in North Dakota, South Dakota, and Nebraska. A combination of crop residues, dedicated bioenergy crops, and coarse grains are the major feedstocks for these biorefineries. A preliminary assessment was carried out by Great River Energy of North Dakota in cooperation with the University of North Dakota's Energy and Environmental Research Center to assess feedstock supplies in the region of Spiritwood, North Dakota (Broekema, 2009). The analysis focused on feedstock supplies for cofiring with coal in a new power facility but also assessed biomass supplies for a cellulosic facility to be constructed in a second stage. The cellulosic facility would use additional biomass and waste heat from the power plant using a proprietary conversion process developed by the Inbicon Company of Denmark. An environmental assessment has not been conducted for cellulosic biofuel production.

Feedstock

The study (Broekema, 2009) concluded that there are significant supplies of biomass available in the area around Spiritwood Station. The most important sources of biomass identified were corn cobs and stover, wheat straw, sugar beet foliage, hay crops, and native grass biomass from CRP land.

Broekema (2009) reported that 83 percent of the land in the region analyzed is farmland and pasture. Soybean, wheat, corn, sunflower, and hay were the principal crops. The remainder of the region had remnant native grasses or perennial grasses on CRP land and mixed vegetation along riparian corridors.

Southeast

Many potential bioenergy feedstocks, including native and nonnative species, have been studied for use in the Southeastern United States. Such feedstocks include cellulosic sources such as canola residue (George et al., 2010), wheat straw (Persson et al., 2010), coastal bermudagrass (Cantrell et al., 2009), and sunn hemp (Cantrell et al., 2010). Carbohydrate-based alternatives to corn-grain ethanol have also been explored, such as kudzu (Sage et al., 2009) and sweet sorghum (Wu et al., 2010). Although many feedstock alternatives exist, most studies and pilot projects relating to biofuel feedstocks in the Southeast have focused on woody biomass and switchgrass (Wright and Turhollow, 2010).

A demonstration-scale cellulosic ethanol facility has been constructed in East Tennessee through the state-sponsored University of Tennessee Biofuels Initiative. Owned by Genera Energy LLC and operated by DuPont Danisco Cellulosic Ethanol (DDCE), the Vonore facility has the capacity to produce 250,000 gallons of ethanol per year. The Vonore facility became operational in January 2010. Because the facility is not a DOE-funded project, an environmental assessment of the facility is not publicly available. The facility meets all environmental permitting requirements. Based on operational experience from the project thus far, DDCE is planning to build a commercial-scale facility with a capacity of 25-50 million gallons of ethanol by 2014.

Feedstock

The Vonore facility will soon be ready to make the change from processing corn cobs to processing switchgrass. A total of 5,162 acres of switchgrass are already in production within an hour's drive of the facility as a result of contracting with 61 farmers in 10 surrounding counties.

Water Resources

Water use is also a significant concern for the environmental sustainability of biomass production in the Southeast (Evans and Cohen, 2009). General circulation models provide contradictory results with respect to future precipitation in the Southeast, but in any case forest biomass and composition changes are projected to differ from forest dynamics expected without climate change (Dale et al., 2010c).

North Central Region

Feedstocks

The Northern Hardwoods Sugar Maple-Beech-Birch community of forests extends from Northern Minnesota through New England and into Maine (Covington, 1980; Drever et al., 2008). These extensive hardwood forests have a lot of potential for woody bioenergy development with extensive existing heat and electricity production using woody biomass already in place (Becker et al., 2009; Jenkins and Sutherland, 2009; Solomon et al., 2009; Volk and Luzadis, 2009). The region has a number of proposed cellulosic ethanol plants of various sizes (Solomon, 2009) and one existing demonstration plant in upstate New York (Checkbiotech.org., 2009). The landscape has historically included high levels of timber harvesting for paper and lumber production. However, timber harvesting has been lower in the past years than before because of the overall decline of these U.S. industries in recent years. Substantial declines in agriculture in the northern reaches of the area, coupled with long-term recovery from widespread clearcutting of earlier eras, create a situation in which forest cover is increasing, particularly in the northernmost areas within the region.

Wildlife and Other Environmental Effects

Increasing woody bioenergy production has the potential to provide new markets for the region's extensive northern hardwoods (Becker et al., 2009). Janowiak and Webster (2010) and Flaspohler et al. (2009) summarize the state of knowledge of potential impacts on soils, hydrology, and biodiversity. They found that some of the greatest impacts could come from extensive removal of forest residues with associated soil compaction and erosion, as well as soil nutrient losses. Carefully planned forest management could minimize negative biodiversity impacts and, in some cases, improve habitat quality (Janowiak and Webster, 2010). A number of states within the region have recently added voluntary biomass harvesting guidelines to minimize negative impacts on soils, biodiversity, and water resources (BURNUP, 2008).

UNCERTAINTIES ABOUT ENVIRONMENTAL EFFECTS OF BIOFUEL PRODUCTION

The environmental effects of corn production have been studied for years and discussed in earlier sections of this chapter. Therefore, the direct environmental effects of expanding use of corn grain for ethanol have been estimated on the basis of increases in planted acreage, changes in rotation to increase corn production, or the proportion of corn diverted to biofuel. A key issue affecting environmental effects from expanding corn-grain ethanol production in the United States are the per-unit product efficiencies achieved across all aspects of the corn-ethanol system (Burney et al., 2010), and whether or not nutrient losses from annual cropping systems can be reduced at the same time (Kitchen et al., 2005; Lerch et al., 2005; NAS-NAE-NRC, 2009). Increasing yield per acre and resource-use efficiency in corn production has the potential of mitigating some of its environmental effects, but the net effects also depend on total planted acreage and locations of planted acres. Business groups and industry associations of corn-grain ethanol producers tend to assume that existing rates of yield progress will be maintained or exceeded as a result of genetic advances and the use of improved genetic tools in the coming decade (Schill, 2007; Monsanto, 2008). However, unexpected changes in weather as a result of global climate change and any associated shifts in weeds, pests, and diseases dynamics, and the opportunity for unforeseen technical advances can affect realized yields in specific locations and result in annual and year-to-year variability (Easterling et al., 2007).

GHG emissions as a result of land-use change are the most uncertain environmental effect of corn-grain ethanol production because of the uncertainty associated with correlating corn-grain ethanol production in the United States with market-mediated land-use change. Although close monitoring of global land-use and land-cover changes and market responses over time can reduce some uncertainties, indirect cause and effect can never be attributed with high certainty.

For cellulosic biofuels, DOE has judged that some of the planned individual biorefineries have acceptable environmental effects locally and receive no-effect determinations. They will be permitted based on successful compliance with local and national environmental assessment requirements as shown earlier. Because those assessments were based on relatively short-term local effects, they cannot be extrapolated to infer the overall environmental effects of meeting RFS2 for several reasons. First, those environmental assessments of biorefineries do not consider the environmental effects over the life cycle of fuels, though they could provide some information on feedstock production and conversion for an attributional LCA. Second, a large number of compliant biorefineries can result in aggregated environmental effects beyond the local scale.

The site-specific nature of cellulosic biofuel production makes a nationwide projection of environmental effects challenging. Unless precise details on how the 16 billion gallons of ethanol-equivalent cellulosic biofuel will be produced (including location of feedstock production, the prior condition of the land [for example, vegetation type], feedstock type to be produced in each location, management practices used, conversion technologies used, volumes of different types of biofuels produced, and so on), estimates of environmental effects of meeting RFS2 will be fraught with large uncertainties. At present, the cellulosic biofuel industry is developing via a piecemeal approach so that where and how feedstock will be produced is uncertain. Therefore, the committee cannot make many quantitative statements about the net effects of producing 16 billion gallons of cellulosic biofuel in the United States for most environmental parameters.

In fact, for some environmental parameters such as GHG emissions, the committee cannot ascertain that producing 16 billion gallons of cellulosic biofuel would result in net GHG benefits because of the large uncertainties associated with indirect land-use change. If only crop and forest residues are used to produce cellulosic biofuels, then GHG emissions from indirect land-use change would be minimal. However, those sources of feedstocks alone are inadequate to produce 16 billion gallons of cellulosic biofuel. If dedicated bioenergy crops are produced on croplands, then the uncertainty associated with indirect land-use change increases.

Some authors have suggested a landscape approach to integrating bioenergy feedstocks into agriculture to increase the likelihood that the development of the biofuel industry will result in net environmental benefits (NAS-NAE-NRC, 2009; Dauber et al., 2010; Karlen, 2010; NRC, 2010b). Landscape planning would provide a basis for careful assessment of various environmental effects and the tradeoffs among effects, especially if conducted within a broad LCA framework.

OPPORTUNITIES TO MINIMIZE NEGATIVE ENVIRONMENTAL EFFECTS

As discussed above, production and use of biofuels to meet RFS2 could provide overall environmental benefits compared to petroleum-based fuels or deplete natural resources and incur negative environmental effects if production is not managed properly (Box 5-3).

BOX 5-3
Sustainability Concerns Associated with Using Forest Resources to Meet RFS2

EISA's definition of renewable biomass from forest resources—that is, forest resources that can be used to produce cellulosic biofuel that can be counted toward RFS2—was selected to reduce the likelihood that "natural" (not planted by humans) forests would be harvested with the primary goal of producing cellulosic bioenergy feedstock. In nonplantation forests, EISA mandates that woody feedstock come from residues ("slash" in the law) or thinnings from state, local, or private forests—explicitly not from federal lands. This definition, in essence, precludes the harvesting for feedstock of nonplantation mature trees, as would be done in a normal timber harvest.

EISA therefore removes about one-third of all U.S. forests from production of woody feedstock that could be used toward RFS2-compliant cellulosic biofuel. In some ways, this could be considered to have minimal effect because timber harvesting on federal lands has already declined drastically over the past couple of decades. Nonindustrial private forest (NIPF) lands have been making up the difference in timber, though U.S. timber harvesting has declined overall in recent years (Adams et al., 2006). Although the mixes of federal, state, county, industrial, and NIPF forestlands vary greatly from state to state (for instance 83 percent of eastern forests are private, whereas only 43 percent of western ones are), these statistics suggest that woody feedstock used to produce cellulosic biofuel into the future will tend to come from NIPFs (USFS, 2001). This raises two primary issues: supply dependability and environmental sustainability.

A major issue with compliance is establishing a process for monitoring "chain of custody" of materials that creates and maintains legally compliant records certifying that feedstock meets these standards. Tracking compliance is possible, though it is arduous and would add to the cost of feedstock procurement. Such a tracking system has been set up to track much of the pulp and lumber in the United States to ensure that it is compliant with the one or more voluntary certification systems to which most major forest and paper companies belong.

The renewable biomass definition raises additional issues with regard to environmental sustainability. Land use is largely regulated at the state and local levels. Likewise, forest management is mostly regulated at the state level and varies widely from state to state (Cubbage et al., 1993). The Southeastern United States is the region with the largest amount of commercial forestry. Most of the western states with significant amounts of valuable timberlands, including Alaska, Idaho, Washington, Oregon, and California, have regulatory forest practice acts governing public and private forest practices that require state permits for most major forest management activities, including timber harvesting. About 50 percent of U.S. forests are in the western states and 57 percent of these are public (USFS, 2001). Much of their forests are federal, and they are subject to a complex set of laws that make them the most restrictively protected in the United States. However, many experts think that environmental protection is not served by restricting all access in fire-threatened forest systems.

The lack of guidelines or standards for sustainable land management practices is a barrier to ensuring a viable biofuel industry that minimizes negative environmental impacts. Although many U.S. forests are subject to voluntary or regulatory guidelines aimed at reducing negative impacts, these guidelines typically focus only on reducing erosion, particularly in the Eastern United States. U.S. forests are generally not subject to mandatory environmental protections that would ensure long-term environmental sustainability of woody biomass feedstock harvesting. Most state and private timber harvesting are required to meet the standards for environmental protection of either the Sustainable Forestry Initiative or Forest Stewardship Council, to which most major paper and forest companies have chosen to belong, or state-specific timber management and timber harvest plans. These standards mostly require companies to adhere to existing laws and guidelines, including voluntary state-level Best Management Practices (BMPs) aimed at reducing nonpoint source pollution caused by erosion from timber management stands and roads (FSCUS, 2010; SFI, 2010). BMPs for harvesting biomass are being developed in many regions of the United States (Ice et al., 2010). All states have recommended BMPs or forest-practice rules as part of silvicultural nonpoint source control programs (Schilling, 2009). Six states have developed specific BMP guidelines directed at biomass harvesting (Maine, Michigan, Minnesota, Missouri, Pennsylvania, and Wisconsin), and others are considering special BMPs (California, Massachusetts, Maryland, Mississippi, and North Carolina) (Evans et al., 2010). Thus, outside of the West, the only form of environmental protection for most state, local, and private forests are voluntary guidelines aimed at reducing nonpoint source pollution with regulatory requirements to avoid only activities that would cause severe negative impacts on wetlands or water quality through erosion.

Determining best management practices for the production of different feedstock types in various regions and developing sustainability standards or certification processes could provide opportunities to enhance environmental benefits and minimize negative environmental effects.

Therefore, the national and international community has been working to select a set of indicators that can be used to measure the environmental effects of increased biofuel production. Indicators are carefully selected categories of measurements that track conditions over time (Cairns et al., 1993), with a purpose of measuring the state of natural resources (including air, water, or land resources), the pressures on them, and the resulting effects on economics and environmental sustainability (Niemi and McDonald, 2004). Indicators need to be repeatable, be statistically valid, measure relevant changes, and be readily monitored (Dale and Beyeler, 2001). A major challenge in selecting and developing a list of indicators for certifying bioenergy sustainability is limitations in data and modeling because the ability to measure and objectively verify critical indicators is limited in many cases (Hecht et al., 2009).

A set of indicators for monitoring environmental effects of biofuels that is complementary and largely based on existing efforts by the Roundtable on Sustainable Biofuels (RSB, 2010), Biomass Research and Development Board (BRDB, 2010), the Global Bioenergy Partnership (GBEP, 2010), the Millennium Ecosystem Assessment (2005), and the National Sustainable Agriculture Information Service (Earles and Williams, 2005) would have to be selected and agreed upon by federal agencies and environmental stakeholder groups. Environmental indicators for biofuels proposed by the groups listed above relate to productivity, GHG emissions, water quality and quantity, air quality, and biodiversity.

Van Dam et al. (2008) reviewed initiatives on biomass sustainability standards and certification and found major differences in the geographic coverage and whether the sustainability standards were voluntary or mandatory. Stakeholder groups that are developing standards and certification systems currently include national and regional governments; companies; nongovernmental organizations; and international organizations and initiatives, such as Biofuels Initiatives and the Roundtable on Sustainable Palm Oil of the United Nations Conference on Trade and Development. The objectives and motivations for certification vary considerably among the stakeholder groups. Van Dam et al. (2008) pointed out that while there is an urgent need for criteria to ensure sustainable production of biomass, some of those criteria can be addressed using existing certification systems, such as forest sustainability certification. The authors suggested that development of certification systems will best be done via an adaptive management process (for example, learning from pilot studies and research) with expansion over time. Furthermore, improved coordination among certification activities is necessary to improve coherence and efficiency in certification of sustainable biomass, to avoid proliferation of redundant or nonaligned standards, and to provide direction in the appropriate approach (van Dam et al., 2008).

To date, the indicators being discussed in these efforts are numerous, and implementing indicators in an assessment process can be costly. Furthermore, there is no agreement among stakeholders as to what indicators should be included in certification systems for bioenergy sustainability (Buchholz et al., 2009). As discussed in another NRC report (2010b, p. 33), "Indicators of sustainability presume the existence of goals and objectives, and yet there is no guarantee that all parties will agree on which sustainability objectives and goals are desirable or most important, particularly if tradeoffs are involved." Thus, indicators are useful measurements toward progress once the sustainability objectives are clearly identified and prioritized (NRC, 2010b).

CONCLUSION

The environmental effects of biofuel production in the United States can be discussed in several contexts. For example, one context includes mitigating the net environmental costs; this chapter provides many specific examples of how biofuel production could result in positive, neutral, or negative environmental outcomes depending on the particular environmental effect of concern, the crop used, the land used to cultivate the crop and its prior use, the management practices used, and other factors including environmental effects from market-mediated land-use and land-cover changes. A separate context is the question of whether achieving RFS2 would provide net environmental benefits or harm compared to using petroleum-based fuels. The committee cannot provide any quantitative answers in most cases or even qualitative answers with certainty in some cases for the following reasons:

- The collective effects of achieving RFS2 will, in large part, depend on where and how the biomass feedstock is grown across the country. Although various models (National Biorefinery Siting Model, USDA, EPA, and others) estimated potential locations of feedstock production, whether farmers would grow bioenergy feedstocks in those locations, the management practices that they would use, and the condition of the lands used before and after bioenergy feedstock production are unknown and not predictable.
- An assessment of the environmental outcome of substituting petroleum-based fuels with the RFS2-mandated biofuels would require a comparison of each environmental effect between biofuels and petroleum-based fuels and a projection of collective effects of the fuel substitution.

The committee's assessment of the environmental effects of achieving RFS2 is summarized below.

GHG Emissions

GHGs are emitted into the atmosphere or stored in soil during different stages of biofuel production. GHG effects of biofuels depend on type of feedstocks grown and the management practices used to grow them, any direct and indirect land-use changes that might be incurred as a result of increasing biofuel production, harvesting and transport of biomass, and the technologies used to convert biomass to fuels. GHG emissions from direct and indirect land-use and land-cover changes are the variables with the highest uncertainty and the greatest effect in many cases throughout the biofuel supply chain. If no direct or indirect land-use or land-cover changes are incurred, biofuels tend to have lower life-cycle GHG emissions than petroleum-based fuels. Feedstocks such as crop and forest residues and municipal solid wastes incur little or no direct and indirect land-use or land-cover changes; therefore, cellulosic biofuels made from those feedstocks are more likely to reduce GHG emissions when care is taken to maintain land productivity and soil carbon storage.

Other cellulosic feedstocks such as herbaceous perennial crops and short-rotation woody crops can contribute to carbon storage in soil particularly if they are planted on land with low carbon content. For example, planting perennial bioenergy crops in place of annual crops could potentially enhance carbon storage in that site. However, planting perennial bioenergy crops on existing cropland can trigger market-mediated land-use changes elsewhere that can result in large GHG emissions. Although RFS2 can levy restrictions to discourage bioenergy feedstock producers from land-clearing or land-cover change in the

United States that would result in net GHG emissions, the policy cannot prevent market-mediated effects on land-use or land-cover changes nor can it control land-use changes outside the United States. Therefore, the extent to which RFS2 contributes to lowering global GHG emissions is uncertain.

Air Quality

The current focus on tailpipe emissions of biofuels compared to petroleum-derived fuels is misguided as it misses the majority of the emissions of air pollutants (other than GHGs) affecting air quality in each of the fuels' life cycles. Overall production and use of ethanol was projected to result in increases in pollutant concentration for ozone and particulate matter than gasoline on a national average, but the local effects could be variable. Those projected air-quality effects from ethanol fuel would be more damaging to human health than those from gasoline use. This is particularly true for corn-grain ethanol. It also showed that the effects from the different fuel options are highly spatially and temporally dependent, thus necessitating a modeling approach that accounts for this variability.

Water Quality

Along the biofuel supply chain, the effect of feedstock production on water quality is less quantifiable than that of fuel conversion. Feedstock production is a nonpoint source discharge; thus, its effect is less certain. Some feedstock types might provide water quality-benefits while others might result in high discharge of sediment and nutrients. Scenarios in which different bioenergy crops are grown in various areas would have to be developed and applied to watershed models to predict changes in water quality resulting from different ways of implementing the RFS2 schedule. Therefore, detailed information on where the bioenergy feedstocks would be grown and how they would be integrated into the existing landscape is necessary to assess the effects of increasing biofuel production on water quality.

Water Quantity and Consumptive Water Use

Consumptive water use over the life cycle of corn-grain ethanol is higher than petroleum-based fuels even if the biofuels are produced from nonirrigated crops. Estimates of consumptive water use for cellulosic biofuels ranges from 2.9 to 1,300 gallons per gallon of gasoline equivalent. Consumptive water use for biofuel produced from switchgrass was estimated to be comparable to that of petroleum-based fuel if the biomass feedstock is not irrigated and if it is converted to fuel by thermochemical conversion. However, biofuels' higher consumptive water use does not necessarily imply that they have more of a negative effect on water resources than petroleum-based fuels because water availability has to be considered in a regional context. For example, a petroleum refinery sited in an arid region with water shortage could be more harmful to its local water resources than a biofuel refinery sited near an aquifer that has rising groundwater level. Therefore, a national assessment of total consumptive water use as a result of meeting RFS2 might not be as useful in assessing effects on water quantity as local and regional assessments. In particular, biorefineries are most likely situated close to sources of bioenergy feedstock production; both biorefinery and feedstock production draw upon local water resources. Regional water availability is particularly important as the number of biorefineries increases in a region. An individual refinery might not pose much stress on a water resource, but multiple refineries could alter the hydrology in a region.

Soil Quality

Whether the environmental effects on soil quality are positive or negative depend, in large part, on the feedstock grown, prior condition of the land, and management practices used. Overharvesting of crop or forest residues can certainly have negative effects on soil quality. In contrast, converting abandoned croplands to herbaceous perennial crops is likely to improve soil quality. Therefore, the effects of increasing biofuel production on soil quality cannot be generalized across the country.

Biodiversity

The effects of achieving RFS2 on biodiversity cannot be readily quantified or qualified because the species affected are largely location-specific and the effects depend on management practices and changes in vegetation cover (including vegetation type and height). Local and regional assessments would be needed to evaluate whether bioeneregy feedstock production would benefit or harm biodiversity.

Overall Environmental Outcome

Production and use of biofuels can be beneficial to some environmental qualities and resource base and have negative effects for others. Thus, the environmental effects of biofuels cannot be focused on one or two environmental parameters (for example, GHG emissions). An assessment of overall environmental outcomes requires a systems approach that considers various environmental effects simultaneously using a suite of indicators. Such assessment would have to be conducted across spatial scales because some effects are localized while others are regional or global. A systems assessment of environmental effects would contribute to developing a biofuel industry that balances tradeoffs and minimizes negative outcomes.

Although using biofuels holds promise to provide net environmental benefits compared to using petroleum-based fuels, the environmental outcome of biofuel production cannot be guaranteed without a landscape and life-cycle vision of where and how the bioenergy feedstocks will be grown to meet the RFS2 consumption mandate. Such landscape and life-cycle vision would contribute to minimizing the potential of negative direct and indirect land-use and land-cover changes, encouraging placement of cellulosic feedstock production in areas that can enhance soil quality or help reduce agricultural nutrient runoffs, anticipating and reducing the potential of groundwater overdraft, and enhancing wildlife habitats. A piecemeal effort to expanding the biofuel industry does not necessarily consider how bioenergy feedstocks could be best integrated into an agricultural landscape to optimize environmental benefits. Without a strategic vision of how RFS2 would be achieved, the overall environmental effects of displacing petroleum-based fuels with 35 billion gallons of ethanol-equivalent biofuels and 1 billion gallons of biodiesel can be positive or negative.

REFERENCES

Abu-Allaban, M., J. Gillies, A. Gertler, R. Clayton, and D. Proffitt. 2007. Motor vehicle contributions to ambient PM_{10} and $PM_{2.5}$ at selected urban areas in the USA. Environmental Monitoring and Assessment 132(1):155-163.

Adams, D.M., R.J. Alig, B.A. McCarl, J.M. Callaway, and S.M. Winnett. 1999. Minimum cost strategies for sequestering carbon in forests. Land Economics 75(3):360-374.

Adams, D.M., R.W. Haynes, and A.J. Daigneault. 2006. Estimated Timber Harvest by U.S. Region and Ownership, 1950-2002. Portland, OR: U.S. Department of Agriculture - Forest Service - Pacific Northwest Research Station.

Adegbidi, H.G., T.A. Volk, E.H. White, L.P. Abrahamson, R.D. Briggs, and D.H. Bickelhaupt. 2001. Biomass and nutrient removal by willow clones in experimental bioenergy plantations in New York State. Biomass and Bioenergy 20(6):399-411.

Aden, A. 2007. Water usage for current and future ethanol production. Southwest Hydrology September/October:22-23.

Adler, P.R., S.J.D. Grosso, and W.J. Parton. 2007. Life-cycle assessment of net greenhouse-gas flux for bioenergy cropping systems. Ecological Applications 17(3):675-691.

Agnetwork. 2010. Grain market outlook: Prospective planting report. Available online at http://www.agnetwork.com/Grain-Market-Outlook--Prospective-Planting-Report/2010-03-29/Article.aspx?oid=1029633&tid=Archive. Accessed September 8, 2010.

Alexander, R.B., R.A. Smith, G.E. Schwarz, E.W. Boyer, J.V. Nolan, and J.W. Brakebill. 2008. Differences in phosphorus and nitrogen delivery to the Gulf of Mexico from the Mississippi River Basin. Environmental Science and Technology 42(3):822-830.

Anderson, L.G. 2009. Ethanol fuel use in Brazil: Air quality impacts. Energy and Environmental Science 2(10):1015-1037.

Anderson-Teixeira, K.J., and E.H. Delucia. 2011. The greenhouse gas value of ecosystems. Global Change Biology 17(1):425-438.

Anderson-Teixeira, K.J., S.C. Davis, M.D. Masters, and E.H. Delucia. 2009. Changes in soil organic carbon under biofuel crops. Global Change Biology Bioenergy 1(1):75-96.

Aneja, V.P., W.H. Schlesinger, and J.W. Erisman. 2009. Effects of agriculture upon the air quality and climate: Research, policy, and regulations. Environmental Science and Technology 43(12):4234-4240.

Anex, R., and R. Lifset. 2009. Post script to the corn ethanol debate. Journal of Industrial Ecology 13(6):996-999.

Anex, R.P., L.R. Lynd, M.S. Laser, A.H. Heggenstaller, and M.M. Liebman. 2007. Potential for enhanced nutrient cycling through coupling of agricultural and bioenergy systems. Crop Science 47:1327-1335.

Applegate, T.J., C. Troche, Z. Jiang, and T. Johnson. 2009. The nutritional value of high-protein corn distillers dried grains for broiler chickens and its effect on nutrient excretion. Poultry Science 88(2):354-359.

Babcock, B.A. 2009. Measuring unmeasurable land-use changes from biofuels. Iowa Ag Review Summer:4-6.

Baker, J.M., T.E. Ochsner, R.T. Venterea, and T.J. Griffis. 2007. Tillage and soil carbon sequestration—What do we really know? Agriculture Ecosystems and Environment 118(1-4):1-5.

Barney, J.N., and J.M. DiTomaso. 2008. Nonnative species and bioenergy: Are we cultivating the next invader? BioScience 58(1):64-70.

Barney, J.N., and J.M. DiTomaso. 2011. Global climate niche estimates for bioenergy crops and invasive species of agronomic origin: Potential problems and opportunities. Plos One 6(3):e17222.

Baskaran, L., H. I. Jager, P. E. Schweizer, and R. Srinivasan. 2009. Progress toward evaluating the sustainability of switchgrass as a bioenergy crop using the SWAT model. Pp. 212-222 in 2009 International SWAT Conference Proceedings. College Station: Texas A&M University System.

Bavin, T.K., T.J. Griffis, J.M. Baker, and R.T. Venterea. 2009. Impact of reduced tillage and cover cropping on the greenhouse gas budget of a maize/soybean rotation ecosystem. Agriculture Ecosystems and Environment 134(3-4):234-242.

Becker, D.R., K. Skog, A. Hellman, K.E. Halvorsen, and T. Mace. 2009. An outlook for sustainable forest bioenergy production in the lake states. Energy Policy 37:5687-5693.

Benke, M.B., X. Hao, P. Caffyn, and T.A. McAllister. 2010. Using manure from cattle fed dried distillers' grains with solubles (DDGS) as fertilizer: Effects on nutrient accumulation in soil and uptake by barley. Agriculture, Ecosystems and Environment 139(4):720-727.

Berndes, G., P. Borjesson, M. Ostwald, and M. Palm. 2008. Multifunctional biomass production systems—An overview with presentation of specific applications in India and Sweden. Biofuels Bioproducts and Biorefining 2(1):16-25.

Bies, L. 2006. The biofuels explosion: Is green energy good for wildlife? Wildlife Society Bulletin 34(4):1203-1205.

Blanco-Canqui, H. 2010. Energy crops and their implications on soil and environment. Agronomy Journal 102(2):403-419.

Blanco-Canqui, H., and R. Lal. 2008. No-tillage and soil-profile carbon sequestration: An on-farm assessment. Soil Science Society of America Journal 72(3):693-701.

Börjesson, P. 2009. Good or bad bioethanol from a greenhouse gas perspective—What determines this? Applied Energy 86(5):589-594.

Börjesson, P., and L.M. Tufvesson. 2011. Agricultural crop-based biofuels—Resource efficiency and environmental performance including direct land use changes. Journal of Cleaner Production 19(2-3):108-120.

Bouwman, A.F., I. Fung, E. Matthews, and J. John. 1993. Global analysis of the potential for N_2O production in natural soils. Global Biogeochemical Cycles 7(3):557-597.

Bouwman, A.F., L.J.M. Boumans, and N.H. Batjes. 2002. Emissions of N_2O and NO from fertilized fields: Summary of available measurement data. Global Biogeochemical Cycles 16(4):1058-1061.

Bouwman, A.F., J.J.M. van Grinsven, and B. Eickhout. 2010. Consequences of the cultivation of energy crops for the global nitrogen cycle. Ecological Applications 20(1):101-109.

Bowyer, C. 2010. Anticipated Indirect Land Use Change Associated with Expanded Use of Biofuels and Bioliquids in the EU—An Analysis of the National Renewable Energy Action Plans. London: Institute of European Environmental Policy.

Boyd, J., and L. Wainger. 2003. Measuring Ecosystem Service Benefits: The Use of Landscape Analysis to Evaluate Environmental Trades and Compensation. Washington, DC: Resources for the Future.

Brauman, A., and G.C. Daily. 2008. Ecosystem services. Pp. 1148-1154 in Human Ecology, S.E. Jørgensen and B.D. Fath, eds. Oxford: Elsevier.

BRDB (Biomass Research and Development Board). 2010. BR&D: Biomass Research and Development. Available online at http://www.usbiomassboard.gov/. Accessed November 9, 2010.

Bremer, V.R., A.J. Liska, T.J. Klopfenstein, G.E. Erickson, H.S. Yang, D.T. Walters, and K.G. Cassman. 2010. Emissions savings in the corn-ethanol life cycle from feeding coproducts to livestock. Journal of Environmental Quality 39(2):472-482.

Bricker, S.B., C.G. Clement, D.E. Pirhalla, S.P. Orlando, and D.R.G. Farrow. 1999. National Estuarine Eutrophication Assessment: Effects of Nutrient Enrichment in the Nation's Estuaries. Silver Spring, MD: National Oceanic and Atmospheric Administration.

Broekema, S. 2009. Feasibility Study of a Biomass Supply for the Spiritwood Industrial Park. Final Report. Maple Grove, MN: Great River Energy.

Brooke, R., G. Fogel, A. Glaser, E. Griffin, and K. Johnson. 2009. Corn Ethanol and Wildlife: How Are Policy and Market Driven Increases in Corn Plantings Affecting Habitat and Wildlife. Washington, DC: National Wildlife Federation.

Buchholz, T., V.A. Luzadis, and T.A. Volk. 2009. Sustainability criteria for bioenergy systems: 848 results from an expert survey. Journal of Cleaner Production 17:S86-S98.

Buddenhagen, C.E., C. Chimera, and P. Clifford. 2009. Assessing biofuel crop invasiveness: A case study. Plos One 4(4).

Burke, I.C., C.M. Yonker, W.J. Parton, C.V. Cole, K. Flach, and D.S. Schimel. 1989. Texture, climate, and cultivation effects on soil organic matter content in U.S. grassland soils. Soil Science Society of America Journal 53:800-805.

Burney, J.A., S.J. Davis, and D.B. Lobell. 2010. Greenhouse gas mitigation by agricultural intensification. Proceedings of the National Academy of Sciences of the United States of America 107(26):12052-12057.

BURNUP (Biomass Utilization and Restoration Network for the Upper Peninsula). 2008. Woody biomass harvesting guidelines. Available online at http://www.upwoodybiomass.org/guidelines_2.asp. Accessed October 30, 2010.

Busse, M.D., P.H. Cochran, W.E. Hopkins, W.H. Johnson, G.M. Riegel, G.O. Fiddler, A.W. Ratcliff, and C.J. Shestak. 2009. Developing resilient ponderosa pine forests with mechanical thinning and prescribed fire in central Oregon's pumice region. Canadian Journal of Forest Research-Revue Canadienne De Recherche Forestiere 39(6):1171-1185.

Cairns, J., P.V. McCormick, and B.R. Niederlehner. 1993. A proposed framework for developing indicators of ecosystem health. Hydrobiologia 263(1):1-44.

Campbell, C.A., H.H. Janzen, K. Paustian, E.G. Gregorich, L. Sherrod, B.C. Liang, and R.P. Zentner. 2005. Carbon storage in soils of the North American Great Plains: Effect of cropping frequency. Agronomy Journal 97(2):349-363.

Campbell, J.E., and E. Block. 2010. Land-use and alternative bioenergy pathways for waste biomass. Environmental Science and Technology 44(22):8665-8669.

Campbell, J.E., D.B. Lobell, and C.B. Field. 2009. Greater transportation energy and GHG offsets from bioelectricity than ethanol. Science 324(5930):1055-1057.

Cantrell, K.B., K.C. Stone, P.G. Hunt, K.S. Ro, M.B. Vanotti, and J.C. Burns. 2009. Bioenergy from coastal bermudagrass receiving subsurface drip irrigation with advance-treated swine wastewater. Bioresource Technology 100(13):3285-3292.

Cantrell, K.B., P.J. Bauer, and K.S. Ro. 2010. Utilization of summer legumes as bioenergy feedstocks. Biomass and Bioenergy 34(12):1961-1967.

Cassman, K.G., S. Wood, P.S. Choo, H.D. Cooper, C. Devendra, J. Dixon, J. Gaskell, S. Khan, R. Lal, L. Lipper, J. Pretty, J. Primavera, N. Ramankutty, E. Viglizzo, K. Wiebe, S. Kadungure, N. Kanbar, Z. Khan, R. Leakey, S. Porter, K. Sebastian, and R. Tharme. 2005. Cultivated systems. Pp. 745-794 in Ecosystems and Human Well-being: Current State and Trends, R. Hassan, R. Scholes, and N. Ash, eds. Washington, DC: Island Press.

Castro, A., and D.H. Wise. 2009. Influence of fine woody debris on spider diversity and community structure in forest leaf litter. Biodiversity and Conservation 18(14):3705-3731.

CENR (Committee on Environment and Natural Resources). 2000. Integrated Assessment of Hypoxia in the Northern Gulf of Mexico. Washington, DC: National Science and Technology Council.

Checkbiotech.org. 2009. Mascoma cellulosic ethanol plant begins operations in Rome, NY. Available online at http://checkbiotech.org/node/24912. Accessed November 5, 2010.

Cherubini, F., and S. Ulgiati. 2010. Crop residues as raw materials for biorefinery systems—A LCA case study. Applied Energy 87(1):47-57.

Chester, M.V., and A. Horvath. 2009. Environmental assessment of passenger transportation should include infrastructure and supply chains. Environmental Research Letters 4(2).

Chiu, Y.-W., B. Walseth, and S. Suh. 2009. Water embodied in bioethanol in the United States. Environmental Science and Technology 43(8):2688-2692.

Christopher, S.F., R. Lal, and U. Mishra. 2009. Regional study of no-till effects on carbon sequestration in Midwestern United States. Soil Science Society of America Journal 73(1):207-216.

Cook, R., S. Phillips, M. Houyoux, P. Dolwick, R. Mason, C. Yanca, M. Zawacki, K. Davidson, H. Michaels, C. Harvey, J. Somers, and D. Luecken. 2011. Air quality impacts of increased use of ethanol under the United States' Energy Independence and Security Act. Atmospheric Environment (In press).

Costanza, R., R. d'Arge, R. de Groot, S. Farber, M. Grasso, B. Hannon, K. Limburg, S. Naeem, R.V. O'Neill, J. Paruelo, R.G. Raskin, P. Sutton, and M. van den Belt. 1997. The value of the world's ecosystem services and natural capital. Nature 387(6630):253-260.

Costello, C., W.M. Griffin, A.E. Landis, and H.S. Matthews. 2009. Impact of biofuel crop production on the formation of hypoxia in the Gulf of Mexico. Environmental Science and Technology 43(20):7985-7991.

Covington, W.W. 1980. Changes in forest floor organic matter and nutrient content following the clear cutting in northern hardwoods. Ecology 62(1):4-48.

Cubbage, F.W., J. O'Laughlin, and C.S. Bullock III. 1993. Forest Resource Policy. New York: John Wiley and Sons.

Dale, V.H., and S.C. Beyeler. 2001. Challenges in the development and use of ecological indicators. Ecological Indicators 1:3-10.

Dale, V.H., K.L. Kline, J. Wiens, and J. Fargione. 2010a. Biofuels: Implications for Land Use and Biodiversity. Washington, DC: Ecological Society of America.

Dale, V.H., C.L. Kling, J.L. Meyer, J. Sanders, H. Stallworth, T. Armitage, D. Wangsness, T. Bianchi, A. Blumberg, W. Boynton, D.J. Conley, W. Crumpton, M.B. Davis, D. Gilbert, R.W. Howarth, R. Lowrance, K. Mankin, J. Opaluch, H. Paerl, K. Reckhow, A.N. Sharpley, T.W. Simpson, C. Snyder, and C. Wright. 2010b. Hypoxia in the Northern Gulf of Mexico; Springer Series on Environmental Management. New York: Springer.

Dale, V.H., M.L. Tharp, K.O. Lannom, and D.G. Hodges. 2010c. Modeling transient response of forests to climate change. Science of the Total Environment 408(8):1888-1901.

Dauber, J., M.B. Jones, and J.C. Stout. 2010. The impact of biomass crop cultivation on temperate biodiversity. Global Change Biology Bioenergy 2(6):289-309.

Daughtry, C.S.T., P.C. Doraiswamy, E.R. Hunt, A.J. Stern, J.E. McMurtrey, and J.H. Prueger. 2006. Remote sensing of crop residue cover and soil tillage intensity. Soil and Tillage Research 91(1-2):101-108.

Davidson, E.A., and I.L. Ackerman. 1993. Changes in soil carbon inventories following cultivation of previously untilled soils. Biogeochemistry 20:161-193.

Davis, A.S., R.D. Cousens, J. Hill, R.N. Mack, D. Simberloff, and S. Raghu. 2010. Screening bioenergy feedstock crops to mitigate invasion risk. Frontiers in Ecology and the Environment 8(10):533-539.

Del Grosso, S.J., A.R. Mosier, W.J. Parton, and D.S. Ojima. 2005. DAYCENT model analysis of past and contemporary soil N_2O and net greenhouse gas flux for major crops in the USA. Soil and Tillage Research 83(1):9-24.

Delucchi, M. 2011. A conceptual framework for estimating the climate impacts of land-use change due to energy crop programs. Biomass and Bioenergy 35(6):2337-2360.

DiTomaso, J.M., J.K. Reaser, C.P. Dionigi, O.C. Doering, E. Chilton, J.D. Schardt, and J.N. Barney. 2010. Biofuel vs. bioinvasion: Seeding policy priorities. Environmental Science and Technology 44(18):6906-6910.

DOE (U.S. Department of Energy). 2005. Design and Construction of a Proposed Fuel Ethanol Plant, Jasper County, Indiana. Golden, CO: U.S. Department of Energy.

DOE (U.S. Department of Energy). 2009. Construction and Operation of a Proposed Cellulosic Ethanol Plant, Range Fuels Soperton Plant, LLC (formerly Range Fuels Inc.), Treutlen County, Georgia. Washington, DC: U.S. Department of Energy.

DOE-EERE (U.S. Department of Energy - Energy Efficiency and Renewable Energy). 2010a. Final Environmental Impact Statement for the Proposed Abengoa Biorefinery Project near Hugoton, Stevens County, Kansas. Golden, CO: U.S. Department of Energy.

DOE-EERE (U.S. Department of Energy - Energy Efficiency and Renewable Energy). 2010b. Final Environmental Impact Statement for the Proposed Abengoa Biorefinery Project near Hugoton, Stevens County, Kansas. Golden, CO: U.S. Department of Energy.

DOE-NETL (U.S. Department of Energy - National Energy Technology Laboratory). 2010. Carbon Sequestration Atlas of the United States and Canada. Washington, DC: U.S. Department of Energy.

Dolan, M.S., C.E. Clapp, R.R. Allmaras, J.M. Baker, and J.A.E. Molina. 2006. Soil organic carbon and nitrogen in a Minnesota soil as related to tillage, residue and nitrogen management. Soil and Tillage Research 89(2):221-231.

Dominguez-Faus, R., S.E. Powers, J.G. Burken, and P.J. Alvarez. 2009. The water footprint of biofuels: A drink or drive issue? Environmental Science and Technology 43(9):3005-3010.

Donner, S.D., and C.J. Kucharik. 2003. Evaluating the impacts of land management and climate variability on crop production and nitrate export across the Upper Mississippi Basin. Global Biogeochemical Cycles 17(3):1085.

Donner, S.D., and C.J. Kucharik. 2008. Corn-based ethanol production compromises goal of reducing nitrogen export by the Mississippi River. Proceedings of the National Academy of Sciences of the United States of America 105(11):4513-4518.

Donner, S.D., M.T. Coe, J.D. Lenters, T.E. Twine, and J.A. Foley. 2002. Modeling the impact of hydrological changes on nitrate transport in the Mississippi River Basin from 1955 to 1994. Global Biogeochemical Cycles 16(3):1043.

Drever, C.R., M.C. Drever, C. Messier, Y. Bergeron, and M. Flannigan. 2008. Fire and the relative roles of weather, climate and landscape characteristics in the Great Lakes-St. Lawrence forest of Canada. Journal of Vegetation Science 19:57-66.

Dumortier, J., D.J. Hayes, M. Carriquiry, F. Dong, X. Du, A. Elobeid, J.F. Fabiosa, and S. Tokgoz. 2010. Sensitivity of Carbon Emission Estimates from Indirect Land-Use Change. Ames: Iowa State University.

Durbin, T.D., J.W. Miller, T. Younglove, T. Huai, and K. Cocker. 2007. Effects of fuel ethanol content and volatility on regulated and unregulated exhaust emissions for the latest technology gasoline vehicles. Environmental Science and Technology 41(11):4059-4064.

Earles, R., and P. Williams. 2005. Sustainable agriculture: An introduction. Available online at http://attra.ncat .org/attra-pub/PDF/sustagintro.pdf. Accessed September 16, 2009.

Easterling, W.E., P.K. Aggarwal, P. Batima, K.M. Brander, L. Erda, S.M. Howden, A. Kirilenko, J. Morton, J.-F. Soussana, J. Schmidhuber, and F.N. Tubiello. 2007. Food, fibre and forest products. Pp. 273-313 in Impacts, Adaptation and Vulnerability. Contribution of Working Group II to the Fourth Assessment Report of the Intergovernmental Panel on Climate Change, M.L. Parry, O.F. Canziani, J.P. Palutikof, P.J.v.d. Linden, and C.E. Hanson, eds. Cambridge: Cambridge University Press.

Edgerton, M. 2010. Corn carbon budgets: Use of "discretionary carbon." Presentation to the Committee on Economic and Environmental Impacts of Increasing Biofuels Production, March 5.

Eisenbies, M.H., E.D. Vance, W.M. Aust, and J.R. Seiler. 2009. Intensive utilization of harvest residues in southern pine plantations: Quantities available and implications for nutrient budgets and sustainable site productivity. BioEnergy Research 2(3):90-98.

Ekvall, T., and A.S.G. Andræ. 2006. Attributional and consequential environmental assessment of the shift to lead-free solders. The International Journal of Life Cycle Assessment 11(5):344-353.

Ekvall, T., and B.P. Weidema. 2004. System boundaries and input data in consequential life cycle inventory analysis. The International Journal of Life Cycle Assessment 9(3):161-171.

Engel, B., I. Chaubey, M. Thomas, D. Saraswat, P. Murphy, and B. Bhaduri. 2010. Biofuels and water quality: Challenges and opportunities for simulation modeling. Biofuels 1(3):463-477.

ENSR AECOM. 2008. Construction and Operation of a Proposed Lignocellulosic Biorefinery, POET Project LIBERTY, LLC, Emmetsburg, Iowa. Washington, DC: U.S. Department of Energy.

EPA (U.S. Environmental Protection Agency). 2010a. EPA Finalizes Regulations for the National Renewable Fuel Standard Program for 2010 and Beyond. Washington, DC: U.S. Environmental Protection Agency.

EPA (U.S. Environmental Protection Agency). 2010b. Fuel-specific lifecycle greenhouse gas emissions results. Available online at http://www.regulations.gov/#!documentDetail;D=EPA-HQ-OAR-2005-0161-3173. Accessed March 3, 2011.

EPA (U.S. Environmental Protection Agency). 2010c. Inventory of U.S. Greenhouse Gas Emissions and Sinks: 1990-2008. Washington, DC: U.S. Environmental Protection Agency.

EPA (U.S. Environmental Protection Agency). 2010d. Renewable Fuel Standard Program (RFS2) Regulatory Impact Analysis. Washington, DC: U.S. Environmental Protection Agency.

Erisman, J., H. van Grinsven, A. Leip, A. Mosier, and A. Bleeker. 2010. Nitrogen and biofuels: An overview of the current state of knowledge. Nutrient Cycling in Agroecosystems 86(2):211-223.

Evans, A.M., R.T. Perschel, and B.A. Kittler. 2010. Revised Assessment of Biomass Harvesting and Retention Guidelines. Santa Fe, NM: Forest Guild.

Evans, J.M., and M.J. Cohen. 2009. Regional water resource implications of bioethanol production in the southeastern United States. Global Change Biology 15(9):2261-2273.

Fargione, J., J. Hill, D. Tilman, S. Polasky, and P. Hawthorne. 2008. Land clearing and the biofuel carbon debt. Science 319(5867):1235-1238.

Fargione, J.E., T.R. Cooper, D.J. Flaspohler, J. Hill, C. Lehman, T. McCoy, S. McLeod, E.J. Nelson, K.S. Oberhauser, and D. Tilman. 2009. Bioenergy and wildlife: Threats and opportunities for grassland conservation. BioScience 59(9):767-777.

Fargione, J.E., R.J. Plevin, and J.D. Hill. 2010. The ecological impacts of biofuels. Annual Review of Ecology and Systematics 41:351-377.

Farrell, A.E., R.J. Plevin, B.T. Turner, A.D. Jones, M. O'Hare, and D.M. Kammen. 2006. Ethanol can contribute to energy and environmental goals. Science 311(5760):506-508.

Fearnside, P.M. 1996. Amazonian deforestation and global warming: Carbon stocks in vegetation replacing Brazil's Amazon forest. Forest Ecology and Management 80(1-3):21-34.

Feng, H., O.D. Rubin, and B.A. Babcock. 2010. Greenhouse gas impacts of ethanol from Iowa corn: Life cycle assessment versus system wide approach. Biomass and Bioenergy 34(6):912-921.

Fight, R.D., and R.J. Barbour. 2005. Financial Analysis of Fuel Treatments. Portland, OR: U.S. Department of Agriculture - Forest Service - Pacific Northwest Research Station.

Fingerman, K.R., M.S. Torn, M.H. O'Hare, and D.M. Kammen. 2010. Accounting for the water impacts of ethanol production. Environmental Research Letters 5(1).

Finnveden, G., M.Z. Hauschild, T. Ekvall, J. Guinee, R. Heijungs, S. Hellweg, A. Koehler, D. Pennington, and S. Suh. 2009. Recent developments in life cycle assessment. Journal of Environmental Management 91(1):1-21.

Fischer, G., E. Hizsnyik, S. Prieler, M. Shah, and H. van Velthuizen. 2009. Biofuels and Food Security. Vienna: The OPEC Fund for International Development.

Flaspohler, D.J., R.E. Froese, and C.R. Webster. 2009. Bioenergy, biomass and biodiversity. Pp. 133-162 in Renewable Energy from Forest Resources in the United States, B.D. Solomon and V.A. Luzadis, eds. New York: Routledge.

Foley, J.A., R. DeFries, G.P. Asner, C. Barford, G. Bonan, S.R. Carpenter, F.S. Chapin, M.T. Coe, G.C. Daily, H.K. Gibbs, J.H. Helkowski, T. Holloway, E.A. Howard, C.J. Kucharik, C. Monfreda, J.A. Patz, I.C. Prentice, N. Ramankutty, and P.K. Snyder. 2005. Global consequences of land use. Science 309(5734):570-574.

Fornara, D.A., and D. Tilman. 2008. Plant functional composition influences rates of soil carbon and nitrogen accumulation. Journal of Ecology 96(2):314-322.

Fortier, J., D. Gagnon, B. Truax, and F. Lambert. 2010. Biomass and volume yield after 6 years in multiclonal hybrid poplar riparian buffer strips. Biomass and Bioenergy 34(7):1028-1040.

Foust, T., A. Aden, A. Dutta, and S. Phillips. 2009. An economic and environmental comparison of a biochemical and a thermochemical lignocellulosic ethanol conversion processes. Cellulose 16(4):547-565.

Frey, H.C., H. Zhai, and N.M. Rouphail. 2009. Regional on-road vehicle running emissions modeling and evaluation for conventional and alternative vehicle technologies. Environmental Science and Technology 43(21):8449-8455.

FSCUS (Forest Stewardship Council United States). 2010. The Forest Stewardship Council. Available online at http://www.fscus.org/. Accessed November 10, 2010.

Gallagher, M.E., W.C. Hockaday, C.A. Masiello, S. Snapp, C.P. McSwiney, and J.A. Baldock. 2011. Biochemical suitability of crop residues for cellulosic ethanol: Disincentives to nitrogen fertilization in corn agriculture. Environmental Science and Technology 45(5):2013-2020.

GAO (U.S. Government Accountability Office). 2009. Biofuels: Potential Effects and Challenges of Required Increases in Production and Use. Washington, DC: U.S. Government Accountability Office.

Gardiner, M.A., J.K. Tuell, R. Isaacs, J. Gibbs, J.S. Ascher, and D.A. Landis. 2010. Implications of three biofuel crops for beneficial arthropods in agricultural landscapes. BioEnergy Research 3(1):6-19.

Garten, C.T., J.L. Smith, D.D. Tyler, J.E. Amonette, V.L. Bailey, D.J. Brice, H.F. Castro, R.L. Graham, C.A. Gunderson, R.C. Izaurralde, P.M. Jardine, J.D. Jastrow, M.K. Kerley, R. Matamala, M.A. Mayes, F.B. Metting, R.M. Miller, K.K. Moran, W.M. Post, R.D. Sands, C.W. Schadt, J.R. Phillips, A.M. Thomson, T. Vugteveen, T.O. West, and S.D. Wullschleger. 2010. Intra-annual changes in biomass, carbon, and nitrogen dynamics at 4-year old switchgrass field trials in west Tennessee, USA. Agriculture Ecosystems and Environment 136(1-2):177-184.

GBEP (Global Bioenergy Partnership). 2010. Second draft of GBEP sustainability criteria and indicators for bioenergy. Available online at http://www.globalbioenergy.org/fileadmin/user_upload/gbep/docs/partners_only/sust_docs/2nd_DRAFT_of_GBEP_Criteria_Indicators_with_TEMPLATES.doc. Accessed November 15, 2010.

Gebber, R., and V.I. Adamchuk. 2010. Precision agriculture and food security. Science 327(5967):828-831.

Geist, H.J., and E.F. Lambin. 2002. Proximate causes and underlying driving forces of tropical deforestation. Bio-Science 52(2):143-150.

Gell, K., J.W. van Groenigen, and M.L. Cayuela. 2011. Residues of bioenergy production chains as soil amendments: Immediate and temporal phytotoxicity. Journal of Hazardous Materials 186(2-3):2017-2025.

George, N., Y. Yang, Z. Wang, R. Sharma-Shivappa, and K. Tungate. 2010. Suitability of canola residue for cellulosic ethanol production. Energy and Fuels 24(8):4454-4458.

Georgescu, M., D.B. Lobell, and C.B. Field. 2011. Direct climate effects of perennial bioenergy crops in the United States. Proceedings of the National Academy of Sciences of the United States of America 108(11):4307-4312.

Gibbs, H.K., M. Johnston, J.A. Foley, T. Holloway, C. Monfreda, N. Ramankutty, and D. Zaks. 2008. Carbon payback times for crop-based biofuel expansion in the tropics: The effects of changing yield and technology. Environmental Research Letters 3(3):034001.

Gilliom, R.J., J.E. Barbash, C.G. Crawford, P.A. Hamilton, J.D. Martin, N. Nakagaki, L.H. Nowell, J.C. Scott, P.E. Stackelberg, G.P. Thelin, and D.M. Wolock. 2007. Pesticides in the Nation's Streams and Ground Water, 1992-2001. Reston, VA: U.S. Geological Survey.

Ginnebaugh, D.L., J.Y. Liang, and M.Z. Jacobson. 2010. Examining the temperature dependence of ethanol (E85) versus gasoline emissions on air pollution with a largely-explicit chemical mechanism. Atmospheric Environment 44(9):1192-1199.

Gollany, H.T., R.W. Rickman, Y. Liang, S.L. Albrecht, S. Machado, and S. Kang. 2011. Predicting agricultural management influence on long-term soil organic carbon dynamics: Implications for biofuel production. Agronomy Journal 103(1):234-246.

Graham, L.A., S.L. Belisle, and C.L. Baas. 2008. Emissions from light duty gasoline vehicles operating on low blend ethanol gasoline and E85. Atmospheric Environment 42(19):4498-4516.

Guo, L.B., and R.M. Gifford. 2002. Soil carbon stocks and land use change: A meta analysis. Global Change Biology 8(4):345-360.

Guzman, J.G., and M. Al-Kaisi. 2010. Landscape position and age of reconstructed prairies effect on soil organic carbon sequestration rate and aggregate associated carbon. Journal of Soil and Water Conservation 65(1):9-21.

Hammerschlag, R. 2006. Ethanol's energy return on investment: A survey of the literature 1990 - Present. Environmental Science and Technology 40(6):1744-1750.

Harto, C., R. Meyers, and E. Williams. 2010. Life cycle water use of low-carbon transport fuels. Energy Policy 38:4933-4944.

Heaton, E., T. Voigt, and S.P. Long. 2004. A quantitative review comparing the yields of two candidate C_4 perennial biomass crops in relation to nitrogen, temperature and water. Biomass and Bioenergy 27(1):21-30.

Hecht, A., D. Shaw, R. Bruins, V. Dale, K. Kline, and A. Chen. 2009. Good policy follows good science: Using criteria and indicators for assessing sustainable biofuel production. Ecotoxicology 18(1):1-4.

Heltman, J., and E. Martinson. 2011. States grapple with new EPA push to adopt strict numeric nutrient limits. Water Policy Report (April 11). Available online at https://environmentalnewsstand.com/Water-Policy-Report/Water-Policy-Report-04/11/2011/menu-id-304.html. Accessed April 18, 2011.

Hertel, T.W., A.A. Golub, A.D. Jones, M. O'Hare, R.J. Plevin, and D.M. Kammen. 2010. Effects of US maize ethanol on global land use and greenhouse gas emissions: Estimating market-mediated responses. BioScience 60(3):223-231.

Hess, P., M. Johnston, B. Brown-Steiner, T. Holloway, J.B. de Andrade, and P. Artaxo. 2009. Air quality issues associated with biofuel production and use. Pp. 169-194 in Biofuels: Environmental Consequences and Interactions with Changing Land Use, R.W. Howarth and S. Bringezu, eds. Ithaca, NY: Cornell University.

Hettinga, W.G., H.M. Junginger, S.C. Dekker, M. Hoogwijk, A.J. McAloon, and K.B. Hicks. 2009. Understanding the reductions in US corn ethanol production costs: An experience curve approach. Energy Policy 37(1):190-203.

Hewitt, C.N., A.R. MacKenzie, P. Di Carlo, C.F. Di Marco, J.R. Dorsey, M. Evans, D. Fowler, M.W. Gallagher, J.R. Hopkins, C.E. Jones, B. Langford, J.D. Lee, A.C. Lewis, S.F. Lim, J. McQuaid, P. Misztal, S.J. Moller, P.S. Monks, E. Nemitz, D.E. Oram, S.M. Owen, G.J. Phillips, T.A.M. Pugh, J.A. Pyle, C.E. Reeves, J. Ryder, J. Siong, U. Skiba, and D.J. Stewart. 2009. Nitrogen management is essential to prevent tropical oil palm plantations from causing ground-level ozone pollution. Proceedings of the National Academy of Sciences of the United States of America 106(44):18447-18451.

Hill, J., E. Nelson, D. Tilman, S. Polasky, and D. Tiffany. 2006. Environmental, economic, and energetic costs and benefits of biodiesel and ethanol biofuels. Proceedings of the National Academy of Sciences of the United States of America 103(30):11206-11210.

Hill, J., S. Polasky, E. Nelson, D. Tilman, H. Huo, L. Ludwig, J. Neumann, H.C. Zheng, and D. Bonta. 2009. Climate change and health costs of air emissions from biofuels and gasoline. Proceedings of the National Academy of Sciences of the United States of America 106(6):2077-2082.

Hochman, G., D. Rajagopal, and D. Zilberman. 2010. The effect of biofuels on crude oil markets. AgBioForum 13(2):112-118.

Hoefnagels, R., E. Smeets, and A. Faaij. 2010. Greenhouse gas footprints of different biofuel production systems. Renewable and Sustainable Energy Reviews 14(7):1661-1694.

Houghton, R.A. 2010. How well do we know the flux of CO_2 from land-use change? Tellus B 62(5):337-351.

Hsu, D.D., D. Inman, G.A. Heath, E.J. Wolfrum, M.K. Mann, and A. Aden. 2010. Life cycle environmental impacts of selected U.S. ethanol production and use pathways in 2022. Environmental Science and Technology 44(13):5289-5297.

Huffaker, R. 2010. Protecting water resources in biofuels production. Water Policy 12:129-134.

Huggins, D.R., R.S. Karow, H.P. Collins, and J.K. Ransom. 2011. Introduction: Evaluating long-term impacts of harvesting crop residues on soil quality. Agronomy Journal 103(1):230-233.

Huo, H., M. Wang, C. Bloyd, and V. Putsche. 2009. Life-cycle assessment of energy use and greenhouse gas emissions of soybean-derived biodiesel and renewable fuels. Environmental Science and Technology 43(3):750-756.

Hutchinson, K.J. 2008. Human impacted water environment classes. M.S. thesis, University of Iowa, Iowa City.

Ice, G.G., E.G. Schilling, and J.G. Vowel. 2010. Trends for forestry best management practices implementation. Journal of Forestry 108(6):267-273.

ISO (International Organization for Standardization). 2006. Environmental Management—Life Cycle Assessment—Principles and Framework. Geneva: International Organization for Standardization.

Izaurralde, R.C., J.R. Williams, W.B. McGill, N.J. Rosenberg, and M.C.Q. Jakas. 2006. Simulating soil C dynamics with EPIC: Model description and testing against long-term data. Ecological Modelling 192(3-4):362-384.

Jackson, R.B., E.G. Jobbagy, R. Avissar, S.B. Roy, D.J. Barrett, C.W. Cook, K.A. Farley, D.C. le Maitre, B.A. McCarl, and B.C. Murray. 2005. Trading water for carbon with biological sequestration. Science 310(5756):1944-1947.

Jacobson, M.Z. 2007. Effects of ethanol (E85) versus gasoline vehicles on cancer and mortality in the United States. Environmental Science and Technology 41(11):4150-4157.

Janowiak, M.K., and C.R. Webster. 2010. Promoting ecological sustainability in woody biomass harvesting. Journal of Forestry 108(1):16-23.

Jenkins, T.L., and J.W. Sutherland. 2009. An integrated supply system for forest biomass. Pp. 92-115 in Renewable Energy from Forest Resources in the United States, B.D. Solomon and V.A. Luzadis, eds. New York: Routledge.

Johnson, G., C. Sheaffer, H.J. Jung, U. Tschirmer, S. Banerjee, K. Petersen, and D.L. Wyse. 2008. Landscape and species diversity: Optimizing the use of land and species for biofuel feedstock production systems. Paper read at the Agronomy Society of America, Crop Science Society of America, and Soil Science Society of America Annual Meeting October 5-9, Houston, TX.

Johnson, J.M.F., A.J. Franzluebbers, S.L. Weyers, and D.C. Reicosky. 2007. Agricultural opportunities to mitigate greenhouse gas emissions. Environmental Pollution 150(1):107-124.

Johnson, K., F.N. Scatena, and Y.D. Pan. 2010. Short- and long-term responses of total soil organic carbon to harvesting in a northern hardwood forest. Forest Ecology and Management 259(7):1262-1267.

Jones, D.L. 2010. Potential air emission impacts of cellulosic ethanol production at seven demonstration refineries in the United States. Journal of the Air and Waste Management Association 60(9):1118-1143.

Jordaan, S.M., D.W. Keith, and B. Stelfox. 2009. Quantifying land use of oil sands production: A life cycle perspective. Environmental Research Letters 4(2):024004.

Kalies, E.L., C.L. Chambers, and W.W. Covington. 2010. Wildlife responses to thinning and burning treatments in southwestern conifer forests: A meta-analysis. Forest Ecology and Management 259(3):333-342.

Kaliyan, N., R.V. Morey, and D.G. Tiffany. 2011. Reducing life cycle greenhouse gas emissions of corn ethanol by integrating biomass to produce heat and power at ethanol plants. Biomass and Bioenergy 35(3):1103-1113.

Kalogo, Y., S. Habibi, H.L. MacLean, and S.V. Joshi. 2006. Environmental implications of municipal solid waste-derived ethanol. Environmental Science and Technology 41(1):35-41.

Karlen, D.L. 2010. A landscape vision for integrating industrial crops into biofuel systems. Paper read at the Association for the Advancement of Industrial Crops Conference September 22, Fort Collins, CO.

Karlen, D.L., R. Lal, R.F. Follett, J.M. Kimble, J.L. Hatfield, J.M. Miranowski, C.A. Cambardella, A. Manale, R.P. Anex, and C.W. Rice. 2010. Crop residues: The rest of the story. Environmental Science and Technology 43(21):8011-8015.

Kendall, A., B. Chang, and B. Sharpe. 2009. Accounting for time-dependent effects in biofuel life cycle greenhouse gas emissions calculations. Environmental Science and Technology 43(18):7142-7147.

Khanal, S.K., M. Rasmussen, P. Shrestha, H. Van Leeuwen, C. Visvanathan, and H. Liu. 2008. Bioenergy and biofuel production from wastes/residues of emerging biofuel industries. Water Environment Research 80(10):1625-1647.

Khanna, M., B. Dhungana, and J. Clifton-Brown. 2008. Costs of producing *Miscanthus* and switchgrass for bioenergy in Illinois. Biomass and Bioenergy 32(6):482-493.

Khanna, M., H. Önal, B. Dhungana, and M. Wander. 2010. Economics of herbaceous bioenergy crops for electricity generation: Implications for greenhouse gas mitigation. Biomass and Bioenergy 35(4):1474-1484.

Kim, S., and B.E. Dale. 2011. Indirect land use change for biofuels: Testing predictions and improving analytical methodologies. Biomass and Bioenergy 35(7):3235-3240.

King, C.W., and M.E. Webber. 2008. Water intensity of transportation. Environmental Science and Technology 42(21):7866-7872.

Kitchen, N.R., K.A. Sudduth, D.B. Myers, R.E. Massey, E.J. Sadler, and R.N. Lerch. 2005. Development of a conservation-oriented precision agriculture system: Crop production assessment and plan of implementation. Journal of Soil and Water Conservation 60:421-430.

Kløverpris, J., H. Wenzel, and P. Nielsen. 2008. Life cycle inventory modelling of land use induced by crop consumption. The International Journal of Life Cycle Assessment 13(1):13-21.

Kravchenko, A.N., and G.P. Robertson. 2011. Whole-profile soil carbon stocks: The danger of assuming too much from analyses of too little. Soil Science Society of America Journal 75(1):235-240.

Kretschmer, B., and S. Peterson. 2010. Integrating bioenergy into computable general equilibrium models—A survey. Energy Economics 32(3):673-686.

Kusiima, J.M., and S.E. Powers. 2010. Monetary value of the environmental and health externalities associated with production of ethanol from biomass feedstocks. Energy Policy 38(6):2785-2796.

Lal, R. 2004. Soil carbon sequestration impacts on global climate change and food security. Science 304:1623-1627.

Lal, R., F. Follett, B.A. Stewart, and J.M. Kimble. 2007. Soil carbon sequestration to mitigate climate change and advance food security. Soil Science Society of America Journal 172:943-956.

Lamb, D.W., P. Frazier, and P. Adams. 2008. Improving pathways to adoption: Putting the right P's in precision agriculture. Computers and Electronics in Agriculture 61(1):4-9.

Lamm, F.R., H.L. Manges, L.R. Stone, A.H. Khan, and D.H. Rogers. 1995. Water requirement of subsurface drip-irrigated corn in Northwest Kansas. Transactions of the American Society of Agricultural and Biological Engineers 38(2):441-448.

Landis, D.A., and B.P. Werling. 2010. Arthropods and biofuel production systems in North America. Insect Science 17(3):220-236.

Landis, D.A., M.M. Gardiner, W. van der Werf, and S.M. Swinton. 2008. Increasing corn for biofuel production reduces biocontrol services in agricultural landscapes. Proceedings of the National Academy of Sciences of the United States of America 105(51):20552-20557.

Lee, J.W., M. Kidder, B.R. Evans, S. Paik, A.C. Buchanan, C.T. Garten, and R.C. Brown. 2010. Characterization of biochars produced from cornstovers for soil amendment. Environmental Science and Technology 44(20):7970-7974.

Lemoine, D.M., R.J. Plevin, A.S. Cohn, A.D. Jones, A.R. Brandt, S.E. Vergara, and D.M. Kammen. 2010. The climate impacts of bioenergy systems depend on market and regulatory policy contexts. Environmental Science and Technology 44(19):7374-7350.

Lerch, R.N., R.J. Kitchen, W.W. Kremer, E.E. Donald, E.J. Alberts, K.A. Saddler, D.B. Sudduth, D.B. Myers, and F. Ghidey. 2005. Development of a conservation-oriented precision agriculture system: Water and soil quality. Journal of Soil and Water Conservation 60:411-421.

Levasseur, A., P. Lesage, M. Margni, L. Deschênes, and R.J. Samson. 2010. Considering time in LCA: Dynamic LCA and its application to global warming impact assessments. Environmental Science and Technology 44(8):3169-3174.

Lewandowski, I., J.M.O. Scurlock, E. Lindvall, and M. Christou. 2003. The development and current status of perennial rhizomatous grasses as energy crops in the US and Europe. Biomass and Bioenergy 25(4):335-361.

Leytem, A.B., P. Kwanyuen, and P. Thacker. 2008. Nutrient excretion, phosphorus characterization, and phosphorus solubility in excreta from broiler chicks fed diets containing graded levels of wheat distillers grains with solubles. Poultry Science 87(12):2505-2511.

Liebig, M.A., M.R. Schmer, K.P. Vogel, and R.B. Mitchell. 2008. Soil carbon storage by switchgrass grown for bioenergy. BioEnergy Research 1(3-4):215-222.

Liska, A.J., and K.G. Cassman. 2009a. Response to Plevin implications for life cycle emissions regulations. Journal of Industrial Ecology 13(4):508-513.

Liska, A.J., and K.G. Cassman. 2009b. Responses to "Comment on 'Response to Plevin: Implications for life cycle emissions regulations'" and "Assessing corn ethanol: relevance and responsibility." Journal of Industrial Ecology 13(6):994-995.

Liska, A.J., H.S. Yang, V.R. Bremer, T.J. Klopfenstein, D.T. Walters, G.E. Erickson, and K.G. Cassman. 2009. Improvements in life cycle energy efficiency and greenhouse gas emissions of corn-ethanol. Journal of Industrial Ecology 13(1):58-74.

Liu, Y., M.A. Evans, and D. Scavia. 2010. Gulf of Mexico hypoxia: Exploring increasing sensitivity to nitrogen loads. Environmental Science and Technology 44(15):5836-5841.

Lu, X.L., and Q.L. Zhuang. 2010. Evaluating climate impacts on carbon balance of the terrestrial ecosystems in the Midwest of the United States with a process-based ecosystem model. Mitigation and Adaptation Strategies for Global Change 15(5):467-487.

Luo, Z., E. Wang, and O.J. Sun. 2010. Can no-tillage stimulate carbon sequestration in agricultural soils? A meta-analysis of paired experiments. Agriculture, Ecosystems and Environment 139:224-231.

Machado, S. 2011. Soil organic carbon dynamics in the Pendleton long-term experiments: Implications for biofuel production in Pacific Northwest. Agronomy Journal 103(1):253-260.

Malcolm, S., and M. Aillery. 2009. Growing crops for biofuels has spillover effects. Amber Waves 7(1):10-15.

Marshall, L. 2009. Biofuels and the Time Value of Carbon: Recommendations for GHG Accounting Protocols. Washington, DC: World Resources Institute.

McCormick, R.L. 2007. The impact of biodiesel on pollutant emissions and public health. Inhalation Toxicology 19(12):1033-1039.

McKechnie, J., S. Colombo, J. Chen, W. Mabee, and H.L. MacLean. 2010. Forest bioenergy or forest carbon? Assessing trade-offs in greenhouse gas mitigation with wood-based fuels. Environmental Science and Technology 45(2):789-795.

McKone, T.E., W.W. Nazaroff, P. Berck, M. Auffhammer, T. Lipman, M.S. Torn, E. Masanet, A. Lobscheid, A. Santero, U. Mishra, A. Barrett, M. Bomberg, K. Fingerman, C. Scown, B. Strogen, and A. Horvath. 2011. Grand challenges for life-cycle assessment of biofuels. Environmental Science and Technology 45(5):1751-1756.

McSwiney, C.P., and G.P. Roberstson. 2005. Nonlinear response of N_2O flux to incremental fertilizer addition in a continuous maize (*Zea mays* L.) cropping system. Global Change Biology 11:1712-1719.

Meehan, T.D., A.H. Hurlbert, and C. Gratton. 2010. Bird communities in future bioenergy landscapes of the Upper Midwest. Proceedings of the National Academy of Sciences of the United States of America 107(43):18533-18538.

Mehta, R.N., M. Chakraborty, P. Mahanta, and P.A. Parikh. 2010. Evaluation of fuel properties of butanol-biodiesel-diesel blends and their impact on engine performance and emissions. Industrial and Engineering Chemistry Research 49(16):7660-7665.

Melamu, R., and H. von Blottnitz. 2011. 2nd generation biofuels a sure bet? A life cycle assessment of how things could go wrong. Journal of Cleaner Production 19(2-3):138-144.

Melillo, J.M., J.M. Reilly, D.W. Kicklighter, A.C. Gurgel, T.W. Cronin, S. Paltsev, B.S. Felzer, X.D. Wang, A.P. Sokolov, and C.A. Schlosser. 2009. Indirect emissions from biofuels: How important? Science 326(5958):1397-1399.

Meyfroidt, P., T.K. Kudel, and E.F. Lambin. 2010. Forest transitions, trade, and the global displacement of land use. Proceedings of the National Academy of Sciences of the United States of America 107(49):20917-20922.

Millar, N., G.P. Robertson, P.R. Grace, R.J. Gehl, and J.P. Hoben. 2010. Nitrogen fertilizer management for nitrous oxide (N_2O) mitigation in intensive corn (Maize) production: An emissions reduction protocol for US Midwest agriculture. Mitigation and Adaptation Strategies for Global Change 15(2):185-204.

Millennium Ecosystem Assessment. 2005. Ecosystems and Human Well Being: Synthesis. Washington, DC: World Resources Institute.

Miller, S.A. 2010. Minimizing land use and nitrogen intensity of bioenergy. Environmental Science and Technology 44(10):3932-3939.

Monsanto. 2008. Monsanto will undertake three-point commitment to double yield in three major crops, make more efficient use of natural resources and improve farmer lives. Available online at http://monsanto.mediaroom.com/index.php?s=43&item=607. Accessed April 13, 2011.

Morais, S., A.A. Martins, and T.M. Mata. 2010. Comparison of allocation approaches in soybean biodiesel life cycle assessment. Journal of the Energy Institute 83:48-55.

Morgan, J.A., R.F. Follett, L.H. Allen, S. del Grosso, J.D. Derner, F. Dijkstra, A. Franzluebbers, R. Fry, K. Paustian, and M.M. Schoeneberger. 2010. Carbon sequestration in agricultural lands of the United States. Journal of Soil and Water Conservation 65(2):6A-13A.

Mubako, S., and C. Lant. 2008. Water resource requirements of corn-based ethanol. Water Resources Research 44(7):W00A02.

Mueller, S. 2010. Detailed Report: 2008 National Dry Mill Corn Ethanol Survey. Chicago: University of Illinois.

Mullins, K.A., W.M. Griffin, and H.S. Matthews. 2010. Policy implications of uncertainty in modeled life-cycle greenhouse gas emissions of biofuels. Environmental Science and Technology 45(1):132-138.

Murray, B.C., B.L. Sohngen, A.J. Sommer, B.M. Depro, K.M. Jones, B.A. McCarl, D. Gillig, B. DeAngelo, and K. Andrasko. 2005. Greenhouse Gas Mitigation Potential in U.S. Forestry and Agriculture. Washington, DC: U.S. Environmental Protection Agency.

Murray, B.C., B. Sohngen, and M.T. Ross. 2007. Economic consequences of consideration of permanence, leakage and additionality for soil carbon sequestration projects. Climatic Change 80(1-2):127-143.

Murray, L., and L. Best. 2010. Bird use of switchgrass fields harvested for bioenergy in the Midwestern U.S. Paper read at the Reshaping Landscapes: Bioenergy and Biodiversity, April 8, Athens, GA.

NAS-NAE-NRC (National Academy of Sciences, National Academy of Engineering, National Research Council). 2009. Liquid Transportation Fuels from Coal and Biomass: Technological Status, Costs, and Environmental Impacts. Washington, DC: National Academies Press.

NCGA (National Corn Growers Association). 2010. Corn production trends. Available online at http://www.ncga. com/corn-production-trends. Accessed September 10, 2010.

Nebraska Corn Board. 2011. Advantage: Nebraska. Available online at http://www.nebraskacorn.org/main-navigation/grain-traders/advantage-nebraska/. Accessed February 11, 2011.

Nelson, N.O., S.C. Agudelo, W.Q. Yuan, and J. Gan. 2011. Nitrogen and phosphorus availability in biochar-amended soils. Soil Science 176(5):218-226.

Nelson, R., M. Langemeier, J. Williams, C. Rice, S. Staggenborg, P. Pfomm, D. Rodgers, D. Wang, and J. Nippert. 2010. Kansas Biomass Resource Assessment: Assessment and Supply of Select Biomass-based Resources. Manhattan: Kansas State University.

Nelson, R.G., J.C. Ascough II, and M.R. Langemeier. 2006. Environmental and economic analysis of switchgrass production for water quality improvement in northeast Kansas. Journal of Environmental Management 79(4):336-347.

Newman, J.K., A.L. Kaleita, and J.M. Laflen. 2010. Soil erosion hazard maps for corn stover management using national resources inventory data and the water erosion prediction project. Journal of Soil and Water Conservation 65(4):211-222.

Ng, T.L., J.W. Eheart, X.M. Cai, and F. Miguez. 2010. Modeling *Miscanthus* in the Soil and Water Assessment Tool (SWAT) to simulate its water quality effects as a bioenergy crop. Environmental Science and Technology 44(18):7138-7144.

Niemi, G.J., and M.E. McDonald. 2004. Application of ecological indicators. Annual Review of Ecology Evolution and Systematics 35:89-111.

Nolan, B.T., K.J. Hitt, and B.C. Ruddy. 2002. Probability of nitrate contamination of recently recharged groundwaters in the conterminous United States. Environmental Science and Technology 36(10):2138-2145.

NRC (National Research Council). 1999. Ozone-Forming Potential of Reformulated Gasoline. Washington, DC: National Academy Press.

NRC (National Research Council). 2003. Cumulative Environmental Effects of Oil and Gas Activities on Alaska's North Slope. Washington, DC: National Academies Press.

NRC (National Research Council). 2008. Water Implications of Biofuels Production in the United States; Committee on Water Implications of Biofuels Production in the United States, National Research Council. Washington, DC: National Academies Press.

NRC (National Research Council). 2010a. Hidden Costs of Energy: Unpriced Consequences of Energy Production and Use. Washington, DC: National Academies Press.

NRC (National Research Council). 2010b. Toward Sustainable Agricultural Systems in the 21st Century. Washington, DC: National Academies Press.

NRC (National Research Council). 2010c. Verifying Greenhouse Gas Emissions: Methods to Support International Climate Agreements. Washington, DC: National Academies Press.

O'Hare, M., R.J. Plevin, J.I. Martin, A.D. Jones, A. Kendall, and E. Hopson. 2009. Proper accounting for time increases crop-based biofuels' greenhouse gas deficit versus petroleum. Environmental Research Letters 4(2009):024001.

O'Hare, M., M. Delucchi, R. Edwards, U. Fritsche, H. Gibbs, T. Hertel, J. Hill, D. Kammen, L. Marelli, D. Mulligan, R. Plevin, and W. Tyner. 2011. Comment on "Indirect land use change for biofuels: Testing predictions and improving analytical methodologies" by Kim and Dale: Statistical reliability and the definition of the indirect land use change (iLUC) issue. Biomass and Bioenergy 35(10):4485-4487.

Odum, E.P. 1969. The strategy of ecosystem development. Science 164(3877):262-270.

Ogle, S.M., F.J. Breidt, M. Easter, S. Williams, and K. Paustian. 2007. An empirically based approach for estimating uncertainty associated with modelling carbon sequestration in soils. Ecological Modelling 205(3-4):453-463.

Ohlrogge, J., D. Allen, B. Berguson, D. DellaPenna, Y. Shachar-Hill, and S. Stymne. 2009. Driving on biomass. Science 324(5930):1019-1020.

Omonode, R.A., D.R. Smith, A. Gál, and T.J. Vyn. 2011. Soil nitrous oxide emissions in corn following three decades of tillage and rotation treatments. Soil Science Society of America Journal 75(1):152-163.

Osterman, L.E., R.Z. Poore, P.W. Swarzenski, and R.E. Turner. 2005. Reconstructing a 180 yr record of natural and anthropogenic induced hypoxia from the sediments of the Louisiana continental shelf. Geology 33(4):329-332.

Overmars, K.P., E. Stehfest, J.P.M. Ros, and A.G. Prins. 2011. Indirect land use change emissions related to EU biofuel consumption: An analysis based on historical data. Environmental Science and Policy 14(3):248-257.

Pang, S.-H., H.C. Frey, and W.J. Rasdorf. 2009. Life cycle inventory energy consumption and emissions for biodiesel versus petroleum diesel fueled construction vehicles. Environmental Science and Technology 43(16):6398-6405.

Parker, N., Q. Hart, P. Tittman, M. Murphy, R. Nelson, K. Skog, E. Gray, A. Schmidt, and B. Jenkins. 2010. National Biorefinery Siting Model: Spatial Analysis and Supply Curve Development. Washington, DC: Western Governors Association.

Parkin, G., P. Weyer, and C. Just. 2007. Riding the bioeconomy wave: Smooth sailing or rough water for the environment and public health. Paper read at the Iowa Water Conference - Water and Bioenergy, March 6, Ames, IA.

Parrish, D.J., and J.H. Fike. 2005. The biology and agronomy of switchgrass for biofuels. Critical Reviews in Plant Sciences 24:423-459.

Pate, R., M. Hightower, C. Cameron, and W. Einfeld. 2007. Overview of Energy-Water Interdependencies and the Emerging Energy Demands on Water Resources. Los Alamos, NM: Sandia National Laboratories.

Paustian, K., E.T. Elliott, and K. Killian. 1997. Modeling soil carbon in relation to management and climate change in some agroecosystems in central North America. Pp. 459-471 in Soil Processes and the Carbon Cycle, R. Lal, J.M. Kimble, R.F. Follett, and B.A. Stewart, eds. Boca Raton, FL: CRC Press.

Perlack, R.D., L.L. Wright, A.F. Turhollow, R.L. Graham, B.J. Stokes, and D.C. Erbach. 2005. Biomass as Feedstock for a Bioenergy and Bioproducts Industry: The Technical Feasibility of a Billion-Ton Annual Supply. Oak Ridge, TN: U.S. Department of Energy, U.S. Department of Agriculture.

Persson, T., A.G.Y. Garcia, J.O. Paz, C.W. Fraisse, and G. Hoogenboom. 2010. Reduction in greenhouse gas emissions due to the use of bio-ethanol from wheat grain and straw produced in the south-eastern USA. The Journal of Agricultural Science 148(05):511-527.

Phillips, S., A. Aden, J. Jechura, D. Dayton, and T. Eggeman. 2007. Thermochemical Ethanol via Indirect Gasification and Mixed Alcohol Synthesis of Lignocellulosic Biomass. Golden, CO: National Renewable Energy Laboratory.

Plevin, R. 2009a. Comment on "Response to Plevin: Implications for life cycle emissions regulations." Journal of Industrial Ecology 13(6):992-993.

Plevin, R.J. 2009b. Modeling corn ethanol and climate. Journal of Industrial Ecology 13(4):495-507.

Plevin, R.J., M. O'Hare, A.D. Jones, M.S. Torn, and H.K. Gibbs. 2010. Greenhouse gas emissions from biofuels' indirect land use change are uncertain but may be much greater than previously estimated. Environmental Science and Technology 44(21):8015-8021.

Powers, S.E., R. Dominguez-Faus, and P.J.J. Alvarez. 2010. The water footprint of biofuel production in the USA. Biofuels 1(2):255-260.

Pradhan, A., D.S. Shrestha, J. Van Gerpen, and J. Duffield. 2008. The energy balance of soybean oil biodiesel production: A review of past studies. Transactions of the American Society of Agricultural and Biological Engineers 51(1):185-194.

Quinn, L.D., D.J. Allen, and J.R. Stewart. 2010. Invasiveness potential of *Miscanthus sinensis*: Implications for bioenergy production in the United States. Global Change Biology Bioenergy 2(6):310-320.

Rabalais, N.N., and R.E. Turner. 2001. Coastal hypoxia: Consequences for living resources and ecosystems. Pp. 1-36 in Coastal and Estuarine Studies, N.N. Rabalais and R.E. Turner, eds. Washington, DC: American Geophysical Union.

Rabalais, N.N., R.E. Turner, and W.J. Wiseman. 2002. Gulf of Mexico hypoxia, aka "The dead zone." Annual Review of Ecology and Systematics 33:235-263.

Rabotyagov, S., T. Campbell, M. Jha, P.W. Gassman, J. Arnold, L. Kurkalova, S. Secchi, H.L. Feng, and C.L. Kling. 2010. Least-cost control of agricultural nutrient contributions to the Gulf of Mexico hypoxic zone. Ecological Applications 20(6):1542-1555.

Raghu, S., R.C. Anderson, C.C. Daehler, A.S. Davis, R.N. Wiedenmann, D. Simberloff, and R.N. Mack. 2006. Adding biofuels to the invasive species fire? Science 313(5794):1742.

Randall, J.M., L.E. Morse, N. Benton, R. Hiebert, S. Lu, and T. Killeffer. 2008. The invasive species assessment protocol: A tool for creating regional and national lists of invasive nonnative plants that negatively impact biodiversity. Invasive Plant Science and Management 1(1):36-49.

Rausch, K.D., and R.L. Belyea. 2006. The future of coproducts from corn processing. Applied Biochemistry and Biotechnology 128(1):47-86.

Ravindranath, N.H., R. Mauvie, J. Fargione, J.G. Canadell, G. Berndes, J. Woods, H. Watson, and J. Sathaye. 2009. Greenhouse gas implications of land use change and land conversion to biofuel crops. Pp. 111-125 in Biofuels: Environmental Consequences and Interactions with Changing Land Use, R.W. Howarth and S. Bringezu, eds. Ithaca, NY: Cornell University.

Reitz, S.R., G.S. Kund, W.G. Carson, P.A. Phillips, and J.T. Trumble. 1999. Economics of reducing insecticide use on celery through low-input pest management strategies. Agriculture Ecosystems and Environment 73(3):185-197.

RFA (Renewable Fuels Association). 2011. Biorefinery locations. Available online at http://www.ethanolrfa.org/bio-refinery-locations/. Accessed January 26, 2011.

Riffell, S., J. Verschuyl, D. Miller, and T.B. Wigley. 2010. Reshaping landscapes: Bioenergy and biodiversity. Paper read at Reshaping Landscapes: Bioenergy and Biodiversity, April 8, Athens, GA.

Robert, P.C. 2002. Precision agriculture: A challenge for crop nutrition management. Plant and Soil 247(1):143-149.

Roberts, K.G., B.A. Gloy, S. Joseph, N.R. Scott, and J. Lehmann. 2010. Life cycle assessment of biochar systems: Estimating the energetic, economic, and climate change potential. Environmental Science and Technology 44(2):827-833.

Robertson, B.A., P.J. Doran, L.R. Loomis, J.R. Robertson, and D.W. Schemske. 2011a. Perennial biomass feedstocks enhance avian diversity. GCB Bioenergy 3(3):235-246.

Robertson, G.P., V.H. Dale, O.C. Doering, S.P. Hamburg, J.M. Melillo, M.M. Wander, W.J. Parton, P.R. Adler, J.N. Barney, R.M. Cruse, C.S. Duke, P.M. Fearnside, R.F. Follett, H.K. Gibbs, J. Goldemberg, D.J. Mladenoff, D. Ojima, M.W. Palmer, A. Sharpley, L. Wallace, K.C. Weathers, J.A. Wiens, and W.W. Wilhelm. 2008. Sustainable biofuels redux. Science 322(5898):49-50.

Robertson, G.P., S.K. Hamilton, W.J. Parton, and S.J.D. Grosso. 2011b. The biogeochemistry of bioenergy landscapes: Carbon, nitrogen, and water considerations. Ecological Applications 21(4):1055-1067.

Robins, J.G. 2010. Cool-season grasses produce more total biomass across the growing season than do warm-season grasses when managed with an applied irrigation gradient. Biomass and Bioenergy 34(4):500-505.

Roth, A.M., D.W. Sample, C.A. Ribic, L. Paine, D.J. Undersander, and G.A. Bartelt. 2005. Grassland bird response to harvesting switchgrass as a biomass energy crop. Biomass & Bioenergy 28(5):490-498.

RSB (Roundtable on Sustainable Biofuels). 2010. Principles and Criteria. Available online at http://rsb.epfl.ch/page-24929.html. Accessed November 15, 2010.

Sage, R.F., H.A. Coiner, D.A. Way, G. Brett Runion, S.A. Prior, H. Allen Torbert, R. Sicher, and L. Ziska. 2009. Kudzu [*Pueraria montana* (Lour.) Merr. Variety *lobata*]: A new source of carbohydrate for bioethanol production. Biomass and Bioenergy 33(1):57-61.

Sahu, M., and R.R. Gu. 2009. Modeling the effects of riparian buffer zone and contour strips on stream water quality. Ecological Engineering 35(8):1167-1177.

Scavia, D., and Y. Liu. 2009. Gulf of Mexico hypoxia forecast and measurement. Available online at http://www.snre.umich.edu/scavia/wp-content/uploads/2009/11/2009_gulf_of_mexico_hypoxic_forecast_and_observation.pdf. Accessed August 17, 2010.

Scavia, D., N.N. Rabalais, R.E. Turner, D. Justic, and W.J. Wiseman, Jr. 2003. Predicting the response of Gulf of Mexico hypoxia to variations in the Mississippi River nitrogen load. Limnology and Oceanography 48:951-956.

Schill, S.R. 2007. 300-bushel corn is coming. Available online at http://www.ethanolproducer.com/articles/3330/300-bushel-corn-is-coming/. Accessed April 13, 2011.

Schilling, E. 2009. Compendium of Forestry Best Management Practices for Controlling Nonpoint Source Pollution in North America. Research Triangle Park, NC: National Council for Air and Stream Improvement.

Schilling, K.E., and J. Spooner. 2006. Effects of watershed-scale land use change on stream nitrate concentrations. Journal of Environmental Quality 35(6):2132-2145.

Schilling, K.E., and C.F. Wolter. 2008. Water Quality Improvement Plan for Raccoon River, Iowa: Total Maximum Daily Load for Nitrate and *Escherichia coli*. Iowa City: Iowa Department of Natural Resources - Geological Survey.

Schmit, T.M., R.N. Boisvert, D. Enahoro, and L.E. Chase. 2009. Optimal dairy farm adjustments to increased utilization of corn distillers dried grains with solubles. Journal of Dairy Science 92(12):6105-6115.

Schnepf, R. 2004. Energy Use in Agriculture: Background and Issues. Washington, DC: Congressional Research Service.

Schweizer, P., H. Jager, and L. Baskaran. 2010. Forecasting changes in water quality and aquatic biodiversity in response to future bioenergy landscapes in the Arkansas-White-Red River basin. Paper read at Reshaping Landscapes: Bioenergy and Biodiversity, April 8, Athens, GA.

Searchinger, T., R. Heimlich, R.A. Houghton, F. Dong, A. Elobeid, J. Fabiosa, S. Tokgoz, D. Hayes, and T.-H. Yu. 2008. Use of U.S. croplands for biofuels increases greenhouse gases through emissions from land-use change. Science 319(5867):1238-1240.

Searchinger, T.D. 2010. Biofuels and the need for additional carbon. Environmental Research Letters 5(2).

Secchi, S., P. Gassman, M. Jha, L. Kurkalova, and C. Kling. 2011. Potential water quality changes due to corn expansion in the Upper Mississippi River Basin. Ecological Applications 21:1068-1084.

SFI (Sustainable Forestry Initiative). 2010. Emerging themes: Bioenergy and carbon. Available online at http://www.sfiprogram.org/forest-conservation/bioenergy-carbon.php. Accessed November 18, 2010.

Shi, A.Z., L.P. Koh, and H.T.W. Tan. 2009. The biofuel potential of municipal solid waste. GCB Bioenergy 1(5):317-320.

Siemens, M.C., and D.E. Wilkins. 2006. Effect of residue management methods on no-till drill performance. Applied Engineering in Agriculture 22(1):51-60.

Simpson, T.W., A.N. Sharpley, R.W. Howarth, H.W. Paerl, and K.R. Mankin. 2008. The new gold rush: Fueling ethanol production while protecting water quality. Journal of Environmental Quality 37(2):318-324.

Singh, A., D. Pant, N.E. Korres, A.-S. Nizami, S. Prasad, and J.D. Murphy. 2010. Key issues in life cycle assessment of ethanol production from lignocellulosic biomass: Challenges and perspectives. Bioresource Technology 101(13):5003-5012.

Snyder, C.S., T.W. Bruulsema, T.L. Jensen, and P.E. Fixen. 2009. Review of greenhouse gas emissions from crop production systems and fertilizer management effects. Agriculture Ecosystems and Environment 133(3-4):247-266.

Sohngen, B., and S. Brown. 2004. Measuring leakage from carbon projects in open economies: a stop timber harvesting project in Bolivia as a case study. Canadian Journal of Forest Research-Revue Canadienne De Recherche Forestiere 34(4):829-839.

Sohngen, B., and R. Sedjo. 2000. Potential carbon flux from timber harvests and management in the context of a global timber market. Climatic Change 44(1-2):151-172.

Solomon, B.D. 2009. Regional economic impacts of cellulosic ethanol development in the North Central states. Pp. 281-298 in Renewable Energy from Forest Resources in the United States, B.D. Solomon and V.A. Luzadis, eds. New York: Routledge.

Solomon, B.D., J.R. Barnes, and K.E. Halvorsen. 2009. From grain to cellulosic ethanol: History, economics and policy. Pp. 49-66 in Renewable Energy from Forest Resources in the United States, B.D. Solomon and V.A. Luzadis, eds. New York: Routledge.

Spatari, S., and H.L. MacLean. 2010. Characterizing model uncertainties in the life cycle of lignocellulose-based ethanol fuels. Environmental Science and Technology 44(22):8773-8780.

Spatari, S., D.M. Bagley, and H.L. MacLean. 2010. Life cycle evaluation of emerging lignocellulosic ethanol conversion technologies. Bioresource Technology 101(2):654-667.

Sprague, L.A., R.M. Hirsch, and B.T. Aulenbach. 2011. Nitrate in the Mississippi River and its tributaries, 1980 to 2008: Are we making progress? Environmental Science and Technology. Available online at http://pubs.acs.org/doi/abs/10.1021/es201221s. Accessed August 31, 2011.

Stavins, R.N., and A.B. Jaffe. 1990. Unintended impacts of public-investments on private decisions—The depletion of forested wetlands. American Economic Review 80(3):337-352.

Stone, K.C., P.G. Hunt, K.B. Cantrell, and K.S. Ro. 2010. The potential impacts of biomass feedstock production on water resource availability. Bioresource Technology 101(6):2014-2025.

Tilman, D., J. Hill, and C. Lehman. 2006. Carbon-negative biofuels from low-input high-diversity grassland biomass. Science 314(5805):1598-1600.

Tolbert, V. 1998. Special issue - Environmental effects of biomass crop production. What do we know? What do we need to know? Guest editorial. Biomass and Bioenergy 14(4):301-306.

Tolbert, V.R., and L.L. Wright. 1998. Environmental enhancement of US biomass crop technologies: Research results to date. Biomass and Bioenergy 15(1):93-100.

Traviss, N., B.A. Thelen, J.K. Ingalls, and M.D. Treadwell. 2010. Biodiesel versus diesel: A pilot study comparing exhaust exposures for employees at a rural municipal facility. Journal of the Air and Waste Management Association 60(9):1026-1033.

Trumble, J.T., W.G. Carson, and G.S. Kund. 1997. Economics and environmental impact of a sustainable integrated pest management program in celery. Journal of Economic Entomology 90(1):139-146.

Turner, B.L., E.F. Lambin, and A. Reenberg. 2007. The emergence of land change science for global environmental change and sustainability. Proceedings of the National Academy of Sciences of the United States of America 104(52):20666-20671.

Turner, R.E., N.N. Rabalais, and D. Justic. 2006. Predicting summer hypoxia in the northern Gulf of Mexico: Riverine N, P, and Si loading. Marine Pollution Bulletin 52(2):139-148.

Tyner, W.E., F. Taheripour, Q. Zhuang, D. Birur, and U. Baldos. 2010. Land Use Changes and Consequent CO_2 Emissions due to US Corn Ethanol Production: A Comprehensive Analysis. West Lafayette, IN: Purdue University.

Uherek, E., T. Halenka, J. Borken-Kleefeld, Y. Balkanski, T. Berntsen, C. Borrego, M. Gauss, P. Hoor, K. Juda-Rezler, J. Lelieveld, D. Melas, K. Rypdal, and S. Schmid. 2010. Transport impacts on atmosphere and climate: Land transport. Atmospheric Environment 44(37):4772-4816.

UNL (University of Nebraska, Lincoln). 2007. Nebraska water map/poster. Available online at http://water.unl.edu/watermap/watermap. Accessed June 11, 2010.

UNL (University of Nebraska, Lincoln). 2009. Groundwater-level changes in Nebraska. Available online at http://snr.unl.edu/data/water/groundwatermaps.asp. Accessed February 10, 2011.

U.S. Congress Office of Technology Assessment. 1987. Technologies to Maintain Biological Diversity. Washington, DC: U.S. Government Printing Office.

USDA-ERS (U.S. Department of Agriculture - Economic Research Service). 2004. Irrigation and Water Use. Available online at http://www.ers.usda.gov/Briefing/WaterUse/. Accessed September 20, 2010.

USDA-ERS (U.S. Department of Agriculture - Economic Research Service). 2010a. Feed grains database: Custom queries. Available online at http://www.ers.usda.gov/Data/FeedGrains/CustomQuery/. Accessed June 24, 2010.

USDA-ERS (U.S. Department of Agriculture - Economic Research Service). 2010b. Feed grains database: Yearbook tables. Available online at http://www.ers.usda.gov/data/FeedGrains/FeedYearbook.aspx. Accessed September 9, 2010.

USDA-ERS (U.S. Department of Agriculture - Economic Research Service). 2010c. Fertilizer use and price. Available online at http://www.ers.usda.gov/Data/FertilizerUse/. Accessed October 26, 2010.

USDA-NASS (U.S. Department of Agriculture - National Agricultural Statistics Service). 2010. Data and statistics: Quick stats. Available online at http://www.nass.usda.gov/Data_and_Statistics/Quick_Stats/index.asp. Accessed June 24, 2010.

USDA-NASS (U.S. Department of Agriculture - National Agricultural Statistics Service). 2009. Irrigated corn for grain, harvested acres: 2007. Available online at http://www.agcensus.usda.gov/Publications/2007/Online_Highlights/Ag_Atlas_Maps/Crops_and_Plants/index.asp. Accessed on June 24, 2010.

USDA-NASS (U.S. Department of Agriculture - National Agricultural Statistics Service). 2011. Prospective Plantings. Washington, DC: U.S. Department of Agriculture.

USDA-NIFA (U.S. Department of Agriculture - National Institute of Food and Agriculture). 2009. Precision, geospatial and sensor technologies: Adoption of precision agriculture. Available online at http://www.csrees.usda.gov/nea/ag_systems/in_focus/precision_if_adoption.html. Accessed November 19, 2010.

USDA-NRCS (U.S. Department of Agriculture - Natural Resources Conservation Service). 2009. Planting and Managing Switchgrass as a Biomass Energy Crop. Washington, DC: U.S. Department of Agriculture.

USFS (U.S. Department of Agriculture - Forest Service). 2001. U.S. Forest Facts and Historical Trends. Washington, DC: U.S. Department of Agriculture.

USGS (U.S. Geological Survey). 2011. WaterWatch. Available online at http://waterwatch.usgs.gov/new/. Accessed February 10, 2011.

van Dam, J., M. Junginger, A. Faaij, I. Jürgens, G. Best, and U. Fritsche. 2008. Overview of recent developments in sustainable biomass certification. Biomass and Bioenergy 32(8):749-780.

van der Voet, E., R.J. Lifset, and L. Luo. 2010. Life-cycle assessment of biofuels, convergence and divergence. Biofuels 1(3):435-449.

Van Groenigen, J.W., G.L. Velthof, O. Oenema, K.J. Van Groenigen, and C. Van Kessel. 2010. Towards an agronomic assessment of N(2)O emissions: A case study for arable crops. European Journal of Soil Science 61(6):903-913.

VanLoocke, A., C.J. Bernache, and T.E. Twine. 2010. The impacts of *Miscanthus* × *giganteus* production on the Midwest US hydrologic cycle. GCB Bioenergy 2:180-191.

Vaquer-Sunyer, R., and C.M. Duarte. 2008. Thresholds of hypoxia for marine biodiversity. Proceedings of the National Academy of Sciences of the United States of America 105(40):15452-15457.

Volk, T.A., and V.A. Luzadis. 2009. Willow biomass production for bioenergy, biofuels and bioproducts in New York. Pp. 238-260 in Renewable Energy from Forest Resources in the United States, B.D. Solomon and V.A. Luzadis, eds. New York: Routledge.

Wang, M.Q., J. Han, Z. Haq, W.E. Tyner, M. Wu, and A. Elgowainy. 2011a. Energy and greenhouse gas emission effects of corn and cellulosic ethanol with technology improvements and land use changes. Biomass and Bioenergy 35(5):1885-1896.

Wang, M., H. Huo, and S. Arora. 2011b. Methods of dealing with co-products of biofuels in life-cycle analysis and consequent results within the U.S. context. Energy Policy 39(10):5726-5736.

Wang, M., M. Wu, and H. Huo. 2007. Life-cycle energy and greenhouse gas emission impacts of different corn ethanol plant types. Environmental Research Letters 2(2):024001.

Wear, D.N., and B.C. Murray. 2004. Federal timber restrictions, interregional spillovers, and the impact on US softwood markets. Journal of Environmental Economics and Management 47(2):307-330.

Webster, C.R., D.J. Flaspohler, R.D. Jackson, T.D. Meehan, and C. Gratton. 2010. Diversity, productivity and landscape-level effects in North American grasslands managed for biomass production. Biofuels 1(3):451-461.

Weidema, B. 2003. Market information in life cycle assessment. Copenhagen: Danish Environmental Protection Agency.

West, T.O., C.C. Brandt, L.M. Baskaran, C.M. Hellwinckel, R. Mueller, C.J. Bernacchi, V. Bandaru, B. Yang, B.S. Wilson, G. Marland, R.G. Nelson, D.G.D. Ugarte, and W.M. Post. 2010. Cropland carbon fluxes in the United States: Increasing geospatial resolution of inventory-based carbon accounting. Ecological Applications 20(4):1074-1086.

Wiegmann, K., K.R. Hennenberg, and U.R. Fritsche. 2008. Degraded Land and Sustainable Bioenergy Feedstock Production. Darmstadt, Germany: Öko-Institut.

Wilhelm, W.W., J.M.E. Johnson, D.L. Karlen, and D.T. Lightle. 2007. Corn stover to sustain soil organic carbon further constrains biomass supply. Agronomy Journal 99(6):1665-1667.

Wilkie, A.C., K.J. Riedesel, and J.M. Owens. 2000. Stillage characterization and anaerobic treatment of ethanol stillage from conventional and cellulosic feedstocks. Biomass and Bioenergy 19(2):63-102.

Williams, P.R.D., D. Inman, A. Aden, and G.A. Heath. 2009. Environmental and sustainability factors associated with next-generation biofuels in the U.S.: What do we really know? Environmental Science and Technology 43(13):4763-4775.

Wise, M., K. Calvin, A. Thomson, L. Clarke, B. Bond-Lamberty, R. Sands, S.J. Smith, A. Janetos, and J. Edmonds. 2009. Implications of limiting CO_2 concentrations for land use and energy. Science 324(5931):1183-1186.

Woodbury, P.B., L.S. Heath, and J.E. Smith. 2006. Land use change effects on forest carbon cycling throughout the southern United States. Journal of Environmental Quality 35(4):1348-1363.

Wright, L., and A. Turhollow. 2010. Switchgrass selection as a "model" bioenergy crop: A history of the process. Biomass and Bioenergy 34(6):851-868.

Wu, J.J. 2000. Slippage effects of the conservation reserve program. American Journal of Agricultural Economics 82(4):979-992.

Wu, M., M. Mintz, M. Wang, and S. Arora. 2009a. Consumptive Water Use in the Production of Ethanol and Petroleum Gasoline. Argonne, IL: Argonne National Laboratory.

Wu, M., M. Mintz, M. Wang, and S. Arora. 2009b. Water consumption in the production of ethanol and petroleum gasoline. Environmental Management 44(5):981-997.

Wu, M., Y. Wu, and M. Wang. 2006. Energy and emission benefits of alternative transportation liquid fuels derived from switchgrass: A fuel life cycle assessment. Biotechnology Progress 22(4):1012-1024.

Wu, X., S. Staggenborg, J.L. Propheter, W.L. Rooney, J. Yu, and D. Wang. 2010. Features of sweet sorghum juice and their performance in ethanol fermentation. Industrial Crops and Products 31(1):164-170.

Wu-Haan, W., W. Powers, R. Angel, and T.J. Applegate. 2010. The use of distillers dried grains plus solubles as a feed ingredient on air emissions and performance from laying hens. Poultry Science 89(7):1355-1359.

Yanowitz, J., and R.L. McCormick. 2009. Effect of E85 on tailpipe emissions from light-duty vehicles. Journal of the Air and Waste Management Association 59(2):172-182.

Yeh, S., S.M. Jordaan, A.R. Brandt, M.R. Turetsky, S. Spatari, and D.W. Keith. 2010. Land use greenhouse gas emissions from conventional oil production and oil sands. Environmental Science and Technology 44(22):8766-8772.

Zhang, X., R.C. Izaurralde, D. Manowitz, T.O. West, W.M. Post, A.M. Thomson, V.P. Bandaruw, J. Nichols, and J.R. Williams. 2010. An integrative modeling framework to evaluate the productivity and sustainability of biofuel crop production systems. Global Change Biology Bioenergy 2(5):258-277.

Zilberman, D., G. Hochman, and D. Rajagopal. 2010. On the inclusion of direct land use in biofuel regulations. University of Illinois Law Review 2011:413-434.

Zimmerman, A.R., B. Gao, and M.Y. Ahn. 2011. Positive and negative carbon mineralization priming effects among a variety of biochar-amended soils. Soil Biology and Biochemistry 43(6):1169-1179.

6

Barriers to Achieving RFS2

The Renewable Fuel Standard as amended under the Energy Security and Independence Act of 2007 (RFS2) mandates that 35 billion gallons of ethanol-equivalent biofuels—15 billion gallons of conventional biofuels, 4 billion gallons of advanced biofuels, and 16 billion gallons of cellulosic biofuels—and 1 billion gallons of biomass-based diesel be consumed in the United States by 2022. As noted in Chapter 2, the United States has the capacity to produce 14.1 billion gallons per year of corn-grain ethanol that can be counted toward conventional biofuel consumption and 2.7 billion gallons per year of biodiesel that can be counted toward biomass-based diesel consumption. Therefore, the committee judges that consumption mandates of those two categories of biofuels will likely be met by 2022. However, cellulosic biofuel is a developing industry, and some formidable barriers could prevent the production and consumption of the combined 20 billion gallons of advanced biofuel[1] and cellulosic biofuel in 2022. Those challenges include producing biomass feedstock and converting it to transportation fuels economically, mitigating environmental effects to meet regulations, social barriers, and constraints to blending ethanol into the fuel supply.

Chapters 4 and 5 describe the potential economic and environmental consequences to inform policymakers of the tradeoffs of meeting RFS2. This chapter discusses barriers to achieving the consumption mandates for advanced and cellulosic biofuels in RFS2. The chapter is organized on the basis of four types of barriers: economic, environmental, policy, and social barriers to achieving RFS2. Economic barriers are ones that maintain the unsubsidized price of biofuels above the price of gasoline. Environmental barriers can be resource limitations or practices or environmental discharges that violate environmental regulations. There could be technical or policy solutions to economic and environmental barriers. For example, a subsidy for biofuels is a policy solution to overcome the economic barrier. Technological improvements that reduce biomass feedstock costs or enhance conversion efficiencies of biomass to fuels reduce cost of production. Better technology also can reduce the environmental effects of

---

[1] The advanced biofuel consumption mandate can be met by biofuels made from cellulosic feedstock, as long as the life-cycle GHG emissions of the fuel product is at least 50 percent lower than that of petroleum-based fuels.

each unit of biofuel produced. Technical barriers often require technical solutions to resolve; most technical challenges are addressed in research and development and demonstrated in pilot plants. However, if the solutions to technical and environmental barriers are too costly, there is still an economic barrier. Policies that could stifle the development of the cellulosic biofuels industry present barriers to achieving RFS2. Social barriers involve potential producers' and consumers' perception of, attitude toward, and acceptance of biofuels.

Should policymakers continue to believe that the consumption mandate of RFS2 is to be met in 2022, the barriers described in this chapter would have to be resolved. Unless these barriers are overcome, the committee concludes that the RFS2 mandate is unlikely to be attained by 2022. Removing barriers to the successful establishment of a cellulosic-biofuel industry at the scale mandated by RFS2 involves identifying potential problems at every stage of the biofuel production process and opportunities to resolve them.

The discovery, extraction, manufacture, distribution, and use of petroleum fuels have been developed and improved for over 150 years. Overcoming the barriers and displacing a significant amount of petroleum with biofuels will require time, innovation, and changes in many fundamental economic, technical, and social processes. Doing so in 11 years would be difficult, costly, and complex. Implementing changes in liquid transportation fuel while avoiding socially unacceptable disruptions requires a deep understanding of each affected component and strong commitment to change. "Drop-in" biofuels that can easily be included in the existing petroleum infrastructure are the simplest way of undertaking a fuel transition. The identification of barriers to transitioning to most other biofuels requires combining knowledge of existing infrastructure with an assessment of the desired properties of future biofuels. In some cases, opportunities to achieve economic, environmental, or social benefits are provided by the need to address particular barriers. As of 2011, the key barrier to achieving RFS2 is economic because technologies for producing cellulosic biofuel are available but not economically viable at a commercial scale,[2] even with the current subsidies and mandates, under 2011 oil prices. Biofuels will only be adopted on a commercial scale if their cost to consumers is competitive with other liquid fuels. Moreover, many barriers identified in this chapter have solutions that are technically feasible, but they could increase the economic and environmental costs of biofuel production. If the economic barrier is removed, environmental and social concerns could pose barriers to producing 20 billion gallons of advanced and cellulosic biofuels in 2022. Although barriers to achieving RFS2 are identified in this chapter, the extent to which each barrier inhibits production and market penetration of biofuels is uncertain. Some barriers might not be obvious or will only be discovered when the technologies of cellulosic biofuel are implemented at a commercial scale. Therefore, commercial-scale demonstrations are critical to proving economic and environmental feasibility.

ECONOMIC BARRIERS

Feedstock Costs

RFS2 cannot be attained without sufficient biomass availability at attractive costs. As discussed in Chapter 4, unless the prices that farmers are paid for bioenergy feedstocks delivered to the biorefinery gate reach $75-$133 per dry ton, farmers are not likely to grow or harvest the necessary amounts of bioenergy feedstock. The price a biorefinery pays for its feedstock is likely to be the largest expense in producing biofuels (Chapter 4). Some reports

---

[2]The National Renewable Energy Laboratory defines a commercial-scale demonstration for biofuel refinery as a facility that has the capacity to process 700 dry tons of feedstock per day.

estimated feedstock costs to be one-third to two-thirds of the cost of corn-grain ethanol and cellulosic biofuel (Foust et al., 2007; NAS-NAE-NRC, 2009; Swanson et al., 2010; Wright et al., 2010). Improving yield; reducing crop loss from drought, pests, and diseases; and reducing costs of harvest, storage, and transportation present opportunities to decrease the costs of bioenergy feedstocks. However, demand from the biofuel market and competition for feedstock with other sectors (for example, bioelectricity generation to meet the state-level Renewable Portfolio Standards) could also drive up feedstock costs and present a barrier to producing biofuels that are cost competitive with petroleum-based fuels (Chapter 3). Furthermore, the exclusion of a large number of forest options by law from the definition of renewable biomass (for example, exclusion of residues from federal forests and from nonplantation forests) poses another limit to biomass supply (Chapter 2).

Storage and Delivery

The year-round operation of biorefineries requires that biomass feedstock produced seasonally be stored until use (Chapter 2). Stored biomass could be susceptible to spoilage so that methods and standards for monitoring feedstock quality would be necessary (DOE-EERE, 2004). Furthermore, the bulky biomass would have to be transported to biorefineries. Cellulosic biomass, regardless of its source, is a high volume, low-density material, and long-distance transport is expensive. Research to address the storage and transport of biomass could improve the economic viability of cellulosic biofuel.

Storage and transportation of biomass feedstock could require additional initial investments and operating costs for producers as well. (See Tables M-5 and M-6 in Appendix M for a list of estimated costs.) Opportunities to address this barrier involve growing multiple feedstocks and having production facilities combine crops grown for feedstock with agricultural, forest, or urban wastes. Several technologies have been suggested as potential solutions to the storage and transportation issues, including torrefaction, liquefaction, and densification (NAS-NAE-NRC, 2009; Sadaka and Negi, 2009; Yan et al., 2009), but implementing those technologies on a commercial scale will require infrastructure and incur additional operational costs, which might not reduce overall production and delivery costs. Studies to solve the transportation and storage problems are under way, including an Idaho National Laboratory analysis of a supply chain of wheat and barley straw to a biorefinery system at a scale of 800,000 dry tons per year (Grant et al., 2006). Corn stover logistics have been analyzed by the National Renewable Energy Laboratory (NREL; Atchison and Hettenhaus, 2003), and Oak Ridge National Laboratory's Bioenergy Feedstock Information Network (ORNL, 2010) provides documents and tools related to logistics for multiple ethanol feedstocks. Those studies provide information on supply logistics with associated costs, identifying opportunities to improve the economic feasibility of harvest, collection, storage, handling, and transport of biomass feedstocks.

Absence of Price Discovery Institutions in Bioenergy Feedstock Markets

The price discovery process is defined as "the process by which buyers and sellers arrive at specific prices and other terms of trade" (Tomek and Robinson, 1983, p. 199).[3] The grain sector relies on a well-developed set of public and private price discovery institutions,

---

[3]"Price discovery" is different from "price determination." Price determination is based on "the theory of pricing and the manner in which economic forces (that is, supply and demand) influence prices under various market structures and lengths of run" (Tomek and Robinson, 1983, p. 213).

but no such institutions exist to facilitate markets in nontraditional feedstock sources. In the absence of such institutions, at least two dimensions of the price discovery process pose potential barriers to the expanded use of biofuels.

First, given the high transportation cost (as discussed earlier) that is likely to exist for such feedstocks, the geographic regions over which such feedstocks are traded is likely to be restricted. A more likely scenario is that a specific geographic region would be served by a few biorefineries. Thus, given the small number of buyers in the region, patterns of spatial price variations are unlikely to communicate price information across regions as in the case of grain markets.

A second dimension of the price-discovery process that could act as a barrier to the expanded use of biofuels is the high cost of information about the quality of feedstock (Hess et al., 2007; Lamb et al., 2007; Anderson et al., 2010). The price of feedstock would be determined, in large part, by the quantity and price of biofuel that can be extracted from that feedstock. To the extent that the quality of feedstock (for example, the yield of biofuel that can be produced from a unit of feedstock) is variable, buyers of feedstock will face uncertainty about the value of that feedstock. If the cost of evaluating feedstock quality is high, this quality uncertainty would, in turn, be reflected in the price that feedstock buyers would offer to feedstock sellers. In particular, buyers faced with a high cost of determining the quality of a good would be expected to discount the price of all such goods, regardless of their high or low quality (Aklerlof, 1970; Stiglitz, 1987, 2002). As a consequence, the expansion of the cellulosic biofuel industry is likely to face a barrier in the high cost of determining the quality of feedstock for biofuels and the high cost of quality information in the price discovery process. However, if the cost of measuring quality is low, a quality incentive program could easily be implemented by biorefineries.

Because of uncertainty about price, quantity, or quality of cellulosic feedstock and the absence of low-cost price-discovery institutions, buyers and sellers will turn to alternative forms of conducting transactions (Williamson, 1996). Market participants are likely to shift to a variety of nonmarket price-discovery mechanisms, such as the use of negotiated contracts, as a means of discovering the information needed by buyers and sellers (Williamson, 1996; Saccomandi, 1998). Such contracts can provide specific expectations regarding price, quantity, quality, delivery timing, or delivery location that are to be met by market participants and the economic incentives or penalties for failing to do so.

Reliance on contracting can provide assurances to buyers and sellers in a new industry such as advanced biofuels. For example, investors in a cellulosic biofuel refinery might be unwilling to invest in new or expanded capacity unless they are certain about the quantity, quality, or price of feedstock available to the refinery. Feedstock producers, on the other hand, would be unlikely to make the investments necessary for feedstock production (for example, specialized harvesting equipment) without the assurance of a long-term buyer commitment. In such cases, contracting can provide the information necessary for the creation and growth of a new industry.

Feedstock Conversion Technologies and Costs

Converting the cellulosic biomass into liquid fuel is the other major cost component. The technical feasibility of conversion has been demonstrated for some time, but lowering the cost to a competitive level is a barrier to achieving RFS2. The first step is finding a technology with high yield, flexibility in terms of feedstocks, and low cost. Both government and companies have sponsored research, development, and demonstration of conversion technologies (see Tables 2-2 and 2-3 in Chapter 2), and there appear to be promising candidates. However, moving a new technology from the laboratory to a commercial operation requires both money and time. Government and private equity can provide the capital, but the time required to commercialize a new technology is difficult to compress. Many potential operational problems and associated costs are only discovered as the process scale is increased. Hettinga et al. (2009) and van den Wall Bake (2009) describe cost reduction and increases in efficiency for corn-grain ethanol in the United States and sugarcane ethanol in Brazil. Technology for corn-grain ethanol for fuel has been developing for over 30 years in the United States. Cost reduction of corn-grain ethanol in the United States was attributed to economies of scale as the ethanol refineries became larger over time, improved ethanol yields, reduced enzyme costs, improved fermentation, better technologies for distillation, dehydration, energy reuse, and automation, and a market developed for coproducts. The cost of conversion at biorefineries decreased by 40-50 percent from the early 1980s to 2005 (Hettinga et al., 2009). Processes for distillation and dehydration are the same for corn-grain and cellulosic ethanol so that cellulosic ethanol (once the sugars have been released) could benefit from the experience gained from corn-grain ethanol. Many economic evaluations for cellulosic ethanol already include cost savings that are assumed to take place as the technology advances (NAS-NAE-NRC, 2009; Tao and Aden, 2009).

Currently, close to half of the commercial companies with secured funding for demonstration of nonfood-based biofuel refineries are planning to use biochemical-based approaches (see Figure 2-18 in Chapter 2) that are roughly analogous to corn-grain ethanol production. However, production of ethanol from corn starch via fermentation is technically simple and efficient compared to production from lignocellulose-rich feedstocks such as herbaceous and woody crops, agricultural and forest residues, and municipal solid waste (MSW) (Table 6-1). Starch is a temporary storage pool for glucose in a plant, and starch is designed to be quickly and easily mobilized by a small number of enzymes. Conversely, cellulose functions as a stable structural component of plant cell walls and is chemically associated with a variety of complex macromolecules, including lignin and hemicellulose, which increase its resistance to physical and biological degradation. Wood consists of about 30-percent lignin, 40-percent cellulose, and 30-percent hemicellulose. Ratios of these macromolecules vary somewhat across the different potential feedstocks (Schnepf, 2010). Ethanol production from cellulosic biomass will not reach the mass efficiency or economic viability of ethanol production from grain unless techniques are developed to break down both cellulose and hemicellulose effectively into sugars (Gírio et al., 2010).

Many herbivores, such as cows, house a complex ecosystem of dozens of species of bacteria and protozoa that efficiently ferment cellulose and hemicellulose over a period of several days. Accomplishing similar efficiency in an industrial setting requires the optimization of a complex series of engineering steps. One complicating factor is that lignin is resistant to enzymatic degradation and protects cellulose and hemicellulose from physical or enzymatic decomposition. Current ethanol production schemes use various pretreatment steps to disassociate lignin and partially hydrolyze hemicellulose. However, this step is expensive and often yields products that are toxic to subsequent enzymatic and fermentation steps.

Technology for producing ethanol from cellulose has been refined to the point of pilot-scale application (Chapter 2), but simultaneous production of ethanol from hemicellulose has proven to be much more difficult. The physiochemical differences between cellulose and hemicellulose compel the use of different and often more complex pretreatment, enzymatic, and fermentation procedures. Thus, the efficiency of conversion of biomass to ethanol is currently low, and economically competitive processes await development of enzymes and organisms for efficiently using hemicelluloses in processes that can be consolidated with the bioprocessing of cellulose. The huge biodiversity of microorganisms and the powerful techniques in biotechnology, particularly microbial evolutionary engineering and

TABLE 6-1 Complexities of Starch-Based Ethanol Production to Biomass-Based Ethanol Production via Fermentation

	Grain Ethanol	Biomass Ethanol	
		Cellulose Fraction	Hemicellulose Fraction
Characteristic of substrate	Corn is about 70% starch, which is composed exclusively of glucose and easily freed from cellular matrix.	Cellulose makes up 35-60% of biomass and is composed exclusively of glucose but is difficult to disassociate from lignin.	Hemicellulose makes up 15-40% of biomass and is composed of numerous hexose and pentose sugars. It is difficult to disassociate from lignin.
Pretreatment	None	Extensive and typically involves application of heat, acids, bases, or oxidizing agents depending on feedstock.	Extensive and typically involves application of heat, acids, bases, or oxidizing agents depending on feedstock; however, the optimal solution for hemicellulose differs from that for cellulose.
Conditioning	Hydration with water	Neutralization of acid or base and removal of toxic products.	Neutralization of acid or base and removal of toxic products.
Release of sugars	Amylase and glucoamylase release glucose.	Three or more different cellulase enzymes release glucose.	A complex cocktail of xylanases and mannases are needed depending on the feedstock. Mannose, galactose, glucose, xylose, and arabinose are the primary end products.
Fermentation of sugars	Glucose is fermented by a single microbial species, often *Saccharomyces cerevisiae*.	Glucose is fermented by a single microbial species, often *S. cerevisiae*.	A mixture of bacteria, yeast and/or fungi is required for fermenting the complex mixture of hexose and pentose sugars.

recombinant DNA, are being applied to address this barrier; however, the time required to develop appropriate solutions and to implement them on a large scale is not readily predictable and will not likely be quick.

Infrastructure Investments for Biorefineries

A major impediment to increasing cellulosic biofuel production is the large capital investment required for commercial production facilities. A 2009 report estimated the capital cost of a cellulosic-ethanol biorefinery with a capacity of 40 million gallons per year to be about $140 million (2007$) (NAS-NAE-NRC, 2009). Capital costs for building cellulosic biorefineries vary by the choice of feedstock, the conversion technology, and the size of biorefinery. For example, $140 million was the estimated cost of a biorefinery if a high-sugar biomass is used as feedstock. The cost would be higher if biomass with high lignin content is the primary feedstock because of the additional cost of the boiler and steam electrical generator for processing large quantities of lignin. Biorefineries benefit from economies of scale so that the larger the biorefinery, the lower the capital cost per unit capacity (NAS-NAE-NRC, 2009).

In a 2002 report, NREL estimated it would cost $197 million (2000$) to build the nth cellulosic-ethanol biorefinery to produce 69.3 million gallons per year of ethanol from 2,200 dry tons per day of feedstock (Aden et al., 2002). In a 2010 report, NREL updated those cost estimates to about $380-$500 million (2007$) to build the nth plant that has a capacity to convert 2,200 dry tons per day of cellulosic feedstock to produce about 50 million gallons per year of ethanol (Kazi et al., 2010).

Capital costs for biorefineries that use gasification to convert biomass to drop-in fuels were estimated to have even higher capital costs than those that use biochemical conversion. One report estimated the capital cost for a biorefinery that uses gasification and Fischer-Tropsch to convert about 4,000 dry tons of biomass per day to be about $600 million (NAS-NAE-NRC, 2009), while another report estimated the capital costs to be about $500 million for a biorefinery that uses 2,200 dry tons of biomass per day (Swanson et al., 2010). NREL also estimated capital cost for a biorefinery that uses fast pyrolysis to convert 2,200 dry tons of biomass to bio-oil followed by upgrading to drop-in fuels. The capital cost was estimated to be about $200 million (Wright et al., 2010). (See Table 4-3 in Chapter 4.) In 2007, the U.S. Department of Energy (DOE) announced funding of six advanced biofuel projects with a total expected production of 130 million gallons per year of ethanol. The total cost of these projects, including both government and industry funding, was estimated at $1.2 billion (2007$).

Depending on the average capacity of cellulosic biorefineries, about 200-350 refineries would have to be built and in operation between now and 2022 to achieve 16 billion gallons of ethanol-equivalent cellulosic biofuels mandated by RFS2 (Anex et al., 2010; Kazi et al., 2010; Swanson et al., 2010; Wright et al., 2010). The number of biorefineries would be even higher if the 4 billion gallons of advanced biofuel are to be met by cellulosic biofuel as well. The total capital costs would be at least $50 billion. Current economics, exclusive of government subsidies and taxes, do not favor the production of biofuels. Biofuel production is only an economically viable business because of the current tax and subsidy structure (Chapter 4). Companies are reluctant to invest capital in a business venture that depends on government subsidies, which can change at any time. To attract investment capital, any subsidy or tax program designed to encourage investment in biofuel production would have to contain provisions that result in rapid or guaranteed payback for the investor. Rapid returns would attract private investment, whereas guaranteed returns would make the fledgling biofuel industry similar to a regulated public utility. The U.S. and global financial crisis at the end of the 2000-2010 period also discouraged investments in biofuels (IEA, 2010). Stable economic and policy conditions are needed to encourage biofuel investment. Sustained high oil prices (for example, $190 per barrel) also would encourage private investment in biorefineries (Chapter 4).

Infrastructure Investments for Fuel Distribution

Developing infrastructure necessary to transport biofuels from biorefinery to point of sale is a barrier to commercial implementation of the RFS2 mandate. Current transport costs for corn-grain ethanol are high. Existing pipeline infrastructure can be used to transport some finished biofuel products to the refineries and blending facilities depending on the fuel properties. If RFS2 is mostly met by ethanol, the need for fuel distribution infrastructure could pose a challenge to market penetration of the fuel.

Pipelines are a cost-effective way of moving large volumes of liquids only if they are to be used for long periods of time. However, ethanol is not compatible with existing petroleum pipelines because of its higher corrosiveness and affinity for water and because it is a better solvent than petroleum products (Farrell et al., 2007; Singh, 2009). Although existing pipelines could be retrofitted to become multipurpose ones that accommodate ethanol transport, they were designed to flow from petroleum refineries to end users. Ethanol transport would require pipelines that link locations where biorefineries exist and are projected to be built to the existing petroleum distribution infrastructure. Alternatively, dedicated ethanol pipelines could be built. The annual operating costs are low, but building a biofuel pipeline system requires a large investment with significant financial risk (DOE, 2010). Also, ethanol volumes would be low compared to petroleum pipelines' throughput. With the use of biofuels currently dependent on short-term government subsidies, private investors are hesitant to invest in dedicated ethanol pipelines. If the government subsidies are removed, then the economic incentive to use ethanol disappears and the value of the pipeline investment is lost. As a result, ethanol might continue to be transported mostly by tanker truck, barge, and rail; each form carries with it inherent cost and risk, including increase in road accidents, spills, and degradation of road surfaces as a result of increased loads. These delivery methods also require unloading racks, possibly new rail sidings or wharf facilities to accept delivery, and manpower to connect the unloading facilities to the delivery vessel. Although ethanol can displace a fraction of the United States' liquid transportation fuels, investment in a fuel delivery and blending infrastructure would be needed.

Infrastructure for refueling would have to be built if an increasing amount of ethanol is used to meet the biofuel consumption mandate (NAS-NAE-NRC, 2009; NRC, 2010). As of 2011, there were about 2,400 E85 refueling stations across the United States (DOE-EERE, 2010).

POLICY BARRIERS

Blend Wall

If the production of fuel ethanol exceeds the amount that can be blended in gasoline, as explained below, then it reaches the so-called "blend wall." Most ethanol in the United States is consumed as a blend of 10-percent ethanol and 90-percent gasoline. If every drop of gasoline-type fuel consumed in U.S. transportation could be blended, then a maximum of about 14 billion gallons of ethanol could be blended. However, most experts believe the effective blend limit[4] is about 9 percent, which is about 12.6 billion gallons, less than current industry production capacity. In 2010, EPA increased the blend limit to 15 percent for vehicles built since 2001. However, even with a blend limit of 15 percent, the blend wall will be reached again around 2014. Thus, the blend wall is a major barrier for increasing ethanol production beyond about 19 billion gallons even if the blend limit is 15 percent.

Appendix N, based on work by Tyner et al. (2011), provides a complete analysis of alternative scenarios for meeting the RFS2 mandate. In that analysis, it becomes clear that production of ethanol from cellulosic feedstocks becomes problematic because of the blend wall. It would require large and rapid investments in fuel dispensers for E85 plus millions of flex-fuel vehicles produced and sold each year. For example, if the blend limit remained at 10 percent, 8.7 million flex-fuel vehicles would have to be sold each year, and 24,000 fueling dispensers would have to be added each year.[5] Even if economic incentives were

---

[4]The proportion of gasoline that actually gets blended with ethanol as a result of infrastructure limitations and total expected gasoline type fuel consumption.

[5]According to Energy Information Administration, the U.S. fuel-flex vehicle (FFV) fleet will grow from 8.0 million vehicles in 2009 to 39 million by 2022. The annual EIA stock of FFVs is adjusted up or down so that the fleet reflects the volume of E85 consumed in the six scenarios reported in Appendix N. From 2010 to 2022, the annual miles driven averaged 12,369 miles while the E85 average fuel efficiency was 17.9 miles per gallon. EPA (2010d) assumes that the average E85 utilization rate will be no more than 40 percent per FFV, given availability, consumer preferences, and price. Thus, a FFV will consume no more than 274 gallons of E85 per year.

provided, accomplishing this level of investment in infrastructure would be a significant challenge. Raising the blend limit to 15 percent and blending all gasoline for transportation with that proportion of ethanol would only alleviate the problem to a small extent.

On the other hand, if cellulosic feedstocks produce hydrocarbons directly via a thermochemical process, the blend wall becomes much less of an issue. Biogasoline and green diesel are not subject to the blend limits; therefore, the blend wall only applies to ethanol.

Another issue with expanding E85 is the challenge of attracting consumers. E85 contains 78 percent of the energy of E10[6] and so would have to be priced at 78 percent or less of the E10 price. If E10 were $3.00 at the pump, E85 could be no more than $2.34 based on the mileage difference and less if consumers considered the cost of more frequent fill-ups. The extent to which this is a challenge depends on the relative prices of gasoline and ethanol. In November 2010, wholesale gasoline was about $2.15 and ethanol about $2.50. Under these conditions, marketing E85 would require very large incentives. At these prices, the required ethanol subsidy would be $0.475 per gallon, which for 16 billion gallons of cellulosic ethanol would be $7.6 billion per year.

Uncertainties in Government Policies

RFS2 does not provide a market guarantee for biofuels, but instead an assurance of a market under most ordinary circumstances. RFS2 as passed by Congress in 2007 was intended to guarantee a market for biofuel producers and thereby eliminate one of the important sources of uncertainty for potential investors in biofuel refineries. The objective is largely achieved for corn-grain ethanol, at least until the blend wall constraint is reached. However, language in the legislation gives EPA the right to waive or defer enforcement of RFS2 under a variety of circumstances. For example, if it is deemed that enforcing RFS2 would result in significant economic dislocation, the EPA administrator has the right to reduce or waive RFS2. Under that provision, the Governor of Texas in 2008 petitioned EPA to waive RFS2 because of high corn prices and the damage to the livestock sector. EPA denied that request, but economic dislocation waivers are still possible. Undoubtedly, uncertainty of enforcement of the mandate is an impediment for private-sector investment. If cellulosic biofuels are likely to be more expensive than fossil fuels as indicated in Chapter 4, the only guarantee of a market is the federal government. If the private investor perceives that the mandate is not iron-clad, then the effect of the mandate policy will be diminished.

For cellulosic biofuel, EPA is "required to set the cellulosic biofuel standard each year based on the volume projected to be available during the following year." The 2011 standard that EPA set is 6.6 million gallons per year of cellulosic biofuel compared to the 250 million gallons per year in RFS2 (EPA, 2010a).

RFS2 is a quantitative mandate regulating minimum usage of renewable fuels in the United States. The mechanism used for enforcing the RFS2 is via renewable identification numbers (RINs) (Thompson et al., 2010). Each batch of biofuel that is produced or imported into the United States that meets the RFS requirements is assigned RINs by EPA. Each fall, EPA converts the overall quantitative RFS level to a share allocation based on fuel market share. For example, if fuel blender A has a fuel market share for fuel type G of 10 percent, and if the RFS for fuel type G is 15 billion gallons ethanol equivalent, then blender A has an obligation to acquire RINs for 1.5 billion gallons ethanol equivalent of biofuel. Blender A can meet the requirement by acquiring 1.5 billion gallons ethanol equivalent of biofuel

---

[6]The energy content of ethanol is two-thirds that of gasoline on a per-gallon basis. Taking into account the energy content of ethanol and gasoline in E10 and E85 blends, E85 contains about 78 percent of the energy of E10 on average.

with its associated RINs and blending the biofuel into fuel type G. Alternatively, blender A could buy the RINs from blender B that has excess RINs; that is, blender B has blended more biofuel than required and has those extra RINs to sell. Blenders can meet the RFS requirement with any combination of actual blending or RIN purchase (Thompson et al., 2010).

Other factors complicate the RIN market. First, a blender can meet up to 20 percent of a given year's (t) requirement with RINs from the previous year (t − 1). However, RINs cannot be carried forward more than one year (Thompson et al., 2010). They are worthless after that. Also, if a blender runs short in year t, the blender can meet the year t obligation in year t + 1. However, the blender is required to meet the full year t and year t + 1 obligation with no borrowing for the next year. Finally, if the cellulosic mandate is waived, as it has been in 2010 and 2011, then blenders can buy RINs from EPA at a price of $3.00 (inflation adjusted) minus wholesale gasoline price or $0.25, whichever is higher (Thompson et al., 2011).

RINs are freely traded among firms, and the RIN price is an indicator of the extent to which RFS is binding. The price rarely goes to zero because there are some transaction costs associated with holding and trading RINs. Ethanol RINs have often traded for about $0.03 per gallon, reflecting this transaction cost. However, the values had been much higher in periods when RFS2 became more stringently binding. Thus, RIN values are one estimate of the cost of RFS2 (Thompson et al., 2011).

There also is an escape clause that permits blenders to buy RINs that are used to track compliance with RFS2 directly from EPA instead of actually purchasing cellulosic biofuels when there is a RFS waiver. When the price of cellulosic biofuel is high relative to gasoline, this provision becomes operative. It was included in RFS2 apparently to provide a relief valve in case the price gap between cellulosic biofuel cost and gasoline got too wide. In other words, Congress did not want to require enforcement of RFS2 if using cellulosic biofuel would increase the price of fuel at the pump substantially. Specifically, blenders can buy RINs from EPA for the higher of $0.25 or $3 minus the wholesale gasoline price. For example, if the biofuel cost were $3.50 and wholesale gasoline were $2.10, it would cost the blender $1.40 to purchase and blend cellulosic biofuel. Alternatively, the blender could purchase RINs for $0.90 from EPA and get no biofuel. Suppose the blender intended to blend 10-percent biofuel. Doing the calculation for 100 gallons, the cost of purchasing and blending the biofuel would be $224 for 100 gallons, and the cost of using 100 gallons of gasoline plus buying ten RINs would be $219. Thus, it would be more attractive for the blender to buy the RINs and forego blending the cellulosic biofuel. While this provision accomplishes its objective of effectively limiting consumer exposure to very high-priced cellulosic biofuels (relative to gasoline), it limits the scope of the RFS market guarantee for potential investors.

The subsidies for biofuels are another source of policy uncertainty. In past legislation, Congress provided biofuel subsidies for a period of time, typically 4-5 years. To the extent investments depend on the subsidies, the uncertainty in subsidy renewal is another impediment for private investment. For example, the subsidy for each gallon of cellulosic biofuel is $1.01 (regardless of the energy content of the biofuel) in 2010, but the subsidy is set to expire in 2012. If an investor were to consider building a biofuel refinery today, not a gallon would be produced before the subsidy expired. Therefore, the subsidy is not likely to have a major effect on investor decisions in cellulosic biofuel. Similarly, the corn-grain ethanol subsidy was set to expire in 2010. A 1-year extension of the $0.45 per gallon subsidy was passed in December 2010. The biodiesel tax incentive, which lapsed in 2009, was extended retroactively until 2011. The corn-grain ethanol subsidy and biodiesel tax incentives demonstrate the barriers created by uncertainty in government policy.

Another area of uncertainty is the Biomass Crop Assistance Program (BCAP), which provides two years of assistance to farmers who participate in the program to provide bioenergy feedstocks for use as biofuels. The congressional appropriation was insufficient to satisfy total demand, and whether the 2-year payment limit will be extended is unknown.

Tariff on imported ethanol is another area of uncertainty. Initially, the import tariff was created to offset the domestic ethanol subsidy, which is available to both domestic and foreign ethanol. However, over time the subsidy has fallen such that there is now a gap between the tariff and the subsidy. The effective tariff (specific plus ad valorem) is about $0.59 per gallon, and the subsidy is $0.45 gallon yielding a net tariff of $0.14 per gallon. Brazil and other countries have argued that the net tariff violates the World Trade Organization rules. How Congress will handle the tariff in the future is unclear.

Nonfederal Laws, Rules, Regulations, and Incentives Affecting Biomass Energy

Some states are implementing or considering low-carbon fuel standards (LCFS) that could affect the development and use of biofuels—for example, California, Oregon, and Northeast and Mid-Atlantic states (California Energy Commission, 2009; NESCAUM, 2010; Oregon Environmental Council, 2010). State-level regulations that are more stringent than national emission standards would create an issue of market fragmentation. Such fragmentation can increase the cost of producing, distributing, and storing a wider variety of fuel blends by reducing the economies of scale and the regional price arbitration that can be achieved when homogeneous national standards are created.

An example of the cost of such "regulatory heterogeneity" (Muehlegger, 2006b) are the national regulations related to air quality in the Clean Air Act Amendments of 1990 and subsequent state-level regulations of a more stringent nature (for example, in California). A variety of estimates have found that the state content regulations, imposed in addition to the national Clean Air Act regulations, can increase the cost of gasoline in the regulating state by $0.030 to $0.045 per gallon (Muehlegger, 2002, 2006b; Chakravorty et al., 2008). (See Chouinard and Perloff [2007]) for a finding that state content regulations created no price effect.) In addition to the absolute cost of refining, distributing, and storing fuels, research suggests that such regulatory heterogeneity can increase fuel price variability on a seasonal basis throughout the year (Davis, 2009) or the variability of prices at times of unexpected supply disruptions (for example, refinery fires or other shutdowns) (Muehlegger, 2002, 2006b).[7]

California is the first state to establish an LCFS and is used in this section as an illustration (California Energy Commission, 2009). Under California's Global Warming Solutions

---

[7]Seasonal variability in fuel prices can increase because the fuel content regulations might not apply throughout the entire year. Thus, traditional seasonal price variations (resulted from regularly observed driving habits through the year) can be exacerbated by variations in regulatory content standards throughout the year (for example, the content regulations are often more stringent during those times of the year in which demand for fuel is the highest because driving activity is the greatest) (Davis, 2009). Price variations due to unexpected supply disruptions can increase with greater regulatory heterogeneity because the number and capacity of refineries capable of producing fuel that satisfies state content regulations may be quite limited or can only be converted to production of fuel that complies with state regulations at significant expense and with a significant time lag. Thus, other refineries may have very limited ability to supply fuel to the market with the heterogeneous regulatory standards, thereby increasing the price movement in that state. In essence, when state regulatory standards proliferate, refineries in other states or regions are less able to absorb a portion of supply disruption and subsequent price increase (Muehlegger, 2002, 2006a). It should also be noted that such regulations can act as a trade barrier that limits imports of gasoline from foreign sources at times of a disruption in supply (Fernandez et al., 2007). Such regulations may also contribute to the exercise of market power by the refineries that serve a specific regulated state or region because the competition from other refineries may be limited by their inability to satisfy the state content regulations, thereby preventing market arbitrage from occurring (Chakravorty et al., 2008). See Chouinard and Perloff (2007) for a finding of no market power effect on prices.

Act (AB32), an LCFS that mandates a reduction in greenhouse-gas (GHG) intensity of transportation fuels was established. The rules governing the LCFS differ in important ways from the federal RFS2 standard and present fuel blenders with different standards for compliance at the state and federal levels. LCFS is a GHG-performance standard that applies to all transportation fuels, including biofuels, compressed natural gas, electricity, and hydrogen. All suppliers of transportation fuels, including blenders, producers, and importers, are required to reduce the average GHG intensity of their fuels compared to GHG intensity of transportation fuels in 2010 (Yeh and Sperling, 2010). The timetable adopted increases the percent GHG reduction required progressively until it reaches 10 percent in 2020. The average GHG intensity of fuels is calculated as the sum of GHGs emitted from the covered fuels divided by the total energy of the fuel. In addition, the act allows trading and banking of GHG credits to encourage technology innovations that would result in low-cost and low-carbon fuels (Yeh and Sperling, 2010). In contrast, RFS2 is a consumption mandate for specific types of biofuels. Some fuels that can contribute to achieving RFS2 might not contribute to meeting LCFS. For example, corn-grain ethanol qualifies as a renewable fuel under RFS2 whether it is blended as E10 or E85. However, under LCFS accounting for GHG reduction (which also includes GHG emissions from indirect land-use change), gasoline blended with 10-percent corn-grain ethanol might not contribute to meeting LCFS's GHG-performance standard (Zhang et al., 2010). Furthermore, the California Air Resources Board uses different approaches for estimating life-cycle GHG emissions than EPA for RFS2. State-level LCFS could be a barrier to meeting RFS2 if multiple states are implementing LCFS and if fuel suppliers use fuels other than biofuels to meet an LCFS target. Having different regulatory agencies using different methods to calculate GHG emission reductions could lead to legal challenges by industry that can delay the implementation of any GHG emission reduction legislation until there is consensus on life-cycle analysis.

ENVIRONMENTAL BARRIERS

Environmental effects of biofuel production can pose a barrier to achieving RFS2 if the effect results in biofuels that do not meet the RFS2 eligibility requirements, if the effect violates environmental regulations, or if a resource limitation constrains the amount of biofuels that can be produced. This section discusses some of the environmental effects at different stages of the supply chain and over the life cycle of biofuel production that could become barriers to producing volumes of biofuels to contribute to the RFS2 consumption mandate.

Life-Cycle GHG Emissions

In addition to the mandated volumes of renewable fuels to be used each year from 2008-2022, RFS2 specifies GHG reduction thresholds for different categories of fuels (Table 6-2). (See also section entitled "Renewable Fuel Standard" in Chapter 1.) That is, biofuels would be required to have life-cycle GHG emissions less than the specified thresholds to qualify as one of the four categories of renewable fuels for meeting the RFS2 consumption mandate.

Uncertainties in life-cycle GHG accounting can pose a barrier to achieving RFS2 because they affect investors' confidence. Although EPA made a ruling on which fuels (that is, fuels produced from which feedstock and in what type of facilities) would meet the GHG reduction threshold, it recognizes the science of GHG accounting is evolving. EPA will reassess the life-cycle GHG determinations (EPA, 2010b) so that the methods and industry data that the agency uses to assess life-cycle GHG emissions could change over time. In

TABLE 6-2 Life-Cycle Greenhouse-gas (GHG) Reduction Thresholds Specified in RFS2

Category	GHG ReductionThreshold[a]
Renewable fuel	20%
Advanced biofuel	50%
Biomass-based diesel	50%
Cellulosic biofuel	60%

[a]GHG reduction threshold is the minimum percent reduction of life-cycle GHG emissions of the 2005 baseline average gasoline or diesel fuel that a renewable fuel replaces.
SOURCE: EPA (2010c).

addition, the ruling on which fuels would meet the GHG reduction threshold could change as empirical data are collected for some parameters that could influence GHG emissions.

As discussed in Chapter 5, GHG emissions from land-use changes globally due to market-mediated effects of U.S. biofuel production are highly uncertain. Although data on land-use changes collected over time can improve knowledge on effects of U.S. biofuel policies on land-use changes worldwide and the precision of life-cycle GHG accounting of biofuels, some uncertainties will always remain because of difficulties in establishing a cause-and-effect relationship between biofuel production and land-use changes from market-mediated effects. Because GHG emissions from land-use changes could span a wide range depending on the actual extent of land-use changes (see Table 5-2 in Chapter 5), investors could not be certain that the biofuels that they plan to produce would meet the GHG reduction threshold set by RFS2. However, GHG emissions in some parts of the biofuel supply chain could decrease as feedstock production and conversion technologies improve.

Air and Water-Quality Effects from Biorefineries

Biorefineries are required to meet the standards of the federal Clean Air Act and the Clean Water Act to obtain permits to operate the facilities. Based on the water and air-quality effects from biorefineries and the environmental impact assessments of planned cellulosic biofuel refineries discussed in Chapter 5, the committee judges that the ability of individual refineries to meet the standards set by those laws is not likely to pose a technical barrier to achieving RFS2. However, in regions that are noncompliant with ambient air-quality standards established by the Clean Air Act, permit requirements can limit the establishment of biorefineries through increased cost for permits and control equipment, delays in the permit process, or outright prohibition. For projects receiving federal funding, a Department of Energy (DOE) Finding and Environmental Assessment would have to be conducted to establish that there will be no adverse impacts with respect to sound, traffic, air quality, water quality, or threatened or endangered species before permits are issued (VeroNews, 2010). (See also section "Regional and Local Environmental Assessments" in Chapter 5.) For example, the central valley of California, where large amounts of biomass are available from agriculture, MSW, and nearby forests, has emission requirements and mandates for Best Available Control Technology that in practice severely limit the capacity to site new industrial facilities (Orta et al., 2010). Fines for water and air-quality permit violations at existing biofuel facilities are relatively commonplace (Beeman, 2007; Smith, 2008; EPA, 2009; Buntjer, 2010; Meersman, 2010; O'Sullivan, 2010). Iowa alone had 394 instances of violations at biofuel facilities during the period 2001-2007 (Beeman, 2007).

There is an additional concern that the industry as a whole could cause a detrimental cumulative impact in large watersheds (Chapter 5). Dominguez-Faus et al. (2009) discussed the possibility that the biofuel industry could be limited by concerns of water quality in high-priority areas such as the Chesapeake Bay or the Mississippi River Basin (Gulf hypoxia) and could cause water shortages in places like the High Plains Aquifer (Ogallala).

Water Use for Irrigating Feedstock and in Biorefineries

Water scarcity in particular regions of the United States could limit the quantity of feedstock that could be grown and the total number of biorefineries that could be built. Although irrigation of bioenergy feedstock in some regions can substantially improve yield (Chapter 2), irrigation is a key factor in determining consumptive water use (Chapter 5).

As an example, the Republican River Basin loses water as it flows through Colorado, Nebraska, and Kansas. Its flow is connected to the Ogallala Aquifer, and the Republican River is subject to drought (UNL, 2011b). As the Ogallala Aquifer loses water (hydraulic head) due to irrigation demands for more corn, the Republican River also becomes at risk to more drought. According to the University of Nebraska water website, "Irrigation development has caused declines of groundwater levels (depth to groundwater from the soil surface) in some areas of the state. The most severely affected areas are the Box Butte area, western end of the Republican River Basin and parts of the Blue River Basin" (UNL, 2011a).

The environmental impact assessments of some of the planned cellulosic refineries discussed in Chapter 5 suggested that biofuel production in those specific facilities would not affect water availability. For example, the Abengoa facility passed the DOE's environmental impact assessment of cumulative impacts on water and groundwater resources, but Abengoa passed because of favorable groundwater availability in that portion of southwest Kansas where the facility is located, a situation that does not exist throughout the state. However, as more biorefineries are built, water availability and consumptive water use would have to be considered locally and regionally to ensure that the water resources will be sustained. A determination of cumulative impact on groundwater availability could be a barrier for future expansion of the industry (Meersman, 2008), but that has yet to be done.

Fuel Certification Requirements

Section 211(f)(1)(A) of the Clean Air Act states, "Effective upon March 31, 1977, it shall be unlawful for any manufacturer of any fuel or fuel additive to first introduce into commerce, or to increase the concentration in use of, any fuel or fuel additive for general use in light duty motor vehicles manufactured after model year 1974 which is not substantially similar to any fuel or fuel additive utilized in the certification of any model year 1975, or subsequent model year, vehicle or engine under section 206" (108 P.L. 201).

EPA regulations prohibit the addition to gasoline of any component that has not been approved for use in gasoline. Ethanol and other aliphatic alcohols and ethers (except methanol) use is approved up to the current blend wall (2.7-percent oxygen by weight). Total oxygen content of fuels that do not contain ethanol or other aliphatic alcohols and ethers can only be 2 percent by weight. Many states have banned the use of methyl tertiary butyl ether (MTBE) or ethyl tertiary butyl ether (ETBE) in gasoline, leaving ethanol as the only commercially available oxygenate for gasoline blending. EPA certification is required before other "new" biofuels, such as those produced by gasification and Fischer-Tropsch or pyrolysis, can become part of the gasoline pool.

This certification is a two-tier system. If the new material is "substantially similar" to gasoline (that is, contains only hydrocarbons and aliphatic alcohols and ethers) or will only be blended at less than 0.25 percent by weight of any fuel batch, then the blender can petition EPA to approve its use based on limited testing (Tier 1 testing). If the potential fuel is not "substantially similar" to current gasoline, then extensive vehicle-based emissions testing is required for approval (Tier 2 testing). EPA has up to 270 days to approve or deny the waiver request. If EPA does not act within 270 days, the waiver is deemed to be granted.

Tier 1 testing requires about 50 gallons for the new material and the cost of producing and testing the fuel can exceed $1 million (NBB, 1998; Scoll and Guerrero, 2006). Tier 2 testing is much more extensive and can take up to a year to complete and cost several million dollars. Larger volumes of fuel are also required. As of 2011, EPA has not determined what level of testing will be required to certify gasification and Fischer-Tropsch fuels, pyrolysis oils, or any other new biofuel for use in gasoline.

In addition to EPA certification for gasoline, ASTM sets fuel property standards that have to be met for any material to be called gasoline, jet fuel, or diesel fuel. These standards currently incorporate the EPA regulations for fuels as well as other performance and marketability standards. The jet fuel standard currently only allows the use of biofuel components that are produced as synthetic paraffinic kerosene at less than 50 percent of the total fuel. The diesel fuel standard currently recognizes and controls the use of fatty acid methyl ester biodiesel, but is silent on other potential blend components as long as their use still allows the blended fuel to meet all other current standards.

SOCIAL BARRIERS

Barriers to achieving RFS2 include social factors that deter producers from growing or harvesting bioenergy feedstocks and that deter consumers from purchasing biofuels. This section is divided into research that investigates barriers faced by farmers and land managers and research that investigates barriers to consumer acceptance and use of biofuels.

Knowledge, Attitudes, and Values of Farmers and Forest Owners

Barriers that farmers and nonindustrial private forest (NIPF) owners face in entering biofuel markets include the lack of information about emerging opportunities to grow and harvest bioenergy feedstocks, logistical barriers to harvesting and transporting bioenergy feedstocks, cultural barriers to introducing new crops into a monoculture landscape, the lack of sufficient economic incentives, and uncertainty around the duration of incentives and policies to support production and harvesting of bioenergy feedstocks. Sociologists and economists have long studied what compels farmers to adopt an innovation, such as a new crop or new technology, and how that innovation spreads or is "diffused" into the broader community. The adoption and diffusion model (Rogers, 2003) attempted to predict the adoption behavior of farmers on the basis of their personal characteristics (education, personality, age, income), the time factor, and the nature of the innovation itself. According to the adoption and diffusion model, innovators are those who are willing to take risks on new crops or new technology; innovators were often found to be independent and not as tightly integrated with their community. Early adopters of the innovation were found to be more integrated in their communities and, because they are often in leadership positions, they help to diffuse the innovation for the late adopters (Rogers, 2003). Adoption and diffusion researchers were also interested in the role that information played in farmer adoption and diffusion behavior—that is, how the type of information and from whom it is received might affect adoption. The adoption and diffusion model has been subsequently criticized for overemphasizing the personal characteristics of the farmer to the neglect of the role of the broader structural, economic, and institutional environment shaping farming decisions, including the policy environment (Buttel et al., 1990).

Economists (Griliches, 1957) have also investigated the economic processes underlying the diffusion of technological innovations such as the spread of hybrid corn. Griliches' foundational research demonstrated that the adoption of new technologies, such as hybrid corn, was not a single event, but was instead a series of developments that occurred at different rates across geographical space over a 20-year time frame (Griliches, 1957). His study shed light on the numerous individual decisions and economic calculations that drove new hybrid corn technology forward, demonstrating that the analysis of spatial patterns in the diffusion of innovation can provide important clues to understanding economic processes. The economics of innovation continues to be an important topic within the fields of economics, business, and the sociology of science.

Economists and other social scientists have begun to research some of the barriers faced by farmers and NIPF owners to entering biofuel markets. A recent study examined factors associated with the potential adoption of *Miscanthus* (a dedicated bioenergy crop) among farmers in Illinois (Villamil et al., 2008). The study concluded that information plays a key role in farmers' consideration of adoption of such a new crop as *Miscanthus*. Researchers found that farmers had different information needs and preferred methods of receiving information when considering adopting *Miscanthus* in different regions of the state. Those farmers with the highest potential for adopting *Miscanthus* were most interested in information related to the agronomy and markets for *Miscanthus*. They preferred to receive the information from farm and agricultural organizations, as opposed to from other farmers.

Another study assessed the willingness of Tennessee farmers to grow switchgrass as an energy crop and the share of their farmland they would be willing to devote to switchgrass (Jensen et al., 2007). Findings showed that only 21 percent of 3,244 farmers who responded to a survey had ever heard of growing switchgrass as a bioenergy crop. About 30 percent said they would be interested in growing switchgrass while 47 percent said they were unsure or did not know if they would be interested. Farmers who had greater off-farm income, higher education levels, and were younger were more willing to convert some of their land to switchgrass production. Farmers with higher net farm income per hectare were less likely to convert a large amount of land, indicating the opportunity cost of planting switchgrass (Jensen et al., 2007). In addition to economic uncertainties, Jensen et al.'s (2007) study indicates the importance of considering farm characteristics and other demographic factors in farmer willingness to enter new markets and grow new crops.

In another study, Song et al. (2009) estimated that the minimum acceptable net return to induce conversion from an annual corn-soybean rotation to a perennial switchgrass crop to be much higher than the risk-free comparative breakeven net return. The reluctance to convert land from traditional row crops to a switchgrass crop was projected to be hindered by the volatility of biofuel prices and the costly reversibility of investment in a switchgrass crop (Song et al., 2009).

Other research investigated the cultural context of introducing different perennial crops (including dedicated bioenergy crops) into the Corn Belt agricultural landscape (Atwell et al., 2009, 2011). The study's rationale was that stakeholder involvement is critical if dedicated bioenergy crops are to be planted on a large scale and that stakeholders' landscape values can affect their decision to plant those crops. Previous research has shown that farm diversification and landscape heterogeneity often are not the cultural norms that define how a well-operated farm would look (Nassauer, 1989; Napier et al., 2000). Atwell et al. (2009) found that most farmers and other stakeholders approved of growing dedicated bioenergy crops, but implementation of these practices was not a priority. They concluded that a shift in community norms about landscapes would be necessary to increase planting of dedicated bioenergy crops.

In a study of the intentions of farmers in the United Kingdom toward producing bioenergy crops for biofuel, Mattison and Norris (2007) surveyed 278 farmers about their interest in growing two bioenergy crops. They found that farmers had positive attitudes toward growing bioenergy crops, but obstacles noted by farmers were inadequate policies to encourage crop production and a lack of infrastructure for biomass processing.

Another study was undertaken with farmers and rural stakeholders in southern Iowa and northeastern Kentucky in 2006 and 2008 (Rossi and Hinrichs, 2011). In-depth interviews were conducted with 48 independent small farmers and stakeholders in two switchgrass bioenergy projects and revealed that farmers were skeptical that switchgrass bioeconomy would bring tangible economic benefits. Their experiences with the switchgrass projects indicated that there were many technological, economic, and logistical barriers yet to be overcome before the biofuel industry could develop further. Although many participants expressed enthusiasm about the potential of cellulosic ethanol to contribute to energy security and rural economic revitalization, they were skeptical that it was economically feasible. In addition to economic uncertainty, some farmers in these studies lacked knowledge and information about growing bioenergy feedstocks and some had concerns about inconsistent policy incentives for producing feedstock for biofuels, and norms and values toward the landscape. All of these factors have been shown to deter farmers from growing or harvesting bioenergy feedstock for biofuels.

Lack of reliable and steady supply of forest resources from NIPFs could be a barrier to achieving RFS. As discussed in Chapter 2, cellulosic biofuels made from feedstock from federal forests is not to be counted toward meeting RFS2. Given that the majority of timber harvested each year has come from NIPFs for the past 50 years (Adams et al., 2006), woody resources for cellulosic biofuel production will likely come from NIPFs. Yet, a large percentage of forest landowners were uncertain or unfamiliar with the idea of producing energy from woody biomass (Joshi and Mehmood, 2011). In particular older landowners, who were about one-quarter of survey respondents, were more uncertain and skeptical about wood-based bioenergy.

NIPFs might not be a reliable year-round source of forest resources for several reasons. First, not all NIPF owners are harvesters or active managers. Most often, owners of large-parcel NIPFs appear most likely to be harvesters or active managers of their forestland (Bliss et al., 1997; Johnson et al., 1997; Best, 2004), are better informed about forest management, and are more receptive to outreach programs administered by forest agencies or university extension programs (Kuhns et al., 1998). Second, many NIPF owners describe nonharvesting objectives, such as aesthetic enjoyment, as their primary reasons for forest ownership (Creighton et al., 2002; Kendra and Hull, 2005; Kilgore et al., 2008). A survey of 4,800 NIPF owners in Arkansas, Florida, and Virginia found similar results (Joshi and Mehmood, 2011). Although timber harvesting is a frequent activity of NIPF owners (Birch, 1994; Butler, 2008), the majority of harvesting activity is for personal, noncommercial uses, such as firewood. Few NIPF owners partake in harvesting for economic gain (Birch, 1994; Butler and Leatherberry, 2004). Third, many NIPF owners make timber harvesting decisions based on short-term financial needs rather than long-term management planning. Thus, uncertainty about the harvesting behavior of NIPF owners will create a barrier to development of a reliable year-round supply of woody resources for cellulosic biofuel. That challenge is already faced by saw and paper mills across the United States.

Consumer Knowledge, Attitudes, and Values about Biofuels

Although some scientists in many countries have positive views of the potential for biofuels and bioenergy, research has shown that these positive views are not always shared by the public (McCormick, 2010). Renewable energy is often viewed favorably by the public, but research shows that often the public is not well informed about biofuels and bioenergy and does not think of biofuels as a form of renewable energy (Rohracher, 2010). Researchers found that this may be related to confusion over the terminology related to bioenergy because of the variety of feedstocks, conversion technologies, products, and markets involved (McCormick, 2010).

Previous research on public opinions on biofuels explored the social and psychological dimensions that shape thinking and behavior toward biofuels (Wegener and Kelly, 2008). In a telephone survey of 1,049 randomly sampled U.S. citizens, researchers found that participants generally had a favorable, but not strong, attitude toward the use of biofuels. Among the participants, 24 percent said they were not well informed about biofuels, such as ethanol. Seventy-one percent were not at all informed about the use of switchgrass to produce ethanol. Only 5 percent identified biofuels as a source of renewable energy. These results suggest that obstacles to public acceptance and adoption of biofuels among the public include a lack of knowledge and experience with biofuels and that policies need to consider the social factors that shape public acceptance and consumer behavior change.

In the U.S. Corn Belt, Delshad et al. (2010) explored public attitudes and knowledge toward biofuel policies and technology. Their findings showed that participants were much more knowledgeable about biofuel technologies than about the policies related to biofuels. Participants were opposed to expanded corn-grain ethanol production because of concerns about rising food prices and environmental impacts of corn-grain ethanol. Research in the Upper Midwest examined consumers' knowledge about climate change and how this knowledge affects their "willingness to pay" more for cellulosic ethanol derived from farm residues, forest residues, and mill waste and municipal waste (Johnson et al., 2011). Findings showed that knowledge about climate change was linked to consumers' willingness to pay more for biofuels, but that consumers only made minor differentiation between sources of biomass in terms of their preferences. Therefore, consumers' attitudes are likely to be a lesser barrier than economics in achieving the RFS2 mandate.

Information and Outreach

As noted above, the lack of information and outreach to farmers and NIPF owners about the emerging energy markets were identified as barriers to the production of biomass for energy. In addition, more information about the price of biomass, the potential environmental and employment benefits, and other opportunities related to biomass production are considered to be important factors in providing incentives for more forest and farm landowners to enter these markets (Joshi and Mehmood, 2011). In addition, there is a recognized need for more education and communication to improve public awareness and acceptance of biofuels (Zoellner et al., 2008; Peck et al., 2009; McCormick, 2010). International efforts have been launched to provide the public with information about biofuels, including initiatives such as the World Bioenergy Association, the Global Sustainable Bioenergy Project, the Global Bioenergy Partnership, the Roundtable on Sustainable Bioenergy, and Bioenergy Promotion. Several educational websites on biomass and bioenergy are currently in development, such as the website developed by the International Energy Agency (IEA, 2011).

CONCLUSION

Because cellulosic biofuel is a developing industry, there are multiple economic, policy, environmental, and social barriers to producing 16-20 billion gallons of ethanol-equivalent advanced and cellulosic biofuels to meet the consumption mandate of RFS2. Resolving most of the barriers is necessary to achieving RFS2, and many of them are interrelated as illustrated by the examples below.

A key barrier to achieving RFS2 is the high cost of producing biofuels compared to petroleum-based fuels and the large capital investments required to put billions of gallons of production capacity in place. As of 2010, biofuel production was contingent on subsidies, tax credits, the import tariff, loan guarantees, RFS2, and similar policies. These policies that provide financial support for biofuels will expire long before 2022 and cannot provide the support necessary for achieving the RFS2 mandate. Uncertainties in policies can affect investors' confidence and discourage investment. In addition, if the cellulosic biofuel produced are mostly ethanol, investments in distribution infrastructure and flex-fuel vehicles would have to be made for such large quantities of ethanol to be consumed in the United States. Given the current blend limit of up to 15-percent ethanol in gasoline, a maximum of 19 billion gallons of ethanol can be consumed unless the number of flex-fuel vehicles increases substantially. However, consumers' willingness to purchase flex-fuel vehicles and use E85 instead of lower blends of ethanol in their vehicles will likely depend on the price of ethanol and their attitude toward biofuels. Producing drop-in biofuels could improve the ability to integrate the mandated volumes of biofuels into U.S. transportation, but would not improve the cost-competitiveness of biofuels with petroleum-based fuels.

Opportunities to reduce the cost of biofuels are to reduce the cost of bioenergy feedstock, which constitutes a large portion of operating costs, and increase the conversion efficiency from biomass to fuels. Research and development to improve the on-farm yield of bioenergy feedstocks through breeding and biotechnology and conversion efficiency in biorefineries would reduce the cost of biofuel production and potentially reduce the environmental effects per unit of biofuel produced. However, a cellulosic-biofuel market will not be realized if farmers and landowners are unaware of the market opportunities for bioenergy feedstocks or are unwilling to participate in that market. If competition for bioenergy feedstocks intensifies because of low supply, the price will likely increase. Given the numerous barriers outlined in this chapter, the committee judges that the consumption mandate for cellulosic biofuel is not likely to be met by 2022 without substantial improvement in technologies in the next few years and a stable economic and policy environment to encourage accelerated demonstration and deployment of cellulosic biofuel.

REFERENCES

Adams, D.M., R.W. Haynes, and A.J. Daigneault. 2006. Estimated Timber Harvest by U.S. Region and Ownership, 1950-2002. Portland, OR: U.S. Department of Agriculture - Forest Service, Pacific Northwest Research Station.

Aden, A., M. Ruth, K. Ibsen, J. Jechura, K. Neeves, J. Sheehan, B. Wallace, L. Montague, A. Slayton, and J. Lukas. 2002. Lignocellulosic Biomass to Ethanol Process Design and Economics Utilizing Co-Current Dilute Acid Prehydrolysis and Enzymatic Hydrolysis for Corn Stover. Golden, CO: National Renewable Energy Laboratory.

Aklerlof, G.A. 1970. The market for "lemons": Quality uncertainty and the market mechanism. Quarterly Journal of Economics 84(3):488-500.

Anderson, W.F., B.S. Dien, H.J.G. Jung, K.P. Vogel, and P.J. Weimer. 2010. Effects of forage quality and cell wall constituents of bermuda grass on biochemical conversion to ethanol. Bioenergy Research 3(3):225-237.

Anex, R.P., A. Aden, F.K. Kazi, J. Fortman, R.M. Swanson, M.M. Wright, J.A. Satrio, R.C. Brown, D.E. Daugaard, A. Platon, G. Kothandaraman, D.D. Hsu, and A. Dutta. 2010. Techno-economic comparison of biomass-to-transportation fuels via pyrolysis, gasification, and biochemical pathways. Fuel 89:S29-S35.

Atchison, J.E., and J.R. Hettenhaus. 2003. Innovative Methods for Corn Stover Collecting, Handling, Storing and Transporting. Golden, CO: National Renewable Energy Laboratory.

Atwell, R., L. Schulte, and L. Westphal. 2009. Landscape, community, countryside: Linking biophysical and social scales in U.S. Corn Belt agricultural landscapes. Landscape Ecology 24(6):791-806.

Atwell, R.C., L.A. Schulte, and L.M. Westphal. 2011. Tweak, adapt, or transform: Policy scenarios in response to emerging bioenergy markets in the U.S. Corn Belt. Ecology and Society 16(1):10.

Beeman, P. 2007. Biofuel plants generate new air, water, soil problems for Iowa. Available online at http://www.desmoinesregister.com/article/20070603/BUSINESS01/706030325/Biofuel-plants-generate-new-air-water-soil-problems-for-Iowa. Accessed January 21, 2011.

Best, C. 2004. Non-governmental organizations: More owners and smaller parcels pose major stewardship challenges. A response to "America's family forest owners." Journal of Forestry 102(7):10-11.

Birch, T.W. 1994. Private forest land owners of the United States. USDA Forest Service Northeast Experiment Station Resource Bulletin NE-134.

Bliss, J.C., S.K. Nepal, R.T. Brooks, and M.D. Larsen. 1997. In the mainstream: Environmental attitudes of mid-south NIPF owners. Southern Journal of Applied Forestry 21(1):37-42.

Buntjer, J. 2010. Heron Lake BioEnergy fined. Daily Globe, December 27. State and Regional News section.

Butler, B.J. 2008. Family Forest Owners of the United States, 2006: A Technical Document Supporting the Forest Service 2010 RPA Assessment. Newtown Square, PA: U.S. Deptartment of Agriculture - Forest Service.

Butler, B.J., and E.C. Leatherberry. 2004. America's family forest owners. Journal of Forestry 102(7):4-14.

Buttel, F.H., O.F. Larson, and G.W. Gillespie, Jr. 1990. The Sociology of Agriculture. New York: Greenwood Press.

California Energy Commission. 2009. Low-carbon fuel standard. Available online at http://www.energy.ca.gov/low_carbon_fuel_standard/index.html. Accessed September 23, 2010.

Chakravorty, U., C. Nauges, and A. Thomas. 2008. Clean air regulation and heterogeneity in U.S. gasoline prices. Journal of Environmental Economics and Management 55(1):106-122.

Chouinard, H.H., and J.M. Perloff. 2007. Gasoline price difference: Taxes, pollution regulations, market power, and market conditions. Berkeley Electronic Journal of Economic Analysis and Policy 7(1):Article 8.

Creighton, J.H., D.M. Baumgartner, and K.A. Blatner. 2002. Ecosystem management and nonindustrial private forest landowners in Washington State, USA. Small-scale Forest Economics, Management, and Policy 1:55-69.

Davis, M.C. 2009. Environmental regulations and the increasing seasonality of gasoline prices. Applied Economic Letters 16(16):1613-1616.

Delshad, A.B., R. Raymond, V. Sawickia, and D.T. Wegenera. 2010. Public attitudes toward political and technological options for biofuels. Energy Policy 38:3414-3425.

DOE (U.S. Department of Energy). 2010. Dedicated Ethanol Pipeline Feasibility Study. Washington, DC: U.S. Department of Energy.

DOE-EERE (U.S. Department of Energy - Energy Efficiency and Renewable Energy). 2004. Biomas bulk processing and storage. Available online at http://www1.eere.energy.gov/biomass/fy04/bulk_processing_storage.pdf. Accessed November 19, 2010.

DOE-EERE (U.S. Department of Energy - Energy Efficiency and Renewable Energy). 2010. Alternative fueling station total counts by state and fuel type. Available online at http://www.afdc.energy.gov/afdc/fuels/stations_counts.html. Accessed April 25, 2011.

Dominguez-Faus, R., S.E. Powers, J.G. Burken, and P.J. Alvarez. 2009. The water footprint of biofuels: A drink or drive issue? Environmental Science and Technology 43(9):3005-3010.

EIA (U.S. Energy Information Administration). 2010. U.S. Natural Gas Pipeline Network, 2009. Available online at http://www.eia.gov/pub/oil_gas/natural_gas/analysis_publications/ngpipeline/ngpipelines_map.html. Accessed November 9, 2010.

EPA (U.S. Environmental Protection Agency). 2009. Ethanol Plant Clean Air Act Enforcement Initiative. Available online at http://www.epa.gov/oecaerth/resources/cases/civil/caa/ethanol/index.html. Accessed February 14, 2011.

EPA (U.S. Environmental Protection Agency). 2010a. EPA Finalizes 2011 Renewable Fuel Standards. Washington, DC: U.S. Environmental Protection Agency.

EPA (U.S. Environmental Protection Agency). 2010b. EPA Lifecycle Analysis of Greenhouse Gas Emissions from Renewable Fuels. Washington, DC: U.S. Environmental Protection Agency.

EPA (U.S. Environmental Protection Agency). 2010c. Regulation of fuels and fuel additives: Changes to renewable fuel standard program; final rule. Federal Register 75(58):14669-15320.

EPA (U.S. Environmental Protection Agency). 2010d. Renewable Fuel Standard Program (RFS2) Regulatory Impact Analysis. Washington, DC: U.S. Environmental Protection Agency.

Farrell, A., D. Sperling, S.M. Arons, A.R. Brandt, M.A. Delucchi, A. Eggert, B.K. Haya, J. Hughes, B.M. Jenkins, A.D. Jones, D.M. Kammen, S.R. Kaffka, C.R. Knittel, D.M. Lemoine, E.W. Martin, M.W. Melaina, J. Ogden, R.J. Plevin, B.T. Turner, R.B. Williams, and C. Yang. 2007. A Low-Carbon Fuel Standard for California. Part 1: Technical Analysis. Davis: University of California, Davis.

Fernandez, A.Z., R.W. Gilmer, and J.L. Story. 2007. Gasoline Content Regulation as a Trade Barrier: Do Boutique Fuels Discourage Fuel Imports? Dallas, TX: Federal Reserve Bank of Dallas.

Foust, T.D., R. Wooley, J. Sheehan, R. Wallace, K. Ibsen, D. Dayton, M. Himmel, J. Ashworth, R. McCormick, M. Melendez, J.R. Hess, K. Kenney, C. Wright, C. Radtke, R. Perlack, J. Mielenz, M. Wang, S. Synder, and T. Werpy. 2007. A National Laboratory Market and Technology Assessment of the 30x30 Scenario. Golden, CO: National Renewable Energy Laboratory.

Gírio, F.M., C. Fonseca, F. Carvalheiro, L.C. Duarte, S. Marques, and R. Bogel-Łukasik. 2010. Hemicelluloses for fuel ethanol: A review. Bioresource Technology 101:4775-4800.

Grant, D., J.R. Hess, K. Kenney, P. Laney, D. Muth, P. Pryfogle, C. Radtke, and C. Wright. 2006. Feasibility of a Producer Owned Ground-Straw Feedstock Supply System for Bioethanol and Other Products. Idaho Falls: Idaho National Laboratory.

Griliches, Z. 1957. Hybrid corn: An exploration of the economics of technological change. Econometrica 48:501-522.

Hess, J.R., C.T. Wright, and K.L. Kenny. 2007. Cellulosic biomass feedstocks and logistics for ethanol production. Biofuels Bioproducts and Biorefining 1(3):181-190.

Hettinga, W.G., H.M. Junginger, S.C. Dekker, M. Hoogwijk, A.J. McAloon, and K.B. Hick. 2009. Understanding the reductions in US corn ethanol production costs: An experience curve approach. Energy Policy 37(1):190-203.

IEA (International Energy Agency). 2010. The Impact of the Financial and Economic Crisis on Global Energy Investment, 2009. Paris: International Energy Agency.

IEA (International Energy Agency). 2011. Education web site on biomass and bioenergy. Available online at http://www.aboutbioenergy.info/. Accessed January 28, 2011.

Jensen, K., C.D. Clark, P. Ellis, B. English, J. Menard, M. Walsh, and D. De La Torre Ugarte. 2007. Farmer willingness to grow switchgrass for energy production. Biomass and Bioenergy 31:773-781.

Johnson, D.M., K.E. Halvorsen, and B.D. Solomen. 2011. Upper midwestern U.S. consumers and ethanol: Knowledge, beliefs and consumption. Biomass and Bioenergy 35(4):1454-1464.

Johnson, R.L., R.J. Alig, E. Moore, and R.J. Moulton. 1997. NIPF landowners view of regulation. Journal of Forestry 95(1):23-28.

Joshi, O., and S.R. Mehmood. 2011. Factors affecting nonindustrial private forest landowners' willingness to supply woody biomass for bioenergy. Biomass and Bioenergy 35(1):186-192.

Kazi, F.K., J. Fortman, R. Anex, G. Kothandaraman, D. Hsu, A. Aden, and A. Dutta. 2010. Techno-Economic Analysis of Biochemical Scenarios for Production of Cellulosic Ethanol. Golden, CO: National Renewable Energy Laboratory.

Kendra, A., and R.B. Hull. 2005. Motivations and behaviors of new forest owners in Virginia. Forest Science 51(2):142-154.

Kilgore, M.A., S. Snyder, S. Taff, and J. Schertz. 2008. Family forest stewardship: Do owners need a financial incentive? Journal of Forestry 106(7):357-362.

Kuhns, M.R., M.W. Brunson, and S.D. Roberts. 1998. Landowners' educational needs and how foresters can respond. Journal of Forestry 96(8):38-43.

Lamb, J.F.S., H.J.G. Jung, C.C. Sheaffer, and D.A. Samac. 2007. Alfalfa leaf protein and stem cell wall polysaccharide yields under hay and biomass management systems. Crop Science 47(4):1407-1415.

Mattison, E.H.A., and K. Norris. 2007. Intentions of UK farmers toward biofuel crop production: Implications for policy targets and land use change. Environmental Science and Technology 41(16):5589-5594.

McCormick, K. 2010. Communicating bioenergy: A growing challenge. Biofuels, Bioproducts and Biorefining 4(5):494-502.

Meersman, T. 2008. Is ethanol tapping too much water? P. 1A in Minneasopolis Star Tribune, January 28, News section.

Meersman, T. 2010. Minnesota ethanol plants' price is pollution. Star Tribune, October 11. Available online at http://www.startribune.com/local/104746614.html?elr=KArks:DCiU1PciUiD3aPc:_Yyc:aUoaEYY_1Pc_bDaEP7U. Accessed January 21, 2011.

Muehlegger, E.J. 2002. The Role of Content Regulation on Pricing and Market Power in Regional Retail and Wholesale Gasoline Markets. Cambridge: Massachusetts Institute of Technology Center for Energy and Environmental Policy Research.

Muehlegger, E.J. 2006a. Gasoline price spikes and regional gasoline content regulation: A structural approach. Available online at http://www.hks.harvard.edu/fs/emuehle/Research%20WP/MuehleggerGasoline3_06.pdf. Accessed September 21, 2010.

Muehlegger, E.J. 2006b. Market effects of regulatory heterogeneity: A study of regional gasoline content regulations. Available online at http://www.hks.harvard.edu/fs/emuehle/Research%20WP/Regulatory%20Heterogeneity.pdf. Accessed September 21, 2010.

Napier, T.L., M. Tucker, and S. McCarter. 2000. Adoption of conservation production systems in three Midwest watersheds. Journal of Soil and Water Conservation 55(2):123-134.

NAS-NAE-NRC (National Academy of Sciences, National Academy of Engineering, National Research Council). 2009. Liquid Transportation Fuels from Coal and Biomass: Technological Status, Costs, and Environmental Impacts. Washington, DC: National Academies Press.

Nassauer, J.I. 1989. Agricultural policy and aesthetic objectives. Journal of Soil and Water Conservation 44(5):384-387.

NBB (National Biodiesel Board). 1998. Summary Results From NBB/USEPA Tier I Health and Environmental Effects Testing for Biodiesel Under the Requirements for USEPA Registration of Fuels and Fuel Additives (40 CFR Part 79, Sec 21 1 (B)(2) and 21 1 (E)): Final Report. Jefferson City, MO: National Biodiesel Board.

NESCAUM (Northeast States for Coordinated Air Use Management). 2010. Low carbon fuels. Available online at http://www.nescaum.org/topics/low-carbon-fuels. Accessed February 4, 2011.

NRC (National Research Council). 2010. Expanding Biofuel Production and the Transition to Advanced Biofuels. Lessons for Sustainability from the Upper Midwest. Washington, DC: National Academies Press.

O'Sullivan, J. 2010. SD ethanol plants fined $225k for water violations. Available online at http://thepostsd.com/2010/12/12/sd-ethanol-plants-fined-225k-for-water-violations/. Accessed February 14, 2011.

Oregon Environmental Council. 2010. Oregon needs a low carbon fuel standard (HB 2186). Available online at http://www.oeconline.org/resources/publications/factsheetarchive/LCFS%20Overview.pdf. Accessed September 23, 2010.

ORNL (Oak Ridge National Laboratory). 2010. Bioenergy Feedstock Information Network. Available online at http://bioenergy.ornl.gov/. Accessed November 18, 2010.

Orta, J., Z. Zhang, K. Koyama, J. McKinney, S. Michael, C. Mitzutani, S. Fromm, G. O'Neill, P. Doughman, M. Leaon, R.d. Mesa, S. Garfield, C. Robinson, and S. Korosec. 2010. 2009 Progress to Plan—Bioenergy Action Plan for California. Sacramento: California Energy Commission.

Peck, P., S.J. Bennett, R. Bissett-Amess, J. Lenhart, and H. Mozaffarian. 2009. Examining understanding, acceptance, and support for the biorefinery concept among EU policy-makers. Biofuels, Bioproducts and Biorefining 3(3):361-383.

Rogers, E.M. 2003. Diffusion of Innovations. 5th ed. New York: Free Press.

Rohracher, H. 2010. Biofuels and their publics: The need for differentiated analyses and strategies. Biofuels 1(1):3-5.

Rossi, A.M., and C.C. Hinrichs. 2011. Hope and skepticism: Farmer and local community views on the socio-economic benefits of agricultural bioenergy. Biomass and Bioenergy 35(4):1418-1428.

Saccomandi, V. 1998. Agricultural Market Economics: A Neo-Institutional Analysis of the Exchange, Circulation and Distribution of Agricultural Products. Assen, Netherlands: Van Gorcum and Company.

Sadaka, S., and S. Negi. 2009. Improvements of biomass physical and thermochemical characteristics via torrefaction process. Environmental Progress and Sustainable Energy 28(3):427-434.

Schnepf, R. 2010. Cellulosic Ethanol: Feedstocks, Conversion Technologies, Economics, and Policy Options. Washington, DC: Congressional Research Service.

Scoll, J., and T. Guerrero. 2006. Legal perspective. An overview of biodiesel registration with the U.S. EPA. Available online at http://www.biodieselmagazine.com/articles/761/legal-perspective. Accessed May 12, 2011.

Singh, R. 2009. Ethanol corrosion in pipelines. Materials Performance 48(5):53-55.

Smith, S. 2008. Pollution violations may test public support for biodiesel. Available online at http://www.biodieselmagazine.com/articles/2383/pollution-violations-may-test-public-support-for-biodiesel. Accessed January 21, 2011.

Song, F., J. Zhao, and S.M. Swinton. 2009. Switching to Perennial Energy Crops under Uncertainty and Costly Reversibility. East Lansing: Michigan State University.

Stiglitz, J.E. 1987. The causes and consequences of the dependence of quality on price. Journal of Economic Literature 25(1):1-48.

Stiglitz, J.E. 2002. Information and the change in the paradigm of economics. American Economic Review 92(3):460-501.

Swanson, R.M., J.A. Satrio, R.C. Brown, A. Platon, and D.D. Hsu. 2010. Techno-Economic Analysis of Biofuels Production Based on Gasification. Golden, CO: National Renewable Energy Laboratory.

Tao, L., and A. Aden. 2009. The economics of current and future biofuels. In Vitro Cellular and Developmental Biology-Plant 45(3):199-217.

Thompson, W., S. Meyer, and P. Westhoff. 2010. The new markets for renewable identification numbers. Applied Economic Perspectives and Policy 32(4):588-603.

Thompson, W., S. Meyer, and P. Westhoff. 2011. What to conclude about biofuel mandates from evolving prices for renewable identification numbers? American Journal of Agricultural Economics 93(2):481-487.

Tomek, W.G., and K.L. Robinson. 1983. Agricultural Product Prices. Ithaca, NY: Cornell University Press.

Tyner, W.E., F. Dooley, and D. Viteri. 2011. Effects of biofuel mandates in a context of ethanol demand constraints, cellulosic biofuel costs, and compliance mechanisms. American Journal of Agricultural Economics 93(2):488-489.

UNL (University of Nebraska, Lincoln). 2011a. UNL water: Agricultural irrigation. Available online at http://water.unl.edu/web/cropswater/home. Accessed January 28, 2011.

UNL (University of Nebraska, Lincoln). 2011b. UNL water: Drought. Available online at http://water.unl.edu/drought. Accessed January 28, 2011.

Van den Wall Bake, J.D., M. Junginger, A. Faaij, T. Poot, and A. Walter. 2009. Explaining the experience curve: Cost reductions of Brazilian ethanol from sugarcane. Biomass and Bioenergy 33(4):644-658.

VeroNews. 2010. Biofuel facility clears permits needed to begin construction on Oslo Rd. Available online at http://www.veronews.com/index.php?option=com_content&view=article&id=11774:biofuel. Accessed January 21, 2011.

Villamil, M.B., A.H. Silvis, and G.A. Bollero. 2008. Potential Miscanthus' adoption in Illinois: Information needs and preferred information channels. Biomass and Bioenergy 32(12):1338-1348.

Wegener, D., and J. Kelly. 2008. Social psychological dimensions of bioenergy development and public acceptance. Bioenergy Research 1(2):107-117.

Williamson, O.E. 1996. The Mechanisms of Governance. Oxford: Oxford University Press.

Wright, M.M., J.A. Satrio, R.C. Brown, D.E. Daugaard, and D.D. Hsu. 2010. Techno-Economic Analysis of Biomass Fast Pyrolysis to Transportation Fuels. Golden, CO: National Renewable Energy Laboratory.

Yan, W., T.C. Acharjee, C.J. Coronella, and V.R. Vasquez. 2009. Thermal pretreatment of lignocellulosic biomass. Environmental Progress and Sustainable Energy 28(3):435-440.

Yeh, S., and D. Sperling. 2010. Low carbon fuel standards: Implementation scenarios and challenges. Energy Policy 38(11):6955-6965.

Zhang, Y., S. Joshi, and H.L. MacLean. 2010. Can ethanol alone meet California's low carbon fuel standard? An evaluation of feedstock and conversion alternatives. Environmental Research Letters 5(January-March):014002.

Zoellner, J., P. Schweizer-Ries, and C. Wemheuer. 2008. Public acceptance of renewable energies: Results from case studies in Germany. Energy Policy 36(11):4136-4141.

APPENDIXES

A

Statement of Task

Using a series of public meetings to gather information and work sessions in which that information is analyzed, the committee will develop a consensus report that includes the following components:

- A quantitative and qualitative description of biofuels currently produced and projected to be produced and consumed by 2022 in the United States under different policy scenarios, including scenarios with and without current Renewable Fuel Standard (RFS) and biofuel tax and tariff policies, and considering a range of future fossil energy and biofuel prices, the impact of a carbon price, and advances in technology. The analysis will include a review of estimates of potential biofuel production levels using RFS-compliant feedstocks from U.S. forests and farmland, including the per-unit cost of that production. The study will assess the effects of current and projected levels of biofuel production, and the incremental impact of additional production, on the number of U.S. acres used for crops, forestry, and other uses, and the associated changes in the price of rural and suburban land.
- A review of model results and other estimates of the relative effects of the RFS, biofuel tax and tariff policy, production costs, and other factors, alone and in combination, on biofuel and petroleum refining capacity, and on the types, amounts and prices of biofuel feedstocks, biofuels, and petroleum-based fuels (including finished motor fuels) produced and consumed in the United States.
- An analysis of the effects of current and projected levels of biofuel production, and the incremental impact of additional production, on U.S. exports and imports of grain crops, forest products and fossil fuels, and on the price of domestic animal feedstocks, forest products, and food grains.
- An analysis of the effect of projected biofuel production on federal revenue and spending, through costs or savings to commodity crop payments, biofuel subsidies, and tariff revenue.

289

- An analysis of the pros and cons of achieving legislated RFS levels, including the impacts of potential shortfalls in feedstock production on the prices of animal feed, food grains, and forest products, and including an examination of the impact of the cellulosic biofuel tax credit established by Sec. 15321 of the Food, Conservation, and Energy Act of 2008 on the regional agricultural and silvicultural capabilities of commercially available forest inventories. This analysis will explore policy options to maintain regional agricultural and silvicultural capacity in the long term, given RFS requirements for annual increases in the volume of renewable fuels, and include recommendations for the means by which the federal government could prevent or minimize adverse impacts of the RFS on the price and availability of animal feedstocks, food and forest products, including options available under current law.
- An analysis of barriers to achieving the RFS requirements.
- An analysis of the impact of current and projected future levels of biofuel production and use, and the incremental impact of additional production, on the environment. The analysis will consider impacts due to changes in land use, fertilizer use, runoff, water use and quality, greenhouse-gas and local pollutant emissions from vehicles utilizing biofuels, use of forestland biomass, and other factors relevant to the full lifecycle of biofuel production and use. The analysis will summarize and evaluate various estimates of the indirect effects of biofuel production on changes in land use and the environmental implications of those effects.
- A comparison of corn ethanol versus other biofuels and renewable energy sources for the transportation sector based on life-cycle analyses, considering cost, energy output, and environmental impacts, including greenhouse-gas emissions.
- Recommendations for additional scientific inquiry related to the items above, and specific areas of interest for future research.

As part of its deliberations, the committee will consider the relevant reports of past NRC committees, the work of relevant current committees, and reports of other organizations, and individual researchers. In addition, the committee will consider the relevant experience and reports of various federal government agencies.

To inform its analysis, the study committee will seek the input of feed grain producers; food animal producers; producers of other food products; energy producers (renewable and petroleum-based fuel producers, fuel blenders); forest owners and forest products manufacturers and users; individuals and entities interested in nutrition, or in the relationship of the environment to energy production; producers and users of renewable fuel feedstocks; users of renewable fuels; and experts in agricultural economics from land grant universities.

B

Biographical Sketches

Lester B. Lave *(chair until May 9, 2011)* was the Harry B. and James H. Higgins Professor of Economics and Finance, a professor of engineering and public policy, and director of the university's Green Design Institute at the Carnegie Mellon University before his death in 2011. Dr. Lave's work focused on environmental quality and risk management, and more specifically on modeling the effects of global climate change, improving social regulations, risk perception and communication, the value of information in tests for carcinogenicity, highway safety, electricity generation and use, and pollution prevention. As the head of the university-wide Green Design Initiative, Dr. Lave collaborated with private businesses and with government agencies such as the U.S. Department of Energy to address the fundamental problems in pollution prevention. A recipient of the Distinguished Achievement Award of the Society for Risk Analysis, Dr. Lave was a member of the Institute of Medicine and served on numerous NRC committees, including the Panel on Energy Efficiency as chair and Committee on America's Energy Future as a member. Dr. Lave received his Ph.D. in economics from Harvard University.

Ingrid C. Burke *(cochair from May 9, 2011)* is director of the Haub School and Ruckelshaus Institute of Environment and Natural Resources at the University of Wyoming. She also is a professor and holds a Wyoming Excellence Chair in the Departments of Botany and Renewable Resources. She is a former professor and University Distinguished Teaching Scholar in the College of Natural Resources at Colorado State University. Dr. Burke is an ecosystem scientist, with particular expertise in carbon and nitrogen cycling of semi-arid ecosystems. She directed the Shortgrass Steppe Long Term Ecological Research team for 6 years, as well as other large interdisciplinary research teams funded by the National Science Foundation, the Environmental Protection Agency, the National Aeronautics and Space Adminstration, and the National Institutes of Health. She was designated a U.S. Presidential Faculty Fellow, has served on the NRC Board on Environmental Science and Toxicology, and was a member of the NRC committee tasked with developing recommendations on *A New Biology for the 21st Century: Ensuring That the United States Leads the Coming Biology Revolution.*

Dr. Burke also serves as an associate editor for the journal *Ecological Applications* and is the new chair of the advisory committee for the Greater Yellowstone National Environmental Observatory Network research site. She received her Ph.D. in botany from the University of Wyoming.

Wallace E. Tyner *(cochair from May 9, 2011)* is the James and Lois Ackerman Professor of Agricultural Economics at Purdue University and co-director of the Purdue Center for Research on Energy Systems and Policy. His research interests are in the areas of energy, agricultural, and natural resource policy analysis, and structural and sectoral adjustment in developing economies. His work in energy economics has encompassed oil, natural gas, coal, oil shale, biomass, ethanol from agricultural sources, and solar energy. Most of his recent work has focused on economic and policy analysis for biofuels, with international work on agricultural trade and policy issues in developing economies. Dr. Tyner received the Agricultural and Applied Economics Association (AAEA) Distinguished Policy Contribution Award in 2005. In 2007 he received the "Energy Patriot Award" from Senator Richard Lugar. In 2009, he was named the Outstanding Graduate Educator in the College of Agriculture, received the College Team award (with colleagues) for biofuel research, and received (with colleagues) the AAEA Quality of Communication award. He teaches a graduate course in benefit-cost analysis. Dr. Tyner is author or co-author of three books: *Energy Resources and Economic Development in India, Western Coal: Promise or Problem* (with R. J. Kalter), and *A Perspective on U.S. Farm Problems and Agricultural Policy* (with Lance McKinzie and Tim Baker). Dr. Tyner has a Ph.D. in economics from the University of Maryland.

Virginia H. Dale is director of the Center for BioEnergy Sustainability, corporate fellow, and group leader of the Landscape Ecology and Regional Analysis Group in the Environmental Sciences Division at Oak Ridge National Laboratory. She is also an adjunct professor in the Department of Ecology and Evolutionary Biology and the Department of Forestry, Wildlife and Fisheries at the University of Tennessee. Dr. Dale's primary research interests are in landscape design for bioenergy, environmental decision-making, land-use change, landscape ecology, and ecological modeling. She has worked on developing tools for resource management, vegetation recovery subsequent to disturbances, effects of climate change on forests, and integrating socioeconomic and ecological models of land-use change. Dr. Dale has served on national scientific advisory boards for five federal agencies (the U.S. Environmental Protection Agency, and the U.S. Departments of Agriculture, Defense, Energy, and Interior). She has also served on several NRC committees. She is editor-in-chief of the journal *Environmental Management* and is on the editorial board for *Ecological Indicators, Ecological Economics,* and the *Journal of Land Use Science*. She was chair of the U.S. Regional Association of the International Association for Landscape Ecology and has served on the scientific review team for The Nature Conservancy. She served on the Executive Committee of the Policy Team for Southern Appalachian Assessment, which won Vice President Gore's Hammer Award. Dr. Dale has a Ph.D. in mathematical ecology from the University of Washington.

Kathleen E. Halvorsen is an associate professor of natural resource policy at Michigan Technological University. She has a joint appointment with the Department of Social Sciences and the School of Forest Resources and Environmental Science. Her research focuses on two main areas. One relates to the development of woody bioenergy in the United States and includes identification of barriers and opportunities related to this development. She views bioenergy as an important tool in the climate change mitigation toolbox. Her other area of research is aimed at understanding relationships to water resources in the United States and Mexico. That research includes risk perceptions of water-borne disease and ecosystem service protection. Over the years, she also has studied public participation and organizational change within the U.S. Department of Agriculture Forest Service. Dr. Halvorsen received her Ph.D. from the University of Washington.

Jason D. Hill is an assistant professor in the Department of Bioproducts and Biosystems Engineering at the University of Minnesota. His research interests include the technological, environmental, economic, and social aspects of sustainable bioenergy production from current and next-generation feedstocks. His work on the life-cycle impacts of transportation biofuels has been published in the journals *Science* and the *Proceedings of the National Academy of Sciences of the United States of America*. His current research focuses on the effects that the expanding global biofuel industry is having on climate change, land use, biodiversity, and human health. Dr. Hill has testified before U.S. Senate committees on the use of diverse prairie biomass for biofuel production and on the greenhouse-gas implications of ethanol and biodiesel. He has also performed independent analysis for the National Renewable Energy Laboratory, NRC, and the U.S. Environmental Protection Agency. Dr. Hill served on the NRC Steering Committee on Expanding Biofuel Production: Sustainability and Transition to Advanced Biofuels. Dr. Hill received his Ph.D. in plant biological sciences from the University of Minnesota.

Stephen R. Kaffka is director of the California Biomass Collaborative and extension specialist in the Department of Plant Sciences at the University of California, Davis. He is chair of the BioEnergy Work Group for the University of California's Division of Agriculture and Natural Resources. From 2003 to 2007, he was director of the Long Term Research on Agricultural Systems Project, in which he led the development of current and new projects focusing on sustainable agriculture. His commodity assignments include sugar and oilseed crops. Since joining the university in 1992, he has also carried out research on water quality and agriculture in the Upper Klamath Basin, and the reuse of saline drainage water for crop, forage, energy biomass feedstocks, and livestock production in salt-affected areas of the San Joaquin Valley. He participates on several advisory committees for the California Energy Commission and California Air Resources Board, including the Bioenergy Interagency Work Group as an ex officio member. He has received meritorious service awards from the American Society of Sugar Beet Technologists and the Soil and Water Conservation Society. He is past president of the California chapter of the American Society of Agronomy and past section leader for the American Society of Agronomy's division on environmental quality. He holds a Ph.D. in agronomy from Cornell University.

Kirk C. Klasing is a professor of animal nutrition in the Department of Animal Science at the University of California, Davis. His research into the impact of nutrition on immunochemistry and disease resistance encompasses three interrelated areas. He examines the impact of an immune response against infectious diseases on growth and reproduction. He is interested in identifying the cytokines and hormones that the immune system releases in order to direct nutritional resources towards defense instead of growth and reproduction. Dr. Klasing strives to quantify the nutritional costs of these immune defenses, and investigates the impact of an animal's diet on the immune response. He also explores the diverse nutritional and immune strategies of carnivorous, nectarivorous, herbivorous, and granivorous animals. Dr. Klasing has served on several NRC committees, including the Committee on Minerals and Toxic Substances in Diets and Water for Animals as chair and the Board on Agriculture and Natural Resources as a current member. He has received the Poultry Science Research Award from the Poultry Science Association, the BioServ Award from the American Institute of Nutrition, and the Lilly Animal Scientist Award. He holds a Ph.D. in nutritional biochemistry from Cornell University.

Stephen J. McGovern has over 35 years of experience in the refining and petrochemical industries. He has been a principal of PetroTech Consultants since 2000, providing consulting services on various refining technologies, including clean fuel projects and refining economics. He has assisted numerous refiners in the evaluation of gasoline and diesel desulfurization technologies, catalytic cracking, and environmental issues. He has provided technical advice to the Defense Advanced Research Projects Agency and commercial enterprises for the production of biofuels. Previously, he was with Mobil Technology Company, where he was involved in process development and refinery technical support. Dr. McGovern has 17 patents and has written many publications. He has lectured, published, and consulted on refining technology and environmental issues. He is a licensed professional engineer in New Jersey and a past director of the Fuels and Petrochemicals Division of the American Institute of Chemical Engineers. He earned a Ph.D. in chemical engineering from Princeton University.

John A. Miranowski is a professor of agricultural economics and of environmental and resource economics at Iowa State University. His background is in natural resource management, agricultural research decision-making, and environmental policy. He served as chair of the Department of Economics from 1995 to 2000. Dr. Miranowski has further expertise in soil conservation, water quality, land management, energy, and global change. He has previously served as director of the Resources and Technology Division of the U.S. Department of Agriculture Economic Research Service, 1984-1994; as executive coordinator of the Secretary of Agriculture's Policy Coordination Council, and special assistant to the deputy secretary of agriculture, 1990-1991; and as the Gilbert F. White Fellow at Resources for the Future, 1981-1982. Dr. Miranowski headed the U.S. delegation to the Organization for Economic Cooperation and Development Joint Working Party on Agriculture and the Environment, 1993-1995. He has served as a member of the Ad Hoc Working Group on Risk Assessment of Federal Coordinating Committee on Science, Education, and Technology, 1990-1992; director of the Executive Board of the Association of Environmental and Resource Economists, 1989-1992; and director of the Executive Board of the American Agricultural Economics Association, 1987-1990. Dr. Miranowski served as a member of the NRC Panel on Alternative Liquid Transportation Fuels, and the Committee on Expanding Biofuel Production—Lessons from the Upper Midwest for Sustainability. He received his Ph.D. in economics from Harvard University.

Aristides A. N. Patrinos is president of Synthetic Genomics. He served on the staff of the U.S. Environmental Protection Agency and joined the U.S. Department of Energy (DOE) in 1988 and led the development of DOE's program in global environmental change. From 1995 to 2006, Dr. Patrinos was the associate director for biological and environmental research in DOE's Office of Science, where he oversaw research activities in the human and microbial genome, structural biology, nuclear medicine, and global environmental change. He also directed the DOE component of the U.S. Human Genome Project and was the DOE representative to the U.S. Climate Change Science Program and the Climate Change Technology Program. He is the recipient of numerous awards and honorary degrees, including three Presidential Rank Awards for meritorious and distinguished service and two Secretary of Energy gold medals. He is a Fellow of the American Association for the Advancement of Science and the American Meteorological Society, and a member of the American Society of Mechanical Engineers and the American Geophysical Union. He has served on the NRC Committees on Strategic Advice on the U.S. Climate Change Science Program and on America's Energy Future. Dr. Patrinos received his Ph.D. in mechanical and astronautical sciences from Northwestern University.

Jerald L. Schnoor is the Allen S. Henry Chair Professor of Environmental Engineering and codirector of the Center for Global and Regional Environmental Research at the University of Iowa. Dr. Schnoor is a member of the National Academy of Engineering for his pioneering work using mathematical models in science-policy decisions. He testified several times before Congress on the environmental effects of acid deposition and the importance of passing the 1990 Clean Air Act. Dr. Schnoor chaired the Board of Scientific Counselors for the U.S. Environmental Protection Agency, Office of Research and Development from 2000-2004. Currently, he is one of three co-directors for the National Science Foundation Project Office on a Collaborative Large-scale Engineering Analysis Network for Environmental Research (CLEANER). As editor-in-chief of *Environmental Science and Technology*, Dr. Schnoor guides the journal in both environmental engineering and environmental science. His research interests are in mathematical modeling of water quality, phytoremediation, and impact of carbon emissions on global change. He conducts research on the aquatic effects modeling of acid precipitation, global change and biogeochemistry, groundwater and hazardous wastes, and exposure risk assessment modeling. Dr. Schnoor has served on several NRC committees including the Committee on Water Implications of Biofuels Production in the United States and the Civil Engineering Peer Committee. Dr. Schnoor received his Ph.D. in civil engineering from the University of Texas.

David Schweikhardt is a professor in the Agricultural, Food, and Resource Economics Department at Michigan State University. He specializes in agricultural and international trade policy. In particular, his work examines the implications of North American Free Trade Agreement and the Uruguay Round Agreement of General Agreement on Tariffs and Trade on U.S. and Michigan agriculture; analysis of U.S. commodity programs; law, economics, and the analysis of changes in legal and economic institutions; political economy of agricultural and trade policy decision-making processes; legal issues in commodity checkoff programs; and labeling of genetically modified food products. Dr. Schweikhardt received a Ph.D. in agricultural economics from Michigan State University.

Theresa L. Selfa is assistant professor in environmental studies at the State University of New York, College of Environmental Science and Forestry (SUNY-ESF). Her current research focuses on biofuel policy and attitudes. Additional research interests include food and agriculture, development, and political ecology. Prior to joining the SUNY-ESF faculty, she was assistant professor of sociology at Kansas State University. She has expertise in rural, environmental, agricultural, and development sociology, with research experience in Brazil, Philippines, Europe, and the United States. She was a postdoctoral associate in Washington State on a project examining alternative agriculture and food systems. She has worked on interdisciplinary water quality projects assessing impacts of farmers' management behavior on water quality in an agricultural watershed in central Kansas and in Devon, England. She is the principal investigator on a study on the impacts of biofuels on rural communities in Kansas and Iowa funded by the U.S. Department of Energy and is a coprincipal investigator on research assessing farmers' land-use decisions regarding advanced biofuel crops funded by the National Science Foundation. Her work has been published in *Society and Natural Resources, Environment and Planning A, Renewable Agriculture and Food Systems*, and *Environmental Science and Policy*. Dr. Selfa received her Ph.D. in development sociology from Cornell University.

Brent Sohngen is a professor in the Department of Agricultural, Environmental, and Development Economics at Ohio State University. His research interest is in modeling land-use and land-cover change, economics of nonpoint source pollution, and valuing environmental change. Projects that he is working on include one on forests, economics, and global climate change, and the global timber market and forestry data project. Dr. Sohngen received his Ph.D. in natural resource and environmental economics from Yale University.

J. Andres Soria is an assistant professor of wood chemistry in the Department of Forest Sciences at the University of Alaska, Fairbanks. He also has an appointment with the School of Applied Environmental Science and Technology at the University of Alaska, Anchorage. He worked as a researcher and instructor at the University of Idaho from 2002 to 2005. Dr. Soria's research involves utilizing waste and undervalued biomass to create products ranging from fuels, additives, and chemical feedstocks from Alaskan species. Dr. Soria also performs research on agricultural byproducts and wastes. He teaches courses in the area of energy and forest products. Dr. Soria's honors include a 2005 Outstanding Graduate Student, Department of Forest Products, University of Idaho; a Stillinger Endowment recipient, University of Idaho, 2002-2005; and a Foster Fellowship recipient, University of Idaho, 2003-2005. He earned his Ph.D. in natural resources from the University of Idaho.

C

Presentations to the Committee

JANUARY 15, 2010

Presentation by Sponsors
Chris Soares, U.S. Department of Treasury
Zia Haq, U.S. Department of Energy (DOE) Energy Efficiency and Renewable Energy (EERE)
Sharlene Weatherwax, DOE Office of Science
Paul Argyropoulos, U.S. Environmental Protection Agency (EPA)
Alan Hecht, EPA
Harry Baumes, U.S. Department of Agriculture (USDA)

Economic Impacts of Increasing Biofuels Production
Bruce McCarl, Texas A&M University

Opportunities and Challenges of Biofuel Production to Improving Environmental Quality
John Sheehan, University of Minnesota

Opportunities and Challenges of Biofuel Production from Woody Biomass to Improving
    Environmental Quality
Marilyn Buford, USDA Forest Service

MARCH 5, 2010

Future Demand for Food
Ron Trostle, USDA Economic Research Service

Climate Effects of Transportation Fuels: Uncertainty and Its Policy Implications
Richard Plevin, University of California, Berkeley

Building Uncertainties into Modeling of Direct, Life-Cycle Greenhouse-Gas Emissions, Net
     Energy and Other Environmental Effects of Biofuels
Garvin Heath, National Renewable Energy Laboratory

Biomass Supply and Infrastructure for Biofuels
Bryan Jenkins, University of California, Davis

Input from Stakeholders' Groups
Geoff Cooper, Renewable Fuels Association
Michael Edgerton, Monsanto, on behalf of the National Corn Growers' Association
Jaime Jonker, National Milk Producers Federation
Al Mannato, American Petroleum Institute
Tim Hogan, National Petrochemical and Refiners Association
Julie Sibbing, National Wildlife Federation

MAY 3, 2010

Update of the "Billion Ton" Study
Robert Perlack, Oak Ridge National Laboratory, and
Bryce Stokes, DOE-EERE

Prospects for Medium Run Biomass Supply in the U.S. South
Robert Abt, North Carolina State University

Input from Stakeholders' Groups
Roger Conway, Growth Energy
Manning Ferraci, National Biodiesel Board
Tom Hance, Gordley Associates on behalf of the American Soybean Association
David Tenny, National Alliance of Forest Owners
Paul Noe, American Forest and Paper Association
Richard Lobb, National Chicken Council
Randy Spronk, National Pork Producers Council
Gregg Doud, National Cattlemen's Beef Association
Joel Brandenberger, National Turkey Federation

JULY 12, 2010

Biofuel Discussion with National Renewable Energy Laboratory Staff
Dale Gardner, Mike Cleary, Helena Chum, Andy Aden, Mark Ruth, and Garvin Heath,
     National Renewable Energy Laboratory

JULY 14, 2010

Research and Development for Improving Yield and Decreasing Environmental Impacts
     of Major Crops
Robb Fraley, Monsanto

OCTOBER 7, 2010

Indirect Land-Use Change
David Zilberman, University of California, Berkeley

World Demand for Food in the Future
David Roland-Holst, University of California, Berkeley

OCTOBER 8, 2010

Biofuels and the Environment: A 2010 Report to Congress
Denice Shaw, Bob Frederick, Caroline Ridley, Stephen LeDuc, EPA

D

Glossary

Alcohol fuels	Fuels that are organic compounds that contain one or more hydroxyl groups (-OH) attached to one or more of the carbon atoms in a hydrocarbon chain. Common alcohol fuels include ethanol, methanol, and butanol.
Algae	A group of aquatic eukaryotic organisms that contain chlorophyll. Algae can be microscopic in size (microalgae) or observable to the eye (macroalgae).
Aliphatic alcohol	An alcohol that contains a hydrocarbon fragment derived from a fully saturated, nonaromatic hydrocarbon.
Anoxia	The absence of dissolved oxygen.
Biodiesel	Diesel fuel consisting of long-chain alkyl esters derived from biological material such as vegetable oils, animal fats, and algal oils.
Biofuel	Fuel derived from biomass.
Biomass	Any organic matter that is available on a renewable or recurring basis, including agricultural crops and trees, wood and wood residues, plants (including aquatic plants), grasses, animal residues, municipal residues, and other residue materials.
Biorefinery	A commercial-scale processing facility that successfully integrates all processes for extracting and converting biomass feedstocks into a spectrum of saleable products.
Carbon sequestration	Net transfer of atmospheric carbon dioxide into long-lived carbon pools.
Cellulose	A polymer of glucose, $(C_6H_{10}O_5)_n$, that forms cell walls of most plants.

301

Commercial demonstration	The National Renewable Energy Laboratory (NREL) defines a commercial demonstration for biofuel refinery as a facility that has the capacity to process 700 dry tons of feedstock per day. In addition, a commercial demonstration facility will be a fully integrated facility that includes all processing steps at a scale sufficient to identify potential operational problems.
Corn stover	Corn stalks, leaves, and cobs that remain after the corn-grain is harvested.
Demonstration facility	NREL defines a demonstration facility for biofuel refinery as one that has the capacity to process 70 dry tons of feedstock per day. A true demonstration facility will be a fully integrated facility that includes all of the processing steps that a commercial-scale plant would have.
Drop-in fuel	Nonpetroleum fuel that is compatible with existing infrastructure for petroleum-based fuels.
Feedstock	Material that can be processed to make fuel, including grains, crop residues, forestry products, plant oils, animal fats, and municipal wastes.
Feedstuff	Nutrient-rich material that can be incorporated into the diet of livestock or other animals.
Green diesel	Hydrogenation product of triglycerides.
Hemicellulose	A matrix of polysaccharides present in almost all plant cell walls with cellulose.
Hydrocarbon fuels	Fuels that are organic compounds that contains primarily carbon and hydrogen and only trace amounts of other atoms such as sulfur, nitrogen, and oxygen. Hydrocarbon fuels include petroleum-based materials such as alkanes, olefins, and aromatics.
Hypoxia	Low dissolved oxygen concentrations, generally less than 2 milligrams per liter.
Land cover	Land cover is the extent and type of physical and biological cover over the surface of land.
Land use	Land use is defined by anthropogenic activities, such as agriculture, forestry, and urban development, that alter land-surface processes including biogeochemistry, hydrology, and biodiversity.
Lignin	A complex polymer that occurs in certain plant cell walls. Lignin binds to cellulose fibers and hardens and strengthens the cell walls of plants.
Lignocellulosic biomass	Plant biomass composed of cellulose, hemicellulose, and lignin.
Pilot demonstration	NREL defines a pilot demonstration for biofuel refinery as a facility that has the capacity to process 1-10 dry tons of feedstock per day. These facilities typically do not include fully integrated processes.
Reid Vapor Pressure	A measure of fuel volatility.
Stumpage	A fee charged by a landowner to companies or operators for the right to harvest timber on that land.

E

Select Acronyms and Abbreviations

BCAP	Biomass Crop Assistance Program
BioBreak	Biofuel breakeven model
BOD	Biological oxygen demand
CBOB	Conventional blendstock for oxygenate blending
CO	Carbon monoxide
CO_2	Carbon dioxide
CPI	Consumer price index
CRP	Conservation Reserve Program
CWD	Coarse woody debris
DDGS	Dried distillers grains with solubles
DOE	U.S. Department of Energy
E10	A blend of up to 10-percent ethanol and the balance petroleum-based gasoline
E15	A blend of up to 15-percent ethanol and the balance petroleum-based gasoline
E85	A blend of up to 85-percent ethanol and the balance petroleum-based gasoline. For the past several years, E85 sold in the United States has averaged about 75-percent ethanol.
EPA	U.S. Environmental Protection Agency
EPAct	Energy Policy Act
EIA	Energy Information Administration
EISA	Energy Independence and Security Act
ETBE	Ethyl tertiary butyl ether

FAME	Fatty acid methyl ester
FAPRI	Food and Agricultural Policy Research Institute
FASOM	Forest and agricultural sector optimization model
FCC	Fluidized catalytic cracking
FFV	Flex-fuel vehicle
F-T	Fischer-Tropsch
GE	General equilibrium
GHG	Greenhouse gas
GMS	Groundwater monitoring system
GTAP	Global Trade Analysis Project
HSPF	Hydrological Simulation Program – Fortran
ILUC	Indirect land-use change
LCFS	Low-carbon fuel standard
LUC	Land-use change
MTBE	Methyl tertiary butyl ether
NASQAN	National Stream-Quality Accounting Network
NAWQA	National Water-Quality Assessment
NEQA	National Environmental Quality Act
NEXRAD	Next-generation radar
NH_3	Ammonia
NIPF	Nonindustrial private forest
NPV	Net present value
NMHCs	Nonmethane hydrocarbons
NO_2	Nitrous oxide
NO_3	Nitrate
NO_x	Nitrous oxides
NPDES	National Pollutant Discharge Elimination System
NWOS	National Woodland Owner Survey
O_3	Ozone
OECD	Organisation for Economic Co-operation and Development
OPEC	Organization of the Petroleum Exporting Countries
PAH	Polycyclic aromatic hydrocarbon
PE	Partial equilibrium
PM	Particulate matter
POLYSYS	Policy analysis system model
QUAL2K	River and stream water quality model
RBOB	Reformulated blendstock for oxygenate blending
REIT	Real estate investment trust

RFS   Renewable Fuel Standard
RINs   Renewable identification numbers
RUSLE   Revised universal soil loss equation
RVP   Reid vapor pressure

SNAP   Supplemental Nutrition Assistance Program
SO_2 Sulfur dioxide
SOC   Soil organic carbon
SOM   Soil organic matter
SPARROW   Spatially referenced regressions on watershed attributes
SRWC   Short-rotation woody crops
SWAT   Soil and water assessment tool

TIMO   Timber Investment Management Organization
TMDL   Total maximum daily loads

USDA   U.S. Department of Agriculture
USGS   U.S. Geological Survey

VEETC   Volumetric ethanol excise tax credit
VOCs   Volatile organic compounds

WASP   Water quality analysis simulation program
WIC   Special Supplemental Nutrition Program for Women, Infants and
               Children
WTA   Willingness to accept
WTP   Willingness to pay

F

Conversion Factors

Mass

1	ounce (oz)	≡	28.3495231	g
1	pound	≡	0.453592	kg
1	(short) ton	≡	0.907185	(metric) tonne

Length

1	foot (ft)	≡	0.3048	m (meter)
1	mile	≡	1.609344	km (kilometer)

Area

1	mi^2	≡	2.589988	km^2
1	acre	≡	0.404685642	hectare (ha)

Volume

1	ft^3	≡	0.028317	m^3
1	gallon	≡	3.785412	liter (L)
1	barrel	≡	158.987295	L

Energy

1	British thermal unit (Btu)	≡	0.001055	megajoule (MJ)

Pressure

1	pounds per square inch (psi)	≡	6,894.76	Pascal (Pa)

Compound units

1	pound per bushel	≡	17.857143	kg/tonne
1	pound per acre	≡	1.120851	kg/ha

1	bushel per acre	≡	0.062768	tonne/ha
1	ton per acre	≡	2.241702	tonne/ha
1	ounce (oz) per gallon	≡	7.489152	g/L
1	ounce per Btu	≡	26,870.16	g/MJ
1	ft^3/acre	≡	0.028317	m^3/ac
1	ft^3/ton	≡	0.031214	m^3/Mg
1	ft^3/Btu	≡	26,839.19	m^3/GJ
1	Btu per gallon	≡	0.000279	MJ/L

G

Petroleum-Based Fuel Economics

Most of the energy currently used in fuel-powered transportation vehicles (for example, cars, trucks, buses, and airplanes) is in the form of liquid fuels derived from petroleum. Liquid fuels are uniquely suited for this service. They have a very high energy density on both a volumetric and weight basis. Vehicles can travel long distances between refueling on relatively small amounts of fuel. Liquid fuels are easy and cheap to transport. Vehicle refueling is fast and safe. The average person is capable of personally refueling his or her car in only a few minutes, with enough gasoline to travel over 250 miles. A commercial jet liner can take on enough fuel to fly halfway around the world in less time than it takes to unload and reload the passengers.

According to the U.S. Bureau of Economic Analysis, total personal consumption expenditures for gasoline, fuel oil, and other energy goods in 2008, when oil prices exceeded $140 per barrel, were only about 4 percent of all personal consumption expenditures, lower than they were for all years between 1950 and 1984 (Figure G-1). The U.S. economy and lifestyle have evolved around the availability of cheap, convenient, liquid transportation fuels.

Although petroleum has been used for thousands of years, the modern petroleum industry really began its rapid development during World War II. It has evolved into a very efficient industry for finding and converting a variety of crude oils into the high quality fuels the market and regulatory bodies demand. The industry has done this while meeting ever tightening emission limits for their production facilities and the fuels they produce.

Gasoline, diesel, and jet fuel are by far the largest volume petroleum-based products. Combined, they account for about 85 percent of consumed petroleum products in the United States. A number of steps are involved in getting these products from the well to the ultimate consumer, including

- Finding the oil-bearing deposits,
- Obtaining the rights to explore for and produce the oil,
- Drilling for the oil and installing facilities to recover the oil,
- Transporting the oil from the well to the refinery,

Gasoline, fuel oil, and other energy goods

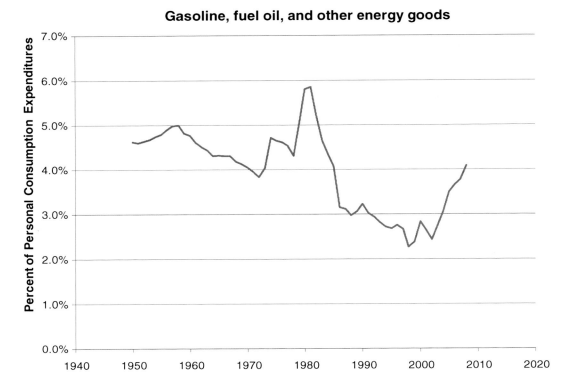

FIGURE G-1 U.S. personal energy consumption expenditures.
DATA SOURCE: EIA (2010).

- Converting (refining) the crude oil into the desired products, and
- Transporting and delivering the products to the final consumer.

Each one of these steps has a monetary cost associated with it. There are also other costs, such as marketing, accounting, research and development, and environmental costs.

COST ELEMENTS

The oil industry is very competitive, and company-specific cost elements are closely guarded from competitors. Publicly owned companies are required to report some information to regulatory agencies and shareholders in filings required by the Securities and Exchange Commission and in their annual reports. This information, however, is usually not sufficient to fully define all of the various cost elements. There is even less information available for privately owned companies or companies owned by foreign governments. Searching through various oil company annual reports, and the information on the Energy Information Administration and other government and non-government websites allows many costs to be defined or at least bracketed.

Crude oil and petroleum product prices have been extremely volatile over the last 15 years. The benchmark West Texas Intermediate crude (WTI), traded on the New York Mercantile Exchange, has gone from less than $14 per barrel to over $140 per barrel while wholesale gasoline prices have gone from $0.32 per gallon to over $3.30 per gallon (Figure G-2).

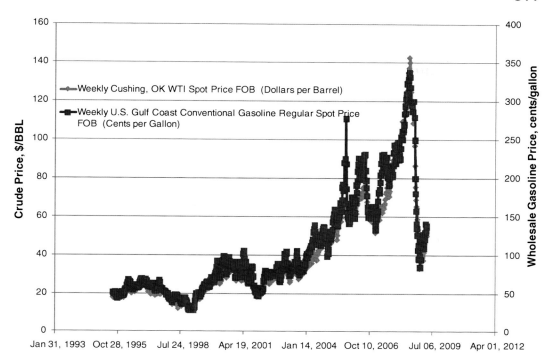

FIGURE G-2 Historic U.S. crude oil and gasoline prices.
DATA SOURCE: EIA (2010).

As expected, there is a very close relationship between crude oil price and gasoline price. There are short-term disconnects, such as the high gasoline prices that occurred during and after Hurricane Katrina in the fall of 2005 and probable gasoline price speculation prior to the summer driving and hurricane seasons the following two years.

The largest single factor affecting the price of gasoline is the price of crude oil. It accounts for over 80 percent of the direct, pretax cost of gasoline and an even larger fraction when its impact on transportation and refining costs are considered.

Crude oil and product transportation costs are the easiest to define. Imported crude oil transportation costs can be defined by comparing free-on-board (FOB) and landed crude prices that are contained in the EIA database. These show average transportation costs of $0.04-$0.05 per gallon for crude coming from the Persian Gulf, but only $0.01-$0.02 per gallon for crude coming from Canada, Mexico, and Venezuela. The elements that make up these transportation costs are capital for the tankers that haul the crude, the manpower to load the crude and operate the tanker, and the fuel to move the tanker between the two ports. The disparity in costs between crude coming from the Persian Gulf and that originating closer to the United States indicates that fuel costs and operating costs are the dominant cost components. Domestic crude transportation costs tend to be lower because the transport distances are shorter and most domestic crude is transported by pipeline.

RELATIONSHIP BETWEEN CRUDE OIL AND PRODUCT PRICES

As shown in Figure G-2, gasoline wholesale spot price closely follows crude oil spot price, with price discontinuities due to natural events such as Hurricane Katrina, which shut down many Gulf Coast refineries. This relationship for gasoline is quantified in Figure G-3 and for diesel fuel in Figure G-4.

Figure G-3 clearly shows that a linear relationship (that is, a constant multiplier such as those frequently used in previous studies and the 2010 Annual Energy Outlook) between gasoline price and crude oil price is not valid as crude price rises above $80 per barrel. Using a linear relationship between gasoline price and crude oil price would over-predict the price of gasoline as crude price rises above $80 per barrel. Figure G-4 shows the relationship between crude price and diesel fuel price. A linear relationship fits the diesel price data better than a second-order relationship fits the gasoline price data.

There is more scatter in the gasoline price data at a given crude price than in the diesel price data. This is due to the annual variations in gasoline demand and gasoline price as shown in Figure G-5. Gasoline demand peaks between late June and early August each year, while the ratio of gasoline price to crude price usually peaks in April, in anticipation of higher summer demand. This cyclical price relationship causes much of the apparent scatter in the price relationship in Figure G-3.

The average price ratio is 0.0264, but the ratio varies from about 0.025 in the winter to 0.030 in the summer. Table G-1 shows the expected gasoline and diesel prices for several different crude prices and pricing relationships.

Table G-1 shows a rather large uncertainty in future gasoline price as crude price increases. This is due both to the seasonal price swings and to the nonlinearity in gasoline price as crude price increases.

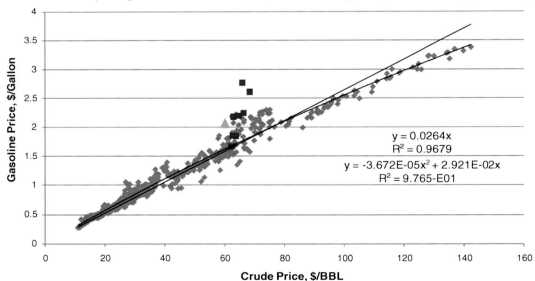

FIGURE G-3 Relationship between gasoline price and crude oil price.
DATA SOURCE: EIA (2010).

Weekly U.S. Gulf Coast No 2 Diesel Low Sulfur Spot Price FOB ($ per Gallon)

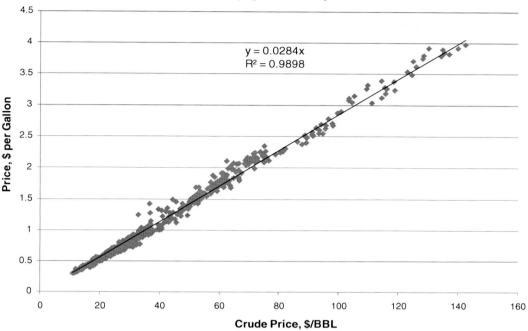

y = 0.0284x
R^2 = 0.9898

FIGURE G-4 Relationship between diesel fuel price and crude oil price.
DATA SOURCE: EIA (2010).

TRUE COST OF CRUDE OIL

The EIA database gives an average U.S. crude acquisition cost of $67.94 per barrel ($1.62 per gallon) for 2007. The 2009 ExxonMobil annual report gives an average capital cost for reserve replacement as less than $23 per barrel for the period from 2005 to 2009. The 2008 ConocoPhillips annual report gives an average crude oil production cost from existing fields of less than $7 per barrel for a similar period. This gives a total crude oil production cost, including all exploration and production costs, of less than $30 per barrel for a major oil company. These costs include lease acquisition costs, but do not include production (severance) taxes, state and national royalty payments, or other income taxes. These tax and royalty payments vary from state to state and country to country. It is beyond the scope of this study to identify all of these government payments, but a few are listed below as examples.

Because of falling oil prices, the Canadian province of Alberta recently reduced its maximum production royalty from 50 percent to 40 percent of the sale price of oil produced in the province (Oil and Gas Jounal, 2010). At $60 per barrel crude, $24 per barrel goes to the province of Alberta. In 2008, ConocoPhillips paid the state of Alaska over $33 per barrel in taxes other than income taxes when their production costs were less than $10 per barrel. Government-owned oil companies, such as Aramco in Saudi Arabia or PDVSA in Venezuela, control the exploration and production of all oil in the country. The difference between oil sale price and exploration and production cost is effectively a tax. Most

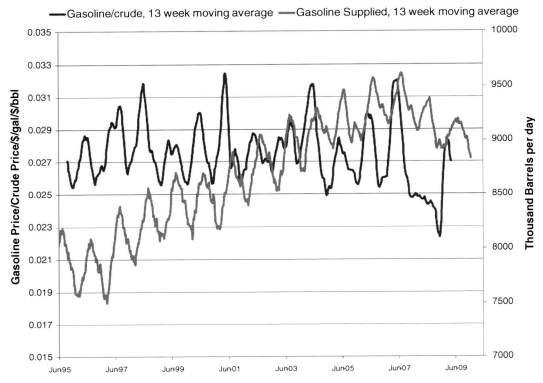

FIGURE G-5 Cyclical nature of gasoline demand and price.
DATA SOURCE: EIA (2010).

TABLE G-1 Expected Pre-Tax Wholesale Gasoline and Diesel Prices

Crude Price, dollars per barrel	50	75	100	150
Product Price, dollars per gallon				
Average Gasoline	1.32	1.98	2.64	3.96
Minimum Gasoline	1.25	1.88	2.50	3.75
Maximum Gasoline	1.50	2.25	3.00	4.50
Second Order Gasoline	1.37	1.98	2.55	3.56
Average Diesel	1.42	2.13	2.84	4.26

oil fields in the Middle East are well defined and rather shallow. Average exploration and production costs for new production from this region are below the world average. If they desire, OPEC countries can lower the price of oil below that, which justifies investment in alternate routes to liquid fuels, such as gas to liquid fuels, coal to liquid fuels, or biofuels.

In the absence of a carbon tax, production of alternative fuels from coal, natural gas, or shale set an upper bound on the future price of crude oil. The report *Liquid Transportation Fuels from Coal and Biomass: Technological Status, Costs, and Environmental Impacts* (NAS-NAE-NRC, 2009) indicated the cost of producing alternate transportation fuels from coal was about $60 per barrel. The 2010 Annual Energy Outlook (EIA, 2010) shows essentially no change (5-percent decline) in the price of coal between 2009 (the last full year of data) and 2022 (in 2008 dollars), while it shows a 78-percent increase in the crude price to over $100 per barrel over the same period and an even higher relative increase in natural gas price. At the same time it predicts only a 6-percent increase in the price of ethanol.

REFERENCES

EIA (Energy Information Administration). 2010. Annual Energy Outlook 2010—With Projections to 2035. Washington, DC: U.S. Department of Energy.

NAS-NAE-NRC (National Academy of Sciences, National Academy of Engineering, National Research Council). 2009. Liquid Transportation Fuels from Coal and Biomass: Technological Status, Costs, and Environmental Impacts. Washington, DC: National Academies Press.

Oil and Gas Journal. 2010. Alberta's royalty retreat. Available online at http://www.ogj.com/articles/print/volume-108/issue-11/General-Interest/editorial-alberta-s-royalty-retreat.html. Accessed March 22, 2010.

H

Ethanol Biorefineries in Operation or Under Construction in the United States in 2010

Company	Location	Feedstock	Nameplate Capacity (million gallons per year)	Operating Production (million gallons per year)	Expansion Capacity of Facilities Under Construction (million gallons per year)
ABE (Advanced BioEnergy, LLC)	Multiple locations	-	198.2	198	
ABE Fairmont	Fairmont, NE	Corn			
ABE South Dakota - Aberdeen	Aberdeen, SD	Corn			
ABE South Dakota - Huron	Huron, SD	Corn			
Abengoa Bioenergy Corp. (total)	Multiple locations	-	378	348	
Abengoa Bioenergy Corp.	Madison, IL	Corn			
Abengoa Bioenergy Corp.	Mt. Vernon, IN	Corn			
Abengoa Bioenergy Corp.	Colwich, KS	Corn/Milo			
Abengoa Bioenergy Corp.	Ravenna, NE	Corn			
Abengoa Bioenergy Corp.	Road O York, NE	Corn			
Abengoa Bioenergy Corp.	Portales, NM	Corn			

Company	Location	Feedstock	Nameplate Capacity (million gallons per year)	Operating Production (million gallons per year)	Expansion Capacity of Facilities Under Construction (million gallons per year)
Absolute Energy, LLC	St. Ansgar, IA	Corn	110	110	
ACE Ethanol, LLC	Stanley, WI	Corn	41	41	
Adkins Energy, LLC	Lena, IL	Corn	40	40	
Ag Energy Resources, Inc.	Benton, IL	Corn			5
AGP	Hastings, NE	Corn	52	52	
Al-Corn Clean Fuel	Claremont, MN	Corn	42	42	
Alchem LLP	Grafton, ND	Corn	10		
AltraBiofuels Coshocton Ethanol LLC	Coshocton, OH	Corn	60		
AltraBiofuels Phoenix Bio Industries	Goshen, CA	Corn	31.5		
Amaizing Energy, LLC	Denison, IA	Corn	55	55	
Appomattox Bio Energy LLC	Hopewell, VA	Corn	65		
Archer Daniels Midland (total)	Multiple locations	-	1450	1450	275
Archer Daniels Midland	Cedar Rapids, IA	Corn			
Archer Daniels Midland	Clinton, IA	Corn			
Archer Daniels Midland	Decatur, IL	Corn			
Archer Daniels Midland	Peoria, IL	Corn			
Archer Daniels Midland	Marshall, MN	Corn			
Archer Daniels Midland	Wallhalla, ND	Corn/Barley			
Archer Daniels Midland	Columbus, NE	Corn			
Arkalon Energy, LLC	Liberal, KS	Corn	110	110	
Aventine Renewable Energy, LLC (total)	Multiple locations	-	244	244	226
Aventine Renewable Energy, LLC	Pekin, IL	Corn			
Aventine Renewable Energy, LLC	Aurora, NE	Corn			

Company	Location	Feedstock	Nameplate Capacity (million gallons per year)	Operating Production (million gallons per year)	Expansion Capacity of Facilities Under Construction (million gallons per year)
Aventine Renewable Energy, LLC	Canton, IL	Corn			
Aventine Renewable Energy, LLC	Aurora, NE	Corn			
Aventine Renewable Energy, LLC	Mount Vernon, IN	Corn			
Badger State Ethanol, LLC	Monroe, WI	Corn	48	48	
Big River Resources Galva, LLC	Galva, IL	Corn	100	100	
Big River Resources, LLC	West Burlington, IA	Corn	100	100	
Big River United Energy	Dyersville, IA	Corn	110	110	
BioFuel Energy - Buffalo Lake Energy, LLC	Fairmont, MN	Corn	115	115	
BioFuel Energy - Pioneer Trail Energy, LLC	Wood River, NE	Corn	115	115	
Bional Clearfield	Clearfield, PA	Corn	110	110	
Blue Flint Ethanol	Underwood, ND	Corn	50	50	
Bonanza Energy, LLC	Garden City, KS	Corn/Milo	55	55	
BP Biofuels North America	Jennings, LA	Sugar Cane Bagasse	1.5	1.5	
Bridgeport Ethanol	Bridgeport, NE	Corn	54	54	
Bunge-Ergon Vicksburg	Vicksburg, MS	Corn	54	54	
Bushmills Ethanol, Inc.	Atwater, MN	Corn	50	50	
Calgren Renewable Fuels, LLC	Pixley, CA	Corn	60	60	
Carbon Green Bioenergy	Lake Odessa, MI	Corn	50	50	
Cardinal Ethanol	Union City, IN	Corn	100	100	
Cargill, Inc.	Eddyville, IA	Corn	35	35	
Cargill, Inc.	Blair, NE	Corn	85	85	
Cascade Grain Products LLC	Clatskanie, OR	Corn	108		

Company	Location	Feedstock	Nameplate Capacity (million gallons per year)	Operating Production (million gallons per year)	Expansion Capacity of Facilities Under Construction (million gallons per year)
Castle Rock Renewable Fuels, LLC	Necedah, WI	Corn	50	50	
Center Ethanol Company	Sauget, IL	Corn	54	54	
Central Indiana Ethanol, LLC	Marion, IN	Corn	40	40	
Central MN Ethanol Coop	Little Falls, MN	Corn	21.5	21.5	
Chief Ethanol	Hastings, NE	Corn	62	62	
Chippewa Valley Ethanol Co.	Benson, MN	Corn	45	45	
Cilion Ethanol	Keyes, CA	Corn			50
Clean Burn Fuels, LLC	Raeford, NC	Corn			60
Commonwealth Agri-Energy, LLC	Hopkinsville, KY	Corn	33	33	
Corn Plus, LLP	Winnebago, MN	Corn	44	44	
Corn, LP	Goldfield, IA	Corn	60	60	
Cornhusker Energy Lexington, LLC	Lexington, NE	Corn	40	40	
Dakota Ethanol, LLC	Wentworth, SD	Corn	50	50	
DENCO II	Morris, MN	Corn	24	24	
Didion Ethanol	Cambria, WI	Corn	40	40	
Dubay Biofuels Greenwood	Greenwood, WI	Cheese Whey			3
E Caruso (Goodland Energy Center)	Goodland, KS	Corn			20
E Energy Adams, LLC	Adams, NE	Corn	50	50	
E3 Biofuels	Mead, NE	Corn	25		
East Kansas Agri-Energy, LLC	Garnett, KS	Corn	35	35	
ESE Alcohol Inc.	Leoti, KS	Seed Corn	1.5	1.5	
Flint Hills Resources LP	Menlo, IA	Corn	110	110	
Flint Hills Resources LP	Shell Rock, IA	Corn	110	110	
Front Range Energy, LLC	Windsor, CO	Corn	40	40	
Gateway Ethanol	Pratt, KS	Corn	55		

Company	Location	Feedstock	Nameplate Capacity (million gallons per year)	Operating Production (million gallons per year)	Expansion Capacity of Facilities Under Construction (million gallons per year)
Gevo	Luverne, MN	Corn	21	21	
Glacial Lakes Energy, LLC - Mina	Mina, SD	Corn	107	107	
Glacial Lakes Energy, LLC	Watertown, SD	Corn	100	100	
Golden Cheese Company of California	Corona, CA	Cheese Whey	5	0	
Golden Grain Energy, LLC	Mason City, IA	Corn	115	115	
Golden Triangle Energy, LLC	Craig, MO	Corn	20	20	
Grain Processing Corp.	Muscatine, IA	Corn	20	20	
Granite Falls Energy, LLC	Granite Falls, MN	Corn	52	52	
Greater Ohio Ethanol, LLC	Lima, OH	Corn	54		
Green Plains Renewable Energy	Lakota, IA	Corn	100	100	
Green Plains Renewable Energy	Riga, MI	Corn	57	57	
Green Plains Renewable Energy	Shenandoah, IA	Corn	65	65	
Green Plains Renewable Energy	Superior, IA	Corn	55	55	
Green Plains Renewable Energy	Bluffton, IN	Corn	115	115	
Green Plains Renewable Energy	Central City, NE	Corn	100	100	
Green Plains Renewable Energy	Ord, NE	Corn	50	50	
Green Plains Renewable Energy	Obion, TN	Corn	115	115	
Guardian Energy	Janesville, MN	Corn	110	110	
Hankinson Renewable Energy, LLC	Hankinson, ND	Corn	110	110	
Hawkeye Renewables, LLC	Fairbank, IA	Corn	110	110	

Company	Location	Feedstock	Nameplate Capacity (million gallons per year)	Operating Production (million gallons per year)	Expansion Capacity of Facilities Under Construction (million gallons per year)
Hawkeye Renewables, LLC	Iowa Falls, IA	Corn	90	90	
Heartland Corn Products	Winthrop, MN	Corn	100	100	
Heron Lake BioEnergy, LLC	Heron Lake, MN	Corn	50	50	
Highwater Ethanol LLC	Lamberton, MN	Corn	55	55	
Homeland Energy	New Hampton, IA	Corn	100	100	
Husker Ag, LLC	Plainview, NE	Corn	75	75	
Idaho Ethanol Processing	Caldwell, ID	Potato Waste	4	4	
Illinois River Energy, LLC	Rochelle, IL	Corn	100	100	
Iroquois Bio-Energy Company, LLC	Rensselaer, IN	Corn	40	40	
KAAPA Ethanol, LLC	Minden, NE	Corn	60	60	
Kansas Ethanol, LLC	Lyons, KS	Corn	55	55	
KL Process Design Group	Upton, WY	Wood Waste	1.5	1.5	
Land O' Lakes	Melrose, MN	Cheese Whey	2.6	2.6	
Levelland/Hockley County Ethanol, LLC	Levelland, TX	Corn	40	40	
LifeLine Foods, LLC	Joseph, MO	Corn	40	40	
Lincolnland Agri-Energy, LLC	Palestine, IL	Corn	48	48	
Lincolnway Energy, LLC	Nevada, IA	Corn	55	55	
Little Sioux Corn Processors, LP	Marcus, IA	Corn	92	92	
Louis Dreyfus Commodities	Grand Junction, IA	Corn	100	100	
Louis Dreyfus Commodities	Norfolk, NE	Corn	45	45	
Marquis Energy, LLC	Hennepin, IL	Corn	100	100	
Marysville Ethanol, LLC	Marysville, MI	Corn	50	50	
Merrick and Company	Aurora, CO	Waste Beer	3	3	

Company	Location	Feedstock	Nameplate Capacity (million gallons per year)	Operating Production (million gallons per year)	Expansion Capacity of Facilities Under Construction (million gallons per year)
Mid America Agri Products/Wheatland	Madrid, NE	Corn	44	44	
Mid-Missouri Energy, Inc.	Malta Bend, MO	Corn	50	50	
Midwest Renewable Energy, LLC	Sutherland, NE	Corn	25	25	
Minnesota Energy	Buffalo Lake, MN	Corn	18		
Nebraska Corn Processing, LLC	Cambridge, NE	Corn	45	45	
NEDAK Ethanol	Atkinson, NE	Corn	44	44	
Nesika Energy, LLC	Scandia, KS	Corn	10	10	
New Energy Corp.	South Bend, IN	Corn	102	102	
North Country Ethanol, LLC	Rosholt, SD	Corn	20	20	
NuGen Energy	Marion, SD	Corn	110	110	
One Earth Energy	Gibson City, IL	Corn	100	100	
Otter Tail Ag Enterprises	Fergus Falls, MN	Corn	57.5	57.5	
Pacific Ethanol	Madera, CA	Corn	40		
Pacific Ethanol	Stockton, CA	Corn	60		
Pacific Ethanol	Burley, ID	Corn	50	50	
Pacific Ethanol	Boardman, OR	Corn	40	40	
Panda Ethanol	Hereford, TX	Corn/Milo			115
Parallel Products	Rancho Cucamonga, CA				
Parallel Products	Louisville, KY	Beverage Waste	5.4	5.4	
Patriot Renewable Fuels, LLC	Annawan, IL	Corn	100	100	
Penford Products	Cedar Rapids, IA	Corn	45	45	
Pinal Energy, LLC	Maricopa, AZ	Corn	55	55	
Pine Lake Corn Processors, LLC	Steamboat Rock, IA	Corn	31	31	
Platinum Ethanol, LLC	Arthur, IA	Corn	110	110	
Plymouth Ethanol, LLC	Merrill, IA	Corn	50	50	

Company	Location	Feedstock	Nameplate Capacity (million gallons per year)	Operating Production (million gallons per year)	Expansion Capacity of Facilities Under Construction (million gallons per year)
POET Biorefining - Alexandria	Alexandria, IN	Corn	68	68	
POET Biorefining - Ashton	Ashton, IA	Corn	56	56	
POET Biorefining - Big Stone	Big Stone City, SD	Corn	79	79	
POET Biorefining - Bingham Lake	Bingham Lake, MN		35	35	
POET Biorefining - Caro	Caro, MI	Corn	53	53	5
POET Biorefining - Chancellor	Chancellor, SD	Corn	110	110	
POET Biorefining - Cloverdale	Cloverdale, IN	Corn	92		
POET Biorefining - Coon Rapids	Coon Rapids, IA	Corn	54	54	
POET Biorefining - Corning	Corning, IA	Corn	65	65	
POET Biorefining - Emmetsburg	Emmetsburg, IA	Corn	55	55	
POET Biorefining - Fostoria	Fostoria, OH	Corn	68	68	
POET Biorefining - Glenville	Albert Lea, MN	Corn	42	42	
POET Biorefining - Gowrie	Gowrie, IA	Corn	69	69	
POET Biorefining - Hanlontown	Hanlontown, IA	Corn	56	56	
POET Biorefining - Hudson	Hudson, SD	Corn	56	56	
POET Biorefining - Jewell	Jewell, IA	Corn	69	69	
POET Biorefining - Laddonia	Laddonia, MO	Corn	50	50	
POET Biorefining - Lake Crystal	Lake Crystal, MN	Corn	56	56	
POET Biorefining - Leipsic	Leipsic, OH	Corn	68	68	
POET Biorefining - Macon	Macon, MO	Corn	46	46	

Company	Location	Feedstock	Nameplate Capacity (million gallons per year)	Operating Production (million gallons per year)	Expansion Capacity of Facilities Under Construction (million gallons per year)
POET Biorefining - Marion	Marion, OH	Corn	68	68	
POET Biorefining - Mitchell	Mitchell, SD	Corn	68	68	
POET Biorefining - North Manchester	North Manchester, IN	Corn	68	68	
POET Biorefining - Portland	Portland, IN	Corn	68	68	
POET Biorefining - Preston	Preston, MN	Corn	46	46	
POET Biorefining - Scotland	Scotland, SD	Corn	11	11	
POET Biorefining- Groton	Groton, SD	Corn	53	53	
Prairie Horizon Agri-Energy, LLC	Phillipsburg, KS	Corn	40	40	
Quad-County Corn Processors	Galva, IA	Corn	30	30	
Range Fuels	Soperton, GA	Woody Biomass			10
Red Trail Energy, LLC	Richardton, ND	Corn	50	50	
Redfield Energy, LLC	Redfield, SD	Corn	50	50	
Reeve Agri-Energy	Garden City, KS	Corn/Milo	12	12	
Renova Energy	Torrington, WY	Corn	5	5	
Show Me Ethanol	Carrollton, MO	Corn	55	55	
Siouxland Energy & Livestock Coop	Sioux Center, IA	Corn	60	60	
Siouxland Ethanol, LLC	Jackson, NE	Corn	50	50	
Southwest Georgia Ethanol, LLC	Camilla, GA	Corn	100	100	
Southwest Iowa Renewable Energy, LLC	Council Bluffs, IA	Corn	110	110	
Sterling Ethanol, LLC	Sterling, CO	Corn	42	42	
Sunoco	Volney, NY	Corn	114	114	
Tate & Lyle	Loudon, TN	Corn	67	67	38
Tharaldson Ethanol	Casselton, ND	Corn/milo	110	110	

Company	Location	Feedstock	Nameplate Capacity (million gallons per year)	Operating Production (million gallons per year)	Expansion Capacity of Facilities Under Construction (million gallons per year)
The Andersons Albion Ethanol LLC	Albion, MI	Corn	55	55	
The Andersons Clymers Ethanol, LLC	Clymers, IN	Corn	110	110	
The Andersons Marathon Ethanol, LLC	Greenville, OH	Corn	110	110	
Trenton Agri Products, LLC	Trenton, NE	Corn	40	40	
United Ethanol	Milton, WI	Corn	52	52	
United WI Grain Producers, LLC	Friesland, WI	Corn	49	49	
Utica Energy, LLC	Oshkosh, WI	Corn	48	48	
Valero Renewable Fuels	Albert City, IA	Corn	110	110	
Valero Renewable Fuels	Charles City, IA	Corn	110	110	
Valero Renewable Fuels	Ft. Dodge, IA	Corn	110	110	
Valero Renewable Fuels	Hartley, IA	Corn	110	110	
Valero Renewable Fuels	Welcome, MN	Corn	110	110	
Valero Renewable Fuels	Albion, NE	Corn	110	110	
Valero Renewable Fuels	Aurora, SD	Corn	120	120	
Valero Renewable Fuels	North Linden, IN	Corn	110	110	
Valero Renewable Fuels	Bloomingburg, OH	Corn	110	110	
Valero Renewable Fuels	Jefferson Junction, WI	Corn	130	130	
Western New York Energy LLC	Shelby, NY		50	50	
Western Plains Energy, LLC	Campus, KS	Corn	45	45	
Western Wisconsin Renewable Energy, LLC	Boyceville, WI	Corn	40	40	

Company	Location	Feedstock	Nameplate Capacity (million gallons per year)	Operating Production (million gallons per year)	Expansion Capacity of Facilities Under Construction (million gallons per year)
White Energy	Russell, KS	Milo/Wheat Starch	48	48	
White Energy	Hereford, TX	Corn/Milo	100	100	
White Energy	Plainview, TX	Corn	110	110	
Wind Gap Farms	Baconton, GA	Brewery Waste	0.4	0.4	
Yuma Ethanol	Yuma, CO	Corn	40	40	

SOURCE: RFA (2010).

REFERENCE

RFA (Renewable Fuels Association). 2010. Biorefinery locations. November 11, 2010. Available online at http://www.ethanolrfa.org/bio-refinery-locations/. Accessed November 17, 2010.

I

Biodiesel Refineries in the United States in 2010

Company	Location	Feedstock	Nameplate Capacity (gallons per year)	Start Date
San Francisco Public Utilities Commission	San Francisco, CA			
Shenandoah Agricultural Products	Clearbrook, VA	Recycled Cooking Oil	300,000	3/15/2010
Allied Renewable Energy, LLC	Birmingham, AL	Soy	15,000,000	5/1/2007
Eagle Biodiesel, Inc.	Bridgeport, AL	Multiple Feedstocks	30,000,000	4/1/2007
Southeastern Biodiesel Solutions, Inc.	Creola, AL			
Delta American Fuel, LLC	Helena, AR	Multiple Feedstocks	40,000,000	3/1/2009
Pinnacle Biofuels, Inc.	Crossett, AR	Multiple Feedstocks	10,000,000	5/1/2008
Amereco Biofuels Corp	Arlington, AZ	Multiple Feedstocks	15,000,000	9/1/2007
Environmental Development Group	Tucson, AZ	Recycled Cooking Oil	1,500,000	Summer 2010
Grecycle Arizona, LLC	Tucson, AZ	Yellow Grease	2,500,000	5/1/2009
Performance Biofuels, LLC	Chandler, AZ	Recycled Cooking Oil	1,500,000	12/1/2008

Company	Location	Feedstock	Nameplate Capacity (gallons per year)	Start Date
Baker Commodities	Los Angeles, CA	Multiple Feedstocks	10,000,000	12/1/2010
Bay Biodiesel, LLC	San Jose, CA	Multiple Feedstocks	3,000,000	3/1/2007
Biodiesel Industries of Ventura, LLC	Port Hueneme, CA	Full Spectrum, including but not limited to yellow grease, jatropha, and algae	3,000,000	8/1/2009
Blue Sky Biofuels	Oakland, CA	Waste Oil	4,000,000	
Community Fuels	Stockton, CA	Multiple Feedstocks	10,000,000	6/1/2008
Crimson Renewable Energy, LP	Bakersfield, CA	Multiple Feedstocks	30,000,000	3Q 2009
Ecolife Biofuels, LLC	San Jacinto, CA	Multiple Feedstocks	1,500,000	4/1/2010
Enviro Fuels Enterprises, LLC	Fresno, CA			3Q 2009
Imperial Valley Biodiesel, LLC	El Centro, CA	Multiple Feedstocks	3,000,000	12/1/2007
Imperial Western Products	Coachella, CA	Multiple Feedstocks	10,500,000	10/1/2001
Manning Beef, LLC	Pico Rivera, CA	Tallow		3Q 2009
New Leaf Biofuel, LLC	San Diego, CA	Used Cooking Oil	1,500,000	12/1/2008
Noil Energy Group	Commerce, CA	Multiple Feedstocks	5,000,000	2Q 2009
Promethean Biofuels Cooperative Corporation	Temecula, CA	Multiple Feedstocks	1,500,000	9/5/2009
Simple Fuels Biodiesel, Inc.	Chilcoot, CA	Waste Oil	1,000,000	3Q 2009
Whole Energy Fuels	Pacifica, CA	Recycled Cooking Oil	3,000,000	2Q 2009
Biofuels of Colorado	Denver, CO	Multiple Feedstocks	15,000,000	
BioDiesel One Ltd	Southington, CT	Multiple Feedstocks	4,000,000	2/1/2009
BioPur Inc.	Bethlehem, CT	Multiple Feedstocks	1,000,000	7/1/2006
Clayton	Clayton, DE`	Multiple Feedstocks	11,000,000	1/1/2010
Agri-Source Fuels, Inc.	Dade City, FL	Multiple Feedstocks	10,000,000	10/1/2007
Genuine Bio-Fuel	Indiantown, FL		3,900,000	1/1/2009
Greenwave Biodiesel	Ft. Lauderdale, FL	Waste Oil	4,000,000	8/1/2010

Company	Location	Feedstock	Nameplate Capacity (gallons per year)	Start Date
Heartland Bio Energy, LLC	Palm Beach County, FL	Multiple Feedstocks	5,000,000	
Johnson Biofuels	Fort Lauderdale, FL	Waste Oil	1,000,000	11/1/2009
Smart Fuels, LLC	Fruitland Park, FL			3Q 2009
Southern Biodiesel Corporation	Miami, FL			4Q 2009
Alterra Bioenergy of Middle Georgia, LLC	Gordon, GA	Multiple Feedstocks	15,000,000	8/1/2007
BullDog BioDiesel	Ellenwood, GA	Multiple Feedstocks	18,000,000	1/1/2008
Down to Earth Energy, Inc.	Monroe, GA	Multiple Feedstocks	2,000,000	8/1/2009
Middle Georgia Biofuels	East Dublin, GA	Poultry Fat, Tallow	1,500,000	4/1/2006
Peach State Labs	Rome, GA	Soy		1/1/2005
Seminole Biodiesel	Bainbridge, GA	Multiple Feedstocks	10,000,000	1/1/2008
AGP	Sergeant Bluff, IA	Soy	30,000,000	8/1/1996
Cargill	Iowa Falls, IA	Soy	37,500,000	6/1/2006
Energy Tec, LLC	Maquoketa, IA	Waste Oil	30,000	9/1/2008
Iowa Renewable Energy, LLC	Washington, IA	Multiple Feedstocks	30,000,000	7/1/2007
Maple River Energy, LLC	Galva, IA	Multiple Feedstocks	5,000,000	5/1/2009
REG Newton, LLC	Newton, IA	Multiple Feedstocks	30,000,000	4/1/2007
REG Ralston, LLC	Ralston, IA	Multiple Feedstocks	12,000,000	3/1/2001
Western Dubuque Biodiesel	Farley, IA	Crude or Refined Vegetable Oils	36,000,000	8/1/2007
Western Iowa Energy, LLC	Wall Lake, IA	Multiple Feedstocks	30,000,000	6/1/2006
Incobrasa Industries, Ltd.	Gilman, IL	Soy	31,000,000	1/1/2007
Midwest Biodiesel Products, Inc.	South Roxanna, IL	Multiple Feedstocks	30,000,000	5/1/2007
REG Danville, LLC	Danville, IL	Multiple Feedstocks	45,000,000	9/1/2008
REG Seneca, LLC	Seneca, IL	Multiple Feedstocks	60,000,000	8/1/2008
Stepan Company	Millsdale, IL	Multiple Feedstocks	22,000,000	1/1/2001
Alternative Fuel Solutions, LLC	Huntington, IN	Multiple Feedstocks	300,000	1/1/2010

Company	Location	Feedstock	Nameplate Capacity (gallons per year)	Start Date
e-biofuels, LLC	Middletown, IN	Multiple Feedstocks	15,000,000	6/1/2007
Heartland Biofuel	Flora, IN	Multiple Feedstocks	500,000	4/1/2006
Integrity Biofuels	Morristown, IN	Multiple Feedstocks	5,000,000	8/1/2006
Louis Dreyfus Agricultural Industries, LLC	Claypool, IN	Soy	80,000,000	1/1/2008
Emergent Green Energy	Minneola, KS	Multiple Feedstocks	2,000,000	3/1/2009
Kansas Biofuels	Wichita, KS	Multiple Feedstocks	1,800,000	10/1/2007
REG Emporia, LLC	Emporia, KS	Multiple Feedstocks	60,000,000	2Q 2010
Griffin Industries	Butler, KY	Multiple Feedstocks	1,750,000	12/1/1998
Owensboro Grain	Owensboro, KY	Soy	50,000,000	1/1/2008
REG New Orleans, LLC	St. Rose, LA	Multiple Feedstocks	60,000,000	2Q 2010
Vanguard Synfuels, LLC	Pollock, LA	Multiple Feedstocks	12,000,000	4/1/2006
Baker Commodities	Billerica, MA	Multiple Feedstocks	15,000,000	12/1/2010
Cape Cod BioFuels	Sandwich, MA	Waste Oil	500,000	9/1/2009
Greenlight Biofuels, LLC	Princess Anne, MD	Multiple Feedstocks	4,000,000	10/1/2007
Maine Standard Biofuel	Portland, ME	Yellow Grease	500,000	1/1/2010
Michigan Biodiesel, LLC	Bangor, MI	Multiple Feedstocks	10,000,000	1/1/2007
Thumb BioEnergy, LLC	Applegate, MI			
TPA Inc.	Warren, MI	Multiple Feedstocks	20,000,000	7/1/2008
Ever Cat Fuels, LLC	Isanti, MN	Multiple Feedstocks	3,000,000	10/1/2009
FUMPA BioFuels	Redwood Falls, MN	Multiple Feedstocks	3,000,000	12/1/2004
Minnesota Soybean Processors	Brewster, MN	Soy	30,000,000	8/1/2005
Soymor	Albert Lea, MN	Refined Vegetable Oils	30,000,000	8/1/2005
AGP	St. Joseph, MO	Soy	29,900,000	9/1/2007
American Energy Producers, Inc.	Tina, MO	Soy	50,000,000	4Q 2010

Company	Location	Feedstock	Nameplate Capacity (gallons per year)	Start Date
Global Fuels, LLC	Dexter, MO	Multiple Feedstocks	3,000,000	4/1/2007
Mid America Biofuels, LLC	Mexico, MO	Soy	30,000,000	12/1/2006
Paseo Cargill Energy, LLC	Kansas City, MO	Soy	37,500,000	4/1/2008
Producers' Choice Soy Energy LLC	Moberly, MO	Soy	5,000,000	6/1/2009
Terra Bioenergy, LLC	St. Joseph, MO	Multiple Feedstocks	30,000,000	4Q 2009
Delta Biofuels, Inc.	Natchez, MS	Multiple Feedstocks	80,000,000	5/1/2007
GreenLight Biofuels, LLC	Meridian, MS			3/1/2009
North Mississippi Biodiesel	New Albany, MS	Soy	7,000,000	10/1/2006
Scott Petroleum Corporation	Greenville, MS	Multiple Feedstocks	20,000,000	10/1/2007
Earl Fisher Bio Fuels	Chester, MT	Canola, Camelina, Safflower, Sunflower	250,000	4/1/2008
Blue Ridge Biofuels	Asheville, NC	Multiple Feedstocks	1,000,000	5/1/2006
Foothills Bio-Energies, LLC	Lenoir, NC	Multiple Feedstocks	5,000,000	9/1/2006
Leland Organic Corporation	Leland, NC	Multiple Feedstocks	30,000,000	9/1/2008
Patriot Biodiesel, LLC	Greensboro, NC	Multiple Feedstocks	1,500,000	12/1/2008
Piedmont Biofuels	Pittsboro, NC	Multiple Feedstocks	4,000,000	11/1/2006
Triangle Biofuels Industries, Inc.	Wilson, NC	Multiple Feedstocks	3,000,000	1/1/2008
ADM	Velva, ND	Canola	85,000,000	8/1/2007
White Mountain Biodiesel, LLC	North Haverhill, NH	Multiple Feedstocks	5,500,000	Summer 2010
Innovation Fuels	Newark, NJ	Multiple Feedstocks	40,000,000	7/1/2004
Rio Valley Biofuels, LLC	Anthony, NM	Multiple Feedstocks	1,000,000	7/1/2006
Bently Biofuels	Minden, NV	Multiple Feedstocks	1,000,000	11/1/2005
Biodiesel of Las Vegas	Las Vegas, NV	Multiple Feedstocks	100,000,000	4Q 2009
Metro Fuel Oil Corp	Brooklyn, NY	Multiple Feedstocks	110,000,000	IQ 2012

Company	Location	Feedstock	Nameplate Capacity (gallons per year)	Start Date
TMT Biofuels, LLC	Port Leyden, NY	Recycled Cooking Oil	250,000	9/1/2008
Ambiol Flex Fuels, LLC	East Toledo, OH	Soy	2,000,000	2/1/2008
Arlington Energy, LLC	Mansfield, OH	Multiple Feedstocks	4,000,000	7/1/2008
Center Alternative Energy Company	Cleveland, OH	Multiple Feedstocks	5,000,000	5/1/2007
Chieftain BioFuels, LLC	Logan, OH	Multiple Feedstocks	3,000,000	
Fina, LLC	Cincinnati, OH	Multiple Feedstocks	60,000,000	12/1/2006
Jatrodiesel Inc.	Miamisburg, OH	Multiple Feedstocks	5,000,000	6/1/2007
Peter Cremer NA	Cincinnati, OH	Soy	30,000,000	10/1/2002
High Plains Bioenergy	Guymon, OK	Multiple Feedstocks	30,000,000	3/1/2008
South East Oklahoma Biodiesel	Valliant, OK	Multiple Feedstocks	5,000,000	11/1/2008
American Biodiesel Energy, Inc.	Erie, PA	Multiple Feedstocks	4,000,000	
Biodiesel of Pennsylvania, Inc.	White Deer, PA	Multiple Feedstocks	1,500,000	3/1/2007
Eagle Bio Diesel	Kane, PA	Multiple Feedstocks	5,000,000	4Q 2009
Keystone BioFuels, Inc.	Shiremanstown, PA		60,000,000	3/1/2006
Lake Erie Biofuels	Erie, PA	Multiple Feedstocks	45,000,000	9/1/2007
Pennsylvania Biodiesel, Inc.	Monaca, PA	Multiple Feedstocks	25,000,000	7/1/2009
Soy Energy, Inc.	New Oxford, PA	Soy	2,500,000	2/1/2007
United Oil Company	Pittsburgh, PA	Multiple Feedstocks	5,000,000	12/1/2005
US Alternative Fuels Corp.	Johnstown, PA			4Q 2008
Lantic Green Energy, LLC	West Greenwich, RI	Waste Oil	500,000	4Q 2008
Newport Biodiesel, LLC	Newport, RI	Recycled Cooking Oil	500,000	1/1/2008
Cateechee Biofuels, LLC	Central, SC	Recycled Cooking Oil	2,000,000	3Q 2009
Ecogy Biofuels, LLC	Estill, SC	Multiple Feedstocks	30,000,000	12/1/2007
Green Valley Biofuels, LLC	Warrenville, SC	Multiple Feedstocks	35,000,000	3Q 2009

Company	Location	Feedstock	Nameplate Capacity (gallons per year)	Start Date
Greenlight Biofuels, LLC	Laurens, SC	Multiple Feedstocks	10,000,000	4Q 2008
Midwest BioDiesel Producers, LLC	Alexandria, SD	Multiple Feedstocks	7,000,000	3/1/2006
Milagro Biofuels of Memphis	Memphis, TN	Multiple Feedstocks	5,000,000	10/1/2006
Southern Alliance for Clean Energy	Knoxville, TN	Recycled Cooking Oil	380,000	7/1/2009
Sullens Biodiesel, LLC	Morrison, TN		2,000,000	
Biodiesel Industries of Greater Dallas-Fort Worth	Denton, TX	Multiple Feedstocks	3,000,000	3/1/2005
Biodiesel of Texas, Inc.	Denton, TX	Multiple Feedstocks	2,000,000	3Q 2009
Direct Fuels	Euless, TX	Multiple Feedstocks	10,000,000	2/1/2008
Green Earth Fuels of Houston, LLC	Galena Park, TX	Multiple Feedstocks	90,000,000	7/1/2007
New Fuel Company	Sanger, TX	Multiple Feedstocks	250,000	4/1/2006
RBF Port Neches, LLC	Port Neches, TX	Multiple Feedstocks	180,000,000	4Q 2008
Red River Biodiesel Ltd.	New Boston, TX	Multiple Feedstocks	15,000,000	5/1/2008
REG Houston, LLC	Seabrook, TX	Multiple Feedstocks	35,000,000	7/1/2008
Texas Biotech, Inc	Arlington, TX			
Texas Green Manufacturing, LLC	Littlefield, TX	Tallow	1,250,000	4/1/2009
The Sun Products Corp	Pasadena, TX	Palm	15,000,000	6/1/2005
VICNRG, LLC	Fort Worth, TX	Multiple Feedstocks	10,000,000	
Washakie Renewable Energy	Plymouth, UT	Multiple Feedstocks	10,000,000	1/1/2009
RECO Biodiesel, LLC	Richmond, VA	Multiple Feedstocks	10,000,000	12/1/2006
Red Birch Energy, Inc.	Bassett, VA	Multiple Feedstocks	2,500,000	6/1/2008
Synergy Biofuels, LLC	Pennington Gap, VA	Multiple Feedstocks	3,000,000	12/1/2008
Virginia Biodiesel Refinery	West Point, VA	Multiple Feedstocks	7,000,000	10/1/2003
General Biodiesel Seattle LLC	Seattle, WA	Multiple Feedstocks	5,000,000	3Q 2009

Company	Location	Feedstock	Nameplate Capacity (gallons per year)	Start Date
Gen-X Energy Group, Inc.	Burbank, WA	Multiple Feedstocks	15,000,000	6/1/2007
Imperium Grays Harbor	Hoquiam, WA	Multiple Feedstocks	100,000,000	8/1/2007
Inland Empire Oilseeds, LLC	Odessa, WA	Canola	8,000,000	11/1/2008
Whole Energy Fuels	Bellingham, WA	Recycled Cooking Oil		1Q 2009
Bio Blend Fuels Inc.	Manitowoc, WI	Multiple Feedstocks	2,600,000	5/1/2009
Sanimax Energy Inc.	Deforest, WI	Multiple Feedstocks	20,000,000	4/1/2007
Sun Power Biodiesel, LLC	Cumberland, WI	Sunflower, Canola	3,000,000	12/1/2009
Walsh Bio Fuels, LLC	Mauston, WI	Multiple Feedstocks	5,000,000	5/1/2007
AC & S, Inc.	Nitro, WV	Soy	3,000,000	12/1/2007

SOURCE: NBB (2010).

REFERENCE

NBB (National Biodiesel Board). 2010. NBB member plant locations. Available online at http://www.biodiesel.org/buyingbiodiesel/plants/. Accessed November 17, 2010.

J

Economic Models Used to Assess the Effects of Biofuel Production in the United States

conomic models are widely used for agricultural and energy policy analysis. For biofuels in the United States, the four main models that have been used are the Food and Agricultural Policy Research Institute (FAPRI) model, the Forest and Agricultural Sector Optimization (FASOM) model, the Global Trade Analysis Project (GTAP) model, and the Policy Analysis System (POLYSYS) model. FAPRI has been developed by Iowa State University and the University of Missouri. FASOM has had many contributors over the years, but development currently is managed by Texas A&M University. GTAP is led by Purdue University and POLYSYS by the University of Tennessee. Each model is large and complicated and has unique strengths and weaknesses. This section provides a general summary of each model.

SUMMARY OF ECONOMIC MODELS

Three of the models (FAPRI, FASOM, and POLYSYS) are partial-equilibrium (PE) models, which means that not all sectors of the economy are included in the model. These PE models focus on the agricultural sector with limited account of other sectors through exogenous information. General-equilibrium (GE) models, such as GTAP, provide coverage of all sectors of the economy, although at a much more aggregate level. GTAP also covers all regions of the world. Thus, GE models capture the interactions among sectors and between product and factor markets. However, GE models, especially global ones like GTAP, cannot model the interactions among detailed sectors of agriculture or regions as PE models can. For many questions, PE models that concentrate on one or a few sectors are appropriate for responding to policy questions regarding the sector(s) of interest. PE models generally are richer in detail regarding cropping practices, land quality and use, and regional variations, and they can permit more in-depth analysis of sector-specific policies. However, the models implicitly assume that what goes on in the agricultural sector will not have a large effect on the rest of the economy and vice versa (that is, the agricultural sector can be analyzed without worrying explicitly about what happens in the rest of the economy).

Forward-looking models assume perfect knowledge of the future; recursive dynamic models assume agents are myopic and operate as if current prices will continue into the future. FASOM is a forward-looking model and solves all years simultaneously, whereas FAPRI and POLYSYS are recursive dynamic models. GTAP can be run either as a comparative static or dynamic model. Comparative static models compute the market equilibrium under one set of conditions; when conditions change, the models compute the new equilibrium without worrying about the path from one equilibrium to the other. The comparative static solution does not have a particular time reference, although it is generally characterized as medium term or about eight years.[1]

The three PE models originally had a heavy focus on agricultural policy, although FASOM was designed to examine competition between forestry and agricultural sectors for land from its early stages. GTAP originally was developed to evaluate the effects of alternative trade policies in international trade negotiations and regional and bilateral trade agreements. Thus, GTAP is more heavily focused on the trade dimension. Specific details of each model follow.

FAPRI

FAPRI is a model of the U.S. agricultural sector, with extensions for the rest of world, designed originally for agricultural policy analysis. The model has been supported by congressional appropriations and is heavily used in providing information to congressional staffers on questions related to the Farm Bill. The FAPRI baseline—projections for the next decade—is used for policy and general outlook work by many others. In fact, the FAPRI baseline is used to help develop the POLYSYS baseline. The U.S. Department of Agriculture Economic Research Service uses the FAPRI baseline for some of their analysis. The model covers all the major U.S. agricultural crops and livestock sectors. It pays strong attention to detailed policy representation. It accounts for cropland use (as well as pasture for the United States and Brazil) and agricultural trade data (especially for the United States), but treats other macroeconomic effects as exogenous inputs to the model. International trade is included with the rest of world divided into regions. Trade adjustment is accounted for directly through supply and demand shifts with no consideration for traditional trade patterns. In other words, it adopts the Heckscher-Ohlin assumption that domestic and imported goods are identical. There are both deterministic and stochastic versions of FAPRI.

FASOM

Before 1996, FASOM was an agricultural sector model, but it was modified in the mid-1990s to include the forestry sector. Since that time the model has been under continuous development. It presently includes detailed regional representation for cropland, pasture, and forestland for the United States. It is connected to the rest of the world by a relatively simple excess supply and demand structure. FASOM has often been used for environmental and greenhouse-gas (GHG) analyses; it was the primary model used by the U.S.

---

[1]The biofuel modeling done by Purdue University to date has been done with the comparative static version of the model. However, others (for example, Massachusetts Institute of Technology, using the Emissions Prediction Policy Analysis model) have used the GTAP database as part of a recursive dynamic simulation model that runs for 100 years or more.

Environmental Protection Agency (EPA) to determine GHG reductions from various bio-fuel pathways. As indicated above, FASOM is the only one of the models that looks forward and solves all years simultaneously.

GTAP

GTAP is both a database and a modeling framework. The GTAP database can be used with other models, and the basic GTAP model can be adapted to particular questions of interest. GTAP adopts the Armington approach to international trade, which means that domestic and imported goods are differentiated. Changes in the structure of international trade are less pronounced than for models with the Heckscher-Ohlin assumption of homogeneous goods. There are 57 sectors plus 3 biofuel sectors in the standard GTAP database. However, most research is done with far fewer regions and sectors. Aggregation tools are available to permit users to specify whatever regional and sector aggregation they prefer. Global land is divided into 18 agroecological zones, and land covers of cropland, pasture, and forest are included. GTAP simulations (using the comparative static approach) essentially estimate the global economic effects from whatever policy or technology shock is of interest. For example, GTAP has been used to estimate the effects of a 15-billion gallon mandate for corn-grain ethanol. To do so, the results of the baseline model run are compared to the results of introducing a shock into the model that requires 15 billion gallons of corn-grain ethanol be produced to satisfy the Renewable Fuel Standard as amended under the Energy Security and Independence Act of 2007 (RFS2) while holding other demands and requirements constant. The results of these two model runs are then compared with respect to corn price, corn exports and imports, other agricultural prices, land-use change, and other relevant economic variables.

POLYSYS

POLYSYS includes eight U.S. crops (corn, sorghum, barley, oats, cotton, rice, wheat, and soybean) and seven livestock sectors (beef, pork, turkeys, broilers, eggs, lamb and mutton, and dairy). The model divides the United States into 305 crop-reporting districts; its current version is able to analyze county-level effects if needed. The POLYSYS simulations are anchored to USDA's published crop-projection baseline, and the simulations provide deviations from that baseline for the shock of interest. Crop and livestock supply is regionally generated, and demand is handled through a system of simultaneous equations. POLYSYS also tracks farm income, government payments, and several environmental variables. For environmental variables, POLYSYS interacts with the Environmental Policy Integrated Climate (EPIC) model (Williams et al., 1990, p. 18). The model can be and commonly is run stochastically to capture the effects of uncertainty and nonlinearities in some of the functions. In recent years, POLYSYS has been modified to include crop residues and dedicated energy crops as biomass feedstock supply options.

LAND COVER IN ECONOMIC MODELS

The models differ in the extent to which types of cover are included. GTAP and FASOM have forestland, pasture, and cropland, although it is predominantly only U.S. land in FASOM. POLYSYS includes cropland and pasture, but not forestland at present, although it may be added in the near future. FAPRI has cropland and pasture for the United States and Brazil but not forestland.

INCORPORATING BIOFUELS INTO ECONOMIC MODELS

All four models were developed before the implementation of RFS2, but each has been modified in recent years to include varying degrees of biofuel coverage. FAPRI and GTAP include only first-generation biofuels (corn-grain ethanol, sugarcane ethanol, and oilseed-based biodiesel). The institutions that run these models plan to expand the database to include second-generation biofuels, and in recent years, FAPRI has increased the level of detail of its representation of Brazil to account for changes due to biofuels. FASOM and POLYSYS include second-generation biofuels from cellulosic feedstocks presently used. FASOM is the only model of the four to include electricity generation from cellulosic feedstocks. Incorporating biofuels into the models is complex. The difficulty of using the models for projections of biofuel feedstock production centers around three areas of insufficient detail.

CAPTURING WORLD MARKET EFFECTS VERSUS AGRICULTURAL PRODUCTION DETAILS

Whether global trade interactions and linkages with other sectors are needed is largely an empirical question. Both GE and PE models have advantages and disadvantages. FAPRI, FASOM, and POLYSYS enter some macro-variables exogenously and consider world trade linkages and flows. The comparative static version of GTAP does not project changes over time; rather, it compares the situation at two points in time given whatever shock has been applied to the model. The PE models normally simulate through time and compare the results with and without a given shock to a predetermined baseline.

CELLULOSIC BIOMASS NOT INCLUDED IN MOST MODELS

Cellulosic biomass crops, including crop residues from grain crops, perennial grasses, bagasse from sugarcane, and woody crops and residues, are being considered as a source of feedstock for liquid transportation fuels. The production and cost estimates that exist vary widely over time, space, and studies. Of the four models being discussed, FASOM includes woody biomass and POLYSYS includes cellulosic biomass crops and woody biomass. GTAP is in the process of adding cellulosic biomass, and FAPRI is in process of including cellulosic crop residue and switchgrass.

REFERENCE

Williams, J.R., P. Dyke, W. Fuchs, V. Benson, O. Rice, and E. Taylor. 1990. EPIC—Erosion/Productivity Impact Calculator: 2. User Manual. Washington, DC: U.S. Department of Agriculture - Agricultural Research Service.

K

BioBreak Model: Assumptions for Willingness to Accept

This appendix provides detail on the assumptions and references used to calculate Willingness to Accept (WTA) in the BioBreak model, discussed in Box 4-2 in Chapter 4.

EQUATION

$$WTA = \{(C_{ES} + C_{Opp})/Y_B + C_{HM} + SF + C_{NR} + C_S + DFC + DVC*D\} - G$$

The supplier's WTA for 1 ton of delivered cellulosic material is equal to the total economic costs the supplier incurs to deliver 1 unit of biomass to the biorefinery less the government incentives received (G) (for example, tax credits and production subsidies). Depending on the type of biomass feedstock, costs include establishment and seeding (C_{ES}), land and biomass opportunity costs (C_{Opp}), harvest and maintenance (C_{HM}), stumpage fees (SF), nutrient replacement (C_{NR}), biomass storage (C_S), transportation fixed costs (DFC), and variable transportation costs calculated as the variable cost per mile (DVC) multiplied by the average hauling distance to the biorefinery (D). Establishment and seeding cost and land and biomass opportunity cost are most commonly reported on a per acre scale. Therefore, the biomass yield per acre (Y_B) is used to convert the per acre costs into per ton costs, and the equation above provides the minimum amount the supplier can accept for the last dry ton of biomass delivered to the biorefinery and still breakeven. The variables in the equation are based on the following assumptions.

WTA PARAMETERS

Nutrient Replacement (C_{NR})

Uncollected cellulosic material adds value to the soil through enrichment and protection against rain, wind, and radiation, thereby limiting erosion that would cause the loss of vital soil nutrients like nitrogen, phosphorus, and potassium. Biomass suppliers will incorporate the costs of soil damage and nutrient loss from biomass collection into the minimum price they are willing to accept. Nutrient replacement cost (C_{NR}) varies by feedstock and harvest technique. After adjusting for 2007 costs, estimates for nutrient replacement costs range from $5 to $21 per ton. Based on the model's baseline oil price ($111 per barrel) and research estimates, nutrient replacement was assumed to have a mean (likeliest) value of $14.20 ($15.20) per ton for stover, $16.20 ($17.20) per ton for switchgrass, $9 per ton for *Miscanthus*, and $6.20 per ton for wheat straw.[1] Alfalfa was assumed to have a 2-year stand with the first-year nutrient costs incorporated into the establishment costs discussed below and a cost of $65 per acre for second-year nutrient application. Given the yield assumptions for second-year alfalfa, this corresponds to approximately $16.25 per ton. Nutrient replacement was assumed unnecessary for woody biomass. The cost of nutrient replacement depends on the natural gas price and is therefore dependent on energy costs. EIA projected natural gas to oil price factor for 2022 was used to scale fertilizer costs at varying oil price levels. At the high oil price ($191 per barrel), nutrient replacement costs increase by about $1.35 per ton. At the low oil price ($52 per barrel), nutrient replacement costs decrease by about $1.00 per ton.

Harvest and Maintenance Costs (C_{HM}) and Stumpage Fees (SF)

Harvest and maintenance cost (C_{HM}) estimates for cellulosic material have varied based on harvest technique and feedstock. Noncustom harvest research estimates range from $14 to $84 per ton for corn stover (McAloon et al., 2000; Aden et al., 2002; Sokhansanj and Turhollow, 2002; Suzuki, 2006; Edwards, 2007; Hess et al., 2007; Perlack, 2007; Brechbill and Tyner, 2008a; Khanna, 2008; Huang et al., 2009), $16 to $58 per ton for switchgrass (Tiffany et al., 2006; Khanna and Dhungana, 2007; Kumar and Sokhansanj, 2007; Brechbill and Tyner, 2008a; Duffy, 2008; Khanna, 2008; Khanna et al., 2008; Perrin et al., 2008; Huang et al., 2009), and $19 to $54 per ton for *Miscanthus* (Khanna and Dhungana, 2007; Khanna, 2008; Khanna et al., 2008), after adjusting for 2007 costs.[2] Estimates for nonspecific biomass range between $15 and $38 per ton (Mapemba et al., 2007, 2008). Woody biomass collection costs up to roadside range between $17 and $50 per ton (USFS, 2003, 2005; BRDI, 2008; Jenkins et al., 2009; Sohngen et al., 2010). Spelter and Toth (2009) find total delivered costs (including transportation) around $58, $66, $75, and $86 per dry ton[3] for woody residue in the Northeast, South, North, and West regions, respectively.[4]

Using the timber harvesting cost simulator outlined in Fight et al. (2006), Sohngen et al. (2010) found harvest costs up to roadside about $25 per dry ton, with a high cost scenario of $34 per dry ton. Based on an oil price of $111 per barrel, the model assumed harvest and maintenance costs have mean (likeliest) values of $44.20 ($47.20), $37.20 ($39.20), $46.20 ($49.20), $33.20 ($34.20), and $27.20 for stover, switchgrass, *Miscanthus*, wheat straw, and

---

[1]For parameters with an assumed skewed distribution in Monte Carlo analysis, the "likeliest" value denotes the value with the highest probability density.

[2]Harvest and maintenance costs were updated using USDA-NASS agricultural fuel, machinery, and labor prices from 1999-2007 (USDA-NASS, 2007a,b).

[3]Based on a conversion rate of 0.59 dry tons per green tons.

[4]Northeast includes Pennsylvania, New Jersey, New York, Connecticut, Massachusetts, Rhode Island, Vermont, New Hampshire, and Maine. South refers to Delaware, Maryland, West Virginia, Virginia, North Carolina, South Carolina, Kentucky, Tennessee, Florida, Georgia, Alabama, Mississippi, Louisiana, Arkansas, Texas, and Oklahoma. States in the North region are Minnesota, Wisconsin, Michigan, Iowa, Missouri, Illinois, Indiana, and Ohio. West includes South Dakota, Wyoming, Colorado, New Mexico, Arizona, Utah, Montana, Idaho, Washington, Oregon, Nevada, and California.

woody biomass. Alfalfa was assumed to be harvested once during the first year and three times during the second year at a cost of $55 per acre per harvest. In addition to harvest costs, suppliers of short-rotation woody crops (SRWC) incur a stumpage fee (SF) with an assumed mean value of $20 per ton. Energy costs affect the cost of harvest through the price of diesel. The relationship between diesel and oil prices was derived using data from 1988-2008. Harvest costs increase by approximately $2.70 per ton at the high oil price and decreases by around $2.00 per ton at the low oil price.

Transportation Costs (DVC, DFC, and D)

Previous research on transportation of biomass has provided two distinct types of cost estimates: (1) total transportation cost and (2) breakdown of variable and fixed transportation costs. Research estimates for total corn stover transportation costs range between $3 per ton and $32 per ton (Aden et al., 2002; Perlack and Turhollow, 2002; Atchison and Hettenhaus, 2003; English et al., 2006; Hess et al., 2007; Perlack, 2007; Brechbill and Tyner, 2008a; Mapemba et al., 2008; Vadas et al., 2008). Total switchgrass and *Miscanthus* transportation costs have been estimated between $14 and $36 per ton (Tiffany et al., 2006; Kumar and Sokhansanj, 2007; Mapemba et al., 2007; Brechbill and Tyner, 2008a; Duffy, 2008; Khanna et al., 2008; Mapemba et al., 2008; Perrin et al., 2008; Vadas et al., 2008), adjusted to 2007 costs.[5] Woody biomass transportation costs are expected to range between $11 and $30 per dry ton (Summit Ridge Investments, 2007; Sohngen et al., 2010). Based on the second method, distance variable cost (DVC) estimates range between $0.09 and $0.60 per ton per mile (Kaylen et al., 2000; Kumar et al., 2003; USFS, 2003; Kumar et al., 2005; USFS, 2005; Searcy et al., 2007; Brechbill and Tyner, 2008a,b; Petrolia, 2008; Huang et al., 2009; Jenkins et al., 2009; Sohngen et al., 2010), while distance fixed cost (DFC) estimates range between $4.80 and $9.80 per ton (Kumar et al., 2003, 2005; Searcy et al., 2007; Petrolia, 2008; Huang et al., 2009), depending on feedstock type. The BioBreak model utilized the latter method of separating fixed and variable transportation costs.

The DFC for corn stover, switchgrass, *Miscanthus,* wheat straw, and second-year alfalfa was assumed to range from $5 to $12 per ton with a mean value of $8.50 per ton. Besides loading and unloading costs, woody biomass requires an on-site chipping fee. Therefore, the DFC for woody biomass was assumed to have a mean value of $10 per ton. The DVC was assumed to follow a skewed distribution to account for future technological progress in transportation of biomass with a mean (likeliest) value of $0.38 ($0.41) per ton per mile for stover, switchgrass, *Miscanthus,* wheat straw, and second-year alfalfa and $0.53 ($0.56) per ton per mile for woody biomass. Energy costs affect the DVC through the price of diesel. The 1988-2008 relationship between diesel and oil prices was used to adjust DVC to each oil price scenario. The DVC increases by approximately $0.07 per ton per mile at the high oil price ($191 per barrel) and decreases by approximately $0.05 per ton per mile at the low oil price ($52 per barrel).

One-way transportation distance (D) has been evaluated up to around 140 miles for woody biomass (USFS, 2003, 2005; Miller and Bender, 2008; Spelter and Toth, 2009) and between 5 and 75 miles (Perlack and Turhollow, 2002, 2003; Atchison and Hettenhaus, 2003; English et al., 2006; Tiffany et al., 2006; Mapemba et al., 2007, 2008; BRDI, 2008; Brechbill and Tyner, 2008a,b; Khanna, 2008; Taheripour and Tyner, 2008; Vadas et al., 2008) for all other feedstocks. In the model, the average hauling distance was calculated using the formulation by French (1960) for a circular supply area with a square road grid provided in Equation (3)

---

[5]Transportation costs were updated using USDA-NASS agricultural fuel prices from 1999-2007 (USDA-NASS, 2007a,b).

below. [6] Average distance (D) is a function of the annual biorefinery biomass demand (BD), annual biomass yield (Y_B) and biomass density (B).

$$D = 0.4789 \sqrt{\frac{BD}{640 * Y_B * B}}$$

(3)

Annual biomass demand was assumed to be consistent with the biorefinery outlined for capital and operating cost distributions (772,000 tons per year). Biomass density was assumed to follow a normal distribution with a mean value of 0.20 for all feedstocks except alfalfa, which has a mean biomass density of 0.15 (McCarl et al., 2000; Perlack and Turhollow, 2002; Popp and Hogan Jr., 2007; Brechbill and Tyner, 2008a,b; Petrolia, 2008; Huang et al., 2009).[7]

Storage Costs (C_S)

Due to the low density of biomass compared to traditional cash crops such as corn and soybean, biomass storage costs (C_S) can vary greatly depending on the feedstock type, harvest technique, and type of storage area. Adjusted for 2007 costs, biomass storage estimates ranged between $2 and $23 per ton. For simulation, BioBreak assumed storage costs follow a skewed distribution for all feedstocks to allow for advancement in storage and densification techniques. The mean (likeliest) cost for woody biomass storage was $11.50 ($12) per ton, while corn stover, switchgrass, *Miscanthus*, wheat straw, and alfalfa storage costs were assumed to have mean (likeliest) values of $10.50 ($11) per ton.

Establishment and Seeding Costs (C_{ES})

Corn stover, wheat straw, and forest residue suppliers were assumed to not incur establishment and seeding costs (C_{ES}), while all other feedstock suppliers must be compensated for their establishment and seeding costs. Costs vary by initial cost, stand length, years to maturity, and interest rate. Stand length for switchgrass ranges between 10 and 20 years with full yield maturity by the third year. *Miscanthus* stand length ranges from 10 to 25 years with full maturity between the second and fifth year. Interest rates used for amortization of establishment costs range between 4 and 8 percent. Amortized cost estimates for switchgrass establishment and seeding, adjusted to 2007 costs, are between $30 and $200 per acre. *Miscanthus* establishment and seeding cost estimates vary widely, based on the assumed level of technology and rhizome costs. James et al. (2010) reported a total rhizome cost (not including equipment and labor) of $8,194 per acre as representative of current costs and $227.61 per acre for a projected cost estimate after technological advancement (2008$). Lewandowski et al. (2003) provided a cost range of $1,206-$2,413 per acre (not updated). Jain et al. (2010) pointed out the benefit of using rhizomes over plugs where the total cost of establishment of rhizomes is about $1,200 per acre in Illinois and $1,215-$1,620 per acre for plugs. Establishment costs for wood also vary by species and location. Cubbage et al. (2010) reported establishment costs of $386-$430 and $520 per acre for yellow pine and Douglas Fir, respectively (2008$).

---

[6]The authors' simplifying assumption of uniform density is maintained.

[7]Although the biomass density for a corn-soybean rotation is assumed to be 0.20, the value used to calculate the average hauling distance for stover from a corn-soybean rotation is 0.10 since only half of the acreage is in corn at any given point in time.

Given research estimates, switchgrass establishment and seeding costs were based on a $250 per acre cost, amortized over 10 years at 10 percent to yield a mean value of $40 per acre in all regions. *Miscanthus* was assumed to have a mean value of $150 per acre per year mean establishment costs based on a total cost of $1,250 per acre amortized over 20 years at 10 percent. Establishment of alfalfa was assumed fixed at $165 per acre, including fertilizer application. Finally, SRWC were assumed to cost $400 per acre to establish and amortized over 15 years at 10 percent to yield a mean value of $52 per acre per year.

Opportunity Costs (C_{Opp})

To provide a complete economic model, the opportunity costs of utilizing biomass for ethanol production were included in BioBreak. Research estimates for the opportunity cost of switchgrass and *Miscanthus* ranged between $70 and $318 per acre (Khanna and Dhungana, 2007; Brechbill and Tyner, 2008a; Khanna et al., 2008; Jain et al., 2010), while estimates for nonspecific biomass opportunity cost ranged between $10 and $76 per acre (Khanna et al., 2008; Mapemba et al., 2008), depending on the harvest restrictions under the Conservation Reserve Program contracts. Opportunity cost of woody biomass was estimated to range between $0 and $30 per ton (USFS, 2003, 2005; Summit Ridge Investments, 2007).

The corn stover harvest activity was developed for a corn-soybean rotation alternative and has no opportunity cost beyond the nutrient replacement cost. A continuous corn alternative, used by 10-20 percent of Corn Belt producers, was developed for corn stover harvest but not included in the BioBreak results presented in this report. The continuous corn production budgets, developed by state extension specialists, are always less profitable than corn-soybean rotation budgets with or without stover harvest. Continuous corn has an associated yield penalty or forgone profit (opportunity costs) relative to the corn-soybean rotation that occurs irrespective of stover harvest. Thus, a comparative analysis of stover harvest with corn-soybean and continuous corn may be misinterpreted.[8]

Given the research estimates for perennial grass opportunity cost, switchgrass, and *Miscanthus* grown on Midwest land were assumed to have mean opportunity costs of $150 per acre on high-quality and $100 per acre on low-quality land. Perennial grasses grown in the Appalachian and South-Central regions were assumed to have lower mean opportunity costs of $75 and $50 per acre. Wheat straw opportunity cost was assumed to follow a distribution with likeliest value of $0 per acre with a range of –$10 to $30 per acre. Negative values for the opportunity costs of wheat straw were based on the potential nuisance cost of wheat straw. Occasionally, straw is burned at harvest to avoid grain planting problems during the following crop season. Forest residue was assumed to have no value in an alternative use and therefore no opportunity cost, and the stumpage fee was assumed to account for the opportunity cost of SRWC. Finally, alfalfa is assumed to have a 2-year stand with first-year harvest sold for hay at a value of $140 per ton while second-year alfalfa was assumed to have 50-percent leaf mass sold for protein value at $160 per ton and the remaining 50 percent used as a biofuel feedstock. Alfalfa opportunity cost (that is, land cost) was assumed fixed at $175 per acre for both years.

---

[8]From the rotation calculator provided by the Iowa State University extension services with a corn price of $4 per bushel, a soybean price of $10 per bushel, and a yield penalty of 7 bushels per acre, the lost net returns to switching from a corn-soybean rotation to continuous corn is about $62 per acre (http://www.extension.iastate.edu/agdm/crops/html/al-20.html). Previous literature that has attributed an opportunity cost to stover based on lost profits from switching from a corn-soybean rotation to continuous corn production has assumed an opportunity cost between $22 and $140 per acre (Khanna and Dhungana, 2007; Duffy, 2010).

Biomass Yield (Y_B)

The final parameter in the BioBreak model is biomass yield per acre of land. Biomass yield is variable in the near and distant future due to technological advancements and environmental uncertainties. Corn-stover yield per acre will vary based on the amount of corn stover that is removable, which depends on soil quality and other topographical characteristics. Harvested corn-stover yield has been estimated between 0.7 to 3.8 tons per acre (Duffy and Nanhou, 2002; Lang, 2002; Perlack and Turhollow, 2002; Sokhansanj and Turhollow, 2002; Atchison and Hettenhaus, 2003; Quick, 2003; Schechinger and Hetten-haus, 2004; Edwards, 2007; Khanna and Dhungana, 2007; Prewitt et al., 2007; BRDI, 2008; Brechbill and Tyner, 2008a; Khanna, 2008; Vadas et al., 2008; Huang et al., 2009; Chen et al., 2010). Potential switchgrass yields range between 0.89 and 17.8 tons per acre (Reynolds et al., 2000; Muir et al., 2001; Bouton, 2002; Kszos et al., 2002; McLaughlin et al., 2002; Talia-ferro, 2002; Vogel et al., 2002; Lewandowski et al., 2003; Ocumpaugh et al., 2003; Parrish et al., 2003; Heaton et al., 2004b; Berdahl et al., 2005; Cassida et al., 2005; Kiniry et al., 2005; McLaughlin and Kszos, 2005; Thomason et al., 2005; Comis, 2006; Fike et al., 2006a,b; Nel-son et al., 2006; Schmer et al., 2006; Shinners et al., 2006; Tiffany et al., 2006; Gibson and Barnhart, 2007; Khanna and Dhungana, 2007; Popp and Hogan, 2007; BRDI, 2008; Brechbill and Tyner, 2008a; Duffy, 2008; Khanna, 2008; Khanna et al., 2008; Perrin et al., 2008; Sand-erson, 2008; Vadas et al., 2008; Walsh, 2008; Huang et al., 2009; Chen et al., 2010; Jain et al., 2010), depending on region, land quality, switchgrass variety, field versus plot trial stud-ies, and harvest technique. On average, *Miscanthus* has significantly higher yield estimates that range between 3.4 and 19.6 tons per acre when yield estimates from both the United States and the European Union are considered (Lewandowski et al., 2000; Clifton-Brown et al., 2001, 2004; Kahle et al., 2001; Clifton-Brown and Lewandowski, 2002; Vargas et al., 2002; Heaton et al., 2004a,b; Khanna and Dhungana, 2007; Stampfl et al., 2007; Christian et al., 2008; Khanna, 2008; Khanna et al., 2008; Smeets et al., 2009). Estimated U.S. *Miscanthus* yields range between 9 and 28 tons per acre (Heaton et al., 2004a,b; Khanna and Dhungana, 2007; Khanna, 2008; Khanna et al., 2008; Chen et al., 2010; Jain et al., 2010). A wheat straw yield of 1 ton per acre was assumed by the Biomass Research and Development Initiative study (BRDI, 2008). For woody biomass, Huang et al. (2009) estimated Aspen wood yield of 0.446 dry tons per acre from a densely forested area in Minnesota, while the BRDI (2008) study assumed short-rotation woody crops yield 5 to 12 tons per acre. Using USDA Forest Service data for Mississippi, the average removal rate of wood residue in 2006 was around 1.1 tons per acre.[9] In a recent study on 2008 wood production costs, Cubbage et al. (2010) estimated an annual yield of 3.6 and 4.3 tons per acre in North Carolina and the Southern United States, respectively. In the same analysis, Douglas Fir was estimated to provide 4 and 5.1 tons per acre annually in Oregon and North Carolina, respectively.

For simulation, the mean yield of corn stover was approximately 2 tons per acre. Switchgrass grown in the Midwest was assumed to have a distribution with a mean (like-liest) value around 4 (3.4) tons per acre on high quality land and 3.1 tons per acre on low quality land.[10] *Miscanthus* grown in the Midwest was assumed to have a mean (likeliest) value of 8.6 (8) tons per acre on high quality land and 7.1 (6) tons per acre on low quality land.[11] Switchgrass grown in the South-Central region has a higher mean yield of about 5.7 tons per acre. For the regions analyzed, the Appalachian region provides the best climatic

---

[9]This value is a lower bound because forestry still had positive net growth over this period.

[10]Plot trials were evaluated at 80 percent of their estimated yield.

[11]This is a significantly lower assumed yield than previous research has assumed or simulated (Khanna and Dhungana, 2007; Khanna et al., 2008; Khanna, 2008; Heaton et al., 2004).

conditions for switchgrass and *Miscanthus* with assumed mean (likeliest) yields of 6 (5) and 8.8 (8) tons per acre, respectively. Wheat straw, forest residues, and SRWC were assumed to be normally distributed with mean yields of 1, 0.5, and 5 tons per acre. First-year alfalfa yield was fixed at 1.25 tons per acre (sold for hay value), while second-year yield was fixed at 4 tons per acre (50-percent leaf mass sold for protein value), resulting in 2 tons per acre of alfalfa for biomass feedstock during the second year.

Biomass Supplier Government Incentives (G)

For biomass supplier government incentives (G), the dollar for dollar matching payments provided in the Food, Conservation, and Energy Act of 2008 (that is, the 2008 Farm Bill) up to $45 per ton of feedstock for collection, harvest, storage, and transportation is used, and it is denoted as "CHST." The CHST payment was considered in the sensitivity analysis rather than the baseline scenario since the payment is a temporary (2-year) program and might not be considered in the supplier's long-run analysis. Although the Bio-Break model is flexible enough to account for any additional biomass supply incentives, the establishment assistance program outlined in the 2008 Farm Bill is not considered because the implementation details were not finalized at the time of the model run.

REFERENCES

Aden, A., M. Ruth, K. Ibsen, J. Jechura, K. Neeves, J. Sheehan, B. Wallace, L. Montague, A. Slayton, and J. Lukas. 2002. Lignocellulosic Biomass to Ethanol Process Design and Economics Utilizing Co-Current Dilute Acid Prehydrolysis and Enzymatic Hydrolysis for Corn Stover. Golden, CO: National Renewable Energy Laboratory.

Atchison, J.E., and J.R. Hettenhaus. 2003. Innovative Methods for Corn Stover Collecting, Handling, Storing and Transporting. Golden, CO: National Renewable Energy Laboratory.

Berdahl, J.D., A.B. Frank, J.M. Krupinsky, P.M. Carr, J.D. Hanson, and H.A. Johnson. 2005. Biomass yield, phenology, and survival of diverse switchgrass cultivars and experimental strains in Western North Dakota. Agronomy Journal 97(2):549-555.

Bouton, J.H. 2002. Bioenergy Crop Breeding and Production Research in the Southeast: Final Report for 1996 to 2001. Oak Ridge, TN: Oak Ridge National Laboratory.

BRDI (Biomass Research and Development Initiative). 2008. Increasing Feedstock Production for Biofuels: Economic Drivers, Environmental Implications, and the Role of Research. Washington, DC: The Biomass Research and Development Board.

Brechbill, S.C., and W.E. Tyner. 2008a. The Economics of Biomass Collection, Transportation, and Supply to Indiana Cellulosic and Electric Utility Facilities. West Lafayette, IN: Purdue University.

Brechbill, S.C., and W.E. Tyner. 2008b. The Economics of Renewable Energy: Corn Stover and Switchgrass. West Lafayette, IN: Purdue University.

Cassida, K.A., J.P. Muir, M.A. Hussey, J.C. Read, B.C. Venuto, and W.R. Ocumpaugh. 2005. Biomass yield and stand characteristics of switchgrass in South Central U.S. environments. Crop Science 45(2):673-681.

Chen, X., H. Huang, M. Khanna, and H. Onal. 2010. Meeting the mandate for biofuels: Implications for land use and food and fuel prices. Paper read at the AAEA, CAES, and WAEA Joint Annual Meeting, July 25-27, Denver, CO.

Christian, D.G., A.B. Riche, and N.E. Yates. 2008. Growth, yield and mineral content of *Miscanthus × giganteus* grown as a biofuel for 14 successive harvests. Industrial Crops and Products 28(3):320-327.

Clifton-Brown, J.C., and I. Lewandowski. 2002. Screening *Miscanthus* genotypes in field trials to optimise biomass yield and quality in southern Germany. European Journal of Agronomy 16(2):97-110.

Clifton-Brown, J.C., S.P. Long, and U. Jørgensen. 2001. *Miscanthus* Productivity. Pp. 46-67 in *Miscanthus* for Energy and Fibre, M.B. Jones and M. Walsh, eds. London: James & James.

Clifton-Brown, J.C., P.F. Stampfl, and M.B. Jones. 2004. *Miscanthus* biomass production for energy in Europe and its potential contribution to decreasing fossil fuel carbon emissions. Global Change Biology 10(4):509-518.

Comis, D. 2006. Switching to switchgrass makes sense. Agricultural Research 54(7):19.

Cubbage, F., S. Koesbandana, P. Mac Donagh, R. Rubilar, G. Balmelli, V.M. Olmos, R. De La Torre, M. Murara, V.A. Hoeflich, H. Kotze, R. Gonzalez, O. Carrero, G. Frey, T. Adams, J. Turner, R. Lord, J. Huang, C. MacIntyre, K. McGinley, R. Abt, and R. Phillips. 2010. Global timber investments, wood costs, regulation, and risk. Biomass and Bioenergy 34(12):1667-1678.

Duffy, M.D. 2008. Estimated Costs for Production, Storage and Transportation of Switchgrass. Ames: Iowa State University.

Duffy, M.D., and V.Y. Nanhou. 2002. Costs of Producing Switchgrass for Biomass in Southern Iowa. Pp. 267-275 in Trends in New Crops and New Uses, J. Janick and A. Whipkey, eds. Alexandria, VA: ASHS Press.

Edwards, W. 2007. Estimating a value for corn stover. Ag Decision Maker (A1-70):4.

English, B.C., D.G. De La Torre Ugarte, K. Jensen, C. Hellwinckel, J. Menard, B. Wilson, R. Roberts, and M. Walsh. 2006. 25% Renewable Energy for the United States By 2025: Agricultural and Economic Impacts. Knoxville: University of Tennessee.

Fight, R.D., B.R. Hartsough, and P. Noordijk. 2006. Users Guide for FRCS: Fuel Reduction Cost Simulator Software. Portland, OR: U.S. Department of Agriculture - Forest Service - Pacific Northwest Research Station.

Fike, J.H., D.J. Parrish, D.D. Wolf, J.A. Balasko, J.J.T. Green, M. Rasnake, and J.H. Reynolds. 2006a. Long-term yield potential of switchgrass-for-biofuel systems. Biomass and Bioenergy 30(3):198-206.

Fike, J.H., D.J. Parrish, D.D. Wolf, J.A. Balasko, J.T. Green Jr., M. Rasnake, and J.H. Reynolds. 2006b. Switchgrass production for the upper southeastern USA: Influence of cultivar and cutting frequency on biomass yields. Biomass and Bioenergy 30(3):207-213.

French, B.C. 1960. Some considerations in estimating assembly cost functions for agricultural processing operations. Journal of Farm Economics 42(4):767-778

Gibson, L., and S. Barnhart. 2007. Switchgrass. AG 0200. Ames: Iowa State University.

Heaton, E.A., S.P. Long, T.B. Voigt, M.B. Jones, and J. Clifton-Brown. 2004a. *Miscanthus* for renewable energy generation: European Union experience and projections for Illinois. Mitigation and Adaptation Strategies for Global Change 9(4):433-451.

Heaton, E.A., T.B. Voigt, and S.P. Long. 2004b. A quantitative review comparing the yields of two candidate C_4 perennial biomass crops in relation to nitrogen, temperature and water. Biomass and Bioenergy 27(1):21-30.

Hess, J.R., C.T. Wright, and K.L. Kenney. 2007. Cellulosic biomass feedstocks and logistics for ethanol production. Biofuels, Bioproducts and Biorefining 1(3):181-190.

Huang, H.-J., S. Ramaswamy, W. Al-Dajani, U. Tschirner, and R.A. Cairncross. 2009. Effect of biomass species and plant size on cellulosic ethanol: A comparative process and economic analysis. Biomass and Bioenergy 33(2):234-246.

Jain, A.K., M. Khanna, M. Erickson, and H. Huang. 2010. An integrated biogeochemical and economic analysis of bioenergy crops in the midwestern United States. Global Change Biology Bioenergy 2(5):217-234.

James, L.K., S.M. Swinton, and K.D. Thelen. 2010. Profitability analysis of cellulosic energy crops compared with corn. Agronomy Journal 102(2):675-687.

Jenkins, B.M., R.B. Williams, N. Parker, P. Tittmann, Q. Hart, M.C. Gildart, S. Kaffka, B.R. Hartsough, and P. Dempster. 2009. Sustainable use of California biomass resources can help meet state and national bioenergy targets. California Agriculture 63(4):168-177.

Kahle, P., S. Beuch, B. Boelcke, P. Leinweber, and H.-R. Schulten. 2001. Cropping of *Miscanthus* in Central Europe: Biomass production and influence on nutrients and soil organic matter. European Journal of Agronomy 15(3):171-184.

Kaylen, M., D.L. Van Dyne, Y.-S. Choi, and M. Blase. 2000. Economic feasibility of producing ethanol from lignocellulosic feedstocks. Bioresource Technology 72(1):19-32.

Khanna, M. 2008. Cellulosic biofuels: Are they economically viable and environmentally sustainable? Choices 23(3):16-23.

Khanna, M., and B. Dhungana. 2007. Economics of Alternative Feedstocks. Pp. 129-146 in Corn-Based Ethanol in Illinois and the U.S.: A Report from the Department of Agricultural and Consumer Economics, University of Illinois. Urbana-Champaign: University of Illinois.

Khanna, M., B. Dhungana, and J. Clifton-Brown. 2008. Costs of producing *Miscanthus* and switchgrass for bioenergy in Illinois. Biomass and Bioenergy 32(6):482-493.

Kiniry, J.R., K.A. Cassida, M.A. Hussey, J.P. Muir, W.R. Ocumpaugh, J.C. Read, R.L. Reed, M.A. Sanderson, B.C. Venuto, and J.R. Williams. 2005. Switchgrass simulation by the ALMANAC model at diverse sites in the southern U.S. Biomass and Bioenergy 29(6):419-425.

Kszos, L.A., S.B. McLaughlin, and M. Walsh. 2002. Bioenergy from switchgrass: Reducing production costs by improving yield and optimizing crop management. Paper read at the Biomass Research and Development Board's Bioenergy 2002—Bioenergy for the Environment conference, September 22-27, Boise, ID.

Kumar, A., and S. Sokhansanj. 2007. Switchgrass (*Panicum vigratum*, L.) delivery to a biorefinery using integrated biomass supply analysis and logistics (IBSAL) model. Bioresource Technology 98(5):1033-1044.

Kumar, A., J.B. Cameron, and P.C. Flynn. 2003. Biomass power cost and optimum plant size in western Canada. Biomass and Bioenergy 24(6):445-464.

Kumar, A., J.B. Cameron, and P.C. Flynn. 2005. Pipeline transport and simultaneous saccharification of corn stover. Bioresource Technology 96(7):819-829.

Lang, B. 2002. Estimating the Nutrient Value in Corn and Soybean Stover. Decorah: Iowa State University.

Lewandowski, I., J.C. Clifton-Brown, J.M.O. Scurlock, and W. Huisman. 2000. *Miscanthus*: European experience with a novel energy crop. Biomass and Bioenergy 19(4):209-227.

Lewandowski, I., J.M.O. Scurlock, E. Lindvall, and M. Christou. 2003. The development and current status of perennial rhizomatous grasses as energy crops in the U.S. and Europe. Biomass and Bioenergy 25(4):335-361.

Mapemba, L.D., F.M. Epplin, C.M. Taliaferro, and R.L. Huhnke. 2007. Biorefinery feedstock production on conservation reserve program land. Review of Agricultural Economics 29(2):227-246.

Mapemba, L.D., F.M. Epplin, R.L. Huhnke, and C.M. Taliaferro. 2008. Herbaceous plant biomass harvest and delivery cost with harvest segmented by month and number of harvest machines endogenously determined. Biomass and Bioenergy 32(11):1016-1027.

McAloon, A., F. Taylor, W. Yee, K. Ibsen, and R. Wooley. 2000. Determining the Cost of Producing Ethanol from Corn Starch and Lignocellulosic Feedstocks. Golden, CO: National Renewable Energy Laboratory.

McCarl, B.A., D.M. Adams, R.J. Alig, and J.T. Chmelik. 2000. Competitiveness of biomass-fueled electrical power plants. Annals of Operations Research 94(1):37-55.

McLaughlin, S.B., and L.A. Kszos. 2005. Development of switchgrass (*Panicum virgatum*) as a bioenergy feedstock in the United States. Biomass and Bioenergy 28(6):515-535.

McLaughlin, S.B., D.G. de la Torre Ugarte, C.T. Garten, L.R. Lynd, M.A. Sanderson, V.R. Tolbert, and D.D. Wolf. 2002. High-value renewable energy from prairie grasses. Environmental Science & Technology 36(10):2122-2129.

Miller, R.O., and B.A. Bender. 2008. Growth and yield of poplar and willow hybrids in the central Upper Peninsula of Michigan. Proceedings of the Short Rotation Crops International Conference: Biofuels, Bioenergy, and Biproducts from Sustainable Agricultural and Forest Crops.

Muir, J.P., M.A. Sanderson, W.R. Ocumpaugh, R.M. Jones, and R.L. Reed. 2001. Biomass production of "Alamo" switchgrass in response to nitrogen, phosphorus, and row spacing. Agronomy Journal 93(4):896-901.

Nelson, R.G., J.C. Ascough II, and M.R. Langemeier. 2006. Environmental and economic analysis of switchgrass production for water quality improvement in northeast Kansas. Journal of Environmental Management 79(4):336-347.

Ocumpaugh, W., M. Hussey, J. Read, J. Muir, F. Hons, G. Evers, K. Cassida, B. Venuto, J. Grichar, and C. Tischler. 2003. Evaluation of Switchgrass Cultivars and Cultural Methods for Biomass Production in the South Central U.S.: Consolidated Report 2002. Oak Ridge, TN: Oak Ridge National Laboratory.

Parrish, D.J., D.D. Wolf, J.H. Fike, and W.L. Daniels. 2003. Switchgrass as a Biofuels Crop for the Upper Southeast: Variety Trials and Cultural Improvements: Final Report for 1997 to 2001. Oak Ridge, TN: Oak Ridge National Laboratory.

Perlack, R.D. 2007. Overview of Plant Feedstock Production for Biofuel: Current Technologies and Challenges, and Potential for Improvement. Presentation to the Panel on Alternative Liquid Transportation Fuels, November 19.

Perlack, R.D., and A.F. Turhollow. 2002. Assessment of Options for the Collection, Handling, and Transport of Corn Stover. Oak Ridge, TN: Oak Ridge National Laboratory.

Perlack, R.D., and A.F. Turhollow. 2003. Feedstock cost analysis of corn stover residues for further processing. Energy 28(14):1395-1403.

Perrin, R.K., K. Vogel, M. Schmer, and R. Mitchell. 2008. Farm-scale production cost of switchgrass for biomass. BioEnergy Research 1(1):91-97.

Petrolia, D.R. 2008. The economics of harvesting and transporting corn stover for conversion to fuel ethanol: A case study for Minnesota. Biomass and Bioenergy 32(7):603-612.

Popp, M., and R. Hogan, Jr. 2007. Assessment of two alternative switchgrass harvest and transport methods. Paper read at the Farm Foundation Conference, April 12-13, St. Louis, MO.

Prewitt, R.M., M.D. Montross, S.A. Shearer, T.S. Stombaugh, S.F. Higgins, S.G. McNeill, and S. Sokhansanj. 2007. Corn stover availability and collection efficiency using typical hay equipment. Transactions of the ASABE 50(3):705-711.

Quick, G.R. 2003. Single-pass corn and stover harvesters: Development and performance. Proceedings of the American Society of Agricultural and Biological Engineers' International Conference on Crop Harvesting and Processing. St. Joseph, MI: American Society of Agricultural and Biological Engineers.

Reynolds, J.H., C.L. Walker, and M.J. Kirchner. 2000. Nitrogen removal in switchgrass biomass under two harvest systems. Biomass and Bioenergy 19(5):281-286.

Sanderson, M.A. 2008. Upland switchgrass yield, nutritive value, and soil carbon changes under grazing and clipping. Agronomy Journal 100(3):510-516.

Schechinger, T.M., and J. Hettenhaus. 2004. Corn Stover Harvesting: Grower, Custom Operator, and Processor Issues and Answers: Report on Corn Stover Harvest Experiences in Iowa and Wisconsin for the 1997-98 and 1998-99 Crop Years. Oak Ridge, TN: Oak Ridge National Laboratory.

Schmer, M.R., K.P. Vogel, R.B. Mitchell, L.E. Moser, K.M. Eskridge, and R.K. Perrin. 2006. Establishment stand thresholds for switchgrass grown as a bioenergy crop. Crop Science 46(1):157-161.

Searcy, E., P. Flynn, E. Ghafoori, and A. Kumar. 2007. The relative cost of biomass energy transport. Applied Biochemistry and Biotechnology 137-140(1):639-652.

Shinners, K.J., G.C. Boettcher, R.E. Muck, P.J. Weimer, and M.D. Casler. 2006. Drying, harvesting and storage characteristics of perennial grasses as biomass feedstocks. Paper read at the 2006 American Society of Agricultural and Biological Engineers Annual Meeting, July 9-12, Portland, OR.

Smeets, E.M.W., I.M. Lewandowski, and A.P.C. Faaij. 2009. The economical and environmental performance of *Miscanthus* and switchgrass production and supply chains in a European setting. Renewable & Sustainable Energy Reviews 13(6-7):1230-1245.

Sohngen, B., J. Anderson, S. Petrova, and K. Goslee. 2010. Alder Springs Biomass Removal Economic Analysis. Arlington VA: Winrock International and U.S. Department of Agriculture - Forest Service.

Sokhansanj, S., and A.F. Turhollow. 2002. Baseline cost for corn stover collection. Applied Engineering in Agriculture 18(5):525-530.

Spelter, H., and D. Toth. 2009. North America's Wood Pellet Sector. Madison, WI: U.S. Department of Agriculture, Forest Service.

Stampfl, P.F., J.C. Clifton-Brown, and M.B. Jones. 2007. European-wide GIS-based modelling system for quantifying the feedstock from *Miscanthus* and the potential contribution to renewable energy targets. Global Change Biology 13(11):2283-2295.

Summit Ridge Investments. 2007. Eastern Hardwood Forest Region Woody Biomass Energy Opportunity. Granville, VT: Summit Ridge Investments, LLC.

Suzuki, Y. 2006. Estimating the Cost of Transporting Corn Stalks in the Midwest. Ames: Iowa State University College of Business, Business Partnership Development.

Taheripour, F., and W.E. Tyner. 2008. Ethanol policy analysis—What have we learned so far? Choices 23(3):6-11.

Taliaferro, C.M. 2002. Breeding and Selection of New Switchgrass Varieties for Increased Biomass Production. Oak Ridge, TN: Oak Ridge National Laboratory.

Thomason, W.E., W.R. Raun, G.V. Johnson, C.M. Taliaferro, K.W. Freeman, K.J. Wynn, and R.W. Mullen. 2005. Switchgrass response to harvest frequency and time and rate of applied nitrogen. Journal of Plant Nutrition 27(7):1199-1226.

Tiffany, D.G., B. Jordan, E. Dietrich, and B. Vargo-Daggett. 2006. Energy and Chemicals from Native Grasses: Production, Transportation and Processing Technologies Considered in the Northern Great Plains. St. Paul: University of Minnesota.

USDA-NASS (U.S. Department of Agriculture - National Agricultural Statistics Service). 2007a. Agricultural Prices 2006 Summary. Washington, DC: U.S. Department of Agriculture - National Agricultural Statistics Service.

USDA-NASS (U.S. Department of Agriculture - National Agricultural Statistics Service). 2007b. Agricultural Prices December 2007. Washington, DC: U.S. Department of Agriculture - National Agricultural Statistics Service.

USFS (U.S. Department of Agriculture - Forest Service). 2003. A Strategic Assessment of Forest Biomass and Fuel Reduction Treatments in Western States. Fort Collins, CO: U.S. Department of Agriculture - Forest Service.

USFS (U.S. Department of Agriculture - Forest Service). 2005. A Strategic Assessment of Forest Biomass and Fuel Reduction Treatments in Western States. Fort Collins, CO: U.S. Department of Agriculture - Forest Service.

Vadas, P., K. Barnett, and D. Undersander. 2008. Economics and energy of ethanol production from alfalfa, corn, and switchgrass in the Upper Midwest, USA. BioEnergy Research 1(1):44-55.

Vargas, L.A., M.N. Andersen, C.R. Jensen, and U. Jørgensen. 2002. Estimation of leaf area index, light interception and biomass accumulation of *Miscanthus sinensis* "Goliath" from radiation measurements. Biomass and Bioenergy 22(1):1-14.

Vogel, K.P., J.J. Brejda, D.T. Walters, and D.R. Buxton. 2002. Switchgrass biomass production in the Midwest USA. Agronomy Journal 94(3):413-420.

Walsh, M. 2008. Switchgrass. Available online at http://bioweb.sungrant.org/Technical/Biomass+Resources/Agricultural+Resources/New+Crops/Herbaceous+Crops/Switchgrass/Default.htm. Accessed September 21, 2010.

L

BioBreak Model Assumptions

	Supplier Breakeven - Parameter Assumptions		
Parameter	Feedstock	Mean Value (Likeliest if Skewed)	Statistical Distribution of Cost Estimates[1]
Nutrient Replacement (C_{NR})	Stover	$13.6/ton ($14.6)	Minimum Extreme
	Switchgrass (All)	$15.6/ton ($16.6)	Minimum Extreme
	Miscanthus (All)	$8.35/ton	Normal
	Wheat Straw	$5.6/ton	Normal
	Short-rotation woody crops (SRWC)	-	-
	Forest Residue	-	-
	Alfalfa (2nd year)	$62.5/acre (~$15.6/ton)	Fixed
Harvest and Maintenance (C_{HM})	Stover	$43/ton ($46)	Minimum Extreme
	Switchgrass (All)	$36/ton ($38)	Minimum Extreme
	Miscanthus (All)	$45/ton ($48)	Minimum Extreme
	Wheat Straw	$32/ton ($33)	Minimum Extreme
	SRWC	$26/ton	Normal
	Forest Residue	$26/ton	Normal
	Alfalfa	$57 per acre per harvest	Fixed
Stumpage Fee (SF)	SRWC	$20/ton	Normal
Distance Fixed Cost (DFC)	Stover	$8.50/ton	Normal
	Switchgrass (All)	$8.50/ton	Normal
	Miscanthus (All)	$8.50/ton	Normal
	Wheat Straw	$8.50/ton	Normal
	SRWC	$10/ton	Normal
	Forest Residue	$10/ton	Normal
	Alfalfa	$8.50/ton	Normal

Supplier Breakeven - Parameter Assumptions			
Parameter	Feedstock	Mean Value (Likeliest if Skewed)	Statistical Distribution of Cost Estimates[1]
Distance Variable Cost (DVC)	Stover	$0.35/ton/mile ($0.38)	Minimum Extreme
	Switchgrass (All)	$0.35/ton/mile ($0.38)	Minimum Extreme
	Miscanthus (All)	$0.35/ton/mile ($0.38)	Minimum Extreme
	Wheat Straw	$0.35/ton/mile ($0.38)	Minimum Extreme
	SRWC	$0.50/ton/mile ($0.53)	Minimum Extreme
	Forest Residue	$0.50/ton/mile ($0.53)	Minimum Extreme
	Alfalfa	$0.35/ton/mile ($0.38)	Minimum Extreme
Distance[2]	Stover (CS)	36 miles	Fixed
	Stover/Alfalfa	26 miles	Fixed
	Alfalfa	43 miles	Fixed
	Switchgrass (MW)	19 miles	Fixed
	Switchgrass (MW_low)	21 miles	Fixed
	Switchgrass (App)	15 miles	Fixed
	Switchgrass (SC)	16 miles	Fixed
	Miscanthus (MW)	13 miles	Fixed
	Miscanthus (MW_low)	14 miles	Fixed
	Miscanthus (App)	13 miles	Fixed
	Wheat Straw	37 miles	Fixed
	SRWC	17 miles	Fixed
	Forest Residue	53 miles	Fixed
Annual Biomass Demand (BD)	All	772,000 tons[3]	Fixed
Yield (Y_B)	Stover (CS)	2.1 tons	Gamma
	Alfalfa (1st year)	1.25 tons	Fixed
	Alfalfa (2nd year)	4 tons	Fixed
	Switchgrass (MW)	4 tons (3.4)	Maximum Extreme
	Switchgrass (MW_low)	3.1 tons	Log Normal
	Switchgrass (App)	6 tons (5)	Maximum Extreme
	Switchgrass (SC)	5.7 tons	Beta
	Miscanthus (MW)	8.6 tons (8)	Maximum Extreme
	Miscanthus (MW_low)	7.1 tons (6)	Maximum Extreme
	Miscanthus (App)	8.8 tons (8)	Maximum Extreme
	Wheat Straw	1 ton	Normal
	SRWC	5 tons	Normal
	Forest Residue	0.5 tons	Normal
Biomass Density (B)	Alfalfa	0.15	Normal
	Other Feedstocks	0.20	Normal
Storage (C_S)	Stover	$10.50/ton ($11)	Minimum Extreme
	Switchgrass (All)	$10.50/ton ($11)	Minimum Extreme
	Miscanthus (All)	$10.50/ton ($11)	Minimum Extreme
	Wheat Straw	$10.50/ton ($11)	Minimum Extreme
	SRWC	$11.50/ton ($12)	Minimum Extreme
	Forest Residue	$11.50/ton ($12)	Minimum Extreme
	Alfalfa	$10.50/ton ($11)	Minimum Extreme
Establishment and Seeding (C_{ES})[4,5]	Stover	-	-
	Switchgrass (All)	$40/acre	Log Normal
	Miscanthus (All)	$150/acre	Log Normal
	Wheat Straw	-	-
	SRWC	$52/acre	Normal
	Forest Residue	-	-
	Alfalfa (1st year w/ fert)	$165/acre	Fixed

Supplier Breakeven - Parameter Assumptions			
Parameter	Feedstock	Mean Value (Likeliest if Skewed)	Statistical Distribution of Cost Estimates[1]
Opportunity Cost (C$_{Opp}$)	Stover (CS)	-	-
	Switchgrass (MW)	$150/acre[6]	Log Normal
	Switchgrass (MW_low)	$100/acre	Log Normal
	Switchgrass (App)	$75/acre	Normal
	Switchgrass (SC)	$50/acre	Normal
	Miscanthus (MW)	$150/acre	Log Normal
	Miscanthus (MW_low)	$100/acre	Log Normal
	Miscanthus (App)	$75/acre	Normal
	Wheat Straw	$1.80/acre ($0)	Maximum Extreme
	SRWC	-	-
	Forest Residue	-	-
	Alfalfa (1st year w/ fert)	$175/acre	Fixed

Processor Breakeven - Parameter Assumptions			
Parameter	Feedstock	Mean Value in Baseline (Likeliest if Skewed)	Distribution
Oil Price (P$_{Oil}$)	All	$52/barrel	Fixed
		$111/barrel	(3 scenarios)
		$191/barrel	
Energy Equivalent Factor (E$_V$)	All	0.68 (0.65)	Maximum Extreme
Tax (T)	All	$1.01/gal	Fixed
Byproduct value (V$_{BP}$)	Stover	$0.16/gal	Fixed
	Switchgrass (All)	$0.18/gal	Fixed
	Miscanthus (All)	$0.18/gal	Fixed
	Wheat Straw	$0.18/gal	Fixed
	SRWC	$0.14/gal	Fixed
	Forest Residue	$0.14/gal	Fixed
	Alfalfa	$0.18/gal	Fixed
Octane (V$_O$)	All	$0.10/gal	Fixed
Capital Cost (C$_I$)	All	$0.91/gal ($0.85)	Maximum Extreme
Non-enzyme Operating Cost	All	$0.36/gal	Fixed
Enzyme Cost	All	$0.50/gal ($0.46)	Minimum Extreme
Yield (Y$_E$)	All – current	70 gal/ton	Normal
	All – future	80 gal/ton	Normal

[1] The cost estimates taken from several published studies cited in Appendix L are not necessarily normally distributed. This column reflects the statistical distribution that the cost observations tended to best fit and that was used in the Monte Carlo process to derive statistical "mean" and "most likely values" if the statistical distribution of costs was skewed.

[2] Average hauling distance is calculated using the formulation by French (1960) for a circular supply area with a square road grid. Technically, the distance is not fixed since it is a function of stochastic parameters including biomass density and yield. French, B. 1960. Some considerations in estimating assembly cost functions for agricultural processing operations. Journal of Farm Economics 62:767-778.

[3] Equivalent to 2,205 tons per day delivered to a biorefinery operating 350 days per year.

[4] Switchgrass establishment seeding cost is amortized over 10 years at 10 percent, *Miscanthus* establishment and seeding cost is amortized over 20 years at 10 percent, and woody biomass is amortized over 15 years at 10 percent. The values presented in the table are annual payments per acre.

[5] All per acre costs are converted to per ton costs using the yield assumptions provided in the table.

[6] Midwest opportunity cost is assumed to be positively correlated with corn yield through stover yield with a correlation of 0.75.

M

Summary of Literature Estimates

Appendix Table M-1 Ethanol Production Research Estimates

Type of Cost	Assumption	Value cited	Value in 2007	Reference
Oil price		$60/barrel		Elobeid et al. (2006)
Ethanol price	Analysis range	$1.50-$3.50/gal		Lambert and Middleton (2010)
	Minimum for industry development	$1.70/gal		
	Historical trend	$P_{oil}/29$		Elobeid et al. (2006)
Energy equivalent factor (EV)		0.667		Elobeid et al. (2006)
		0.667		Tokgoz et al. (2007)
Tax credit	Corn	$0.45/gal	$0.45/gal	2008 Farm Bill
	Cellulosic	$1.01/gal	$1.01/gal	2008 Farm Bill
Byproduct credit	Cellulosic		$0.14-0.21/gal*	Aden et al. (2002)
			$0.16/gal*	Khanna and Dhungana (2007)
		2.61 kWh/gal		Aden (2008)
		$0.12/gal**		Khanna (2008)
	Rank from low to high excess electricity	Aspen wood Corn stover Poplar Switchgrass		Huang et al. (2009)
	Corn	$0.48/gal		Khanna (2008)

continued

Appendix Table M-1 Continued

Type of Cost	Assumption	Value cited	Value in 2007	Reference
Investment cost	69.3 MMGY	$197.4 million		Aden et al. (2002)
(Cellulosic biorefineries)	55.5 MMGY	$231.7 million	$231.7 million	Aden (2008)
	50 MMGY	$294 million		Wright and Brown (2007)
	100 MMGY	$400 million		Taheripour and Tyner (2008)
	Stover (69.6) Switchgrass (64)	$202.2 million (0.46/gal if 10-10) $212.1 million	$0.50[1]	Huang et al. (2009)
	Hybrid poplar (68)	(0.53/gal if 10-10) $203.3 million	$0.58	
	Aspen wood (86)	(0.50/gal if 10-10) $187 million	$0.545	
		(0.34/gal if 10-10)	$0.37	
		$0.55/gallon	$0.55/gal	Jiang and Swinton (2008)
Other costs	Partial variable costs	$0.11/gal		Aden et al. (2002)
	"Other" costs	$0.11/gal		Aden et al. (2002)
	Total non-feedstock costs	$1.48/gal		Chen et al. (2010)
Enzyme cost		$0.07-0.20/gal		Aden et al. (2002)
		$0.32/gal	$0.32/gal	Aden (2008)
	2012 target	$0.10/gal		Aden (2008)
		$0.14-0.18/gal		Bothast (2005)
		$0.18/gal		Jha et al. (Presentation)
		$0.40-$1.00/gal	$0.40-$1.00/gal	Industry Source
		$0.10-0.25/gal		Tiffany et al. (2006)
Operating costs	Stover Switchgrass (crop) Switchgrass (grass) Hybrid poplar Aspen wood	$1.42/gal[2] $1.73/gal $1.86/gal $1.83/gal $1.56/gal	$1.58/gal $1.92/gal $2.06/gal $2.03/gal $1.73/gal	Huang et al. (2009)
		$1.10/gal	$1.10/gal	Jiang and Swinton (2008)
Ethanol yield (Gal/dry ton)	Stover	87.9		Aden et al. (2002)
	Stover (current)	71.9		Aden (2008)
	Stover (theoretical)	112.7		Aden (2008)
	Stover (2012 target)	90		Aden (2008)
	Stover Switchgrass Miscanthus (Nth generation plant)	79.2		Khanna and Dhungana (2007)

Appendix Table M-1 Continued

Type of Cost	Assumption	Value cited	Value in 2007	Reference
	Stover	72		McAloon et al. (2000)
	Stover	70 gal/raw ton		Tokgoz et al. (2007)
	Stover	70		Petrolia (2008)
	Stover	96		Comis (2006)
	Switchgrass (range)	60-140		Crooks (2006)
	Switchgrass (typical)	80-90		
	Stover	80		Perlack and Turhollow (2002)
	Stover *Miscanthus* Switchgrass (Nth generation plant)	79.2		Khanna (2008)
	Switchgrass	80-90		BRDI (2008)
	Switchgrass (theoretical)	110		
	Woody	89.5		
		80-120		Atchison and Hettenhaus (2003)
	Stover (base)	67.8		Tiffany et al. (2006)
	Stover (future)	89.7		
	Stover	89.8		Huang et al. (2009)
	Switchgrass	82.7		
	Hybrid poplar	88.2		
	Aspen wood	111.4		
	Switchgrass Stover	54.4		Jiang and Swinton (2008)
	Cellulosic (Nth plant)	79		Chen et al. (2010)
	Stover (current) Stover (projected)	67.4		Sheehan et al. (2003)
		89.8		
	Stover (current) Stover (theoretical)	79.2		Wallace et al. (2005)
		107		
Optimal plant size	Cellulosic	2,294-4,408		Huang et al. (2009)
Online days		350		Aden et al. (2002)
		350		Huang et al. (2009)

[1]Updated using building materials price index.
[2]Updated using machinery price index.
*Updated using EIA (2008).
**Not updated since author did not provide year of estimate.

Appendix Table M-2 Harvest and Maintenance[1]

Type of Feedstock	Type of Cost	Cited Cost per ton ($)	Cost per ton (2007$)	Reference
Corn stover	Baling, stacking and grinding	26	45	Hess et al. (2007)
Corn stover	Collection	31-36	66-77	McAloon et al. (2000)
Corn stover	Collection	35-46	64-84	McAloon et al. (2000)
Corn stover	Collection	17.70	17.70	Perlack (2007) Presentation
Corn stover	Up to Storage	20-21	36-39	Sokhansanj and Turhollow (2002)
Corn stover		28	36	Suzuki (2006)
Corn stover	Baling and staging	26	47	Aden et al. (2002)
Corn stover	Harvest	14	14	Edwards (2007)
Corn stover	Custom Harvest Bale Rake and Bale Shred, Rake, and Bale	7.47 8.84 10.70	7.47 8.84 10.70	Brechbill and Tyner (2008a)
Corn stover	Harvest	35.41-36.58	35.41-36.58	Khanna (2008)
Corn stover	Combine, Shred, Bale and Stack	19.16	24.33	Haung et al. (2009)
Corn stover	Harvest and Bale	7.26	7.26	Lamert and Middleton (2010)
Corn stover	Harvest cost	19.6	36	Jiang and Swinton (2008)
Corn stover or Switchgrass	Move to fieldside	2	2	Brechbill and Tyner (2008a)
Switchgrass	Collection	12-22	16-28	Kumar and Sokhansanj (2007)
Switchgrass	Harvest	32	32	Duffy (2007)
Switchgrass	Harvest	35	58	Khanna et al. (2008)
Switchgrass	Harvest, maintenance and establishment	123.5/acre	210/acre	Khanna and Dhungana (2007)
Switchgrass	Harvest	15	26	Perrin et al. (2008)
Switchgrass	Custom Harvest Bale Rake and Bale Shred, Rake and Bale	2.01 3.09 4.79	2.01 3.09 4.79	Brechbill and Tyner (2008a)
Switchgrass	Harvest	27.8-34.72	27.8-34.72	Khanna (2008)
Switchgrass`	Harvest (square bales)	21.86	27.8	Huang et al. (2009)
Switchgrass	Weed control	9.36/acre	9.36/acre[2]	University of Tennessee switchgrass budget (2008)

Appendix Table M-2 Continued

Type of Feedstock	Type of Cost	Cited Cost per ton ($)	Cost per ton (2007$)	Reference
	Mow, rake, bale, equipment, repair, interest, operating capital	242.92/acre	242.92/acre	
Switchgrass	Maintenance and fertilization			Mooney et al. (2009)
	0 lb N/acre	17.23/acre	17.23/acre	
	60 lb N/acre	46.5/acre	46.5/acre	
	120 lb N/acre	72.7/acre	72.7/acre	
	180 lb N/acre	99/acre	99/acre	
Switchgrass	Harvest cost (function of yield)			Mooney et al. (2009)
	7.7 tons/acre	200/acre	200/acre	
	12.5 tons/acre	311.85/acre	311.85/acre	
	2.4 tons/acre	79/acre	79/acre	
	7.2 tons/acre	190/acre	190/acre	
Switchgrass	Total production cost	54.4	54.4	Jiang and Swinton (2008)
Prairie grasses (include switchgrass)	Harvest	17.7-19.3		Tiffany et al. (2006)
Miscanthus	Harvest	33	54	Khanna et al. (2008)
Miscanthus	Harvest, maintenance, and establishment	301/acre	512/acre	Khanna and Dhungana (2007)
Miscanthus	Harvest	18.72-32.65	18.72-32.65	Khanna (2008)
Straw	Harvest and bale	7.26	7.26	Lamert and Middleton (2010)
Nonspecific		10-30	15-45	Mapemba et al. (2007)
Nonspecific		23	38	Mapemba et al. (2008)

Hybrid poplar and Aspen wood	Logging cost
	  Range
	  Assumed
	Chipping cost
	  Range
	  Assumed (Minnesota)
		14-28
		14.5
		
		12-27
		12.7
			17.8-34.6
			18.4
			
			15.2-34.3
			16.1
				Huang et al. (2009)

Aspen wood	Stumpage	51.9	66	Huang et al. (2009)
Woody biomass	Cut and extract to roadside	35-87[3]		USFS (2003, 2005)
Woody biomass	Roadside	40-46	40-46	BRDI (2008)
Woody biomass	Stumpage	4	4	BRDI (2008)
Short-rotation woody	Harvest/collection	17-29/acre	17-29/acre	BRDI (2008)

continued

Appendix Table M-2 Continued

Type of Feedstock	Type of Cost	Cited Cost per ton ($)	Cost per ton (2007$)	Reference
Woody (slash)	Collect and transport 2.8 m	24.32/Gt 31.17/Dt	24.32/Gt 31.17/Dt	Han et al. (2010)
Woody biomass	Up to roadside and on truck	25/Dt	25/Dt*	Sohngen et al. (2010)
Woody biomass	Up to roadside and on truck (high)	34/Dt	34/Dt (2009$)	Sohngen et al. (2010)
	Delivered cost range	34-65	34-65 (2009$)	
Woody biomass	Up to roadside	30-50	30-50[4]	Jenkins et al. (2009)
Woody residues	Delivered cost			
	West	56/GMT	86/Dt* (2008$)	Spelter and Toth (2009)
	North	49/GMT	75/Dt* (2008$)	Spelter and Toth (2009)
	South	42/GMT	66/Dt* (2008$)	Spelter and Toth (2009)
	Northeast	38/GMT	58/Dt* (2008$)	Spelter and Toth (2009)

[1]Harvest and maintenance costs were updated using USDA-NASS agricultural fuel, machinery, and labor prices from 1999-2007 (USDA-NASS, 2007a,b).
[2]Values are in 2008$.
[3]Price not updated.
[4]This value was based on a summary of the literature and therefore does not have a relevant year for cost.
*Assume a conversion of 0.59 for green tons to dry tons.

Appendix Table M-3 Nutrient and Replacement[1]

Type of Feedstock	Type of Cost	Cited Cost per ton ($)	Cost per ton (2007$)	Reference
Corn stover		10.2	14.1	Hoskinson et al. (2007)
Corn stover		4.6	8.4	Khanna and Dhungana (2007)
Corn stover		7	14.4	Aden et al. (2002)
Corn stover		4.2	4.2	Petrolia (2008)
Corn stover		10	21	Perlack and Turhollow (2003)
Corn stover		6.4-12.2[2]		Atchison and Hettenhaus (2003)
Corn stover	Whole plant harvest	9.7	13.3	Karlen and Birrell (2007)
Corn stover	Cob and top 50% harvest	9.5	13.1	Karlen and Birrell (2007)
Corn stover	Bottom 50% harvest	10.1	13.9	Karlen and Birrell (2007)
Corn stover		15.64	15.64	Brechbill and Tyner (2008a)
Corn stover		7.26	10	Huang et al. (2009)
Corn stover		6.5	13.7	Jiang and Swinton (2008)
Corn stover	Replace N, P, K	21.70	21.70 (2009$)	Karlen (2010)

Appendix Table M-3 Continued

Type of Feedstock	Type of Cost	Cited Cost per ton ($)	Cost per ton (2007$)	Reference
Corn stover or Straw		11.13	15.40	Lambert and Middleton (2010)
Switchgrass		6.7	12.1	Perrin et al. (2008)
Switchgrass		10.8	19.77	Khanna et al. (2008)
Switchgrass	Fertilizer, equipment, labor	83.86/acre	83.86/acre[3]	University of Tennessee (2008)
Miscanthus		2.5	4.6	Khanna et al. (2008)
Miscanthus		4.20	7.73	Cost using average fertilizer rates from literature summarized in Khanna et al. (2008) and updated Khanna et al. (2008) costs

[1]Nutrient and replacement costs were updated using USDA-NASS agricultural fertilizer prices from 1999-2007 (USDA-NASS, 2007a,b).
[2]Price not updated.
[3]Value in 2008$.

Appendix Table M-4 Distance

Distance (miles)	Type	Reference
46-134	Round-trip	Mapemba et al. (2007)
22-62	One-way	Perlack and Turhollow (2003)
22-61	One-way	Perlack and Turhollow (2002)
50	Round-trip	Khanna et al. (2008)
50	One-way max	English et al. (2006)
50	One-way	Vadas et al. (2008)
100	One-way (wood)	USFS (2003, 2005)
10-50	One-way	Atchison and Hettenhaus (2003)
5-50	One-way	Brechbill and Tyner (2008a,b)
16.6-47	One-way average	Perlack and Turhollow (2002)
50	One-way max	Taheripour and Tyner (2008)
75	One-way max	BRDI (2008)
50	One-way	Tiffany et al. (2006)
83	One-way (wood)	Sohngen et al. (2010)
46-138	One-way range (wood)	Sohngen et al. (2010)
50	One-way (wood)	Spelter and Toth (2009)

Appendix Table M-5 Transportation Cost[1]

Type of Feedstock	Type of Cost	Cost cited ($)	Cost (2007$)	Reference
Corn stover	Per ton	8.85	12.5	English et al. (2006)
Corn stover	Per ton	10.25	27	Hess et al. (2007)
Corn stover	DVC[2]	0.15	0.35	Kaylen et al. (2000)
Corn stover	Max DVC for positive NPV	0.28	0.66	Kaylen et al. (2000)
Corn stover	Per ton	10.8	10.8	Perlack (2007)
Corn stover	Per ton	13	31	Aden et al. (2002)
Corn stover	Per ton	4.2-10.5	11-27.7	Perlack and Turhollow (2002)
Corn stover	DVC	0.08-0.29	0.17-0.63	Kumar et al. (2005)
	DFC[3]	4.5	9.8	
	DFC range	0-6	0-13.3	
Corn stover	DVC	0.18	0.32	Searcy et al. (2007)
	DFC	4	7.3	
Corn stover	DVC	0.16	0.38	Kumar et al. (2003)
	DFC	3.6	8.6	
Corn stover	10 miles	3.4		Atchison and Hettenhaus (2003)
	15 miles	5.1		
	30 miles	10.2		
	40 miles	13.5		
	50 miles	17[4]		
Corn stover	DVC			Petrolia (2008)
	0-25 miles	0.13-0.23	0.13-0.23	
	25-100 miles	0.10-0.19	0.10-0.19	
	>100 miles	0.09-0.16	0.09-0.16	
	DFC square bales	1.70	1.70	
	DFC round bales	3.10	3.10	
Corn stover	Per ton	10.9	13.8	Vadas et al. (2008)
Corn stover	DFC	6.9	9.71	Huang et al. (2009)
	DVC	0.16	0.23	
Corn stover or Switchgrass	Average DVC	0.20	0.20	Brechbill and Tyner (2008a,b)
Corn stover or Switchgrass	Custom loading	1.15	1.15	Brechbill and Tyner (2008a)
	Custom DVC	0.28	0.28	
	Owned DVC	0.12	0.12	
	Custom per ton			
	10 miles	3.92	3.92	
	20 miles	6.69	6.69	
	30 miles	9.46	9.46	
	40 miles	12.23	12.23	
	50 miles	15	15	
Corn stover	Own equipment (per ton)			Brechbill and Tyner (2008a)
	10 miles	3.31-6.18	3.31-6.18[5]	
	20 miles	4.65-7.52	4.65-7.52	
	30 miles	5.99-8.86	5.99-8.86	
	40 miles	7.33-7.71	7.33-7.71	
	50 miles	8.67-9.05	8.67-9.05	

Appendix Table M-5 Continued

Type of Feedstock	Type of Cost	Cost cited ($)	Cost (2007$)	Reference
Switchgrass	Own equipment (per ton)			Brechbill and Tyner (2008a)
	10 miles	3.13-3.93	3.13-3.93[6]	
	20 miles	4.47-5.27	4.47-5.27	
	30 miles	5.81-6.61	5.81-6.61	
	40 miles	7.15-7.95	7.15-7.95	
	50 miles	8.49-9.29	8.49-9.29	
Switchgrass	Per ton	14.75	14.75	Duffy (2007)
Switchgrass	Per ton	19.2-23	27-32.4	Kumar and Sokhansanj (2007)
Switchgrass	Per ton	13	28	Perrin et al. (2008)
Switchgrass	Per ton	10.9	13.8	Vadas et al. (2008)
Switchgrass	DFC	3.39	4.78	Huang et al. (2009)
	DVC	0.16	0.23	
Switchgrass	Stage and load	19.15/acre	19.15/acre (2008$)	UT (2008)
Native prairie (include switchgrass)	Per ton	4[7]		Tiffany et al. (2006)
Switchgrass or *Miscanthus*	Per ton for 50 miles	7.9	17.1	Khanna et al. (2008)
Nonspecific	Per ton	7.4-19.3	13.7-35.6	Mapemba et al. (2007)
Nonspecific	Per ton	14.5	31.5	Mapemba et al. (2008)
Hybrid poplar and Aspen wood	DFC	4.13	5.8	Huang et al. (2009)
	DVC	0.16	0.23	
Woody biomass	Per ton	11-22	11-22	Summit Ridge Investments (2007)
Woody biomass	DVC	0.2-0.6 Used 0.35[8]		USFS (2003, 2005)
Woody biomass	DVC	0.22	0.22	Sohngen et al. (2010)
Wood	DVC	0.20-0.60	0.20-0.60	Jenkins et al. (2009)

[1]Transportation costs were updated using USDA-NASS agricultural fuel prices from 1999-2007 (USDA-NASS, 2007a,b).
[2]DVC is distance variable cost in per ton per mile.
[3]DFC is distance fixed cost per ton.
[4]Prices not updated.
[5]Authors used 2006 wages and March 2008 fuel costs.
[6]Authors used 2006 wages and March 2008 fuel costs.
[7]Price not updated.
[8]Price not updated.

Appendix Table M-6 Storage[1]

Type of Feedstock	Type of Cost	Cited cost per ton ($)	Cost per ton (2007$)	Reference
Corn stover		4.44	5.64	Hess et al. (2007)
Corn stover	Round bales Square bales	6.82 12.93	6.82 12.93	Petrolia (2008)
Corn stover		4.39-21.95	4.39-21.95	Khanna (2008)
Stover or switchgrass	Square bales	7.25	7.9	Huang et al. (2009)
Switchgrass		16.67	16.67	Duffy (2007)
Switchgrass		4.14	5.18	Khanna et al. (2008)
Switchgrass		4.43-21.68	4.43-21.68	Khanna (2008)
Miscanthus		4.40	5.50	Khanna et al. (2008)
Miscanthus		4.64-23.45	4.64-23.45	Khanna (2008)
Nonspecific		2	2.18	Mapemba et al. (2008)
Hybrid poplar or Aspen wood	Keep on stump until needed	0	0	Huang et al. (2009)

[1]Storage costs were updated using USDA-NASS Agricultural building material prices from 1999-2007 (USDA-NASS, 2007a,b).

Appendix Table M-7 Establishment and Seeding[1]

Type of Feedstock	Type of Cost	Land rent included	Cited cost per acre ($)	Cost per acre (2007$)	Reference
Switchgrass		Yes	200	200	Duffy (2007)
Switchgrass		No Yes	25.76 85.46	46 153	Perrin et al. (2008)

Switchgrass	PV per ton
	10 yr PV per acre
	Amortized
	  4% over 10 years
	  8% over 10 years
		No	7.21/ton
			142.3
			
			17.3
			20.7
				12.6/ton
				249
				
				30.25
				36.25
					Khanna et al. (2008)

Switchgrass		Yes	72.5-110	88.5-134	Vadas et al. (2008)
Switchgrass	Grassland Cropland (includes fertilizer)	No	134 161	180 216	Huang et al. (2009)
Switchgrass	Prorated Establishment and Reseeding (10 years)		45.69	45.69[2]	UT (2008)
Switchgrass	Plots with seeding:				
	2.5 lb/acre	No	150	150	Mooney et al. (2009)
	5 lb/acre	No	202.6	202.6	
	7.5 lb/acre	No	255	255	
	10 lb/acre	No	306.6	306.6	

Appendix Table M-7 Continued

Type of Feedstock	Type of Cost	Land rent included	Cited cost per acre ($)	Cost per acre (2007$)	Reference
	12.5 lb/acre	No	$359	$359	
Switchgrass	Seed and fertilizer cost per acre (no equip/machinery)	No	$171	$171 (2008$)	James et al. (2010)
Miscanthus	PV per ton	No	2.29/ton	4/ton	Khanna et al. (2008)
	20 yr PV per acre		261	457	
	Amortized				
	4% over 20 years		19	33.2	
	8% over 20 years		26.20	45.87	
Miscanthus	Total	No	1,206-2,413		Lewandowski et al. (2003)
	Amortized				
	4% over 20 years		88-175	176-350	
	8% over 20 years		121-242	242-484	
Miscanthus	Total rhizome cost per acre (no equip/labor)	No	8,194	8,194 (2008$)	James et al. (2010)
Miscanthus	Total rhizome cost per acre – projected (no equipment/labor)	No	227.61	227.61 (2008$)	James et al. (2010)
Miscanthus	Plugs	No	3,000-4,000/ha	1,215-1,619/ac	Jain et al. (2010)
Miscanthus	Rhizomes in Illinois	No	2,957/ha	1,197/ac	Jain et al. (2010)
Hybrid poplar	Total cutting cost per acre	No	242	242 (2008$)	James et al. (2010)
Hybrid poplar	Includes nutrients (cropland)	No	35	47	Huang et al. (2009)
Timber	Yellow pine (South average)		386	386 (2008$)	Cubbage et al. (2010)
Timber	Yellow pine (NC)		430	430 (2008$)	Cubbage et al. (2010)
Timber	Douglas fir (NC, OR)		520	520 (2008$)	Cubbage et al. (2010)

[1]Establishment and Seeding costs were updated using USDA-NASS agricultural fuel and seed prices from 1999-2007 (USDA-NASS, 2007a,b).

[2]Value in 2008$.

Appendix Table M-8 Opportunity Cost[1]

Type of Feedstock	Type of Cost	Cited cost per acre ($)	Cost per acre (2007$)	Reference
Corn stover	Feed value less harvest and nutrient cost	24/ton	24/ton	Edwards (2007)
	@ 2.4 tons/acre	57/acre	57/acre	
Corn stover	Lost profits	22-58	22-58	Khanna and Dhungana (2007)
Corn stover	Lost profits when switch to continuous corn	94-140		Scenarios derived using Duffy (2010)
Switchgrass	Cash rents	70 ($14/ton)	70 ($14/ton)	Brechbill and Tyner (2008a)
Switchgrass	Lost profits	78-231	78-231	Khanna and Dhungana (2007)
Switchgrass	Cash rental rate – alternative land use (TN)	68	68	Mooney et al. (2009)
Switchgrass	Forgone profits per ton	46-103/Mt	42-93/ton	Jain et al. (2010)
Switchgrass or *Miscanthus*	Forgone profits – Michigan	366/ha	148/ac	Jain et al. (2010)
Switchgrass or *Miscanthus*	Forgone profits – Illinois	785/ha	318/ac	Jain et al. (2010)
Switchgrass or *Miscanthus*	Lost profits	78	76	Khanna et al. (2008)
Miscanthus	Forgone profits per ton	19-103/Mt	17-93/ton	Jain et al. (2010)
Miscanthus	Lost profits	78-231	78-231	Khanna and Dhungana (2007)
Nonspecific	Lost Conservation Reserve Program (CRP) payments if harvest every year	35	36	Mapemba et al. (2008)
Nonspecific	Lost CRP payments if harvest once every 3 years	10.1	10.4	Mapemba et al. (2008)
Nonspecific	Non-CRP land crops	10/ton	10.3/ton	Mapemba et al. (2008)
Nonspecific		78	76	Khanna et al. (2008)
Woody biomass	Alternative use	0-25	0-25	Summit Ridge Investments (2007)
Woody biomass	Chip value	30/ton	30/ton[2]	USFS (2003, 2005)

[1]Opportunity costs were updated using USDA-NASS agricultural land rent prices from 1999-2007 (USDA-NASS, 2007a,b).

[2]Price not updated since no year was provided for initial estimate.

Appendix Table M-9 Yield

Biomass Type	Assumptions	Estimated Yield (tons acre^{-1})	Location	Reference
Corn stover	Soil tolerance	2.02	IL	Khanna and Dhungana (2007)
Corn stover		2.4	IA	Edwards (2007)
Corn stover	2000-2005 mean	2.31-3	WI	Vadas et al. (2008)
Corn stover		2-3.8		Atchison and Hettenhaus (2003)
Corn stover	130 bu/acre yield 170 bu/acre yield 200 bu/acre yield	0-2.6 0-3.6 0-4.3		Atchison and Hettenhaus (2003)
Corn stover	Bale Rake and bale Shred, rake, and bale	1.62 2.23 2.98	IN	Brechbill and Tyner (2008a)
Corn stover		1.1		Perlack and Turhollow (2002)
Corn stover	Produced Delivered	3.6 1.5	Midwest	Sokhansanj and Turhollow (2002)
Corn stover	Produced Delivered	2.4-4 1.8-1.9	IL	Khanna (2008)
Corn stover		3		BRDI (2008)
Corn stover	Total produced 125 bu/acre 140 bu/acre >140 bu/acre	 3.5 3.92 4		Lang (2002)
Corn stover	Total produced Removable	4.2 2.94	IA	Quick (2003)
Corn stover	Collected	0.8-2.2	KY	Prewitt et al. (2003)
Corn stover	Collected (trial)	1.25-1.5	IA, WI	Schechinger and Hettenhaus (2004)
Corn stover	Four scenarios (assumed)	1.5, 3, 4, and 6	IA	Duffy and Nanhou (2001)
Corn stover	Produced	2.54	MN	Haung et al. (2009)
Corn stover	Previous study	1.6	MI	James et al. (2010)
Corn stover	Produced (150 bu/ac)	2.93	Corn Belt	Jiang and Swinton (2008)
Corn stover	Harvested (50%)	1.46	Corn Belt	Jiang and Swinton (2008)
Corn stover	No-till	0.67	Average	Chen et al. (2010)
Switchgrass	Field Trials	2.58	IA, IL	Khanna and Dhungana (2007)
Switchgrass		4	IA	Duffy (2007)

continued

Appendix Table M-9 Continued

Biomass Type	Assumptions	Estimated Yield (tons acre^{-1})	Location	Reference
Switchgrass	Farm-scale	2.23 (5 year average) (Range = 1.7-2.7) 3.12 (10 year average) (Range = 2.6-3.5)	SD, NE	Perrin et al. (2008)
Switchgrass	Delivered yield (years 3-10)	3.13	IL	Khanna et al. (2008)
Switchgrass	Peak yield	4.2	IL	Khanna et al. (2008)
Switchgrass	10 year PV	19.74	IL	Khanna et al. (2008)
Switchgrass	Nitrogen level	4-5.8	Upper Midwest	Vadas et al. (2008)
Switchgrass	Research blocks	7.14 (average) 9.8 (best)	Southern and Mid-Atlantic	Lewandowski et al. (2003)
Switchgrass	Plot trials	3.6-8.9 (previous) 2.3-4 (own)	United States Northern	Shinners et al. (2006)
Switchgrass	Plot trials	6.33 4.64-8.5	SE	Fike et al. (2006)
Switchgrass	Field trials Mean Strains: Dacotah ND3743 Summer Sunburst Trailblazer Shawnee OK NU-2 Cave-in-Rock	1.12-4.1 1.11-4.22 0.91-3.92 1.18-4.38 1.43-5.57 1.15-4.88 1.06-4.5 0.89-4.18 0.97-4.27	ND	Berdahl et al. (2005)
Switchgrass	Plot trials	5.2-5.6 4.7-5	IA NE	Vogel et al. (2002)
Switchgrass	Field trials Mean Range	0.5-3.2 0-6.4	Northern Great Plains	Schmer et al. (2006)
Switchgrass	Peer-reviewed articles	4.46		Heaton et al. (2004a)

Appendix Table M-9 Continued

Biomass Type	Assumptions	Estimated Yield (tons acre^{-1})	Location	Reference
Switchgrass	Farm trials (avg)			McLaughlin and Kszos (2005)
	Alamo (1 cut)	6.2	VA, TN, WV, KY, NC	
	Alamo (1 cut)	6-8.5	TX, AR, LA	
	Alamo (1 cut)	5.4	IA	
	Alamo (1 cut)	5.8-7.2	AL, GA	
	Alamo (2 cut)	7	VA, TN, WV, KY, NC	
	Alamo (2 cut)	7.2-10.3	AL	
	Kanlow (1 cut)	6.2	VA, TN, WV, KY, NC	
	Kanlow (1 cut)	5.8	IA	
	Kanlow (1 cut)	5.2-7	AL, GA	
	Kanlow (1 cut)	9.2	NE	
	Kanlow (2 cut)	6.9-8.1	AL	
	Cave-in-rock (1 cut)	7.3	NE	
	Rockwell (1 cut)	4.2	KS	
	Shelter (1 cut)	4.2	KS	
	Sunburst (1 cut)	4.9	ND	
	Trailblazer (1 cut)	4.4	ND	
	Best	12.2	VA, TN, WV, KY, NC	
	Alamo (1 cut)	11	TX, AR, LA	
	Alamo (1 cut)	7.8	IA	
	Alamo (1 cut)	15.4	AL	
	Alamo (1 cut)	11.3	VA, TN, WV, KY, NC	
	Alamo (2 cut)	15.4	Al	
	Alamo (2 cut)	10.4	VA, TN, WV, KY, NC	
	Kanlow (1 cut)	11	AL, GA	
	Kanlow (1 cut)	6.2	ND	
	Sunburst (1 cut)	5.4	ND	
	Trailblazer (1 cut)			
Switchgrass	U.S. average	4.2		McLaughlin et al. (2002)
Switchgrass		5	IN	Brechbill and Tyner (2008a)
Switchgrass	Alamo	5.35-6.9	18 sites	Walsh (2008)
	Kanlow	5.2-6.9	18 sites	
	Max one year	15.4	AL	
Switchgrass	Delivered	2.3-2.5	IL	Khanna (2008)
Switchgrass		4.2-10.3		BRDI (2008)
Switchgrass	First	0	AR	Popp and Hogan (2007)
	Second year	3		
	Third+ year	5		
Switchgrass	Assumptions			Kszos et al. (2002)
	Lake states	4.8		
	Corn Belt	5.98		
	Southeast	5.49		
	Appalachian	5.84		
	North Plains	3.47		
	South Plains	4.3		
	Northeast	4.87		
Switchgrass		4	Northern Plains	Tiffany et al. (2006)

continued

Appendix Table M-9 Continued

Biomass Type	Assumptions	Estimated Yield (tons acre^{-1})	Location	Reference
Switchgrass		7-16	Southeast	Comis (2006)
		5-6	Western Corn Belt	
		1-4	ND	
Switchgrass		1-4	IA	Gibson and Barnhart (2007)
		2-6.4		
Switchgrass	Cropland and grassland	4.9	MN	Huang et al. (2009)
Switchgrass	80% of *Miscanthus* 2004			Smeets et al. (2009)
		6.7	Poland	
		7.1	Hungary	
		5.4	United Kingdom	
		9	Italy	
		5.8	Lithuania	
	2030 (1.5% increase/ year)	9.4	Poland	
		9.8	Hungary	
		7.6	United Kingdom	
		12	Italy	
		8	Lithuania	
	3 experiments on loss	3.8-6.7	Italy	Monti et al. (2009)
	Sustainable yield (124 kg N/acre)	6.7	United States	
Switchgrass	One cut	5.8	OK	Thomason et al. (2005)
	Two cut	5.6		
	Three cut	7.3		
	Max yield (2 harvests)	16.4		
Switchgrass	Predicted yields		KS	Nelson et al. (2006)
	0-200 lbs/acre N	2.5-5.9		
	100 lbs/acre N	4.6		
Switchgrass	Max (Alamo)	10	TX	Muir et al. (2001)
	Average (2 sites)	4.8-6.5		
Switchgrass	3 years of data (avg)	5.5	LA	Kiniry et al. (2005)
		7.7	AR	
		8.3-10	TX	
	7 years of data (avg)	6.6	TX	
Switchgrass	Cave-in-Rock (2 cut)	2.8	PA	Sanderson (2008)
	Shawnee (2 cut)	2.7		
	Trailblazer (2 cut)	2.6		
	Mean (2 cut)	2.7		
	Cave-in-Rock (3 cut)	3.2		
	Shawnee (3 cut)	3.2		
	Trailblazer (3 cut)	3.2		
	Mean (3 cut)	3.2		

Appendix Table M-9 Continued

Biomass Type	Assumptions	Estimated Yield (tons acre^{-1})	Location	Reference
Switchgrass	One-cut range	5-9	TN	Reynolds et al. (2000)
	Two-cut range	6.8-10.3		
	Cave-in-rock (2 cut)	8.7		
	Alamo (2 cut)	8.9		
	Kanlow (2 cut)	8.2		
	Shelter (2 cut)	8.1		
Switchgrass	Alamo (1 cut)	5.4-5.9	TX, Upper South	Lewandowski et al. (2003)
		11.6	AL	
	Alamo (2 cut)	15.4	AL	
	Kanlow (1 cut)	4.5-5.5	TX, Upper South	
		8.3	AL	
	Kanlow (2 cut)	10.3	AL	
	Kanlow (3-4 years)	5	Britain	
	Cave-in-Rock (1 cut)	2.4-4.2	TX, Upper South	
		4.2	AL	
	Cave-in-Rock (2 cut)	4.6	AL	
	Cave-in-Rock (3-6 years)	4.7	Britain	
Switchgrass	Alamo (3-4 years)	4.9-8.8	TX	Cassida et al. (2005b)
	Caddo (3-4 years)	2.2-2.7		
	Alamo (3 years)	4.8	LA	
	Caddo (3 years)	0.5		
	Alamo (3 years)	7.5	AR	
	Caddo (3 years)	3.3		
Switchgrass	Kanlow (avg)	5.9	AL	Bouton et al. (2002)
	Alamo (avg)	6.0		
Switchgrass	Cave-in-rock (1 cut)	3.9-7.3	Southeast (6)	Fike et al. (2006b)
	Shelter (1 cut)	3.7-6.8		
	Alamo (1 cut)	4.8-9.8		
	Kanlow (1 cut)	5.4-9.5		
	Cave-in-rock (2 cut)	5.8-9.5		
	Shelter (2 cut)	4.9-9.1		
	Alamo (2 cut)	6-10		
	Kanlow (2 cut)	6-9.5		
Switchgrass	Alamo (1 cut)	1.2-9	Texas	Ocumpaugh et al. (2003)
	Alamo (2 cut)	1.3-8.6		
Switchgrass	Upland (1 cut)	4.8-5.3		Parrish et al. (2003)
	Upland (2 cut)	6.5-6.7		
	Lowland (1 cut)	6.6-7		
	Lowland (2 cut)	6.8-7.3		
Switchgrass	Alamo	1.6	KS	Taliaferro (2002)
		2.8	AR	
		2.8	VA	
		2.8	OK	
	Kanlow	1.4	KS	
		2.9	AR	
		2.5	VA	
		2.8	OK	

continued

Appendix Table M-9 Continued

Biomass Type	Assumptions	Estimated Yield (tons acre^-1)	Location	Reference
Switchgrass	Cave-in-rock 3 year average	2.2	Northern Illinois	Pyter et al. (2007)
		5.2	Central Illinois	
		2.7	Southern Illinois	
Switchgrass	POLYSYS assumption	4.87	Northeast	De La Torre Ugarte et al. (2003)
		5.84	Appalachian	
		5.98	Corn Belt	
		4.8	Lake states	
		5.49	Southeast	
		4.30	Southern Plains	
		3.47	Northern Plains	
Switchgrass	Calibrated values for 2008 (assumed 2% growth following 2008)	3.5-6.5	Appalachian	Marshall and Sugg (2010)
		5.16-6.4	Corn Belt	
		3.8-6.5	Delta states	
		4.5-6.0	Lake states	
		3	Mountain states	
		4.8-6.0	Northern Plains	
		3.2-6.2	Northeast	
		3.5-6.3	Southern Plains	
		4.4-6.5	Southeast	
Switchgrass	Assumption	4	Southern MI	James et al. (2010)
Switchgrass	Previous Literature	4.46-6.69		Reijnders (2010)
Switchgrass	Plots – varying seed and nitrogen	3.8-7.9	TN	Mooney et al. (2009)
Switchgrass	One year max – plot	10.2	TN	Mooney et al. (2009)
Switchgrass	Plots	4	IA	Lemus et al. (2002)
Switchgrass	Assumption (prev studies)	3.6	Corn Belt	Jiang and Swinton (2008)
Switchgrass	Simulated (MISCANMOD)	3.8	U.S. Average	Chen et al. (2010)
Switchgrass	Average model yield (range)	6.8 (3.6-17.8)	Midwest	Jain et al. (2010)
Switchgrass	Farm-gate yield (annualized yield after losses)	3.75-4.2	Midwest	Jain et al. (2010)
Switchgrass	Average observed peak yield	6.6	Midwest	Jain et al. (2010)
Wheat straw		1		BRDI (2008)

Appendix Table M-9 Continued

Biomass Type	Assumptions	Estimated Yield (tons acre^{-1})	Location	Reference
Wheat Straw	Estimated	0.27	Average	Chen et al. (2010)
Miscanthus	Simulated	8.9	IL	Khanna and Dhungana (2007)
Miscanthus		14.5 avg 12-17 range 114.58 (20 year PV)	IL	Khanna et al. (2008)
Miscanthus	Potential Delivered	12-18 8.1-8.5	IL	Khanna (2008)
Miscanthus	3 year average	9.8	Northern Illinois	Pyter et al. (2007)
		15.5	Central Illinois	
		15.8	Southern Illinois	
	1 year	14.1	Urbana, Illinois	
Miscanthus	Field experiment	5.71 (14 year) 3.43-11.73 (3 year)	EU	Christian et al. (2008)
Miscanthus		1.8-19.6	EU	Lewandowski et al. (2003)
Miscanthus	Projection	13.36 (mean) 10.93-17.81		Heaton et al. (2004b)
Miscanthus	Peer-reviewed articles	10	U.S. and EU	Heaton et al. (2004a)
Miscanthus	3 year state average	13.2	IL	Heaton et al. (2008)
Miscanthus	3 year max state average	17	IL	Heaton et al. (2008)
Miscanthus	Assumption	10	MI	James et al. (2010)
Miscanthus	Peak Delayed	7.5-17.2 4.3-11.6	EU	Clifton-Brown et al. (2004)
Miscanthus	Autumn yields without irrigation	4.5-11.15	EU	Lewandowski et al. (2000)
Miscanthus	Yield range (high-end irrigated)	0.9-19.6	EU	Lewandowski et al. (2000)
Miscanthus	Modeled harvestable yield	6.2-9.4	EU	Stampfl et al. (2007)
Miscanthus	Above ground Mean harvested	6.6-14.9 5.2	Germany	Kahle et al. (2001)
Miscanthus	First year average First year max First year min Second year average Second year max Third year max	0.85 2.6 0.16 3.8 12 18.2	EU	Clifton-Brown et al. (2001)
Miscanthus	1996 (drought) 1997	3.4 5.9	Denmark	Vargas et al. (2002)

continued

Appendix Table M-9 Continued

Biomass Type	Assumptions	Estimated Yield (tons acre^{-1})	Location	Reference
Miscanthus	First year average	0.85	Germany	Clifton-Brown and Lewandowski (2002)
	First year max	1.34		
	Second year average	2.8		
	Second year max	4.3		
	Third year average	7.3		
	Third year max	11.4		
Miscanthus	Assumption	9.81	Southern MI	James et al. (2010)
Miscanthus	Previous literature	4.46-5.8		Reijnders (2010)
Miscanthus	Simulated (MISCANMOD)	11.6	U.S. average	Chen et al. (2010)
Miscanthus	Average model yield (range)	19 (0-27.7)	Midwest	Jain et al. (2010)
Miscanthus	Farm-gate yield (annualized yield after losses)	6.3-8.6	Midwest	Jain et al. (2010)
Miscanthus	Average observed peak yield	16.6	Midwest	Jain et al. (2010)
Hybrid poplar		3.5-5.3	Lake states	Huang et al. (2009)
	Assumption	3.43-4	MN	
		4		
Poplar	10 year average (best growing taxa)	3.7	Upper MI	Miller and Bender (2008)
Poplar	Assumption	5	Southern MI	James et al. (2010)
Hybrid poplar	POLYSYS assumption	3.99	NE	De La Torre Ugarte et al. (2003)
		3.56	Appalachian	
		4.63	Corn Belt	
		4.41	Lake states	
		4.50	Southeast	
		3.75	Southern Plains	
		3.83	Northern Plains	
		5.73	Pacific Northwest	
Willow	10 year average (best taxa)	3.4	Upper Michigan	Miller and Bender (2008)
Willow	POLYSYS assumption	4.9	Northeast	De La Torre Ugarte et al. (2003)
		4.50	Appalachian	
		4.70	Corn Belt	
		4.60	Lake states	
Aspen wood		0.446 (dry)	MN	Huang et al. (2009)
SRWC		5-12		BRDI (2008)
Woody biomass	Stock	4.6-39		USFS (2003, 2005)

Appendix Table M-9 Continued

Biomass Type	Assumptions	Estimated Yield (tons acre^{-1})	Location	Reference
Wood residue	2006 average removal rate in Mississippi (lower bound)	1.1	Mississippi	USDA Forest Service data
Yellow pine	15 m^3/hectare/yr	4.3 (2.3 – 4)[1]	Southern U.S.	Cubbage et al. (2010)
Yellow pine	12.5 m^3/hectare/yr	3.6 (2 – 3.3)	North Carolina	Cubbage et al. (2010)
Douglas fir	14 m^3/hectare/yr	4 (3.3)	Oregon	Cubbage et al. (2010)
Douglas fir	18 m^3/hectare/yr	5.1 (4.25)	North Carolina	Cubbage et al. (2010)
Sorghum	Previous literature	16.41		Reijnders (2010)

[1]The first value is derived using a general conversion factor of 0.64 dry metric tons per cubic meter (DMT/m^3) for softwoods. The yields in parentheses are based on conversion factors provided by engineeringtoolbox.com of 0.35-0.60 DMT/m^3 and 0.53 DMT/m^3 for Yellow Pine and Douglas Fir, respectively. (Accessed September 15, 2010) http://www.engineeringtoolbox.com/wood-density-d_40.html.

Appendix Table M-10 Interest Rate

Details	Rate	Reference
	8%	Brechbill and Tyner (2008a,b), Brechbill et al. (2008)
	7.5%	Quick (2003)
	7.5%	Sokhansanj and Turhollow (2002)
Establishment and seeding	8%	Duffy and Nanhou (2001)
Operating expenses	9%	Duffy and Nanhou (2001)
Real discount rate	4%	Popp and Hogan (2007)
Farmer's real opportunity cost of machinery	5%	James et al. (2010)
Real discount rate (PV calc)	6.5%	de la Torre Ugarte et al. (2003)
Nominal interest rate	8%	Mooney et al. (2009)
Real discount rate	5.4%	Mooney et al. (2009)
Establishment and seeding	4%	Jain et al. (2010)

Appendix Table M-11 Stand Length

Crop	Length	Reference
Switchgrass	10 years	Brechbill et al. (2008)
Switchgrass	10 years	Duffy and Nanhou (2001)
Switchgrass	12 years	Popp and Hogan (2007)
Switchgrass	20 years	Tiffany et al. (2006)
Switchgrass	10 years	Khanna (2008)
Switchgrass	10 years	Khanna et al. (2008)
Switchgrass	10 years	Khanna and Dhungana (2007)
Switchgrass	10+ years	Lewandowski et al. (2003)
Switchgrass	10+ years	Fike et al. (2006)
Switchgrass	10 years	James et al. (2010)
Switchgrass	10 years	de la Torre Ugarte et al. (2003)
Switchgrass	10 years	Mooney et al. (2009)
Switchgrass	5 years[1]	Mooney et al. (2009)
Switchgrass	10 years	Miller and Bender (2008)
Switchgrass	10 years	Jain et al. (2010)
Miscanthus	20 years	Khanna (2008)
Miscanthus	20 years	Khanna et al. (2008)
Miscanthus	20 years	Khanna and Dhungana (2007)
Miscanthus	20-25 years	Lewandowski et al. (2003)
Miscanthus	10 years	James et al. (2010)
Miscanthus	15 years	Jain et al. (2010)
Miscanthus	10 years (sensitivity)	Jain et al. (2010)
Short-rotation poplar	10 years	James et al. (2010)
Poplar	10 year analysis	Miller and Bender (2008)
Poplar	6-10 years	de la Torre Ugarte et al. (2003)
Willow	10-year analysis	Miller and Bender (2008)
Willow	22 years	de la Torre Ugarte et al. (2003)
Yellow pine (South U.S.)	30 years	Cubbage et al. (2010)
Yellow pine (NC)	23 years	Cubbage et al. (2010)
Douglas fir	45 years	Cubbage et al. (2010)

[1]Based on the assumption that it will be optimal to replace with improved seed and contracts.

Appendix Table M-12 Yield Maturity Rate

Type of Feedstock	Year 1	Year 2	Year 3	Reference
Switchgrass	20-35%	60-75%	100%	Walsh (2008)
Switchgrass	No harvest	-	-	
Switchgrass	30%	67%	100%	Kszos et al. (2002)
Switchgrass	0	60%	100%	Popp and Hogan (2007)
Switchgrass	~33%	~66%	100%	McLaughlin and Kszos (2005)
Switchgrass	Max at 3 years			James et al. (2010)
Switchgrass	30%	67%	100%	de la Torre Ugarte et al. (2003)
Switchgrass	14% of 3rd year	59% of 3rd year		Mooney et al. (2009)
Switchgrass	30-100%	67-100%	100%	Jain et al. (2010)
Miscanthus	2-5 years for full	-	-	Heaton et al. (2004)
Miscanthus	Max at 4 years	-	-	Atkinson (2009)
Miscanthus	2 years in warm climate 3 years in cooler climates	-	-	Clifton-Brown et al. (2001)
Miscanthus	Max at 3 years	-	-	James et al. (2010)
Miscanthus	0	40-50%	100%	Jain et al. (2010)
Willow	60% in year 4, 100% after	-	-	de la Torre Ugarte et al. (2003)
Timber	5-year establishment period	-	-	Cubbage et al. (2010)

REFERENCES

Aden, A. 2008. Biochemical Production of Ethanol from Corn Stover: 2007 State of Technology Model. Golden, CO: National Renewable Energy Laboratory.

Aden, A., M. Ruth, K. Ibsen, J. Jechura, K. Neeves, J. Sheehan, B. Wallace, L. Montague, A. Slayton, and J. Lukas. 2002. Lignocellulosic Biomass to Ethanol Process Design and Economics Utilizing Co-Current Dilute Acid Prehydrolysis and Enzymatic Hydrolysis for Corn Stover. Golden, CO: National Renewable Energy Laboratory.

Atchison, J.E., and J.R. Hettenhaus. 2003. Innovative Methods for Corn Stover Collecting, Handling, Storing and Transporting. Golden, CO: National Renewable Energy Laboratory.

Banowetz, G.M., A. Boateng, J.J. Steiner, S.M. Griffith, V. Sethi, and H. El-Nashaar. 2008. Assessment of straw biomass feedstock resources in the Pacific Northwest. Biomass & Bioenergy 32:629-634.

Berdahl, J.D., A.B. Frank, J.M. Krupinsky, P.M. Carr, J.D. Hanson, and H.A. Johnson. 2005. Biomass yield, phenology, and survival of diverse switchgrass cultivars and experimental strains in Western North Dakota. Agronomy Journal 97(2):549-555.

Bothast, R.J. 2005. Cellulosic Ethanol Technology Assessment. Available online at http://www.farmfoundation. org/news/articlefiles/949-rodbothast.pdf. Accessed September 16, 2011.

Bouton, J.H. 2002. Bioenergy Crop Breeding and Production Research in the Southeast: Final Report for 1996 to 2001. Oak Ridge, TN: Oak Ridge National Laboratory.

BRDI (Biomass Research and Development Initiative). 2008. Increasing Feedstock Production for Biofuels: Economic Drivers, Environmental Implications, and the Role of Research. Washington, DC: The Biomass Research and Development Board.

Brechbill, S.C., and W.E. Tyner. 2008a. The Economics of Renewable Energy: Corn Stover and Switchgrass. West Lafayette, IN: Purdue University.

Brechbill, S.C., and W.E. Tyner. 2008b. The Economics of Biomass Collection, Transportation, and Supply to Indiana Cellulosic and Electric Utility Facilities. West Lafayette, IN: Purdue University.

Cassida, K.A., J.P. Muir, M.A. Hussey, J.C. Read, B.C. Venuto, and W.R. Ocumpaugh. 2005. Biomass yield and stand characteristics of switchgrass in South Central U.S. environments. Crop Science 45(2):673-681.

Chen, X., H. Huang, M. Khanna, and H. Onal. 2010. Meeting the mandate for biofuels: Implications for land use and food and fuel prices. Paper presented at the AAEA, CAES, and WAEA Joint Annual Meeting, July 25-27, Denver, CO.

Christian, D.G., A.B. Riche, and N.E. Yates. 2008. Growth, yield and mineral content of *Miscanthus* × *giganteus* grown as a biofuel for 14 successive harvests. Industrial Crops and Products 28(3):320-327.

Clifton-Brown, J.C., and I. Lewandowski. 2002. Screening *Miscanthus* genotypes in field trials to optimise biomass yield and quality in southern Germany. European Journal of Agronomy 16(2):97-110.

Clifton-Brown, J.C., S.P. Long, and U. Jørgensen. 2001. *Miscanthus* Productivity. Pp. 46-47 in *Miscanthus* for Energy and Fibre, M.B. Jones and M. Walsh, eds. London: James & James.

Clifton-Brown, J.C., P.F. Stampfl, and M.B. Jones. 2004. *Miscanthus* biomass production for energy in Europe and its potential contribution to decreasing fossil fuel carbon emissions. Global Change Biology 10(4):509-518.

Comis, D. 2006. Switching to switchgrass makes sense. Agricultural Research 54(7):19.

Crooks, A. 2006. On the road to energy independence: How soon will cellulosic ethanol be a factor? Rural Cooperatives 73(5).

Cubbage, F., S. Koesbandana, P. Mac Donagh, R. Rubilar, G. Balmelli, V.M. Olmos, R. De La Torre, M. Murara, V.A. Hoeflich, H. Kotze, R. Gonzalez, O. Carrero, G. Frey, T. Adams, J. Turner, R. Lord, J. Huang, C. MacIntyre, K. McGinley, R. Abt, and R. Phillips. 2010. Global timber investments, wood costs, regulation, and risk. Biomass and Bioenergy 34(12):1667-1678.

de la Torre Ugarte, D.G., M.E. Walsh, H. Shapouri, and S.P. Slinsky. 2003. The Economic Impacts of Bioenergy Crop Production on U.S. Agriculture. Washington, DC: U.S. Department of Agriculture.

Duffy, M.D. 2008. Estimated Costs for Production, Storage and Transportation of Switchgrass. Ames: Iowa State University.

Duffy, M.D. 2009. Estimated costs of crop production in Iowa—2010. Ag Decision Maker (A1-20): 13p.

Duffy, M.D., and V.Y. Nanhou. 2002. Costs of Producing Switchgrass for Biomass in Southern Iowa. Pp. 267-275 in Trends in New Crops and New Uses, J. Janick and A. Whipkey, eds. Alexandria, VA: ASHS Press.

Edwards, W. 2007. Estimating a value for corn stover. Ag Decision Maker (A1-70): 4p.

Elobeid, A., S. Tokgoz, D.J. Hayes, B.A. Babcock, and C.E. Hart. 2006. The Long-Run Impact of Corn-Based Ethanol on the Grain, Oilseed, and Livestock Sectors: A Preliminary Assessment. Ames: Iowa State University.

EngineeringToolbox.com. 2010. Wood Densities: Density of wood as apple, ash, cedar, elm and more. Available online at http://www.engineeringtoolbox.com/wood-density-d_40.html. Accessed September 15, 2010.

English, B.C., D.G. De La Torre Ugarte, K. Jensen, C. Hellwinckel, J. Menard, B. Wilson, R. Roberts, and M. Walsh. 2006. 25% Renewable Energy for the United States By 2025: Agricultural and Economic Impacts. Knoxville: University of Tennessee.

Fike, J.H., D.J. Parrish, D.D. Wolf, J.A. Balasko, J.T. Green, Jr., M. Rasnake, and J.H. Reynolds. 2006a. Switchgrass production for the upper southeastern USA: Influence of cultivar and cutting frequency on biomass yields. Biomass and Bioenergy 30(3):207-213.

Fike, J.H., D.J. Parrish, D.D. Wolf, J.A. Balasko, J.J.T. Green, M. Rasnake, and J.H. Reynolds. 2006b. Long-term yield potential of switchgrass-for-biofuel systems. Biomass and Bioenergy 30(3):198-206.

Gibson, L., and S. Barnhart. 2007. Switchgrass. AG 0200. Ames: Iowa State University.

Han, H.-S., J. Halbrook, F. Pan, and L. Salazar. 2010. Economic evaluation of a roll-off trucking system removing forest biomass resulting from shaded fuelbreak treatments. Biomass and Bioenergy 34(7):1006-1016.

Heaton, E.A., T.B. Voigt, and S.P. Long. 2004a. A quantitative review comparing the yields of two candidate C_4 perennial biomass crops in relation to nitrogen, temperature and water. Biomass and Bioenergy 27(1):21-30.

Heaton, E.A., S.P. Long, T.B. Voigt, M.B. Jones, and J. Clifton-Brown. 2004b. *Miscanthus* for renewable energy generation: European Union experience and projections for Illinois. Mitigation and Adaptation Strategies for Global Change 9(4):433-451.

Heaton, E.A., F.G. Dohleman, and S.P. Long. 2008. Meeting US biofuel goals with less land: The potential of *Miscanthus*. Global Change Biology 14(9):2000-2014.

Hess, J.R., C.T. Wright, and K.L. Kenney. 2007. Cellulosic biomass feedstocks and logistics for ethanol production. Biofuels, Bioproducts and Biorefining 1(3):181-190.

Hoskinson, R.L., D.L. Karlen, S.J. Birrell, C.W. Radtke, and W.W. Wilhelm. 2007. Engineering, nutrient removal, and feedstock conversion evaluations of four corn stover harvest scenarios. Biomass and Bioenergy 31(2-3):126-136.

Huang, H.-J., S. Ramaswamy, W. Al-Dajani, U. Tschirner, and R.A. Cairncross. 2009. Effect of biomass species and plant size on cellulosic ethanol: A comparative process and economic analysis. Biomass and Bioenergy 33(2):234-246.

Jain, A.K., M. Khanna, M. Erickson, and H. Huang. 2010. An integrated biogeochemical and economic analysis of bioenergy crops in the midwestern United States. Global Change Biology Bioenergy 2(5):217-234.

James, L.K., S.M. Swinton, and K.D. Thelen. 2010. Profitability analysis of cellulosic energy crops compared with corn. Agronomy Journal 102(2):675-687.

Jenkins, B.M., R.B. Williams, N. Parker, P. Tittmann, Q. Hart, M.C. Gildart, S. Kaffka, B.R. Hartsough, and P. Dempster. 2009. Sustainable use of California biomass resources can help meet state and national bioenergy targets. California Agriculture 63(4):168-177.

Jiang, Y., and S.M. Swinton. 2008. Market Interactions, Farmer Choices, and the Sustainability of Growing Advanced Biofuels. East Lansing: Michigan State University.

Kahle, P., S. Beuch, B. Boelcke, P. Leinweber, and H.-R. Schulten. 2001. Cropping of *Miscanthus* in Central Europe: Biomass production and influence on nutrients and soil organic matter. European Journal of Agronomy 15(3):171-184.

Karlen, D.L. 2010. Corn stover feedstock trials to support predictive modeling. Global Change Biology Bioenergy 2(5):235-247.

Kaylen, M., D.L. Van Dyne, Y.-S. Choi, and M. Blase. 2000. Economic feasibility of producing ethanol from lignocellulosic feedstocks. Bioresource Technology 72(1):19-32.

Khanna, M. 2008. Cellulosic biofuels: Are they economically viable and environmentally sustainable? Choices 23(3):16-23.

Khanna, M., and B. Dhungana. 2007. Economics of alternative feedstocks. Pp. 129-146 in Corn-Based Ethanol in Illinois and the U.S.: A Report from the Department of Agricultural and Consumer Economics, University of Illinois. Urbana-Champaign: University of Illinois.

Khanna, M., B. Dhungana, and J. Clifton-Brown. 2008. Costs of producing *Miscanthus* and switchgrass for bioenergy in Illinois. Biomass and Bioenergy 32(6):482-493.

Kiniry, J.R., K.A. Cassida, M.A. Hussey, J.P. Muir, W.R. Ocumpaugh, J.C. Read, R.L. Reed, M.A. Sanderson, B.C. Venuto, and J.R. Williams. 2005. Switchgrass simulation by the ALMANAC model at diverse sites in the southern U.S. Biomass and Bioenergy 29(6):419-425.

Krissek, G. 2008. Future Opportunities and Challenges for Ethanol Production and Technology. Available online at http://www.farmfoundation.org/news/articlefiles/378-krissek%202-5-28.pdf. Accessed February 13, 2011.

Kszos, L.A., S.B. McLaughlin, and M. Walsh. 2002. Bioenergy from Switchgrass: Reducing production costs by improving yield and optimizing crop management. Paper presented at the Biomass Research and Development Board's Bioenergy 2002—Bioenergy for the Environment Conference, September 22-27, Boise, ID.

Kumar, A., and S. Sokhansanj. 2007. Switchgrass (*Panicum vigratum*, L.) delivery to a biorefinery using integrated biomass supply analysis and logistics (IBSAL) model. Bioresource Technology 98(5):1033-1044.

Kumar, A., J.B. Cameron, and P.C. Flynn. 2003. Biomass power cost and optimum plant size in western Canada. Biomass and Bioenergy 24(6):445-464.

Kumar, A., J.B. Cameron, and P.C. Flynn. 2005. Pipeline transport and simultaneous saccharification of corn stover. Bioresource Technology 96(7):819-829.

Lambert, D.K., and J. Middleton. 2010. Logistical design of a regional herbaceous crop residue-based ethanol production complex. Biomass and Bioenergy 34(1):91-100.

Lang, B. 2002. Estimating the Nutrient Value in Corn and Soybean Stover. Decorah: Iowa State University.

Lemus, R., E.C. Brummer, K.J. Moore, N.E. Molstad, C.L. Burras, and M.F. Barker. 2002. Biomass yield and quality of 20 switchgrass populations in southern Iowa, USA. Biomass and Bioenergy 23(6):433-442.

Lewandowski, I., J.C. Clifton-Brown, J.M.O. Scurlock, and W. Huisman. 2000. *Miscanthus*: European experience with a novel energy crop. Biomass and Bioenergy 19(4):209-227.

Lewandowski, I., J.M.O. Scurlock, E. Lindvall, and M. Christou. 2003. The development and current status of perennial rhizomatous grasses as energy crops in the U.S. and Europe. Biomass and Bioenergy 25(4):335-361.

Mapemba, L.D., F.M. Epplin, C.M. Taliaferro, and R.L. Huhnke. 2007. Biorefinery feedstock production on conservation reserve program land. Review of Agricultural Economics 29(2):227-246.

Mapemba, L.D., F.M. Epplin, R.L. Huhnke, and C.M. Taliaferro. 2008. Herbaceous plant biomass harvest and delivery cost with harvest segmented by month and number of harvest machines endogenously determined. Biomass and Bioenergy 32(11):1016-1027.

Marshall, L., and Z. Sugg. 2010. Fields of Fuel: Market and Environmental Implications of Switching to Grass for U.S. Transport. WRI Policy Note. Washington, DC: World Resources Institute.

McAloon, A., F. Taylor, W. Yee, K. Ibsen, and R. Wooley. 2000. Determining the Cost of Producing Ethanol from Corn Starch and Lignocellulosic Feedstocks. Golden, CO: National Renewable Energy Laboratory.

McLaughlin, S.B., and L.A. Kszos. 2005. Development of switchgrass (*Panicum virgatum*) as a bioenergy feedstock in the United States. Biomass and Bioenergy 28(6):515-535.

Miller, R.O., and B.A. Bender. 2008. Growth and yield of poplar and willow hybrids in the central Upper Peninsula of Michigan. In Proceedings of the Short Rotation Crops International Conference: Biofuels, Bioenergy, and Biproducts from Sustainable Agricultural and Forest Crops, R.S. Zalesny Jr., R. Mitchell and J. Richardson, eds. August 19-21, Bloomington, MN: U.S. Forest Service.

Monti, A., S. Fazio, and G. Venturi. 2009. The discrepancy between plot and field yields: Harvest and storage losses of switchgrass. Biomass and Bioenergy 33(5):841-847.

Mooney, D.F., R.K. Roberts, B.C. English, D.D. Tyler, and J.A. Larson. 2009. Yield and breakeven price of "Alamo" switchgrass for biofuels in Tennessee. Agronomy Journal 101(5):1234-1242.

Muir, J.P., M.A. Sanderson, W.R. Ocumpaugh, R.M. Jones, and R.L. Reed. 2001. Biomass production of "Alamo" switchgrass in response to nitrogen, phosphorus, and row spacing. Agronomy Journal 93(4):896-901.

Nelson, R.G., J.C. Ascough II, and M.R. Langemeier. 2006. Environmental and economic analysis of switchgrass production for water quality improvement in northeast Kansas. Journal of Environmental Management 79(4):336-347.

Ocumpaugh, W., M. Hussey, J. Read, J. Muir, F. Hons, G. Evers, K. Cassida, B. Venuto, J. Grichar, and C. Tischler. 2003. Evaluation of Switchgrass Cultivars and Cultural Methods for Biomass Production in the South Central U.S.: Consolidated Report 2002. Oak Ridge, TN: Oak Ridge National Laboratory.

Perlack, R.D. 2007. Overview of Plant Feedstock Production for Biofuel: Current Technologies and Challenges, and Potential for Improvement. Presentation to the Panel on Alternative Liquid Transportation Fuels, November 19.

Perlack, R.D., and A.F. Turhollow. 2002. Assessment of Options for the Collection, Handling, and Transport of Corn Stover. Oak Ridge, TN: Oak Ridge National Laboratory.

Perlack, R.D., and A.F. Turhollow. 2003. Feedstock cost analysis of corn stover residues for further processing. Energy 28(14):1395-1403.

Perrin, R.K. 2008. Ethanol and food prices: Preliminary assessment. Lincoln: University of Nebraska.

Perrin, R.K., K. Vogel, M. Schmer, and R. Mitchell. 2008. Farm-scale production cost of switchgrass for biomass. BioEnergy Research 1(1):91-97.

Petrolia, D.R. 2008. The economics of harvesting and transporting corn stover for conversion to fuel ethanol: A case study for Minnesota. Biomass and Bioenergy 32(7):603-612.

Popp, M., and R. Hogan, Jr. 2007. Assessment of two alternative switchgrass harvest and transport methods. Paper presented at the Farm Foundation Conference, April 12-13, St. Louis, MO.

Prewitt, R.M., M.D. Montross, S.A. Shearer, T.S. Stombaugh, S.F. Higgins, S.G. McNeill, and S. Sokhansanj. 2007. Corn stover availability and collection efficiency using typical hay equipment. Transactions of the American Society of Agricultural and Biological Engineers 50(3):705-711.

Pyter, R., T. Voigt, E. Heaton, F. Dohleman, and S. Long. 2007. Giant *Miscanthus*: Biomass Crop for Illinois. Pp. 39-42 in Proceedings of the 6th National New Crops Symposium: Creating Markets for Economic Development of New Crops, J. Janick and A. Whipkey, eds. October 14-18, Alexandria, VA.

Quick, G.R. 2003. Single-pass corn and stover harvesters: Development and performance. In Proceedings of the American Society of Agricultural and Biological Engineers' International Conference on Crop Harvesting and Processing. February 9-11, at Louisville, KY.

Reijnders, L. 2010. Transport biofuel yields from food and lignocellulosic C_4 crops. Biomass and Bioenergy 34(1): 152-155.

Reynolds, J.H., C.L. Walker, and M.J. Kirchner. 2000. Nitrogen removal in switchgrass biomass under two harvest systems. Biomass and Bioenergy 19(5):281-286.

Sanderson, M.A. 2008. Upland switchgrass yield, nutritive value, and soil carbon changes under grazing and clipping. Agronomy Journal 100(3):510-516.

Schechinger, T.M., and J. Hettenhaus. 2004. Corn Stover Harvesting: Grower, Custom Operator, and Processor Issues and Answers: Report on Corn Stover Harvest Experiences in Iowa and Wisconsin for the 1997-98 and 1998-99 Crop Years. Oak Ridge, TN: Oak Ridge National Laboratory.

Schmer, M.R., K.P. Vogel, R.B. Mitchell, L.E. Moser, K.M. Eskridge, and R.K. Perrin. 2006. Establishment stand thresholds for switchgrass grown as a bioenergy crop. Crop Science 46(1):157-161.

Searcy, E., P. Flynn, E. Ghafoori, and A. Kumar. 2007. The relative cost of biomass energy transport. Applied Biochemistry and Biotechnology 137-140(1):639-652.

Sheehan, J., A. Aden, K. Paustian, K. Killian, J. Brenner, M. Walsh, and R. Nelson. 2003. Energy and environmental aspects of using corn stover for fuel ethanol. Journal of Industrial Ecology 7(3-4):117-146.

Shinners, K.J., G.C. Boettcher, R.E. Muck, P.J. Weimer, and M.D. Casler. 2006. Drying, Harvesting and Storage Characteristics of Perennial Grasses as Biomass Feedstocks. Paper presented at the 2006 American Society of Agricultural and Biological Engineers Annual Meeting, July 9-12, Portland, OR.

Smeets, E.M.W., I.M. Lewandowski, and A.P.C. Faaij. 2009. The economical and environmental performance of *Miscanthus* and switchgrass production and supply chains in a European setting. Renewable & Sustainable Energy Reviews 13(6-7):1230-1245.

Sohngen, B., J. Anderson, S. Petrova, and K. Goslee. 2010. Alder Springs Biomass Removal Economic Analysis. Arlington, VA: Winrock International and U.S. Department of Agriculture - Forest Service.

Sokhansanj, S., and A.F. Turhollow. 2002. Baseline cost for corn stover collection. Applied Engineering in Agriculture 18(5):525-530.

Spelter, H., and D. Toth. 2009. North America's Wood Pellet Sector. Madison, WI: U.S. Department of Agriculture - Forest Service.

Summit Ridge Investments. 2007. Eastern Hardwood Forest Region Woody Biomass Energy Opportunity. Granville, VT: Summit Ridge Investments, LLC and U.S. Department of Agriculture - Forest Service.

Stampfl, P.F., J.C. Clifton-Brown, and M.B. Jones. 2007. European-wide GIS-based modelling system for quantifying the feedstock from *Miscanthus* and the potential contribution to renewable energy targets. Global Change Biology 13(11):2283-2295.

Suzuki, Y. 2006. Estimating the Cost of Transporting Corn Stalks in the Midwest. Ames: Iowa State University College of Business, Business Partnership Development

Taheripour, F., and W.E. Tyner. 2008. Ethanol policy analysis—What have we learned so far? Choices. 23 (3):6-11.

Taliaferro, C.M. 2002. Breeding and Selection of New Switchgrass Varieties for Increased Biomass Production. Oak Ridge, TN: Oak Ridge National Laboratory

Thomason, W.E., W.R. Raun, G.V. Johnson, C.M. Taliaferro, K.W. Freeman, K.J. Wynn, and R.W. Mullen. 2005. Switchgrass response to harvest frequency and time and rate of applied nitrogen. Journal of Plant Nutrition 27(7):1199-1226.

Tiffany, D.G., B. Jordan, E. Dietrich, and B. Vargo-Daggett. 2006. Energy and Chemicals from Native Grasses: Production, Transportation and Processing Technologies Considered in the Northern Great Plains. St. Paul: University of Minnesota.

Tokgoz, S., A. Elobeid, J. Fabiosa, D.J. Hayes, B.A. Babcock, T.-H.E. Yu, F. Dong, C.E. Hart, and J.C. Beghin. 2007. Emerging Biofuels: Outlook of Effects on U.S. Grain, Oilseed, and Livestock Markets. Ames: Iowa State University.

USDA-NASS (U.S. Department of Agriculture - National Agricultural Statistics Service). 2007a. Agricultural Prices: December 2007. Washington, DC: U.S. Department of Agriculture - National Agricultural Statistics Service.

USDA-NASS (U.S. Department of Agriculture - National Agricultural Statistics Service). 2007b. Agricultural Prices 2006 Summary. Washington, DC: U.S. Department of Agriculture - National Agricultural Statistics Service.

USFS (U.S. Department of Agriculture - Forest Service). 2003. A Strategic Assessment of Forest Biomass and Fuel Reduction Treatments in Western States. Fort Collins, CO: U.S. Department of Agriculture - Forest Service.

USFS (U.S. Department of Agriculture - Forest Service). 2005. A Strategic Assessment of Forest Biomass and Fuel Reduction Treatments in Western States. Fort Collins, CO: U.S. Department of Agriculture - Forest Service.

UT (University of Tennessee). 2009. Guideline Switchgrass Establishment and Annual Production Budgets over Three Year Planning Horizon: Estimated Production Expenses as of July 14, 2009. Knoxville: University of Tennessee.

Vadas, P., K. Barnett, and D. Undersander. 2008. Economics and energy of ethanol production from alfalfa, corn, and switchgrass in the Upper Midwest, USA. BioEnergy Research 1(1):44-55.

Vargas, L.A., M.N. Andersen, C.R. Jensen, and U. Jørgensen. 2002. Estimation of leaf area index, light interception and biomass accumulation of *Miscanthus sinensis* "Goliath" from radiation measurements. Biomass and Bioenergy 22(1):1-14.

Vogel, K.P., J.J. Brejda, D.T. Walters, and D.R. Buxton. 2002. Switchgrass biomass production in the Midwest USA. Agronomy Journal 94(3):413-420.

Wallace, R., K. Ibsen, A. McAloon, and W. Yee. 2005. Feasibility Study for Co-Locating and Integrating Ethanol Production Plants from Corn Starch and Lignocellulosic Feedstocks. Golden, CO: National Renewable Energy Laboratory.

Walsh, M. 2008. Switchgrass. May 31, 2008. Available online at http://bioweb.sungrant.org/Technical/Biomass+Resources/Agricultural+Resources/New+Crops/Herbaceous+Crops/Switchgrass/Default.htm. Accessed September 21, 2010.

Wright, M.M., and R.C. Brown. 2007. Comparative economics of biorefineries based on the biochemical and thermochemical platforms. Biofuels, Bioproducts and Biorefining 1(1):49-56.

N

Blend Wall

Total national consumption of gasoline in the United States was about 140 billion gallons per year in 2009 and is expected to fall over time as a result of increasing fuel economy standards (Tyner and Viteri, 2010). As of 2010, if every drop of gasoline were blended as E10, the maximum ethanol that could be absorbed would be 14 billion gallons. In reality, blending 10-percent ethanol into gasoline is not feasible in all regions and seasons. Most experts consider about 9 percent to be the effective maximum, which amounts to about 12.6 billion gallons per year of ethanol blended (Tyner et al., 2008). U.S. ethanol production capacity already exceeds this level. Thus, the nation's ability to consume ethanol as fuel has reached a limit called the blend wall.

This physical constraint is the biggest issue facing U.S. ethanol industry in 2010. If the blending limit of 10 percent is maintained, the ethanol industry cannot grow; indeed, it could not even operate its productive capacity of over 13 billion gallons in 2010. The blend wall partially explains why about 2 billion gallons of annual capacity was shut down during much of 2009, and about 1 billion gallons of capacity remained inoperative in 2010. It also explains why ethanol prices during much of the 2009 were driven mainly by corn prices, instead of gasoline prices as it was before 2008-2009.

In 2010, the relatively low U.S. price for ethanol has led to some ethanol exports. With the world sugar price at its highest level since 1995, Brazil allocated relatively more sugarcane to sugar and less to ethanol, so the Brazilian ethanol price in the summer of 2010 was higher than the U.S. price. However, in late summer, corn price started increasing significantly, and ethanol price increased in step. Although ethanol exports occurred, the exports were a tiny fraction of the total U.S. production.

The basic economics of the blend wall are depicted in Figure N-1. Moving from left to right down the demand curve, once the blend wall is reached, the price plummets from the market equilibrium (with subsidy) at P^* (or P_m without subsidy) to the intersection of the supply curve and the blend wall P^{BW}. Ethanol becomes priced on a breakeven basis with corn, which was the situation in the first three quarters of 2009. Markets picked up in the

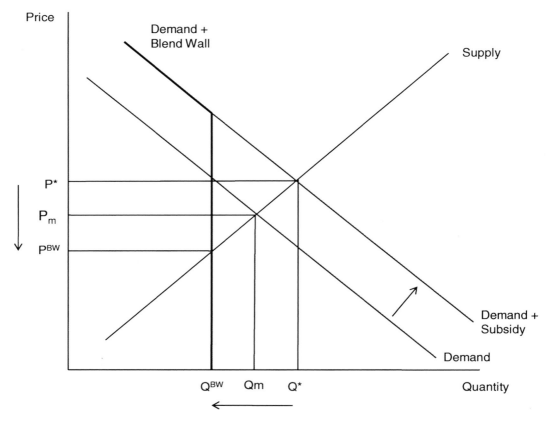

FIGURE N-1 Economics of the ethanol blend wall.
NOTE: P* = market price with subsidy; P_m = market price without subsidy; P^{BW} = price at the blend wall; Q* = quantity with subsidy; Q_m = quantity without subsidy; Q^{BW} = quantity at the blend wall.

fourth quarter as more ethanol can be blended in winter months than summer months due to summer evaporative emissions constraints. However, in spring 2010, ethanol pricing returned to breakeven with corn. Given the fact that gasoline demand is expected to decrease because of higher fuel-economy standards, the blend wall becomes a severe constraint to future ethanol growth and a barrier to achieving the legislative mandates of the Energy Independence and Security Act (110 P.L. 104).

Ethanol production and use are facing two opposing realities: the Renewable Fuel Standard (RFS2) that mandates increasing consumption of biofuels each year from 2005 to 2022, and a physical blend wall that makes it difficult for ethanol consumption to grow beyond present levels. An ethanol industry support and lobby group called Growth Energy petitioned EPA to increase the blending limit from 10 to 15 percent. In October 2010, EPA approved the 15-percent limit for cars built in 2007 or later. In the following January, EPA extended the waiver to model year 2001 to 2006 light-duty vehicles. But even an increase of the blending limit to 15 percent only buys some time (about 4 years) before the blend wall is reached again and only so long as ethanol remains the primary biofuel.

In the following analysis, the consequences of seven alternative RFS2 pathways were analyzed:

- The blend limit remains at 10 percent (E10), and all biofuel is ethanol.
- The blend limit is increased to 15 percent (E15), and all biofuel is ethanol.
- The blend limit is 10 percent (E10), and all cellulosic biofuel is thermochemically produced biogasoline or equivalent. The physical properties of thermochemical biofuels are identical with gasoline, and thus, it can be blended with gasoline at any percentage.
- The blend limit is 15 percent (E15), and all cellulosic biofuel is thermochemically produced biogasoline or equivalent.
- The blend limit is 10 percent (E10), and cellulosic technology is so expensive that EPA waives the cellulosic part of RFS.
- The blend limit is 15 percent (E15), and cellulosic technology is so expensive that EPA waives the cellulosic part of RFS.
- A regional strategy is used to emphasize use of E85 in the Midwest where most of the ethanol is produced.

For each of the scenarios, the total net present value (NPV) of installing the flex-fuel vehicles (FFVs) and pumps was calculated using a real social discount rate of 10 percent and an average inflation of 3 percent per year. The cost of installing E85 fuel dispensers depends on the type of tank installed (new underground tank or conversion of existing tank). Between 30 to 60 percent of the E85 installations involve new tanks, while the others convert a current tank (Moriarty et al., 2009). The typical gas station has 3.3 tanks, one for regular, midgrade, and premium. If a tank is converted, the station loses the revenue stream from one blend. However, some stations, especially at convenience stores, lack the space to add a new tank (Moriarty et al., 2009). Cost estimates for new tanks range from $50,000 to $200,000, with a mean at $74,418. The average cost of a tank retrofit is about $21,244. Thus, the weighted average cost of installing a tank is $45,000. In addition, it costs an additional $100 per vehicle to produce a FFV instead of a standard vehicle (Corts, 2010). Other infrastructure costs are not included so that the cost estimates provided here are clearly underestimates of total cost.

The first alternative of maintaining the blending limit at 10 percent and producing only ethanol as a biofuel is clearly out of question. It would require massive increases in E85, with accompanying huge increases in FFVs (Tyner and Viteri, 2010). Annual sales of FFVs would need to be at least 8.7 million cars per year compared with a cumulative total of 7.9 million on the road today. The total FFVs needed by 2022 would be 121.5 million. It would also require installation of 24,277 E85 fueling pumps per year compared with a cumulative total of 2,100 operating today. A cumulative total of 158,000 stations would need to add flex-fuel pumps. The total cost of E85 pump installation and FFVs around the whole United States have a NPV of $11.13 billion for this scenario. Furthermore, E85 would have to be priced no more than 78 percent of E10 blend gasoline because of the mileage difference. (See the discussion of the regional strategy below for more on E85 pricing.) The bottom line is that this scenario is not likely to be feasible, and EPA would be forced to waive RFS2 at some point. The time profile of E10 and E85 for this scenario is illustrated in Figure N-2.

The second alternative of a 15-percent blend limit with only ethanol as biofuel is less restrictive than the first but suffers similar problems over the longer term. The higher blend limit essentially extends the time before the blend wall is reached but does not solve the problem. E15 consumption would grow from 13.1 billion gallons per year in 2010 to 17.5 billion gallons per year in 2022, as the continued growth in E85 once again crowds out the use of the lower blend fuel (Figure N-2). By 2022, there needs to be about 90.4 million FFVs on the roads, served by 236,208 E85 gas dispensers. The total NPV value cost

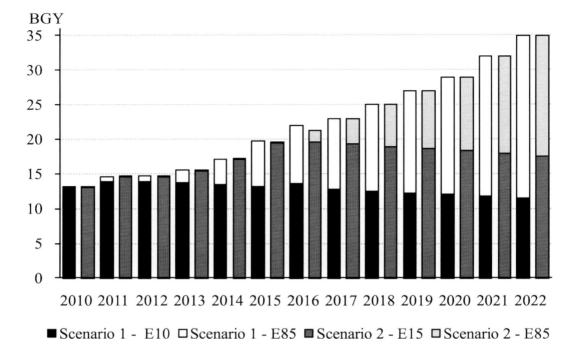

FIGURE N-2 Projected consumption of E10 and E85 under scenario 1 in which the ethanol can be used as E85 blend in FFVs or E10 in conventional vehicles, and under scenario 2 in which ethanol can be used as E85 blend in FFVs or E15 in conventional vehicles.

of installation for E85 pumps and FFVs is $8.0 billion for this scenario. Thus, compared to the E10 scenario, the adoption of an E15 blending limit would reduce the consumption of E85 by 6 billion gallons per year in 2022 and lessen the demand for FFVs and E85 pumps. This change would save a NPV of $3.1 billion. The time profile of scenario 2 also is illustrated in Figure N-2.

The third scenario combines E10 and E85 with the thermochemical refining of cellulosic biomass. Because the biofuel produced using thermochemical conversion can have similar physical properties as gasoline, it can be consumed and blended with gasoline at any percentage. Thermochemical conversion also produces green diesel, but this section focuses on the gasoline component, and ethanol is a gasoline substitute. Ethanol is still blended as an oxygenate in E10 ethanol, and relatively small volumes of E85 are produced (7.6 billion gallons per year in 2022 instead of 23.5 billion gallons per year in scenario 1) (Figure N-3). This scenario would require fewer FFVs, approximately 39 million being needed on the road by 2022, and about 76,100 E85 dispensers. The NPV of E85 pump installation and more FFVs falls to $3.7 billion. Compared to Scenario 1, the adoption of thermochemical biofuels avoids $7.4 billion in E85 related investment. The time path of the different fuels for scenario 3 is illustrated in Figure N-3.

Scenario 4 combines the use of thermochemically produced biogasoline with increasing the ethanol blend rate to E15. The growth of E85 is limited, with only 90 million gallons of E85 in the market by 2022 (Figure N-3). As such, there is also limited demand for increasing the market penetration of FFVs. It is assumed that no additional E85 vehicles are produced and the FFV fleet contracts to 8.5 million by 2022 as vehicles are retired. The number of E85

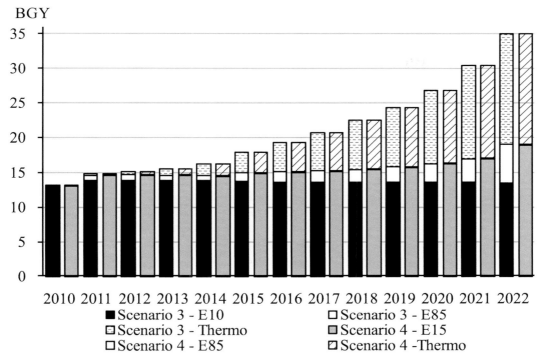

FIGURE N-3 Projected consumption of E10 and E85 under scenario 3 in which the ethanol can only be used as E85 blend in FFVs and as E10 blended with thermochemically produced biogasoline and under scenario 4 in which the ethanol can be used as E85 blend in FFVs and as E15 blended with thermochemically produced biogasoline.

dispensers does not change from 2010 because of low E85 usage rates. This leads to a NPV of $554 million for replacement FFVs.

The fifth scenario assumes E10 and E85 as in scenario 1 but also assumes that cellulosic technology does not become feasible. As a result, the 16 billion gallon cellulosic biofuel component of RFS2 is waived. By 2011, all gasoline will be blended with a 10-percent ethanol mix, and by 2012, E85 consumption will start growing in small amounts until it reaches 7.6 billion gallons in 2022. The effect of the RFS2 waiver on the number of FFVs and E85 stations is the same as the use of thermochemical biofuels in scenario 3. Again, 39.2 million FFVs are needed, as well as 76,100 E85 dispensers. The NPV for pump installation and new FFVs is the same as scenario 3 ($3.7 billion). In addition, the investment cost of thermochemical plants is avoided.

The sixth scenario is analogous to scenario 5 except that the blend level is E15 instead of E10. Again no cellulosic biofuel is required because it is waived. The results are the same as scenario 4 except that no cellulosic biofuels are produced. The only cost is FFV replacement cost of $554 billion.

The seventh scenario is to make E85 more attractive and economically viable by focusing the E85 marketing and infrastructure investments in the Midwest where most of the ethanol is produced. If E85 could reach substantial penetration in the Midwest, it is argued that the blend wall might not be reached. Ethanol will not have to be transported far because it is distributed and used close to where it is produced. This strategy could lower transport and distribution cost of ethanol by as much as $0.15-$0.20 per gallon, and enable the E85 blend to come closer to competitive pricing. For 16 billion gallons of cellulosic ethanol, that savings could be as much as $3.2 billion. Competitive pricing is key for this or any strategy to promote E85. E85 (with 74-percent ethanol) contains about 78 percent of the energy of E10, the normal competitive fuel. So if the retail price of E10 is $2.80, the retail price of E85 could be no more than $2.18 to be competitive with E10. When ethanol is up against the blend wall as at present, its price can be substantially below gasoline as it is priced essentially breakeven with corn. In July 2010, wholesale prices were about $2.10 for gasoline and $1.70 for ethanol. It is possible, but unlikely, that E85 could become competitive under these conditions. In August 2010, the price of corn increased substantially due to world market conditions; consequently, ethanol price increased as well such that, by the end of the month, ethanol and gasoline were close to each other in price. In November 2010, gasoline was about $2.15 and ethanol about $2.50. Under these conditions, it would be extremely difficult to market E85.

If E85 penetrates the market substantially, then the blend wall is effectively surpassed. In that case, RFS2 becomes binding. With a binding RFS2, the mandate is demanding more ethanol than the market would otherwise produce (Figure N-4). The price, P^{RFS}, is higher

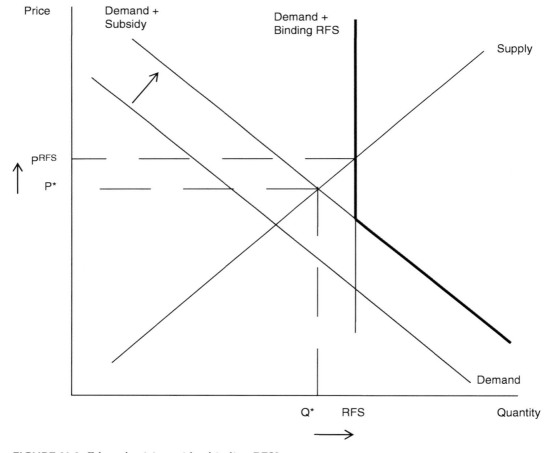

FIGURE N-4 Ethanol pricing with a binding RFS2.
NOTE: P^{RFS} = price at RFS; P^* = market price with subsidy; Q^* = quantity with subsidy.

than the market plus subsidy price. In other words, there is an economic rent attached to the binding mandate. It is unlikely in this situation that E85 could be competitive with E10 in the market place even with lower transportation and distribution costs in the Midwest. Over time companies could use cross-subsidization to lower the price of E85 and increase that of E10, but that strategy requires that the flex-vehicle fleet and E85 dispensers be in place, which takes years to occur.

Scaling up FFV production and service station dispensing facilities to saturate the Midwest market will be a large task. In fact, there are not enough cars in the Midwest to satisfy the E85 demand even if all cars were FFVs. Also, some have argued that because E85 customers would spend much more time refueling than E10 customers (more frequent trips to the pump), customers might demand an even larger price discount for E85.

The analyses demonstrate that ethanol is not likely to be the only biofuel in the U.S. market. The blend wall becomes a near impenetrable barrier to meeting RFS2. If the thermochemical production processes become viable, then RFS2 can be met with a combination of ethanol from corn and sugarcane, and hydrocarbon fuels from cellulose.

REFERENCES

Corts, K.S. 2010. Building out alternative fuel retail infrastructure: Government fleet spillovers in E85. Journal of Environmental Economics and Management 59:219-234.

Moriarty, K., C. Johnson, T. Sears, and P. Bergeron. 2009. E85 Dispenser Study. Golden, CO: National Renewable Energy Laboratory.

Tyner, W.E., and D. Viteri. 2010. Implications of blending limits on the U.S. ethanol and biofuels markets. Biofuels 1(2):251-253.

Tyner, W.E., F. Dooley, C. Hurt, and J. Quear. 2008. Ethanol pricing issues for 2008. Industrial Fuels and Power February: pp.50-57.

O

Safety and Quality of Biofuel Coproducts as Animal Feed

Safety of biofuel coproducts, such as distillers grains from corn-grain ethanol production, as animal feedstuffs can pose a barrier to meeting the Renewable Fuel Standard (RFS2) because whether those biofuel products meet the GHG reduction threshold of RFS depends in part on GHG credits from coproducts. The safety concerns include health and welfare of the animals consuming the coproducts and the safety of the foods that are derived from these animals. Both of these issues are affected by the presence of antibiotic residues and mycotoxins in distillers grains and the potential increase in fecal shedding of *Escherichia coli* O157 in cattle that were given distillers grains as part of their ration.

In corn-grain or sugar-based ethanol production, bacterial contamination during the fermentation is a concern (Skinner and Leathers, 2004). Bacterial contaminants compete with the ethanol-producing yeast for sugars and micronutrients, and they produce organic acids that inhibit yeast, thereby reducing ethanol yield. Antibiotics, including virginiamycin, erythromycin, and tylosin, are sometimes added to control or prevent bacterial contamination in biorefineries. Administering these antibiotics to animals is strictly regulated by the Food and Drug Administration (FDA) and the U.S. Department of Agriculture (USDA), especially immediately prior to slaughter or to egg-laying hens and lactating cattle. When coproducts containing antibiotics are inadvertently fed to livestock, residues in meat, milk, or eggs could result in condemnation of products or, if not discovered, unacceptably high levels in human foods. FDA is concerned about the potential animal and human health hazards from antibiotic residues in distillers grains used as animal feed. In 2009, FDA announced that it would conduct a nationwide survey to determine the extent and levels of antibiotic residues in distillers grains produced in the United States (FDA, 2009). The outcome of the survey could resolve whether antibiotic residue in corn-grain ethanol coproducts would be a barrier to achieving RFS2. Alternative to antibiotics such as stabilized chlorine dioxide also can be used to control or prevent bacterial contamination.

Corn grain might be contaminated by mycotoxins (toxins produced by fungi). These mycotoxins are typically concentrated by about two to three fold when the corn grain is converted to distillers grains because starch comprises about two-thirds of the grain and its removal by fermentation results in the enrichment of mycotoxins in the distillers grains (Whitlow, 2008). Mycotoxins of particular concerns are aflatoxins and fumonisin. Aflatoxin is carcinogenic and affects the liver (Wild and Gong, 2010). Fumonisins have been reported to induce liver and kidney tumors in rodents and identified as possibly carcinogenic to humans. Both mycotoxins affect growth and are immunosuppressive in animals (Wild and Gong, 2010).

One study assessed aflatoxins, deoxynivalenol, fumonisins, T-2 toxin, and zearalenone in samples of distillers grains from 20 ethanol refineries in the Midwestern United States (Zhang et al., 2009). That study found that none of the samples had aflatoxins or deoxynivalenol levels that exceed FDA guidelines for use as animal feed and that less than 10 percent of the samples had fumonisin levels that exceed FDA guideline for feeding equids and rabbits. However, the level of mycotoxins in corn depends on the weather and the amount of insect damage sustained by the plants and therefore is likely to vary from year to year. In a survey of dried distillers grain (DDG) samples from 2009-2010 corn crops in Indiana, Siegel (2010) found that 20 percent of the DDG had mycotoxin levels that were too high to be used as animal feed. These contaminated DDG were mostly disposed of by applying to land as fertilizer.

Another concern of using distillers grains as part of animal feed is its potential contribution to increased prevalence of *Escherichia coli* O157 in cattle. Prevalence of *E. coli* O157 in cattle could be a food safety concern. Jacob et al. (2008a,b) compared the prevalence of *E. coli* 157 in feces of cattle that were fed diets with wet or dried distillers grains to those without distillers grains at all. They found an increase in *E.coli* O157 prevalence in batch cultures of ruminal and fecal fermentation of cattle fed DDG (Jacob et al., 2008a). However, the effect of feeding wet distillers grains on *E. coli* O157 prevalence in cattle was inconclusive (Jacob et al., 2008b). Edrington et al. (2010) also did not observe any effect of feeding wet distillers grains on *E. coli* O157 in feedlot cattle.

In addition to food safety, the nutritional quality of DDG could be a concern if they are to be included in animal diets. Variations in DDG composition affect nutritional quality and market value. Samples of DDG from dry grind ethanol biorefineries in the upper Midwest were found to have consistent fat content but variable protein content that ranged from 260 to 380 g/kg of dry matter (Belyea et al., 2010). In general, including DDG in animal diets does not appear to affect meat and carcass quality of broilers, pigs, and heifers (Xu et al., 2007, 2010; Corzo et al., 2009; Depenbusch et al., 2009). However, finishing pigs fed with a diet of over 20 percent DDG could have fat quality that does not meet the standard of pork processors (Xu et al., 2010). High levels of fat in DDG cause milk fat depression in dairy cattle and limit the inclusion rates in dairy feeds. New technologies that remove the fat from DDG promise to circumvent this problem. This high variability in protein content and quality diminishes the value of DDG as a feedstuff, especially for poultry and pigs.

Use of a large proportion of DDG in animal diet also raises environmental concerns. Inclusion of DDG in poultry diets was shown to increase nitrogen and phosphorus levels in poultry excreta. Moreover, the solubility of excreted phosphorus in poultry fed with DDG is higher than that of poultry without DDG in its diet (Leytem et al., 2008). Another study reported high phosphorus excretion in dry cows and heifers that were fed with DDG (Schmit et al., 2009). Disposal of the manure with high nutrient content is an environmental concern.

REFERENCES

Belyea, R.L., K.D. Rausch, T.E. Clevenger, V. Singh, D.B. Johnston, and M.E. Tumbleson. 2010. Sources of variation in composition of DDGS. Animal Feed Science and Technology 159(3-4):122-130.

Corzo, A., M.W. Schilling, R.E. Loar, V. Jackson, S. Kin, and V. Radhakrishnan. 2009. The effects of feeding distillers dried grains with solubles on broiler meat quality. Poultry Science 88(2):432-439.

Depenbusch, B.E., C.M. Coleman, J.J. Higgins, and J.S. Drouillard. 2009. Effects of increasing levels of dried corn distillers grains with solubles on growth performance, carcass characteristics, and meat quality of yearling heifers. Journal of Animal Science 87(8):2653-2663.

Edrington, T.S., J.C. MacDonald, R.L. Farrow, T.R. Callaway, R.C. Anderson, and D.J. Nisbet. 2010. Influence of wet distiller's grains on prevalence of *Escherichia coli* O157:H7 and salmonella in feedlot cattle and antimicrobial susceptibility of generic *Escherichia coli* isolates. Foodborne Pathogens and Disease 7(5):605-608.

FDA (Food and Drug Adminstration). 2009. FY 2010 Nationwide Survey of Distillers Grains for Antibiotic Residues. Available online at http://www.fda.gov/AnimalVeterinary/Products/AnimalFoodFeeds/Contaminants/ucm190907.htm. Accessed September 6, 2010.

Jacob, M.E., J.T. Fox, J.S. Drouillard, D.G. Renter, and T.G. Nagaraja. 2008a. Effects of dried distillers' grain on fecal prevalence and growth of *Escherichia coli* O157 in batch culture fermentations from cattle. Applied and Environmental Microbiology 74(1):38-43.

Jacob, M.E., J.T. Fox, S.K. Narayanan, J.S. Drouillard, D.G. Renter, and T.G. Nagaraja. 2008b. Effects of feeding wet corn distillers grains with solubles with or without monensin and tylosin on the prevalence and antimicrobial susceptibilities of fecal foodborne pathogenic and commensal bacteria in feedlot cattle. Journal of Animal Science 86(5):1182-1190.

Leytem, A.B., P. Kwanyuen, and P. Thacker. 2008. Nutrient excretion, phosphorus characterization, and phosphorus solubility in excreta from broiler chicks fed diets containing graded levels of wheat distillers grains with solubles. Poultry Science 87(12):2505-2511.

Schmit, T.M., R.N. Boisvert, D. Enahoro, and L.E. Chase. 2009. Optimal dairy farm adjustments to increased utilization of corn distillers dried grains with solubles. Journal of Dairy Science 92(12):6105-6115.

Seigel, V. 2010. Corn dried distillers grains and mycotoxins. Paper presented at the 14th Distillers Grains Symposium, May 12-13, Indianapolis, IN.

Skinner, K.A., and T.D. Leathers. 2004. Bacterial contaminants of fuel ethanol production. Journal of Industrial Microbiology & Biotechnology 31(9):401-408.

Whitlow, L. 2008. Mycotoxins cause concerns. Available online at http://ethanolproducer.com/dgq/article.jsp?article_id=1233&article_title=Mycotoxins+cause+concerns. Accessed September 7, 2010.

Wild, C.P., and Y.Y. Gong. 2010. Mycotoxins and human disease: A largely ignored global health issue. Carcinogenesis 31(1):71-82.

Xu, G., S.K. Baidoo, L.J. Johnston, J.E. Cannon, and G.C. Shurson. 2007. Effects of adding increasing levels of corn dried distillers grains with solubles (DDGS) to corn-soybean meal diets on growth performance and pork quality of growing-finishing pigs. Journal of Animal Science 85:76.

Xu, G., S.K. Baidoo, L.J. Johnston, D. Bibus, J.E. Cannon, and G.C. Shurson. 2010. Effects of feeding diets containing increasing content of corn distillers dried grains with solubles to grower-finisher pigs on growth performance, carcass composition, and pork fat quality. Journal of Animal Science 88(4):1398-1410.

Zhang, Y.H., J. Caupert, P.M. Imerman, J.L. Richard, and G.C. Shurson. 2009. The occurrence and concentration of mycotoxins in US distillers dried grains with solubles. Journal of Agricultural and Food Chemistry 57(20):9828-9837.